Probability and Statistics (Continued)

LINDVALL · Lectures on the Coupling Method
MANTON, WOODBURY, and TOLLEY · Statistical Applications Using Fuzzy Sets
MARDIA · The Art of Statistical Science: A Tribute to G. S. Watson
MORGENTHALER and TUKEY · Configural Polysampling: A Route to Practical Robustness
MUIRHEAD · Aspects of Multivariate Statistical Theory
OLIVER and SMITH · Influence Diagrams, Belief Nets and Decision Analysis
*PARZEN · Modern Probability Theory and Its Applications
PRESS · Bayesian Statistics: Principles, Models, and Applications
PUKELSHEIM · Optimal Experimental Design
PURI and SEN · Nonparametric Methods in General Linear Models
PURI, VILAPLANA, and WERTZ · New Perspectives in Theoretical and Applied Statistics
RAO · Asymptotic Theory of Statistical Inference
RAO · Linear Statistical Inference and Its Applications, *Second Edition*
RAO and SHANBHAG · Choquet-Deng Type Functional Equations and Applications to Stochastic Models
RENCHER · Methods of Multivariate Analysis
ROBERTSON, WRIGHT, and DYKSTRA · Order Restricted Statistical Inference
ROGERS and WILLIAMS · Diffusions, Markov Processes, and Martingales, Volume I: Foundations, *Second Edition;* Volume II: Îto Calculus
ROHATGI · An Introduction to Probability Theory and Mathematical Statistics
ROSS · Stochastic Processes
RUBINSTEIN · Simulation and the Monte Carlo Method
RUBINSTEIN and SHAPIRO · Discrete Event Systems: Sensitivity Analysis and \ Stochastic Optimization by the Score Function Method
RUZSA and SZEKELY · Algebraic Probability Theory
SCHEFFE · The Analysis of Variance
SEBER · Linear Regression Analysis
SEBER · Multivariate Observations
SEBER and WILD · Nonlinear Regression
SERFLING · Approximation Theorems of Mathematical Statistics
SHORACK and WELLNER · Empirical Processes with Applications to Statistics
SMALL and McLEISH · Hilbert Space Methods in Probability and Statistical Inference
STAPLETON · Linear Statistical Models
STAUDTE and SHEATHER · Robust Estimation and Testing
STOYANOV · Counterexamples in Probability
STYAN · The Collected Papers of T. W. Anderson: 1943–1985
TANAKA · Time Series Analysis: Nonstationary and Noninvertible Distribution Theory
THOMPSON and SEBER · Adaptive Sampling
WHITTAKER · Graphical Models in Applied Multivariate Statistics
YANG · The Construction Theory of Denumerable Markov Processes

Applied Probability and Statistics

ABRAHAM and LEDOLTER · Statistical Methods for Forecasting
AGRESTI · Analysis of Ordinal Categorical Data
AGRESTI · Categorical Data Analysis
AGRESTI · An Introduction to Categorical Data Analysis
ANDERSON and LOYNES · The Teaching of Practical Statistics
ANDERSON, AUQUIER, HAUCK, OAKES, VANDAELE, and WEISBERG · Statistical Methods for Comparative Studies
*ARTHANARI and DODGE · Mathematical Programming in Statistics
ASMUSSEN · Applied Probability and Queues
*BAILEY · The Elements of Stochastic Processes with Applications to the Natural Sciences
BARNETT · Interpreting Multivariate Data

*Now available in a lower priced paperback edition in the Wiley Classics Library.

Applied Probability and Statistics (Continued)

BARNETT and LEWIS · Outliers in Statistical Data, *Second Edition*
BARTHOLOMEW, FORBES, and McLEAN · Statistical Techniques for Manpower Planning, *Second Edition*
BATES and WATTS · Nonlinear Regression Analysis and Its Applications
BECHHOFER, SANTNER, and GOLDSMAN · Design and Analysis of Experiments for Statistical Selection, Screening, and Multiple Comparisons
BELSLEY · Conditioning Diagnostics: Collinearity and Weak Data in Regression
BELSLEY, KUH, and WELSCH · Regression Diagnostics: Identifying Influential Data and Sources of Collinearity
BERRY, CHALONER, and GEWEKE · Bayesian Analysis in Statistics and Econometrics: Essays in Honor of Arnold Zellner
BHAT · Elements of Applied Stochastic Processes, *Second Edition*
BHATTACHARYA and WAYMIRE · Stochastic Processes with Applications
BIEMER, GROVES, LYBERG, MATHIOWETZ, and SUDMAN · Measurement Errors in Surveys
BIRKES and DODGE · Alternative Methods of Regression
BLOOMFIELD · Fourier Analysis of Time Series: An Introduction
BOLLEN · Structural Equations with Latent Variables
BOULEAU · Numerical Methods for Stochastic Processes
BOX · R. A. Fisher, the Life of a Scientist
BOX and DRAPER · Empirical Model-Building and Response Surfaces
BOX and DRAPER · Evolutionary Operation: A Statistical Method for Process Improvement
BOX, HUNTER, and HUNTER · Statistics for Experimenters: An Introduction to Design, Data Analysis, and Model Building
BROWN and HOLLANDER · Statistics: A Biomedical Introduction
BUCKLEW · Large Deviation Techniques in Decision, Simulation, and Estimation
BUNKE and BUNKE · Nonlinear Regression, Functional Relations and Robust Methods: Statistical Methods of Model Building
CHATTERJEE and HADI · Sensitivity Analysis in Linear Regression
CHATTERJEE and PRICE · Regression Analysis by Example, *Second Edition*
CLARKE and DISNEY · Probability and Random Processes: A First Course with Applications, *Second Edition*
COCHRAN · Sampling Techniques, *Third Edition*
*COCHRAN and COX · Experimental Designs, *Second Edition*
CONOVER · Practical Nonparametric Statistics, *Second Edition*
CONOVER and IMAN · Introduction to Modern Business Statistics
CORNELL · Experiments with Mixtures, Designs, Models, and the Analysis of Mixture Data, *Second Edition*
COX · A Handbook of Introductory Statistical Methods
*COX · Planning of Experiments
COX, BINDER, CHINNAPPA, CHRISTIANSON, COLLEDGE, and KOTT · Business Survey Methods
CRESSIE · Statistics for Spatial Data, *Revised Edition*
DANIEL · Applications of Statistics to Industrial Experimentation
DANIEL · Biostatistics: A Foundation for Analysis in the Health Sciences, *Sixth Edition*
DAVID · Order Statistics, *Second Edition*
*DEGROOT, FIENBERG, and KADANE · Statistics and the Law
*DEMING · Sample Design in Business Research
DILLON and GOLDSTEIN · Multivariate Analysis: Methods and Applications
DODGE and ROMIG · Sampling Inspection Tables, *Second Edition*
DOWDY and WEARDEN · Statistics for Research, *Second Edition*
DRAPER and SMITH · Applied Regression Analysis, *Second Edition*
DUNN · Basic Statistics: A Primer for the Biomedical Sciences, *Second Edition*
DUNN and CLARK · Applied Statistics: Analysis of Variance and Regression, *Second Edition*

*Now available in a lower priced paperback edition in the Wiley Classics Library.

STOCHASTIC PROCESSES

Second Edition

STOCHASTIC PROCESSES

Second Edition

Sheldon M. Ross
University of California, Berkeley

JOHN WILEY & SONS, INC.

ACQUISITIONS EDITOR Angela Battle
MARKETING MANAGER Debra Riegert
SENIOR PRODUCTION EDITOR Tony VenGraitis
MANUFACTURING MANAGER Dorothy Sinclair
TEXT AND COVER DESIGN A Good Thing, Inc.
PRODUCTION COORDINATION Elm Street Publishing Services, Inc.

This book was set in Times Roman by Bi-Comp, Inc. and printed and bound by Courier/Stoughton. The cover was printed by Phoenix Color.

Recognizing the importance of preserving what has been written, it is a policy of John Wiley & Sons, Inc. to have books of enduring value published in the United States printed on acid-free paper, and we exert our best efforts to that end.

The paper in this book was manufactured by a mill whose forest management programs include sustained yield harvesting of its timberlands. Sustained yield harvesting principles ensure that the number of trees cut each year does not exceed the amount of new growth.

Library of Congress Cataloging-in-Publication Data:
Ross, Sheldon M.
Stochastic processes/Sheldon M. Ross.—2nd ed.
p. cm.
Includes bibliographical references and index.
ISBN 978-0-471-12062-9
ISBN 0-471-12062-6 (cloth:alk. paper)
1. Stochastic processes. I. Title.
QA274.R65 1996
519.2—dc20 95-38012
CIP

Printed in the United States of America

12 13 14 15

On March 30, 1980, a beautiful six-year-old girl died.
This book is dedicated to the memory of

Nichole Pornaras

Preface to the First Edition

This text is a nonmeasure theoretic introduction to stochastic processes, and as such assumes a knowledge of calculus and elementary probability. In it we attempt to present some of the theory of stochastic processes, to indicate its diverse range of applications, and also to give the student some probabilistic intuition and insight in thinking about problems. We have attempted, wherever possible, to view processes from a probabilistic instead of an analytic point of view. This attempt, for instance, has led us to study most processes from a sample path point of view.

I would like to thank Mark Brown, Cyrus Derman, Shun-Chen Niu, Michael Pinedo, and Zvi Schechner for their helpful comments.

SHELDON M. ROSS

Preface to the Second Edition

The second edition of *Stochastic Processes* includes the following changes:

(i) Additional material in Chapter 2 on compound Poisson random variables, including an identity that can be used to efficiently compute moments, and which leads to an elegant recursive equation for the probability mass function of a nonnegative integer valued compound Poisson random variable;

(ii) A separate chapter (Chapter 6) on martingales, including sections on the Azuma inequality; and

(iii) A new chapter (Chapter 10) on Poisson approximations, including both the Stein-Chen method for bounding the error of these approximations and a method for improving the approximation itself.

In addition, we have added numerous exercises and problems throughout the text. Additions to individual chapters follow:

In Chapter 1, we have new examples on the probabilistic method, the multivariate normal distribution, random walks on graphs, and the complete match problem. Also, we have new sections on probability inequalities (including Chernoff bounds) and on Bayes estimators (showing that they are almost never unbiased). A proof of the strong law of large numbers is given in the Appendix to this chapter.

New examples on patterns and on memoryless optimal coin tossing strategies are given in Chapter 3.

There is new material in Chapter 4 covering the mean time spent in transient states, as well as examples relating to the Gibb's sampler, the Metropolis algorithm, and the mean cover time in star graphs.

Chapter 5 includes an example on a two-sex population growth model.

Chapter 6 has additional examples illustrating the use of the martingale stopping theorem.

Chapter 7 includes new material on Spitzer's identity and uses it to compute mean delays in single-server queues with gamma-distributed interarrival and service times.

Chapter 8 on Brownian motion has been moved to follow the chapter on martingales to allow us to utilize martingales to analyze Brownian motion.

Chapter 9 on stochastic order relations now includes a section on associated random variables, as well as new examples utilizing coupling in coupon collecting and bin packing problems.

We would like to thank all those who were kind enough to write and send comments about the first edition, with particular thanks to He Sheng-wu, Stephen Herschkorn, Robert Kertz, James Matis, Erol Pekoz, Maria Rieders, and Tomasz Rolski for their many helpful comments.

SHELDON M. ROSS

Contents

CHAPTER 1

Preliminaries

1.1 PROBABILITY

A basic notion in probability theory is *random experiment:* an experiment whose outcome cannot be determined in advance. The set of all possible outcomes of an experiment is called the *sample space* of that experiment, and we denote it by S.

An *event* is a subset of a sample space, and is said to occur if the outcome of the experiment is an element of that subset. We shall suppose that for each event E of the sample space S a number $P(E)$ is defined and satisfies the following three axioms*:

Axiom (1) $0 \leq P(E) \leq 1$.

Axiom (2) $P(S) = 1$.

Axiom (3) For any sequence of events $E_1, E_2, \ldots$ that are mutually exclusive, that is, events for which $E_iE_j = \phi$ when $i \neq j$ (where ϕ is the null set),

$$P\left(\bigcup_{i=1}^{\infty} E_i\right) = \sum_{i=1}^{\infty} P(E_i).$$

We refer to $P(E)$ as the probability of the event E.

Some simple consequences of axioms (1), (2), and (3) are:

1.1.1. If $E \subset F$, then $P(E) \leq P(F)$.

1.1.2. $P(E^c) = 1 - P(E)$ where E^c is the complement of E.

1.1.3. $P(\bigcup_1^n E_i) = \sum_1^n P(E_i)$ when the E_i are mutually exclusive.

1.1.4. $P(\bigcup_1^\infty E_i) \leq \sum_1^\infty P(E_i)$.

The inequality (1.1.4) is known as *Boole's inequality.*

* Actually $P(E)$ will only be defined for the so-called measurable events of S. But this restriction need not concern us.

An important property of the probability function P is that it is continuous. To make this more precise, we need the concept of a limiting event, which we define as follows: A sequence of events $\{E_n, n \geq 1\}$ is said to be an *increasing* sequence if $E_n \subset E_{n+1}$, $n \geq 1$ and is said to be *decreasing* if $E_n \supset E_{n+1}$, $n \geq 1$. If $\{E_n, n \geq 1\}$ is an increasing sequence of events, then we define a new event, denoted by $\lim_{n\to\infty} E_n$ by

$$\lim_{n\to\infty} E_n = \bigcup_{i=1}^{\infty} E_i \qquad \text{when } E_n \subset E_{n+1},\ n \geq 1.$$

Similarly if $\{E_n, n \geq 1\}$ is a decreasing sequence, then define $\lim_{n\to\infty} E_n$ by

$$\lim_{n\to\infty} E_n = \bigcap_{i=1}^{\infty} E_i, \qquad \text{when } E_n \supset E_{n+1},\ n \geq 1.$$

We may now state the following:

PROPOSITION 1.1.1

If $\{E_n, n \geq 1\}$ is either an increasing or decreasing sequence of events, then

$$\lim_{n\to\infty} P(E_n) = P\left(\lim_{n\to\infty} E_n\right).$$

Proof Suppose, first, that $\{E_n, n \geq 1\}$ is an increasing sequence, and define events F_n, $n \geq 1$ by

$$F_1 = E_1,$$

$$F_n = E_n\left(\bigcup_{1}^{n-1} E_i\right)^c = E_n E_{n-1}^c, \qquad n > 1.$$

That is, F_n consists of those points in E_n that are not in any of the earlier E_i, $i < n$. It is easy to verify that the F_n are mutually exclusive events such that

$$\bigcup_{i=1}^{\infty} F_i = \bigcup_{i=1}^{\infty} E_i \quad \text{and} \quad \bigcup_{i=1}^{n} F_i = \bigcup_{i=1}^{n} E_i \qquad \text{for all } n \geq 1.$$

Thus

$$P\left(\bigcup_1^\infty E_i\right) = P\left(\bigcup_i^\infty F_i\right)$$

$$= \sum_1^\infty P(F_i) \qquad \text{(by Axiom 3)}$$

$$= \lim_{n\to\infty} \sum_1^n P(F_i)$$

$$= \lim_{n\to\infty} P\left(\bigcup_1^n F_i\right)$$

$$= \lim_{n\to\infty} P\left(\bigcup_1^n E_i\right)$$

$$= \lim_{n\to\infty} P(E_n),$$

which proves the result when $\{E_n, n \geq 1\}$ is increasing.

If $\{E_n, n \geq 1\}$ is a decreasing sequence, then $\{E_n^c, n \geq 1\}$ is an increasing sequence; hence,

$$P\left(\bigcup_1^\infty E_n^c\right) = \lim_{n\to\infty} P(E_n^c).$$

But, as $\bigcup_1^\infty E_n^c = (\bigcap_1^\infty E_n)^c$, we see that

$$1 - P\left(\bigcap_1^\infty E_n\right) = \lim_{n\to\infty} [1 - P(E_n)],$$

or, equivalently,

$$P\left(\bigcap_1^\infty E_n\right) = \lim_{n\to\infty} P(E_n),$$

which proves the result.

Example 1.1(a) Consider a population consisting of individuals able to produce offspring of the same kind. The number of individuals initially present, denoted by X_0, is called the size of the zeroth generation. All offspring of the zeroth generation constitute the first generation and their number is denoted by X_1. In general, let X_n denote the size of the nth generation.

Since $X_n = 0$ implies that $X_{n+1} = 0$, it follows that $P\{X_n = 0\}$ is increasing and thus $\lim_{n\to\infty} P\{X_n = 0\}$ exists. What does it represent? To answer this use Proposition 1.1.1 as follows:

$$\begin{aligned}\lim_{n\to\infty} P\{X_n = 0\} &= P\left\{\lim_{n\to\infty}\{X_n = 0\}\right\}\\ &= P\left\{\bigcup_n \{X_n = 0\}\right\}\\ &= P\{\text{the population ever dies out}\}.\end{aligned}$$

That is, the limiting probability that the nth generation is void of individuals is equal to the probability of eventual extinction of the population.

Proposition 1.1.1 can also be used to prove the Borel-Cantelli lemma.

PROPOSITION 1.1.2

The Borel-Cantelli Lemma
Let $E_1, E_2, \ldots$ denote a sequence of events. If

$$\sum_{i=1}^{\infty} P(E_i) < \infty,$$

then

$$P\{\text{an infinite number of the } E_i \text{ occur}\} = 0.$$

Proof The event that an infinite number of the E_i occur, called the $\limsup_{i\to\infty} E_i$, can be expressed as

$$\limsup_{i\to\infty} E_i = \bigcap_{n=1}^{\infty} \bigcup_{i=n}^{\infty} E_i.$$

This follows since if an infinite number of the E_i occur, then $\bigcup_{i=n}^{\infty} E_i$ occurs for each n and thus $\bigcap_{n=1}^{\infty} \bigcup_{i=n}^{\infty} E_i$ occurs. On the other hand, if $\bigcap_{n=1}^{\infty} \bigcup_{i=n}^{\infty} E_i$ occurs, then $\bigcup_{i=n}^{\infty} E_i$ occurs for each n, and thus for each n at least one of the E_i occurs where $i \geq n$; and, hence, an infinite number of the E_i occur.

As $\bigcup_{i=n}^{\infty} E_i$, $n \geq 1$, is a decreasing sequence of events, it follows from Proposition 1.1.1 that

$$P\left(\bigcap_{n=1}^{\infty} \bigcup_{i=n}^{\infty} E_i\right) = P\left(\lim_{n\to\infty} \bigcup_{i=n}^{\infty} E_i\right)$$

$$= \lim_{n\to\infty} P\left(\bigcup_{i=n}^{\infty} E_i\right)$$

$$\leq \lim_{n\to\infty} \sum_{i=n}^{\infty} P(E_i)$$

$$= 0,$$

and the result is proven.

EXAMPLE 1.1(B) Let $X_1, X_2, \ldots$ be such that

$$P\{X_n = 0\} = 1/n^2 = 1 - P\{X_n = 1\}, \qquad n \geq 1.$$

If we let $E_n = \{X_n = 0\}$, then, as $\sum_n^\infty P(E_n) < \infty$, it follows from the Borel-Cantelli lemma that the probability that X_n equals 0 for an infinite number of n is equal to 0. Hence, for all n sufficiently large, X_n must equal 1, and so we may conclude that, with probability 1,

$$\lim_{n\to\infty} X_n = 1.$$

For a converse to the Borel-Cantelli lemma, independence is required.

PROPOSITION 1.1.3

Converse to the Borel-Cantelli Lemma
If $E_1, E_2, \ldots$ are independent events such that

$$\sum_{n=1}^{\infty} P(E_n) = \infty,$$

then

$$P\{\text{an infinite number of the } E_n \text{ occur}\} = 1.$$

Proof

$$P\{\text{an infinite number of the } E_n \text{ occur}\} = P\left\{\lim_{n\to\infty} \bigcup_{i=n}^{\infty} E_i\right\}$$

$$= \lim_{n\to\infty} P\left(\bigcup_{i=n}^{\infty} E_i\right)$$

$$= \lim_{n\to\infty} \left[1 - P\left(\bigcap_{i=n}^{\infty} E_i^c\right)\right].$$

Now,

$$P\left(\bigcap_{i=n}^{\infty} E_i^c\right) = \prod_{i=n}^{\infty} P(E_i^c) \qquad \text{(by independence)}$$

$$= \prod_{i=n}^{\infty} (1 - P(E_i))$$

$$\leq \prod_{i=n}^{\infty} e^{-P(E_i)} \qquad \text{(by the inequality } 1 - x \leq e^{-x}\text{)}$$

$$= \exp\left(-\sum_{n}^{\infty} P(E_i)\right)$$

$$= 0 \quad \text{since } \sum_{i=n}^{\infty} P(E_i) = \infty \text{ for all } n.$$

Hence the result follows.

EXAMPLE 1.1(c) Let $X_1, X_2, \ldots$ be independent and such that

$$P\{X_n = 0\} = 1/n = 1 - P\{X_n = 1\}, \qquad n \geq 1.$$

If we let $E_n = \{X_n = 0\}$, then as $\sum_{n=1}^{\infty} P(E_n) = \infty$ it follows from Proposition 1.1.3 that E_n occurs infinitely often. Also, as $\sum_{n=1}^{\infty} P(E_n^c) = \infty$ it also follows that E_n^c also occurs infinitely often. Hence, with probability 1, X_n will equal 0 infinitely often and will also equal 1 infinitely often. Hence, with probability 1, X_n will not approach a limiting value as $n \to \infty$.

1.2 Random Variables

Consider a random experiment having sample space S. A *random variable* X is a function that assigns a real value to each outcome in S. For any set of real numbers A, the probability that X will assume a value that is contained in the set A is equal to the probability that the outcome of the experiment is contained in $X^{-1}(A)$. That is,

$$P\{X \in A\} = P(X^{-1}(A)),$$

where $X^{-1}(A)$ is the event consisting of all points $s \in S$ such that $X(s) \in A$.

The *distribution function* F of the random variable X is defined for any real number x by

$$F(x) = P\{X \leq x\} = P\{X \in (-\infty, x]\}.$$

We shall denote $1 - F(x)$ by $\bar{F}(x)$, and so

$$\bar{F}(x) = P\{X > x\}.$$

A random variable X is said to be *discrete* if its set of possible values is countable. For discrete random variables,

$$F(x) = \sum_{y \leq x} P\{X = y\}.$$

A random variable is called *continuous* if there exists a function $f(x)$, called the *probability density function*, such that

$$P\{X \text{ is in } B\} = \int_B f(x)\, dx$$

for every set B. Since $F(x) = \int_{-\infty}^{x} f(x)\, dx$, it follows that

$$f(x) = \frac{d}{dx} F(x).$$

The *joint distribution function* F of two random variables X and Y is defined by

$$F(x, y) = P\{X \leq x, Y \leq y\}.$$

The distribution functions of X and Y,

$$F_X(x) = P\{X \leq x\} \quad \text{and} \quad F_Y(y) = P\{Y \leq y\},$$

can be obtained from $F(x, y)$ by making use of the continuity property of the probability operator. Specifically, let y_n, $n \geq 1$, denote an increasing sequence converging to ∞. Then as the events $\{X \leq x, Y \leq y_n\}$, $n \geq 1$, are increasing and

$$\lim_{n\to\infty}\{X \leq x, Y \leq y_n\} = \bigcup_{n=1}^{\infty} \{X \leq x, Y \leq y_n\} = \{X \leq x\},$$

it follows from the continuity property that

$$\lim_{n\to\infty} P\{X \leq x, Y \leq y_n\} = P\{X \leq x\},$$

or, equivalently,

$$F_X(x) = \lim_{y\to\infty} F(x, y).$$

Similarly,

$$F_Y(y) = \lim_{x\to\infty} F(x, y).$$

The random variables X and Y are said to be *independent* if

$$F(x, y) = F_X(x)F_Y(y)$$

for all x and y.

The random variables X and Y are said to be *jointly continuous* if there exists a function $f(x, y)$, called the *joint probability density function*, such that

$$P\{X \text{ is in } A, Y \text{ is in } B\} = \int_A \int_B f(x, y)\, dy\, dx$$

for all sets A and B.

The joint distribution of any collection $X_1, X_2, \ldots, X_n$ of random variables is defined by

$$F(x_1, \ldots, x_n) = P\{X_1 \leq x_1, \ldots, X_n \leq x_n\}.$$

Furthermore, the n random variables are said to be independent if

$$F(x_1, \ldots, x_n) = F_{X_1}(x_1)F_{X_2}(x_2) \cdots F_{X_n}(x_n),$$

where

$$F_{X_i}(x_i) = \lim_{\substack{x_j\to\infty \\ j\neq i}} F(x_1, \ldots, x_n).$$

1.3 Expected Value

The *expectation* or *mean* of the random variable X, denoted by $E[X]$, is defined by

(1.3.1)
$$E[X] = \int_{-\infty}^{\infty} x\, dF(x)$$
$$= \begin{cases} \int_{-\infty}^{\infty} xf(x)\, dx & \text{if } X \text{ is continuous} \\ \sum_x xP\{X = x\} & \text{if } X \text{ is discrete} \end{cases}$$

provided the above integral exists.

Equation (1.3.1) also defines the expectation of any function of X, say $h(X)$. Since $h(X)$ is itself a random variable, it follows from (1.3.1) that

$$E[h(X)] = \int_{-\infty}^{\infty} x\, dF_h(x),$$

where F_h is the distribution function of $h(X)$. However, it can be shown that this is identical to $\int_{-\infty}^{\infty} h(x)\, dF(x)$. That is,

(1.3.2)
$$E[h(X)] = \int_{-\infty}^{\infty} h(x)\, dF(x).$$

The variance of the random variable X is defined by

$$\begin{aligned} \text{Var } X &= E[(X - E[X])^2] \\ &= E[X^2] - E^2[X]. \end{aligned}$$

Two jointly distributed random variables X and Y are said to be uncorrelated if their covariance, defined by

$$\begin{aligned} \text{Cov}(X, Y) &= E[(X - EX)(Y - EY)] \\ &= E[XY] - E[X]E[Y] \end{aligned}$$

is zero. It follows that independent random variables are uncorrelated. However, the converse need not be true. (The reader should think of an example.)

An important property of expectations is that the expectation of a sum of random variables is equal to the sum of the expectations.

(1.3.3)
$$E\left[\sum_{i=1}^{n} X_i\right] = \sum_{i=1}^{n} E[X_i].$$

The corresponding property for variances is that

$$\text{(1.3.4)} \qquad \operatorname{Var}\left[\sum_{i=1}^{n} X_i\right] = \sum_{i=1}^{n} \operatorname{Var}(X_i) + 2\sum_{i<j}\sum \operatorname{Cov}(X_i, X_j).$$

Example 1.3(a) ***The Matching Problem.*** At a party n people put their hats in the center of a room where the hats are mixed together. Each person then randomly selects one. We are interested in the mean and variance of X—the number that select their own hat.

To solve, we use the representation

$$X = X_1 + X_2 + \cdots + X_n,$$

where

$$X_i = \begin{cases} 1 & \text{if the } i\text{th person selects his or her own hat} \\ 0 & \text{otherwise.} \end{cases}$$

Now, as the ith person is equally likely to select any of the n hats, it follows that $P\{X_i = 1\} = 1/n$, and so

$$E[X_i] = 1/n,$$

$$\operatorname{Var}(X_i) = \frac{1}{n}\left(1 - \frac{1}{n}\right) = \frac{n-1}{n^2}.$$

Also

$$\operatorname{Cov}(X_i, X_j) = E[X_i X_j] - E[X_i]E[X_j].$$

Now,

$$X_i X_j = \begin{cases} 1 & \text{if the } i\text{th and } j\text{th party goers both select their own hats} \\ 0 & \text{otherwise,} \end{cases}$$

and thus

$$\begin{aligned} E[X_i X_j] &= P\{X_i = 1, X_j = 1\} \\ &= P\{X_i = 1\}P\{X_j = 1 \mid X_i = 1\} \\ &= \frac{1}{n}\frac{1}{n-1}. \end{aligned}$$

Hence,

$$\text{Cov}(X_i, X_j) = \frac{1}{n(n-1)} - \left(\frac{1}{n}\right)^2 = \frac{1}{n^2(n-1)}.$$

Therefore, from (1.3.3) and (1.3.4),

$$E[X] = 1$$

and

$$\begin{aligned}\text{Var}(X) &= \frac{n-1}{n} + 2\binom{n}{2}\frac{1}{n^2(n-1)} \\ &= 1.\end{aligned}$$

Thus both the mean and variance of the number of matches are equal to 1. (See Example 1.5(f) for an explanation as to why these results are not surprising.)

Example 1.3(b) ***Some Probability Identities.*** Let $A_1, A_2, \ldots, A_n$ denote events and define the indicator variables $I_j, j = 1, \ldots, n$ by

$$I_j = \begin{cases} 1 & \text{if } A_j \text{ occurs} \\ 0 & \text{otherwise.} \end{cases}$$

Letting

$$N = \sum_{j=1}^{n} I_j,$$

then N denotes the number of the A_j, $1 \le j \le n$, that occur. A useful identity can be obtained by noting that

$$(1.3.5) \qquad (1-1)^N = \begin{cases} 1 & \text{if } N = 0 \\ 0 & \text{if } N > 0. \end{cases}$$

But by the binomial theorem,

$$\begin{aligned}(1.3.6) \quad (1-1)^N &= \sum_{i=0}^{N} \binom{N}{i}(-1)^i \\ &= \sum_{i=0}^{n} \binom{N}{i}(-1)^i \qquad \text{since } \binom{m}{i} = 0 \text{ when } i > m.\end{aligned}$$

Hence, if we let

$$I = \begin{cases} 1 & \text{if } N > 0 \\ 0 & \text{if } N = 0, \end{cases}$$

then (1.3.5) and (1.3.6) yield

$$1 - I = \sum_{i=0}^{n} \binom{N}{i} (-1)^i$$

or

$$I = \sum_{i=1}^{n} \binom{N}{i} (-1)^{i+1}. \tag{1.3.7}$$

Taking expectations of both sides of (1.3.7) yields

$$E[I] = E[N] - E\left[\binom{N}{2}\right] + \cdots + (-1)^{n+1} E\left[\binom{N}{n}\right]. \tag{1.3.8}$$

However,

$$\begin{aligned} E[I] &= P\{N > 0\} \\ &= P\{\text{at least one of the } A_i \text{ occurs}\} \\ &= P\left(\bigcup_1^n A_i\right) \end{aligned}$$

and

$$E[N] = E\left[\sum_{j=1}^{n} I_j\right] = \sum_{j=1}^{n} P(A_j),$$

$$\begin{aligned} E\left[\binom{N}{2}\right] &= E[\text{number of pairs of the } A_j \text{ that occur}] \\ &= E\left[\sum\sum_{i<j} I_i I_j\right] \\ &= \sum\sum_{i<j} E[I_i I_j] \\ &= \sum\sum_{i<j} P(A_i A_j), \end{aligned}$$

and, in general, by the same reasoning,

$$E\left[\binom{N}{i}\right] = E[\text{number of sets of size } i \text{ that occur}]$$

$$= E\left[\sum\sum_{j_1<j_2<\cdots<j_i}\sum I_{j_1}I_{j_2}\cdots I_{j_i}\right]$$

$$= \sum\sum_{j_1<j_2<\cdots<j_i}\sum P(A_{j_1}A_{j_2}\cdots A_{j_i}).$$

Hence, (1.3.8) is a statement of the well-known identity

$$P\left(\bigcup_{i=1}^{n} A_i\right) = \sum_{i=1}^{n} P(A_i) - \sum_{i<j}\sum P(A_iA_j) + \sum\sum_{i<j<k}\sum P(A_iA_jA_k)$$
$$- \cdots + (-1)^{n+1}P(A_1A_2\cdots A_n).$$

Other useful identities can also be derived by this approach. For instance, suppose we want a formula for the probability that exactly r of the events $A_1, \ldots, A_n$ occur. Then define

$$I_r = \begin{cases} 1 & \text{if } N = r \\ 0 & \text{otherwise} \end{cases}$$

and use the identity

$$\binom{N}{r}(1-1)^{N-r} = I_r$$

or

$$I_r = \binom{N}{r}\sum_{i=0}^{N-r}\binom{N-r}{i}(-1)^i$$

$$= \sum_{i=0}^{n-r}\binom{N}{r}\binom{N-r}{i}(-1)^i$$

$$= \sum_{i=0}^{n-r}\binom{N}{r+i}\binom{r+i}{r}(-1)^i.$$

Taking expectations of both sides of the above yields

$$E[I_r] = \sum_{i=0}^{n-r}(-1)^i\binom{r+i}{r}E\left[\binom{N}{r+i}\right]$$

or

(1.3.9) $P\{\text{exactly } r \text{ of the events } A_1, \ldots, A_n \text{ occur}\}$

$$= \sum_{i=0}^{n-r} (-1)^i \binom{r+i}{r} \sum_{j_1<j_2<\cdots<j_{r+i}} \cdots \sum P(A_{j_1} A_{j_2} \cdots A_{j_{r+i}}).$$

As an application of (1.3.9) suppose that m balls are randomly put in n boxes in such a way that, independent of the locations of the other balls, each ball is equally likely to go into any of the n boxes. Let us compute the probability that exactly r of the boxes are empty. By letting A_i denote the event that the ith box is empty, we see from (1.3.9) that

$$P\{\text{exactly } r \text{ of the boxes are empty}\}$$
$$= \sum_{i=0}^{n-r} (-1)^i \binom{r+i}{r} \binom{n}{r+i} \left(1 - \frac{r+i}{n}\right)^m,$$

where the above follows since $\sum_{j_1<\cdots<j_{r+i}}$ consists of $\binom{n}{r+i}$ terms and each term in the sum is equal to the probability that a given set of $r + i$ boxes is empty.

Our next example illustrates what has been called the *probabilistic method*. This method, much employed and popularized by the mathematician Paul Erdos, attempts to solve deterministic problems by first introducing a probability structure and then employing probabilistic reasoning.

Example 1.3(c) A graph is a set of elements, called nodes, and a set of (unordered) pairs of nodes, called edges. For instance, Figure 1.3.1 illustrates a graph with the set of nodes $N = \{1, 2, 3, 4, 5\}$ and the set of edges $E = \{(1, 2), (1, 3), (1, 5), (2, 3), (2, 4), (3, 4), (3, 5)\}$. Show that for any graph there is a subset of nodes A such that at least one-half of the edges have one of their nodes in A and the other in A^c. (For instance, in the graph illustrated in Figure 1.3.1 we could take $A = \{1, 2, 4\}$.)

Solution. Suppose that the graph contains m edges, and arbitrarily number them as $1, 2, \ldots, m$. For any set of nodes B, if we let $C(B)$ denote the number of edges that have exactly one of their nodes in B, then the problem is to show that $\max_B C(B) \geq m/2$. To verify this, let us introduce probability by randomly choosing a set of nodes S so that each node of the graph is independently in S with probability 1/2. If we now let X denote the number of edges in

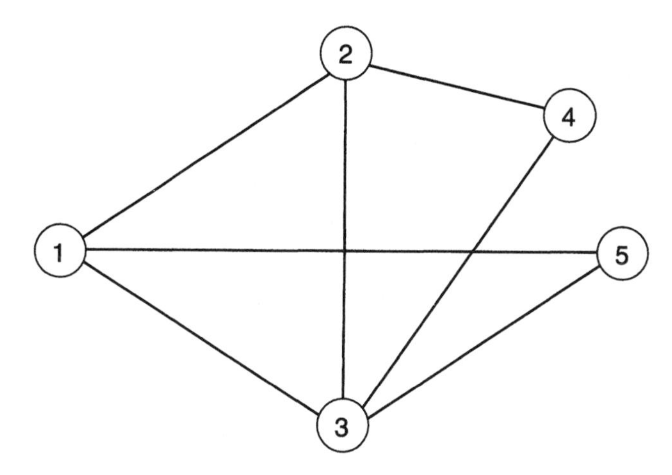

Figure 1.3.1. A graph.

the graph that have exactly one of their nodes in S, then X is a random variable whose set of possible values is all of the possible values of $C(B)$. Now, letting X_i equal 1 if edge i has exactly one of its nodes in S and letting it be 0 otherwise, then

$$E[X] = E\left[\sum_{i=1}^{m} X_i\right] = \sum_{i=1}^{m} E[X_i] = m/2.$$

Since at least one of the possible values of a random variable must be at least as large as its mean, we can thus conclude that $C(B) \geq m/2$ for some set of nodes B. (In fact, provided that the graph is such that $C(B)$ is not constant, we can conclude that $C(B) > m/2$ for some set of nodes B.)

Problems 1.9 and 1.10 give further applications of the probabilistic method.

1.4 Moment Generating, Characteristic Functions, and Laplace Transforms

The *moment generating function* of X is defined by

$$\begin{aligned}\psi(t) &= E[e^{tX}] \\ &= \int e^{tX}\, dF(x).\end{aligned}$$

All the moments of X can be successively obtained by differentiating ψ and then evaluating at $t = 0$. That is,

$$\psi'(t) = E[Xe^{tX}].$$
$$\psi''(t) = E[X^2e^{tX}]$$
$$\vdots$$
$$\psi^n(t) = E[X^ne^{tX}].$$

Evaluating at $t = 0$ yields

$$\psi^n(0) = E[X^n], \qquad n \geq 1.$$

It should be noted that we have assumed that it is justifiable to interchange the differentiation and integration operations. This is usually the case.

When a moment generating function exists, it uniquely determines the distribution. This is quite important because it enables us to characterize the probability distribution of a random variable by its generating function.

Example 1.4(a) Let X and Y be independent normal random variables with respective means μ_1 and μ_2 and respective variances

Table 1.4.1

Discrete Probability Distribution	Probability Mass Function, $p(x)$	Moment Generating Function, $\psi(t)$	Mean	Variance
Binomial with parameters n, p, $0 \leq p \leq 1$	$\binom{n}{x} p^x(1-p)^{n-x}$ $x = 0, 1, \ldots, n$	$(pe^t + (1-p))^n$	np	$np(1-p)$
Poisson with parameter $\lambda > 0$	$e^{-\lambda}\frac{\lambda^x}{x!}$, $x = 0, 1, 2, \ldots$	$\exp\{\lambda(e^t - 1)\}$	λ	λ
Geometric with parameter $0 \leq p \leq 1$	$p(1-p)^{x-1}$, $x = 1, 2, \ldots$	$\frac{pe^t}{1-(1-p)e^t}$	$\frac{1}{p}$	$\frac{1-p}{p^2}$
Negative binomial with parameters r, p	$\binom{x-1}{r-1} p^r(1-p)^{x-r}$, $x = r, r+1, \ldots$	$\left(\frac{pe^t}{1-(1-p)e^t}\right)^r$	$\frac{r}{p}$	$\frac{r(1-p)}{p^2}$

σ_1^2 and σ_2^2. The moment generating function of their sum is given by

$$\begin{aligned}\psi_{X+Y}(t) &= E[e^{t(X+Y)}] \\ &= E[e^{tX}]E[e^{tY}] \qquad \text{(by independence)} \\ &= \psi_X(t)\psi_Y(t) \\ &= \exp\{(\mu_1 + \mu_2)t + (\sigma_1^2 + \sigma_2^2)t^2/2\},\end{aligned}$$

where the last equality comes from Table 1.4.2. Thus the moment generating function of $X + Y$ is that of a normal random variable with mean $\mu_1 + \mu_2$ and variance $\sigma_1^2 + \sigma_2^2$. By uniqueness, this is the distribution of $X + Y$.

As the moment generating function of a random variable X need not exist, it is theoretically convenient to define the *characteristic function* of X by

$$\phi(t) = E[e^{itX}], \qquad -\infty < t < \infty,$$

where $i = \sqrt{-1}$. It can be shown that ϕ always exists and, like the moment generating function, uniquely determines the distribution of X.

We may also define the joint moment generating of the random variables $X_1, \ldots, X_n$ by

$$\psi(t_1, \ldots, t_n) = E\left[\exp\left\{\sum_{j=1}^{n} t_j X_j\right\}\right],$$

or the joint characteristic function by

$$\phi(t_1, \ldots, t_n) = E\left[\exp\left\{i\sum_{j=1}^{n} t_j X_j\right\}\right].$$

It may be proven that the joint moment generating function (when it exists) or the joint characteristic function uniquely determines the joint distribution.

Example 1.4(b) The Multivariate Normal Distribution. Let $Z_1, \ldots, Z_n$ be independent standard normal random variables. If for some constants a_{ij}, $1 \le i \le m$, $1 \le j \le n$, and μ_i, $1 \le i \le m$,

$$\begin{aligned}X_1 &= a_{11}Z_1 + \cdots + a_{1n}Z_n + \mu_1, \\ X_2 &= a_{21}Z_1 + \cdots + a_{2n}Z_n + \mu_2, \\ &\vdots \\ X_i &= a_{i1}Z_1 + \cdots + a_{in}Z_n + \mu_i, \\ &\vdots \\ X_m &= a_{m1}Z_1 + \cdots + a_{mn}Z_n + \mu_m\end{aligned}$$

Table 1.4.2

Continuous Probability Distribution	Probability Density Function, $f(x)$	Moment Generating Function, $\phi(t)$	Mean	Variance
Uniform over (a, b)	$\frac{1}{b-a}, a < x < b$	$\frac{e^{tb} - e^{ta}}{t(b-a)}$	$\frac{a+b}{2}$	$\frac{(b-a)^2}{12}$
Exponential with parameter $\lambda > 0$	$\lambda e^{-\lambda x}, x \geq 0$	$\frac{\lambda}{\lambda - t}$	$\frac{1}{\lambda}$	$\frac{1}{\lambda^2}$
Gamma with parameters (n, λ), $\lambda > 0$	$\frac{\lambda e^{-\lambda x}(\lambda x)^{n-1}}{(n-1)!}, x \geq 0$	$\left(\frac{\lambda}{\lambda - t}\right)^n$	$\frac{n}{\lambda}$	$\frac{n}{\lambda^2}$
Normal with parameters (μ, σ^2)	$\frac{1}{\sqrt{2\pi}\,\sigma} e^{-(x-\mu)^2/2\sigma^2}, -\infty < x < \infty$	$\exp\left\{\mu t + \frac{\sigma^2 t^2}{2}\right\}$	μ	σ^2
Beta with parameters a, b, $a > 0$, $b > 0$	$cx^{a-1}(1-x)^{b-1}, 0 < x < 1$ $c = \frac{\Gamma(a+b)}{\Gamma(a)\Gamma(b)}$		$\frac{a}{a+b}$	$\frac{ab}{(a+b)^2(a+b+1)}$

then the random variables $X_1, \ldots, X_m$ are said to have a multivariate normal distribution.

Let us now consider

$$\psi(t_1, \ldots, t_m) = E[\exp\{t_1 X_1 + \cdots + t_m X_m\}],$$

the joint moment generating function of $X_1, \ldots, X_m$. The first thing to note is that since $\sum_{i=1}^{m} t_i X_i$ is itself a linear combination of the independent normal random variables $Z_1, \ldots, Z_n$ it is also normally distributed. Its mean and variance are

$$E\left[\sum_{i=1}^{m} t_i X_i\right] = \sum_{i=1}^{m} t_i \mu_i$$

and

$$\begin{aligned}\operatorname{Var}\left(\sum_{i=1}^{m} t_i X_i\right) &= \operatorname{Cov}\left(\sum_{i=1}^{m} t_i X_i, \sum_{j=1}^{m} t_j X_j\right) \\ &= \sum_{i=1}^{m}\sum_{j=1}^{m} t_i t_j \operatorname{Cov}(X_i, X_j).\end{aligned}$$

Now, if Y is a normal random variable with mean μ and variance σ^2 then

$$E[e^Y] = \psi_Y(t)|_{t=1} = e^{\mu + \sigma^2/2}.$$

Thus, we see that

$$\psi(t_1, \ldots, t_m) = \exp\left\{\sum_{i=1}^{m} t_i \mu_i + 1/2 \sum_{i=1}^{m}\sum_{j=1}^{m} t_i t_j \operatorname{Cov}(X_i, X_j)\right\},$$

which shows that the joint distribution of $X_1, \ldots, X_m$ is completely determined from a knowledge of the values of $E[X_i]$ and $\operatorname{Cov}(X_i, X_j)$, $i, j = 1, \ldots, m$.

When dealing with random variables that only assume nonnegative values, it is sometimes more convenient to use *Laplace transforms* rather than characteristic functions. The Laplace transform of the distribution F is defined by

$$\tilde{F}(s) = \int_0^\infty e^{-sx}\, dF(x).$$

This integral exists for complex variables $s = a + bi$, where $a \geq 0$. As in the case of characteristic functions, the Laplace transform uniquely determines the distribution.

We may also define Laplace transforms for arbitrary functions in the following manner: The Laplace transform of the function g, denoted $\tilde{g}$, is defined by

$$\tilde{g}(s) = \int_0^\infty e^{-sx}\, dg(x)$$

provided the integral exists. It can be shown that $\tilde{g}$ determines g up to an additive constant.

1.5 Conditional Expectation

If X and Y are discrete random variables, the conditional probability mass function of X, given $Y = y$, is defined, for all y such that $P\{Y = y\} > 0$, by

$$P\{X = x \mid Y = y\} = \frac{P\{X = x, Y = y\}}{P\{Y = y\}}.$$

The conditional distribution function of X given $Y = y$ is defined by

$$F(x \mid y) = P\{X \le x \mid Y = y\}$$

and the conditional expectation of X given $Y = y$, by

$$E[X \mid Y = y] = \int x\, dF(x \mid y) = \sum_x x P\{X = x \mid Y = y\}.$$

If X and Y have a joint probability density function $f(x, y)$, the conditional probability density function of X, given $Y = y$, is defined for all y such that $f_Y(y) > 0$ by

$$f(x \mid y) = \frac{f(x, y)}{f_Y(y)},$$

and the conditional probability distribution function of X, given $Y = y$, by

$$F(x \mid y) = P\{X \le x \mid Y = y\} = \int_{-\infty}^{x} f(x \mid y)\, dx.$$

The conditional expectation of X, given $Y = y$, is defined, in this case, by

$$E[X \mid Y = y] = \int_{-\infty}^{\infty} x f(x \mid y)\, dx.$$

Thus all definitions are exactly as in the unconditional case except that all probabilities are now conditional on the event that $Y = y$.

Let us denote by $E[X|Y]$ that function of the random variable Y whose value at $Y = y$ is $E[X|Y = y]$. An extremely useful property of conditional expectation is that for all random variables X and Y

$$E[X] = E[E[X|Y]] = \int E[X|Y = y]\, dF_Y(y) \tag{1.5.1}$$

when the expectations exist.

If Y is a discrete random variable, then Equation (1.5.1) states

$$E[X] = \sum_y E[X|Y = y]P\{Y = y\},$$

While if Y is continuous with density $f(y)$, then Equation (1.5.1) says

$$E[X] = \int_{-\infty}^{\infty} E[X|Y = y]f(y)\, dy.$$

We now give a proof of Equation (1.5.1) in the case where X and Y are both discrete random variables.

Proof of (1.5.1) *when X and Y Are Discrete* To show

$$E[X] = \sum_y E[X|Y = y]P\{Y = y\}.$$

We write the right-hand side of the above as

$$\begin{aligned}
\sum_y E[X|Y = y]P\{Y = y\} &= \sum_y \sum_x xP\{X = x|Y = y\}P\{Y = y\} \\
&= \sum_y \sum_x xP\{X = x, Y = y\} \\
&= \sum_x x \sum_y P\{X = x, Y = y\} \\
&= \sum_x xP\{X = x\} \\
&= E[X].
\end{aligned}$$

and the result is obtained.

Thus from Equation (1.5.1) we see that $E[X]$ is a weighted average of the conditional expected value of X given that $Y = y$, each of the terms $E[X|Y = y]$ being weighted by the probability of the event on which it is conditioned.

Example 1.5(a) ***The Sum of a Random Number of Random Variables.*** Let $X_1, X_2, \ldots$ denote a sequence of independent and identically distributed random variables; and let N denote a nonnegative integer valued random variable that is independent of the sequence $X_1, X_2, \ldots$. We shall compute the moment generating function of $Y = \sum_1^N X_i$ by first conditioning on N. Now

$$\begin{aligned}
E\left[\exp\left\{t\sum_1^N X_i\right\}\middle| N=n\right]
&= E\left[\exp\left\{t\sum_1^n X_i\right\}\middle| N=n\right] \\
&= E\left[\exp\left\{t\sum_1^n X_i\right\}\right] \qquad \text{(by independence)} \\
&= (\Psi_X(t))^n,
\end{aligned}$$

where $\Psi_X(t) = E[e^{tX}]$ is the moment generating function of X. Hence,

$$E\left[\exp\left\{t\sum_1^N X_i\right\}\middle| N\right] = (\psi_X(t))^N$$

and so

$$\psi_Y(t) = E\left[\exp\left\{t\sum_1^N X_i\right\}\right] = E[(\psi_X(t))^N].$$

To compute the mean and variance of $Y = \sum_1^N X_i$, we differentiate $\psi_Y(t)$ as follows:

$$\begin{aligned}
\psi_Y'(t) &= E[N(\psi_X(t))^{N-1}\psi_X'(t)], \\
\psi_Y''(t) &= E[N(N-1)(\psi_X(t))^{N-2}(\psi_X'(t))^2 + N(\psi_X(t))^{N-1}\psi_X''(t)].
\end{aligned}$$

Evaluating at $t = 0$ gives

$$E[Y] = E[NE[X]] = E[N]E[X]$$

and

$$\begin{aligned}
E[Y^2] &= E[N(N-1)E^2[X] + NE[X^2]] \\
&= E[N]\operatorname{Var}(X) + E[N^2]E^2[X].
\end{aligned}$$

Hence,

$$\operatorname{Var}(Y) = E[Y^2] - E^2[Y]$$
$$= E[N]\operatorname{Var}(X) + E^2[X]\operatorname{Var}(N).$$

Example 1.5(b) A miner is trapped in a mine containing three doors. The first door leads to a tunnel that takes him to safety after two hours of travel. The second door leads to a tunnel that returns him to the mine after three hours of travel. The third door leads to a tunnel that returns him to his mine after five hours. Assuming that the miner is at all times equally likely to choose any one of the doors, let us compute the moment generating function of X, the time when the miner reaches safety.

Let Y denote the door initially chosen. Then

$$E[e^{tX}] = \tfrac{1}{3}(E[e^{tX} \mid Y = 1] + E[e^{tX} \mid Y = 2] + E[e^{tX} \mid Y = 3]). \tag{1.5.2}$$

Now given that $Y = 1$, it follows that $X = 2$, and so

$$E[e^{tX} \mid Y = 1] = e^{2t}.$$

Also, given that $Y = 2$, it follows that $X = 3 + X'$, where X' is the number of additional hours to safety after returning to the mine. But once the miner returns to his cell the problem is exactly as before, and thus X' has the same distribution as X. Therefore,

$$E[e^{tX} \mid Y = 2] = E[e^{t(3+X')}]$$
$$= e^{3t}E[e^{tX}].$$

Similarly,

$$E[e^{tX} \mid Y = 3] = e^{5t}E[e^{tX}].$$

Substitution back into (1.5.2) yields

$$E[e^{tX}] = \tfrac{1}{3}(e^{2t} + e^{3t}E[e^{tX}] + e^{5t}E[e^{tX}])$$

or

$$E[e^{tX}] = \frac{e^{2t}}{3 - e^{3t} - e^{5t}}.$$

Not only can we obtain expectations by first conditioning upon an appropriate random variable, but we may also use this approach to compute probabilities. To see this, let E denote an arbitrary event and define the indicator random variable X by

$$X = \begin{cases} 1 & \text{if } E \text{ occurs} \\ 0 & \text{if } E \text{ does not occur.} \end{cases}$$

It follows from the definition of X that

$$E[X] = P(E)$$
$$E[X \mid Y = y] = P(E \mid Y = y) \qquad \text{for any random variable } Y.$$

Therefore, from Equation (1.5.1) we obtain that

$$P(E) = \int P(E \mid Y = y)\, dF_Y(y).$$

Example 1.5(c) Suppose in the matching problem, Example 1.3(a), that those choosing their own hats depart, while the others (those without a match) put their selected hats in the center of the room, mix them up, and then reselect. If this process continues until each individual has his or her own hat, find $E[R_n]$ where R_n is the number of rounds that are necessary.

We will now show that $E[R_n] = n$. The proof will be by induction on n, the number of individuals. As it is obvious for $n = 1$ assume that $E[R_k] = k$ for $k = 1, \ldots, n - 1$. To compute $E[R_n]$, start by conditioning on M, the number of matches that occur in the first round. This gives

$$E[R_n] = \sum_{i=0}^{n} E[R_n \mid M = i] P\{M = i\}.$$

Now, given a total of i matches in the initial round, the number of rounds needed will equal 1 plus the number of rounds that are required when $n - i$ people remain to be matched with their hats. Therefore,

$$\begin{aligned}
E[R_n] &= \sum_{i=0}^{n} (1 + E[R_{n-i}]) P\{M = i\} \\
&= 1 + E[R_n] P\{M = 0\} + \sum_{i=1}^{n} E[R_{n-i}] P\{M = i\} \\
&= 1 + E[R_n] P\{M = 0\} + \sum_{i=1}^{n} (n - i) P\{M = i\} \\
&\qquad \text{(by the induction hypothesis)} \\
&= 1 + E[R_n] P\{M = 0\} + n(1 - P\{M = 0\}) - E[M] \\
&= E[R_n] P\{M = 0\} + n(1 - P\{M = 0\}) \\
&\qquad \text{(since } E[M] = 1\text{)}
\end{aligned}$$

which proves the result.

EXAMPLE 1.5(D) Suppose that X and Y are independent random variables having respective distributions F and G. Then the distribution of $X + Y$—which we denote by $F * G$, and call the *convolution* of F and G—is given by

$$\begin{aligned}(F * G)(a) &= P\{X + Y \leq a\} \\ &= \int_{-\infty}^{\infty} P\{X + Y \leq a \mid Y = y\}\, dG(y) \\ &= \int_{-\infty}^{\infty} P\{X + y \leq a \mid Y = y\}\, dG(y) \\ &= \int_{-\infty}^{\infty} F(a - y)\, dG(y).\end{aligned}$$

We denote $F * F$ by F_2 and in general $F * F_{n-1} = F_n$. Thus F_n, the n-fold convolution of F with itself, is the distribution of the sum of n independent random variables each having distribution F.

EXAMPLE 1.5(E) ***The Ballot Problem.*** In an election, candidate A receives n votes and candidate B receives m votes, where $n > m$. Assuming that all orderings are equally likely, show that the probability that A is always ahead in the count of votes is $(n - m)/(n + m)$.

Solution. Let $P_{n,m}$ denote the desired probability. By conditioning on which candidate receives the last vote counted we have

$$\begin{aligned}P_{n,m} = P\{&A \text{ always ahead} \mid A \text{ receives last vote}\}\frac{n}{n+m} \\ &+ P\{A \text{ always ahead} \mid B \text{ receives last vote}\}\frac{m}{n+m}.\end{aligned}$$

Now it is easy to see that, given that A receives the last vote, the probability that A is always ahead is the same as if A had received a total of $n - 1$ and B a total of m votes. As a similar result is true when we are given that B receives the last vote, we see from the above that

$$P_{n,m} = \frac{n}{n+m} P_{n-1,m} + \frac{m}{m+n} P_{n,m-1}. \tag{1.5.3}$$

We can now prove that

$$P_{n,m} = \frac{n-m}{n+m}$$

by induction on $n + m$. As it is obviously true when $n + m = 1$—that is, $P_{1,0} = 1$—assume it whenever $n + m = k$. Then when

$n + m = k + 1$ we have by (1.5.3) and the induction hypothesis

$$P_{n,m} = \frac{n}{n+m}\frac{n-1-m}{n-1+m} + \frac{m}{m+n}\frac{n-m+1}{n+m-1}$$
$$= \frac{n-m}{n+m}.$$

The result is thus proven.

The ballot problem has some interesting applications. For example, consider successive flips of a coin that always lands on "heads" with probability p, and let us determine the probability distribution of the first time, after beginning, that the total number of heads is equal to the total number of tails. The probability that the first time this occurs is at time $2n$ can be obtained by first conditioning on the total number of heads in the first $2n$ trials. This yields

$$P\{\text{first time equal} = 2n\}$$
$$= P\{\text{first time equal} = 2n \mid n \text{ heads in first } 2n\}\binom{2n}{n}p^n(1-p)^n.$$

Now given a total of n heads in the first $2n$ flips, it is easy to see that all possible orderings of the n heads and n tails are equally likely and thus the above conditional probability is equivalent to the probability that in an election in which each candidate receives n votes, one of the candidates is always ahead in the counting until the last vote (which ties them). But by conditioning on whoever receives the last vote, we see that this is just the probability in the ballot problem when $m = n - 1$. Hence,

$$P\{\text{first time equal} = 2n\} = P_{n,n-1}\binom{2n}{n}p^n(1-p)^n$$
$$= \frac{\binom{2n}{n}p^n(1-p)^n}{2n-1}.$$

Example 1.5(f) ***The Matching Problem Revisited.*** Let us reconsider Example 1.3(a) in which n individuals mix their hats up and then randomly make a selection. We shall compute the probability of exactly k matches.

First let E denote the event that no matches occur, and to make explicit the dependence on n write $P_n = P(E)$. Upon conditioning on whether or not the first individual selects his or her own hat—call these events M and M^c—we obtain

$$P_n = P(E) = P(E \mid M)P(M) + P(E \mid M^c)P(M^c).$$

Clearly, $P(E|M) = 0$, and so

$$P_n = P(E|M^c)\frac{n-1}{n}. \tag{1.5.4}$$

Now, $P(E|M^c)$ is the probability of no matches when $n - 1$ people select from a set of $n - 1$ hats that does not contain the hat of one of them. This can happen in either of two mutually exclusive ways. Either there are no matches and the extra person does not select the extra hat (this being the hat of the person that chose first), or there are no matches and the extra person does select the extra hat. The probability of the first of these events is P_{n-1}, which is seen by regarding the extra hat as "belonging" to the extra person. Since the second event has probability $[1/(n - 1)]P_{n-2}$, we have

$$P(E|M^c) = P_{n-1} + \frac{1}{n-1}P_{n-2}$$

and thus, from Equation (1.5.4),

$$P_n = \frac{n-1}{n}P_{n-1} + \frac{1}{n}P_{n-2},$$

or, equivalently,

$$P_n - P_{n-1} = -\frac{1}{n}(P_{n-1} - P_{n-2}). \tag{1.5.5}$$

However, clearly

$$P_1 = 0, \qquad P_2 = \tfrac{1}{2}.$$

Thus, from Equation (1.5.5),

$$P_3 - P_2 = -\frac{(P_2 - P_1)}{3} = -\frac{1}{3!} \quad \text{or} \quad P_3 = \frac{1}{2!} - \frac{1}{3!},$$

$$P_4 - P_3 = -\frac{(P_3 - P_2)}{4} = \frac{1}{4!} \quad \text{or} \quad P_4 = \frac{1}{2!} - \frac{1}{3!} + \frac{1}{4!},$$

and, in general, we see that

$$P_n = \frac{1}{2!} - \frac{1}{3!} + \frac{1}{4!} - \cdots + \frac{(-1)^n}{n!}.$$

To obtain the probability of exactly k matches, we consider any fixed group of k individuals. The probability that they, and only they, select their own hats is

$$\frac{1}{n}\frac{1}{n-1}\cdots\frac{1}{n-(k-1)}P_{n-k}=\frac{(n-k)!}{n!}P_{n-k},$$

where P_{n-k} is the conditional probability that the other $n - k$ individuals, selecting among their own hats, have no matches. As there are $\binom{n}{k}$ choices of a set of k individuals, the desired probability of exactly k matches is

$$\binom{n}{k}\frac{(n-k)!}{n!}P_{n-k}=\frac{\dfrac{1}{2!}-\dfrac{1}{3!}+\cdots+\dfrac{(-1)^{n-k}}{(n-k)!}}{k!},$$

which, for n large, is approximately equal to $e^{-1}/k!$.

Thus for n large the number of matches has approximately the Poisson distribution with mean 1. To understand this result better recall that the Poisson distribution with mean λ is the limiting distribution of the number of successes in n independent trials, each resulting in a success with probability p_n, when $np_n \to \lambda$ as $n \to \infty$. Now if we let

$$X_i=\begin{cases}1 & \text{if the } i\text{th person selects his or her own hat}\\ 0 & \text{otherwise,}\end{cases}$$

then the number of matches, $\sum_{i=1}^{n} X_i$, can be regarded as the number of successes in n trials when each is a success with probability $1/n$. Now, whereas the above result is not immediately applicable because these trials are not independent, it is true that it is a rather weak dependence since, for example,

$$P\{X_i = 1\} = 1/n$$

and

$$P\{X_i = 1 \mid X_j = 1\} = 1/(n-1), \qquad j \neq i.$$

Hence we would certainly hope that the Poisson limit would still remain valid under this type of weak dependence. The results of this example show that it does.

Example 1.5(g) ***A Packing Problem.*** Suppose that n points are arranged in linear order, and suppose that a pair of adjacent points is chosen at random. That is, the pair $(i, i + 1)$ is chosen with probability $1/(n - 1)$, $i = 1, 2, \ldots, n - 1$. We then continue to randomly choose pairs, disregarding any pair having a point

1 2 3 4 5 6 7 8

Figure 1.5.1

previously chosen, until only isolated points remain. We are interested in the mean number of isolated points.

For instance, if $n = 8$ and the random pairs are, in order of appearance, (2, 3), (7, 8), (3, 4), and (4, 5), then there will be two isolated points (the pair (3, 4) is disregarded) as shown in Figure 1.5.1.

If we let

$$I_{i,n} = \begin{cases} 1 & \text{if point } i \text{ is isolated} \\ 0 & \text{otherwise,} \end{cases}$$

then $\sum_{i=1}^{n} I_{i,n}$ represents the number of isolated points. Hence

$$E[\text{number of isolated points}] = \sum_{i=1}^{n} E[I_{i,n}]$$
$$= \sum_{i=1}^{n} P_{i,n},$$

where $P_{i,n}$ is defined to be the probability that point i is isolated when there are n points. Let

$$P_n \equiv P_{n,n} = P_{1,n}.$$

That is, P_n is the probability that the extreme point n (or 1) will be isolated. To derive an expression for $P_{i,n}$, note that we can consider the n points as consisting of two contiguous segments, namely,

$$1, 2, \ldots, i \quad \text{and} \quad i, i+1, \ldots, n.$$

Since point i will be vacant if and only if the right-hand point of the first segment and the left-hand point of the second segment are both vacant, we see that

$$P_{i,n} = P_i P_{n-i+1}. \tag{1.5.6}$$

Hence the $P_{i,n}$ will be determined if we can calculate the corresponding probabilities that extreme points will be vacant. To derive an expression for the P_n condition on the initial pair—say $(i, i + 1)$—and note that this choice breaks the line into two independent segments—$1, 2, \ldots, i - 1$ and $i + 2, \ldots, n$. That is, if the initial pair is $(i, i+1)$, then the extreme point n will be isolated if the extreme point of a set of $n-i-1$ points is isolated. Hence we have

$$P_n = \sum_{i=1}^{n-1} \frac{P_{n-i-1}}{n-1} = \frac{P_1 + \cdots + P_{n-2}}{n-1}$$

or

$$(n-1)P_n = P_1 + \cdots + P_{n-2}.$$

Substituting $n-1$ for n gives

$$(n-2)P_{n-1} = P_1 + \cdots + P_{n-3}$$

and subtracting these two equations gives

$$(n-1)P_n - (n-2)P_{n-1} = P_{n-2}$$

or

$$P_n - P_{n-1} = -\frac{P_{n-1} - P_{n-2}}{n-1}.$$

Since $P_1 = 1$ and $P_2 = 0$, this yields

$$P_3 - P_2 = -\frac{P_2 - P_1}{2} = \frac{1}{2} \quad \text{or} \quad P_3 = \frac{1}{2!}$$

$$P_4 - P_3 = -\frac{P_3 - P_2}{3} = -\frac{1}{3!} \quad \text{or} \quad P_4 = \frac{1}{2!} - \frac{1}{3!},$$

and, in general,

$$P_n = \frac{1}{2!} - \frac{1}{3!} + \cdots + \frac{(-1)^{n-1}}{(n-1)!} = \sum_{j=0}^{n-1} \frac{(-1)^j}{j!}, \qquad n \geq 2.$$

Thus, from (1.5.6),

$$P_{i,n} = \begin{cases} \displaystyle\sum_{j=0}^{n-1} \frac{(-1)^j}{j!} & i = 1, n \\ 0 & i = 2, n-1 \\ \displaystyle\sum_{j=0}^{i-1} \frac{(-1)^j}{j!} \sum_{j=0}^{n-i} \frac{(-1)^j}{j!} & 2 < i < n-1. \end{cases}$$

For i and $n - i$ large we see from the above that $P_{i,n} \approx e^{-2}$, and, in fact, it can be shown from the above that $\sum_{i=1}^{n} P_{i,n}$—the expected number of vacant points—is approximately given by

$$\sum_{i=1}^{n} P_{i,n} \approx (n+2)e^{-2} \qquad \text{for large } n.$$

EXAMPLE 1.5(H) ***A Reliability Example.*** Consider an n component system that is subject to randomly occurring shocks. Suppose that each shock has a value that, independent of all else, is chosen from a distribution G. If a shock of value x occurs, then each component that was working at the moment the shock arrived will, independently, instantaneously fail with probability x. We are interested in the distribution of N, the number of necessary shocks until all components are failed.

To compute $P\{N > k\}$ let E_i, $i = 1, \ldots, n$, denote the event that component i has survived the first k shocks. Then

$$\begin{aligned} P\{N > k\} &= P\left(\bigcup_{1}^{n} E_i\right) \\ &= \sum_i P(E_i) - \sum_{i<l} P(E_i E_l) \\ &\quad + \cdots + (-1)^{n+1} P(E_1 E_2 \cdots E_n). \end{aligned}$$

To compute the above probability let p_j denote the probability that a given set of j components will all survive some arbitrary shock. Conditioning on the shock's value gives

$$\begin{aligned} p_j &= \int P\{\text{set of } j \text{ survive} \mid \text{value is } x\}\, dG(x) \\ &= \int (1-x)^j\, dG(x). \end{aligned}$$

Since

$$\begin{aligned} P(E_i) &= p_1^k, \\ P(E_i E_l) &= p_2^k, \ldots, P(E_1 \cdots E_n) = p_n^k, \end{aligned}$$

we see that

$$P\{N > k\} = np_1^k - \binom{n}{2} p_2^k + \binom{n}{3} p_3^k \cdots (-1)^{n+1} p_n^k.$$

The mean of N can be computed from the above as follows:

$$
\begin{aligned}
E[N] &= \sum_{k=0}^{\infty} P\{N > k\} \\
&= \sum_{k=0}^{\infty} \sum_{i=1}^{n} \binom{n}{i} (-1)^{i+1} p_i^k \\
&= \sum_{i=1}^{n} \binom{n}{i} (-1)^{i+1} \sum_{k=0}^{\infty} p_i^k \\
&= \sum_{i=1}^{n} \binom{n}{i} \frac{(-1)^{i+1}}{1 - p_i}.
\end{aligned}
$$

The reader should note that we have made use of the identity $E[N] = \sum_{k=0}^{\infty} P\{N > k\}$, valid for all nonnegative integer-valued random variables N (see Problem 1.1).

Example 1.5(i) ***Classifying a Poisson Number of Events.*** Suppose that we are observing events, and that N, the total number that occur, is a Poisson random variable with mean λ. Suppose also that each event that occurs is, independent of other events, classified as a type j event with probability p_j, $j = 1, \ldots, k$, $\sum_{j=1}^{k} p_j = 1$. Let N_j denote the number of type j events that occur, $j = 1, \ldots, k$, and let us determine their joint probability mass function.

For any nonnegative integers n_j, $j = 1, \ldots, k$, let $n = \sum_{j=1}^{k} n_j$. Then, since $N = \sum_j N_j$, we have that

$$
\begin{aligned}
&P\{N_j = n_j, j = 1, \ldots, k\} \\
&\quad = P\{N_j = n_j, j = 1, \ldots, k \mid N = n\} P\{N = n\} \\
&\qquad + P\{N_j = n_j, j = 1, \ldots, k \mid N \neq n\} P\{N \neq n\} \\
&\quad = P\{N_j = n_j, j = 1, \ldots, k \mid N = n\} P\{N = n\}.
\end{aligned}
$$

Now, given that there are a total of $N = n$ events it follows, since each event is independently a type j event with probability p_j, $1 \leq j \leq k$, that $N_1, N_2, \ldots, N_k$ has a multinomial distribution with parameters n and $p_1, p_2 \ldots, p_k$. Therefore,

$$
\begin{aligned}
P\{N_j = n_j, j = 1, \ldots, k\} &= \frac{n!}{n_1! n_2! \cdots n_k!} p_1^{n_1} p_2^{n_2} \cdots p_k^{n_k} e^{-\lambda} \lambda^n / n! \\
&= \prod_j e^{-\lambda p_j} \frac{(\lambda p_j)^{n_j}}{n_j!}.
\end{aligned}
$$

Thus we can conclude that the N_j are independent Poisson random variables with respective means λp_j, $j = 1, \ldots, k$.

Conditional expectations given that $Y = y$ satisfy all of the properties of ordinary expectations, except that now all probabilities are conditioned on the event that $\{Y = y\}$. Hence, we have that

$$E\left[\sum_{i=1}^{n} X_i \mid Y = y\right] = \sum_{i=1}^{n} E[X_i \mid Y = y]$$

implying that

$$E\left[\sum_{i=1}^{n} X_i \mid Y\right] = \sum_{i=1}^{n} E[X_i \mid Y].$$

Also, from the equality $E[X] = E[E[X \mid Y]]$ we can conclude that

$$E[X \mid W = w] = E[E[X \mid W = w, Y] \mid W = w]$$

or, equivalently,

$$E[X \mid W] = E[E[X \mid W, Y] \mid W].$$

Also, we should note that the fundamental result

$$E[X] = E[E[X \mid Y]]$$

remains valid even when Y is a random vector.

1.5.1 Conditional Expectations and Bayes Estimators

Conditional expectations have important uses in the Bayesian theory of statistics. A classical problem in this area arises when one is to observe data $\boldsymbol{X} = (X_1, \ldots, X_n)$ whose distribution is determined by the value of a random variable θ, which has a specified probability distribution (called the prior distribution). Based on the value of the data $\boldsymbol{X}$ a problem of interest is to estimate the unseen value of θ. An estimator of θ can be any function $d(\boldsymbol{X})$ of the data, and in Bayesian statistics one often wants to choose $d(\boldsymbol{X})$ to minimize $E[(d(\boldsymbol{X}) - \theta)^2 \mid \boldsymbol{X}]$, the conditional expected squared distance between the estimator and the parameter. Using the facts that

(i) conditional on $\boldsymbol{X}$, $d(\boldsymbol{X})$ is a constant; and

(ii) for any random variable W, $E[(W - c)^2]$ is minimized when $c = E[W]$

it follows that the estimator that minimizes $E[(d(\boldsymbol{X}) - \theta)^2 | \boldsymbol{X}]$, called the *Bayes estimator*, is given by

$$d(\boldsymbol{X}) = E[\theta | \boldsymbol{X}].$$

An estimator $d(\boldsymbol{X})$ is said to be an *unbiased* estimator of θ if

$$E[d(\boldsymbol{X}) | \theta] = \theta.$$

An important result in Bayesian statistics is that the only time that a Bayes estimator is unbiased is in the trivial case where it is equal to θ with probability 1. To prove this, we start with the following lemma.

Lemma 1.5.1

For any random variable Y and random vector $\mathbf{Z}$

$$E[(Y - E[Y|\mathbf{Z}])E[Y|\mathbf{Z}]] = 0.$$

Proof

$$E[YE[Y|\mathbf{Z}]] = E[E[YE[Y|\mathbf{Z}]|\mathbf{Z}]]$$
$$= E[E[Y|\mathbf{Z}]E[Y|\mathbf{Z}]]$$

where the final equality follows because, given $\mathbf{Z}$, $E[Y|\mathbf{Z}]$ is a constant and so $E[YE[Y|\mathbf{Z}]|\mathbf{Z}] = E[Y|\mathbf{Z}]E[Y|\mathbf{Z}]$. Since the final equality is exactly what we wanted to prove, the lemma follows.

PROPOSITION 1.5.2

If $P\{E[\theta|\boldsymbol{X}] = \theta\} \neq 1$ then the Bayes estimator $E[\theta|\boldsymbol{X}]$ is not unbiased.

Proof Letting $Y = \theta$ and $\mathbf{Z} = \boldsymbol{X}$ in Lemma 1.5.1 yields that

$$E[(\theta - E[\theta|\boldsymbol{X}])E[\theta|\boldsymbol{X}]] = 0. \tag{1.5.7}$$

Now let $Y = E[\theta|\boldsymbol{X}]$ and suppose that Y is an unbiased estimator of θ so that $E[Y|\theta] = \theta$. Letting $\boldsymbol{Z} = \theta$ we obtain from Lemma 1.5.1 that

$$E[(E[\theta|\boldsymbol{X}] - \theta)\theta] = 0. \tag{1.5.8}$$

Upon adding Equations (1.5.7) and (1.5.8) we obtain that

$$E[(\theta - E[\theta|\boldsymbol{X}])E[\theta|\boldsymbol{X}]] + E[(E[\theta|\boldsymbol{X}] - \theta)\theta] = 0.$$

or,

$$E[(\theta - E[\theta|X])E[\theta|X] + (E[\theta|X] - \theta)\theta] = 0$$

or,

$$-E[(\theta - E[\theta|X])^2] = 0$$

implying that, with probability 1, $\theta - E[\theta|X] = 0$.

1.6 The Exponential Distribution, Lack of Memory, and Hazard Rate Functions

A continuous random variable X is said to have an *exponential distribution* with parameter λ, $\lambda > 0$, if its probability density function is given by

$$f(x) = \begin{cases} \lambda e^{-\lambda x} & x \geq 0 \\ 0 & x < 0, \end{cases}$$

or, equivalently, if its distribution is

$$F(x) = \int_{-\infty}^{x} f(y)\, dy = \begin{cases} 1 - e^{-\lambda x} & x \geq 0 \\ 0 & x < 0. \end{cases}$$

The moment generating function of the exponential distribution is given by

$$E[e^{tX}] = \int_0^\infty e^{tx} \lambda e^{-\lambda x}\, dx = \frac{\lambda}{\lambda - t}. \tag{1.6.1}$$

All the moments of X can now be obtained by differentiating (1.6.1), and we leave it to the reader to verify that

$$E[X] = 1/\lambda, \qquad \text{Var}(X) = 1/\lambda^2.$$

The usefulness of exponential random variables derives from the fact that they possess the memoryless property, where a random variable X is said to be without memory, or *memoryless*, if

$$P\{X > s + t \mid X > t\} = P\{X > s\} \qquad \text{for } s, t \geq 0. \tag{1.6.2}$$

If we think of X as being the lifetime of some instrument, then (1.6.2) states that the probability that the instrument lives for at least $s + t$ hours, given that it has survived t hours, is the same as the initial probability that it lives for at least s hours. In other words, if the instrument is alive at time t, then the distribution of its remaining life is the original lifetime distribution. The condition (1.6.2) is equivalent to

$$\bar{F}(s+t) = \bar{F}(s)\bar{F}(t),$$

and since this is satisfied when F is the exponential, we see that such random variables are memoryless.

Example 1.6(a) Consider a post office having two clerks, and suppose that when A enters the system he discovers that B is being served by one of the clerks and C by the other. Suppose also that A is told that his service will begin as soon as either B or C leaves. If the amount of time a clerk spends with a customer is exponentially distributed with mean $1/\lambda$, what is the probability that, of the three customers, A is the last to leave the post office?

The answer is obtained by reasoning as follows: Consider the time at which A first finds a free clerk. At this point either B or C would have just left and the other one would still be in service. However, by the lack of memory of the exponential, it follows that the amount of additional time that this other person has to spend in the post office is exponentially distributed with mean $1/\lambda$. That is, it is the same as if he was just starting his service at this point. Hence, by symmetry, the probability that he finishes before A must equal $\frac{1}{2}$.

Example 1.6(b) Let $X_1, X_2, \ldots$ be independent and identically distributed continuous random variables with distribution F. We say that a record occurs at time n, $n > 0$, and has value X_n if $X_n > \max(X_1, \ldots, X_{n-1})$, where $X_0 = -\infty$. That is, a record occurs each time a new high is reached. Let τ_i denote the time between the ith and the $(i + 1)$th record. What is its distribution?

As a preliminary to computing the distribution of τ_i, let us note that the record times of the sequence $X_1, X_2, \ldots$ will be the same as for the sequence $F(X_1), F(X_2), \ldots$, and since $F(X)$ has a uniform $(0, 1)$ distribution (see Problem 1.2), it follows that the distribution of τ_i does not depend on the actual distribution F (as long as it is continuous). So let us suppose that F is the exponential distribution with parameter $\lambda = 1$.

To compute the distribution of τ_i, we will condition on R_i the ith record value. Now $R_1 = X_1$ is exponential with rate 1. R_2 has the distribution of an exponential with rate 1 given that it is greater than R_1. But by the lack of memory property of the exponential this means that R_2 has the same distribution as R_1 plus an independent exponential with rate 1. Hence R_2 has the same distribution as the sum of two independent exponential random variables with rate 1. The same argument shows that R_i has the same distribution as the sum of i independent exponentials with rate 1. But it is well known (see Problem 1.29) that such a random variable has the gamma distribution with parameters $(i, 1)$. That is, the density of R_i is given by

$$f_{R_i}(t) = \frac{e^{-t}t^{i-1}}{(i-1)!}, \qquad t \geq 0.$$

Hence, conditioning on R_i yields

$$\begin{aligned} P\{\tau_i > k\} &= \int_0^\infty P\{\tau_i > k \mid R_i = t\} \frac{e^{-t}t^{i-1}}{(i-1)!}\,dt \\ &= \int_0^\infty (1 - e^{-t})^k e^{-t} \frac{t^{i-1}}{(i-1)!}\,dt, \qquad i \geq 1, \end{aligned}$$

where the last equation follows since if the ith record value equals t, then none of the next k values will be records if they are all less than t.

It turns out that not only is the exponential distribution "memoryless," but it is the unique distribution possessing this property. To see this, suppose that X is memoryless and let $\bar{F}(x) = P\{X > x\}$. Then

$$\bar{F}(s + t) = \bar{F}(s)\bar{F}(t).$$

That is, $\bar{F}$ satisfies the functional equation

$$g(s + t) = g(s)g(t).$$

However, the only solutions of the above equation that satisfy any sort of reasonable condition (such as monotonicity, right or left continuity, or even measurability) are of the form

$$g(x) = e^{-\lambda x}$$

for some suitable value of λ. [A simple proof when g is assumed right continuous is as follows: Since $g(s + t) = g(s)g(t)$, it follows that $g(2/n) = g(1/n + 1/n) = g^2(1/n)$. Repeating this yields $g(m/n) = g^m(1/n)$. Also $g(1) = g(1/n + \cdots + 1/n) = g^n(1/n)$. Hence, $g(m/n) = (g(1))^{m/n}$, which implies, since g is right continuous, that $g(x) = (g(1))^x$. Since $g(1) = g^2(1/2) \geq 0$, we obtain $g(x) = e^{-\lambda x}$, where $\lambda = -\log(g(1))$]. Since a distribution function is always right continuous, we must have

$$\bar{F}(x) = e^{-\lambda x}.$$

The memoryless property of the exponential is further illustrated by the failure rate function (also called the hazard rate function) of the exponential distribution.

Consider a continuous random variable X having distribution function F and density f. The *failure* (or *hazard*) *rate* function $\lambda(t)$ is defined by

$$\lambda(t) = \frac{f(t)}{\bar{F}(t)}. \tag{1.6.3}$$

To interpret $\lambda(t)$, think of X as being the lifetime of some item, and suppose that X has survived for t hours and we desire the probability that it will not survive for an additional time dt. That is, consider $P\{X \in (t, t+dt) \mid X > t\}$. Now

$$\begin{aligned} P\{X \in (t, t+dt) \mid X > t\} &= \frac{P\{X \in (t, t+dt), X > t\}}{P\{X > t\}} \\ &= \frac{P\{X \in (t, t+dt)\}}{P\{X > t\}} \\ &\approx \frac{f(t)\,dt}{\bar{F}(t)} \\ &= \lambda(t)\,dt. \end{aligned}$$

That is, $\lambda(t)$ represents the probability intensity that a t-year-old item will fail.

Suppose now that the lifetime distribution is exponential. Then, by the memoryless property, it follows that the distribution of remaining life for a t-year-old item is the same as for a new item. Hence $\lambda(t)$ should be constant. This checks out since

$$\lambda(t) = \frac{\lambda e^{-\lambda t}}{e^{-\lambda t}} = \lambda.$$

Thus, the failure rate function for the exponential distribution is constant. The parameter λ is often referred to as the *rate* of the distribution. (Note that the rate is the reciprocal of the mean, and vice versa.)

It turns out that the failure rate function $\lambda(t)$ uniquely determines the distribution F. To prove this, we note that

$$\lambda(t) = \frac{-\dfrac{d}{dt}\bar{F}(t)}{\bar{F}(t)}.$$

Integration yields

$$\log \bar{F}(t) = -\int_0^t \lambda(t)\,dt + k$$

or

$$\bar{F}(t) = c \exp\left\{-\int_0^t \lambda(t)\,dt\right\}.$$

Letting $t = 0$ shows that $c = 1$ and so

$$\bar{F}(t) = \exp\left\{-\int_0^t \lambda(t)\,dt\right\}.$$

1.7 Some Probability Inequalities

We start with an inequality known as Markov's inequality.

Lemma 1.7.1 Markov's Inequality

If X is a nonnegative random variable, then for any $a > 0$

$$P\{X \geq a\} \leq E[X]/a.$$

Proof Let $I\{X \geq a\}$ be 1 if $X \geq a$ and 0 otherwise. Then, it is easy to see since $X \geq 0$ that

$$aI\{X \geq a\} \leq X.$$

Taking expectations yields the result.

PROPOSITION 1.7.2 Chernoff Bounds

Let X be a random variable with moment generating function $M(t) = E[e^{tX}]$. Then for $a > 0$

$$P\{X \geq a\} \leq e^{-ta}M(t) \qquad \text{for all } t > 0$$

$$P\{X \leq a\} \leq e^{-ta}M(t) \qquad \text{for all } t < 0.$$

Proof For $t > 0$

$$P\{X \geq a\} = P\{e^{tX} \geq e^{ta}\} \leq E[e^{tX}]e^{-ta}$$

where the inequality follows from Markov's inequality. The proof for $t < 0$ is similar.

Since the Chernoff bounds hold for all t in either the positive or negative quadrant, we obtain the best bound on $P\{X \geq a\}$ by using that t that minimizes $e^{-ta}M(t)$.

Example 1.7(a) ***Chernoff Bounds for Poisson Random Variables.*** If X is Poisson with mean λ, then $M(t) = e^{\lambda(e^t-1)}$. Hence, the Chernoff bound for $P\{X \geq j\}$ is

$$P\{X \geq j\} \leq e^{\lambda(e^t-1)-tj}.$$

The value of t that minimizes the preceding is that value for which $e^t = j/\lambda$. Provided that $j/\lambda > 1$, this minimizing value will be positive and so we obtain in this case that

$$P\{X \geq j\} \leq e^{\lambda(j/\lambda-1)}(\lambda/j)^j = e^{-\lambda}(\lambda e)^j/j^j, \qquad j > \lambda.$$

Our next inequality relates to expectations rather than probabilities.

PROPOSITION 1.7.3 Jensen's Inequality

If f is a convex function, then

$$E[f(X)] \geq f(E[X])$$

provided the expectations exist.

Proof We will give a proof under the supposition that f has a Taylor series expansion. Expanding about $\mu = E[X]$ and using the Taylor series with a remainder formula yields

$$\begin{aligned} f(x) &= f(\mu) + f'(\mu)(x - \mu) + f''(\xi)(x - \xi)^2/2 \\ &\geq f(\mu) + f'(\mu)(x - \mu) \end{aligned}$$

since $f''(\xi) \geq 0$ by convexity. Hence,

$$f(X) \geq f(\mu) + f'(\mu)(X - \mu).$$

Taking expectations gives that

$$E[f(X)] \geq f(\mu) + f'(\mu)E[X - \mu] = f(\mu).$$

1.8 Limit Theorems

Some of the most important results in probability theory are in the form of limit theorems. The two most important are:

Strong Law of Large Numbers*

If $X_1, X_2, \ldots$ are independent and identically distributed with mean μ, then

$$P\left\{\lim_{n\to\infty}(X_1 + \cdots + X_n)/n = \mu\right\} = 1.$$

Central Limit Theorem

If $X_1, X_2, \ldots$ are independent and identically distributed with mean μ and variance σ^2, then

$$\lim_{n\to\infty} P\left\{\frac{X_1 + \cdots + X_n - n\mu}{\sigma\sqrt{n}} \leq a\right\} = \int_{-\infty}^{a} \frac{1}{\sqrt{2\pi}} e^{-x^2/2}\, dx.$$

Thus if we let $S_n = \sum_{i=1}^n X_i$, where $X_1, X_2, \ldots$ are independent and identically distributed, then the Strong Law of Large Numbers states that, with probability 1, S_n/n will converge to $E[X_i]$; whereas the central limit theorem states that S_n will have an asymptotic normal distribution as $n \to \infty$.

1.9 Stochastic Processes

A *stochastic process* $\underline{\mathbf{X}} = \{X(t), t \in T\}$ is a collection of random variables. That is, for each t in the *index set* T, $X(t)$ is a random variable. We often interpret t as time and call $X(t)$ the state of the process at time t. If the index set T is a countable set, we call $\underline{\mathbf{X}}$ a discrete-time stochastic process, and if T is a continuum, we call it a continuous-time process.

Any realization of $\underline{\mathbf{X}}$ is called a *sample path*. For instance, if events are occurring randomly in time and $X(t)$ represents the number of events that occur in $[0, t]$, then Figure 1.9.1 gives a sample path of $\underline{\mathbf{X}}$ which corresponds

* A proof of the Strong Law of Large Numbers is given in the Appendix to this chapter.

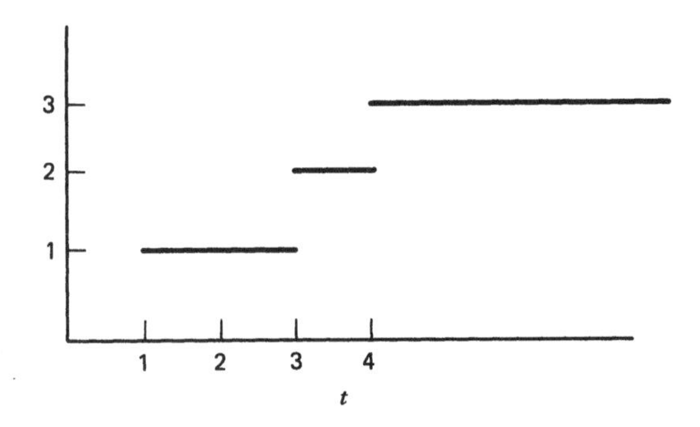

Figure 1.9.1. A sample path of $X(t)$ = number of events in $[0, t]$.

to the initial event occurring at time 1, the next event at time 3 and the third at time 4, and no events anywhere else.

A continuous-time stochastic process $\{X(t), t \in T\}$ is said to have *independent increments* if for all $t_0 < t_1 < t_2 < \cdots < t_n$, the random variables

$$X(t_1) - X(t_0), X(t_2) - X(t_1), \ldots, X(t_n) - X(t_{n-1})$$

are independent. It is said to possess *stationary* increments if $X(t + s) - X(t)$ has the same distribution for all t. That is, it possesses independent increments if the changes in the processes' value over nonoverlapping time intervals are independent; and it possesses stationary increments if the distribution of the change in value between any two points depends only on the distance between those points.

Example 1.9(a) Consider a particle that moves along a set of $m + 1$ nodes, labelled $0, 1, \ldots, m$, that are arranged around a circle (see Figure 1.9.2). At each step the particle is equally likely to move one position in either the clockwise or counterclockwise direction. That is, if X_n is the position of the particle after its nth step then

$$P\{X_{n+1} = i + 1 | X_n = i\} = P\{X_{n+1} = i - 1 | X_n = i\} = 1/2$$

where $i + 1 \equiv 0$ when $i = m$, and $i - 1 \equiv m$ when $i = 0$. Suppose now that the particle starts at 0 and continues to move around according to the above rules until all the nodes $1, 2, \ldots, m$ have been visited. What is the probability that node i, $i = 1, \ldots, m$, is the last one visited?

Solution. Surprisingly enough, the probability that node i is the last node visited can be determined without any computations. To do so, consider the first time that the particle is at one of the two

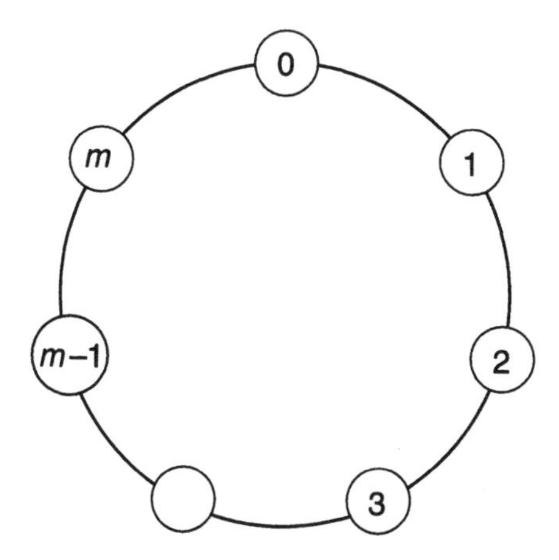

Figure 1.9.2. Particle moving around a circle.

neighbors of node i, that is, the first time that the particle is at one of the nodes $i - 1$ or $i + 1$ (with $m + 1 \equiv 0$). Suppose it is at node $i - 1$ (the argument in the alternative situation is identical). Since neither node i nor $i + 1$ has yet been visited it follows that i will be the last node visited if, and only if, $i + 1$ is visited before i. This is so because in order to visit $i + 1$ before i the particle will have to visit all the nodes on the counterclockwise path from $i - 1$ to $i + 1$ before it visits i. But the probability that a particle at node $i - 1$ will visit $i + 1$ before i is just the probability that a particle will progress $m - 1$ steps in a specified direction before progressing one step in the other direction. That is, it is equal to the probability that a gambler who starts with 1 unit, and wins 1 when a fair coin turns up heads and loses 1 when it turns up tails, will have his fortune go up by $m - 1$ before he goes broke. Hence, as the preceding implies that the probability that node i is the last node visited is the same for all i, and as these probabilities must sum to 1, we obtain

$$P\{i \text{ is the last node visited}\} = 1/m, \qquad i = 1, \ldots, m.$$

Remark The argument used in the preceding example also shows that a gambler who is equally likely to either win or lose 1 on each gamble will be losing n before he is winning 1 with probability $1/(n + 1)$; or equivalently

$$P\{\text{gambler is up 1 before being down } n\} = \frac{n}{n+1}.$$

Suppose now we want the probability that the gambler is up 2 before being down n. Upon conditioning upon whether he reaches up 1 before down n we obtain that

$$P\{\text{gambler is up 2 before being down } n\}$$
$$= P\{\text{up 2 before down } n \mid \text{up 1 before down } n\}\frac{n}{n+1}$$
$$= P\{\text{up 1 before down } n+1\}\frac{n}{n+1}$$
$$= \frac{n+1}{n+2}\frac{n}{n+1} = \frac{n}{n+2}.$$

Repeating this argument yields that

$$P\{\text{gambler is up } k \text{ before being down } n\} = \frac{n}{n+k}.$$

Example 1.9(B) Suppose in Example 1.9(A) that the particle is not equally likely to move in either direction but rather moves at each step in the clockwise direction with probability p and in the counterclockwise direction with probability $q = 1 - p$. If $.5 < p < 1$ then we will show that the probability that state i is the last state visited is a strictly increasing function of i, $i = 1, \ldots, m$.

To determine the probability that state i is the last state visited, condition on whether $i - 1$ or $i + 1$ is visited first. Now, if $i - 1$ is visited first then the probability that i will be the last state visited is the same as the probability that a gambler who wins each 1 unit bet with probability q will have her cumulative fortune increase by $m - 1$ before it decreases by 1. Note that this probability does not depend on i, and let its value be P_1. Similarly, if $i + 1$ is visited before $i - 1$ then the probability that i will be the last state visited is the same as the probability that a gambler who wins each 1 unit bet with probability p will have her cumulative fortune increase by $m - 1$ before it decreases by 1. Call this probability P_2, and note that since $p > q$, $P_1 < P_2$. Hence, we have

$$\begin{aligned} P\{i \text{ is last state}\} &= P_1 P\{i-1 \text{ before } i+1\} \\ &\quad + P_2(1 - P\{i-1 \text{ before } i+1\}) \\ &= (P_1 - P_2)P\{i-1 \text{ before } i+1\} + P_2. \end{aligned}$$

Now, since the event that $i - 1$ is visited before $i + 1$ implies the event that $i - 2$ is visited before i, it follows that

$$P\{i-1 \text{ before } i+1\} < P\{i-2 \text{ before } i\},$$

and thus we can conclude that

$$P\{i-1 \text{ is last state}\} < P\{i \text{ is last state}\}.$$

EXAMPLE 1.9(c) A graph consisting of a central vertex, labeled 0, and rays emanating from that vertex is called a star graph (see Figure 1.9.3). Let r denote the number of rays of a star graph and let ray i consist of n_i vertices, for $i = 1, \ldots, r$. Suppose that a particle moves along the vertices of the graph so that it is equally likely to move from whichever vertex it is presently at to any of the neighbors of that vertex, where two vertices are said to be neighbors if they are joined by an edge. Thus, for instance, when at vertex 0 the particle is equally likely to move to any of its r neighbors. The vertices at the far ends of the rays are called *leafs*. What is the probability that, starting at node 0, the first leaf visited is the one on ray i, $i = 1, \ldots, r$?

Solution. Let L denote the first leaf visited. Conditioning on R, the first ray visited, yields

$$P\{L = i\} = \sum_{j=1}^{r} \frac{1}{r} P\{L = i \mid \text{first ray visited is } j\}. \tag{1.9.1}$$

Now, if j is the first ray visited (that is, the first move of the particle is from vertex 0 to its neighboring vertex on ray j) then it follows, from the remark following Example 1.9(A), that with probability $1/n_j$ the particle will visit the leaf at the end of ray j before returning to 0 (for this is the complement of the event that the gambler will be up 1 before being down $n - 1$). Also, if it does return to 0 before reaching the end of ray j, then the problem in essence begins anew. Hence, we obtain upon conditioning whether the particle reaches the end of ray j before returning to 0 that

$$P\{L = i \mid \text{first ray visited is } i\} = 1/n_i + (1 - 1/n_i)P\{L = i\}$$
$$P\{L = i \mid \text{first ray visited is } j\} = (1 - 1/n_j)P\{L = i\}, \qquad \text{for } j \neq i.$$

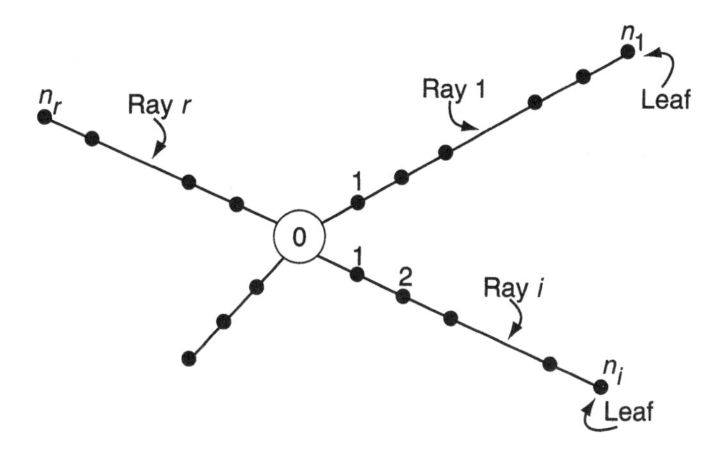

Figure 1.9.3. A star graph.

Substituting the preceding into Equation (1.9.1) yields that

$$rP\{L = i\} = 1/n_i + \left(r - \sum_j 1/n_j\right) P\{L = i\}$$

or

$$P\{L = i\} = \frac{1/n_i}{\sum_j 1/n_j}, \qquad i = 1, \ldots, r.$$

Problems

1.1. Let N denote a nonnegative integer-valued random variable. Show that

$$E[N] = \sum_{k=1}^{\infty} P\{N \geq k\} = \sum_{k=0}^{\infty} P\{N > k\}.$$

In general show that if X is nonnegative with distribution F, then

$$E[X] = \int_0^\infty \overline{F}(x)\, dx$$

and

$$E[X^n] = \int_0^\infty n x^{n-1} \overline{F}(x)\, dx.$$

1.2. If X is a continuous random variable having distribution F show that:

(a) $F(X)$ is uniformly distributed over $(0, 1)$;

(b) if U is a uniform $(0, 1)$ random variable, then $F^{-1}(U)$ has distribution F, where $F^{-1}(x)$ is that value of y such that $F(y) = x$.

1.3. Let X_n denote a binomial random variable with parameters (n, p_n), $n \geq 1$. If $np_n \to \lambda$ as $n \to \infty$, show that

$$P\{X_n = i\} \to e^{-\lambda}\lambda^i/i! \qquad \text{as } n \to \infty.$$

1.4. Compute the mean and variance of a binomial random variable with parameters n and p.

1.5. Suppose that n independent trials—each of which results in either outcome $1, 2, \ldots, r$ with respective probabilities $p_1, p_2, \ldots, p_r$—are performed, $\sum_1^r p_i = 1$. Let N_i denote the number of trials resulting in outcome i.

(a) Compute the joint distribution of $N_1, \ldots, N_r$. This is called the *multinomial* distribution.

(b) Compute $\text{Cov}(N_i, N_j)$.

(c) Compute the mean and variance of the number of outcomes that do not occur.

1.6. Let $X_1, X_2, \ldots$ be independent and identically distributed continuous random variables. We say that a record occurs at time n, $n > 0$ and has value X_n if $X_n > \max(X_1, \ldots, X_{n-1})$, where $X_0 \equiv -\infty$.

(a) Let N_n denote the total number of records that have occurred up to (and including) time n. Compute $E[N_n]$ and $\text{Var}(N_n)$.

(b) Let $T = \min\{n: n > 1 \text{ and a record occurs at } n\}$. Compute $P\{T > n\}$ and show that $P\{T < \infty\} = 1$ and $E[T] = \infty$.

(c) Let T_y denote the time of the first record value greater than y. That is,

$$T_y = \min\{n: \quad X_n > y\}.$$

Show that T_y is independent of X_{T_y}. That is, the time of the first value greater than y is independent of that value. (It may seem more intuitive if you turn this last statement around.)

1.7. Let X denote the number of white balls selected when k balls are chosen at random from an urn containing n white and m black balls. Compute $E[X]$ and $\text{Var}(X)$.

1.8. Let X_1 and X_2 be independent Poisson random variables with means λ_1 and λ_2.

(a) Find the distribution of $X_1 + X_2$.

(b) Compute the conditional distribution of X_1 given that $X_1 + X_2 = n$.

1.9. A round-robin tournament of n contestants is one in which each of the $\binom{n}{2}$ pairs of contestants plays each other exactly once, with the outcome of any play being that one of the contestants wins and the other loses. Suppose the players are initially numbered $1, 2, \ldots, n$. The permutation $i_1, \cdots, i_n$ is called a Hamiltonian permutation if i_1 beats i_2, i_2 beats i_3, $\cdots$, and i_{n-1} beats i_n. Show that there is an outcome of the round-robin for which the number of Hamiltonians is at least $n!/2^{n-1}$.
(*Hint:* Use the probabilistic method.)

1.10. Consider a round-robin tournament having n contestants, and let k, $k < n$, be a positive integer such that $\binom{n}{k}[1 - 1/2^k]^{n-k} < 1$. Show that it is possible for the tournament outcome to be such that for every set of k contestants there is a contestant who beat every member of this set.

1.11. If X is a nonnegative integer-valued random variable then the function $P(z)$, defined for $|z| \leq 1$ by

$$P(z) = E[z^X] = \sum_{j=0}^{\infty} z^j P\{X = j\},$$

is called the probability generating function of X.

(a) Show that

$$\frac{d^k}{dz^k} P(z)_{|z=0} = k! P\{X = k\}.$$

(b) With 0 being considered even, show that

$$P\{X \text{ is even}\} = \frac{P(-1) + P(1)}{2}.$$

(c) If X is binomial with parameters n and p, show that

$$P\{X \text{ is even}\} = \frac{1 + (1 - 2p)^n}{2}.$$

(d) If X is Poisson with mean λ, show that

$$P\{X \text{ is even}\} = \frac{1 + e^{-2\lambda}}{2}.$$

(e) If X is geometric with parameter p, show that

$$P\{X \text{ is even}\} = \frac{1 - p}{2 - p}.$$

(f) If X is a negative binomial random variable with parameters r and p, show that

$$P\{X \text{ is even}\} = \frac{1}{2}\left[1 + (-1)^r \left(\frac{p}{2 - p}\right)^r\right].$$

1.12. If $P\{0 \leq X \leq a\} = 1$, show that

$$\text{Var}(X) \leq a^2/4.$$

1.13. Consider the following method of shuffling a deck of n playing cards, numbered 1 through n. Take the top card from the deck and then replace it so that it is equally likely to be put under exactly k cards, for $k = 0, 1, \ldots, n - 1$. Continue doing this operation until the card that was initially on the bottom of the deck is now on top. Then do it one more time and stop.

(a) Suppose that at some point there are k cards beneath the one that was originally on the bottom of the deck. Given this set of k cards explain why each of the possible $k!$ orderings is equally likely to be the ordering of the last k cards.

(b) Conclude that the final ordering of the deck is equally likely to be any of the $N!$ possible orderings.

(c) Find the expected number of times the shuffling operation is performed.

1.14. A fair die is continually rolled until an even number has appeared on 10 distinct rolls. Let X_i denote the number of rolls that land on side i. Determine:

(a) $E[X_1]$.

(b) $E[X_2]$.

(c) the probability mass function of X_1.

(d) the probability mass function of X_2.

1.15. Let F be a continuous distribution function and let U be a uniform (0, 1) random variable.

(a) If $X = F^{-1}(U)$, show that X has distribution function F.

(b) Show that $-\log(U)$ is an exponential random variable with mean 1.

1.16. Let $f(x)$ and $g(x)$ be probability density functions, and suppose that for some constant c, $f(x) \le cg(x)$ for all x. Suppose we can generate random variables having density function g, and consider the following algorithm.

Step 1: Generate Y, a random variable having density function g.

Step 2: Generate U, a uniform (0, 1) random variable.

Step 3: If $U \le \dfrac{f(Y)}{cg(Y)}$ set $X = Y$. Otherwise, go back to Step 1.

Assuming that successively generated random variables are independent, show that:

(a) X has density function f.

(b) the number of iterations of the algorithm needed to generate X is a geometric random variable with mean c.

1.17. Let $X_1, \ldots, X_n$ be independent and identically distributed continuous random variables having distribution F. Let $X_{i,n}$ denote the ith smallest of $X_1, \ldots, X_n$ and let $F_{i,n}$ be its distribution function. Show that:

(a) $F_{i,n}(x) = F(x)F_{i-1,n-1}(x) + \overline{F}(x)F_{i,n-1}(x)$.

(b) $F_{i,n-1}(x) = \dfrac{i}{n} F_{i+1,n}(x) + \dfrac{n-i}{n} F_{i,n}(x)$.

(*Hints:* For part (a) condition on whether $X_n \le x$, and for part (b) start by conditioning on whether X_n is among the i smallest of $X_1, \ldots, X_n$.)

1.18. A coin, which lands on heads with probability p, is continually flipped. Compute the expected number of flips that are made until a string of r heads in a row is obtained.

1.19. An urn contains a white and b black balls. After a ball is drawn, it is returned to the urn if it is white; but if it is black, it is replaced by a white ball from another urn. Let M_n denote the expected number of white balls in the urn after the foregoing operation has been repeated n times.

(a) Derive the recursive equation

$$M_{n+1} = \left(1 - \frac{1}{a+b}\right) M_n + 1.$$

(b) Use part (a) to prove that

$$M_n = a + b - b\left(1 - \frac{1}{a+b}\right)^n.$$

(c) What is the probability that the $(n+1)$st ball drawn is white?

1.20. A Continuous Random Packing Problem. Consider the interval $(0, x)$ and suppose that we pack in this interval random unit intervals—whose left-hand points are all uniformly distributed over $(0, x-1)$—as follows. Let the first such random interval be I_1. If $I_1, \ldots, I_k$ have already been packed in the interval, then the next random unit interval will be packed if it does not intersect any of the intervals $I_1, \ldots, I_k$, and the interval will be denoted by I_{k+1}. If it does intersect any of the intervals $I_1, \ldots, I_k$, we disregard it and look at the next random interval. The procedure is continued until there is no more room for additional unit intervals (that is, all the gaps between packed intervals are smaller than 1). Let $N(x)$ denote the number of unit intervals packed in $[0, x]$ by this method.

For instance, if $x = 5$ and the successive random intervals are (.5, 1.5), (3.1, 4.1), (4, 5), (1.7, 2.7), then $N(5) = 3$ with packing as follows:

Let $M(x) = E[N(x)]$. Show that M satisfies

$$M(x) = 0, \qquad x < 1,$$

$$M(x) = \frac{2}{x-1}\int_0^{x-1} M(y)\,dy + 1, \qquad x > 1.$$

1.21. Let $U_1, U_2, \ldots$ be independent uniform (0, 1) random variables, and let N denote the smallest value of n, $n \geq 0$, such that

$$\prod_{i=1}^{n} U_i \geq e^{-\lambda} > \prod_{i=1}^{n+1} U_i, \qquad \text{where } \prod_{i=1}^{0} U_i \equiv 1.$$

Show that N is a Poisson random variable with mean λ.
(*Hint:* Show by induction on n, conditioning on U_1, that $P\{N = n\} = e^{-\lambda}\lambda^n/n!$.)

1.22. The conditional variance of X, given Y, is defined by

$$\mathrm{Var}(X|Y) = E[(X - E[X|Y])^2|Y].$$

Prove the conditional variance formula; namely,

$$\mathrm{Var}(X) = E[\mathrm{Var}(X|Y)] + \mathrm{Var}(E[X|Y]).$$

Use this to obtain $\mathrm{Var}(X)$ in Example 1.5(B) and check your result by differentiating the generating function.

1.23. Consider a particle that moves along the set of integers in the following manner. If it is presently at i then it next moves to $i + 1$ with probability p and to $i - 1$ with probability $1 - p$. Starting at 0, let α denote the probability that it ever reaches 1.

(a) Argue that

$$\alpha = p + (1 - p)\alpha^2.$$

(b) Show that

$$\alpha = \begin{cases} 1 & \text{if } p \geq 1/2 \\ p/(1-p) & \text{if } p < 1/2. \end{cases}$$

(c) Find the probability that the particle ever reaches n, $n > 0$.

(d) Suppose that $p < 1/2$ and also that the particle eventually reaches n, $n > 0$. If the particle is presently at i, $i < n$, and n has not yet been reached, show that the particle will next move to $i + 1$ with probability $1 - p$ and to $i - 1$ with probability p. That is, show that

$$P\{\text{next at } i + 1 \mid \text{at } i \text{ and will reach } n\} = 1 - p.$$

(Note that the roles of p and $1 - p$ are interchanged when it is given that n is eventually reached.)

1.24. In Problem 1.23, let $E[T]$ denote the expected time until the particle reaches 1.

(a) Show that

$$E[T] = \begin{cases} 1/(2p - 1) & \text{if } p > 1/2 \\ \infty & \text{if } p \leq 1/2. \end{cases}$$

(b) Show that, for $p > 1/2$,

$$\text{Var}(T) = \frac{4p(1-p)}{(2p-1)^3}.$$

(c) Find the expected time until the particle reaches n, $n > 0$.

(d) Find the variance of the time at which the particle reaches n, $n > 0$.

1.25. Consider a gambler who on each gamble is equally likely to either win or lose 1 unit. Starting with i show that the expected time until the gambler's fortune is either 0 or k is $i(k - i)$, $i = 0, \ldots, k$.
(*Hint:* Let M_i denote this expected time and condition on the result of the first gamble.)

1.26. In the ballot problem compute the probability that A is never behind in the count of the votes.

1.27. Consider a gambler who wins or loses 1 unit on each play with respective possibilities p and $1 - p$. What is the probability that, starting with n units, the gambler will play exactly $n + 2i$ games before going broke? (*Hint:* Make use of ballot theorem.)

1.28. Verify the formulas given for the mean and variance of an exponential random variable.

1.29. If $X_1, X_2, \ldots, X_n$ are independent and identically distributed exponential random variables with parameter λ, show that $\sum_1^n X_i$ has a gamma distribution with parameters (n, λ). That is, show that the density function of $\sum_1^n X_i$ is given by

$$f(t) = \lambda e^{-\lambda t}(\lambda t)^{n-1}/(n-1)!, \qquad t \geq 0.$$

1.30. In Example 1.6(A) if server i serves at an exponential rate λ_i, $i = 1, 2$, compute the probability that Mr. A is the last one out.

1.31. If X and Y are independent exponential random variables with respective means $1/\lambda_1$ and $1/\lambda_2$, compute the distribution of $Z = \min(X, Y)$. What is the conditional distribution of Z given that $Z = X$?

1.32. Show that the only continuous solution of the functional equation

$$g(s + t) = g(s) + g(t)$$

is $g(s) = cs$.

1.33. Derive the distribution of the ith record value for an arbitrary continuous distribution F (see Example 1.6(B)).

1.34. If X_1 and X_2 are independent nonnegative continuous random variables, show that

$$P\{X_1 < X_2 \mid \min(X_1, X_2) = t\} = \frac{\lambda_1(t)}{\lambda_1(t) + \lambda_2(t)},$$

where $\lambda_i(t)$ is the failure rate function of X_i.

1.35. Let X be a random variable with probability density function $f(x)$, and let $M(t) = E[e^{tX}]$ be its moment generating function. The tilted density function f_t is defined by

$$f_t(x) = \frac{e^{tx} f(x)}{M(t)}.$$

Let X_t have density function f_t.

(a) Show that for any function $h(x)$

$$E[h(X)] = M(t) E[\exp\{-tX_t\} h(X_t)].$$

(b) Show that, for $t > 0$,

$$P\{X > a\} \leq M(t) e^{-ta} P\{X_t > a\}.$$

(c) Show that if $E[X_{t^*}] = a$ then

$$\min_t M(t) e^{-ta} = M(t^*) e^{-t^* a}.$$

1.36. Use Jensen's inequality to prove that the arithmetic mean is at least as large as the geometric mean. That is, for nonnegative x_i, show that

$$\sum_{i=1}^{n} x_i/n \geq \left(\prod_{i=1}^{n} x_i\right)^{1/n}.$$

1.37. Let $X_1, X_2, \ldots$ be a sequence of independent and identically distributed continuous random variables. Say that a peak occurs at time n if $X_{n-1} < X_n > X_{n+1}$. Argue that the proportion of time that a peak occurs is, with probability 1, equal to 1/3.

1.38. In Example 1.9(A), determine the expected number of steps until all the states $1, 2, \ldots, m$ are visited.
(*Hint:* Let X_i denote the number of additional steps after i of these states have been visited until a total of $i + 1$ of them have been visited, $i = 0, 1, \ldots, m - 1$, and make use of Problem 1.25.)

1.39. A particle moves along the following graph so that at each step it is equally likely to move to any of its neighbors.

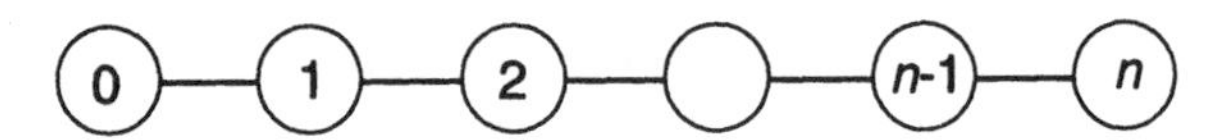

Starting at 0 show that the expected number of steps it takes to reach n is n^2.
(*Hint:* Let T_i denote the number of steps it takes to go from vertex $i - 1$ to vertex i, $i = 1, \ldots, n$. Determine $E[T_i]$ recursively, first for $i = 1$, then $i = 2$, and so on.)

1.40. Suppose that $r = 3$ in Example 1.9(C) and find the probability that the leaf on the ray of size n_1 is the last leaf to be visited.

1.41. Consider a star graph consisting of a central vertex and r rays, with one ray consisting of m vertices and the other $r - 1$ all consisting of n vertices. Let P_r denote the probability that the leaf on the ray of m vertices is the last leaf visited by a particle that starts at 0 and at each step is equally likely to move to any of its neighbors.

(a) Find P_2.

(b) Express P_r in terms of P_{r-1}.

1.42. Let $Y_1, Y_2, \ldots$ be independent and identically distributed with

$$P\{Y_n = 0\} = \alpha$$

$$P\{Y_n > y\} = (1 - \alpha)e^{-y}, \qquad y > 0.$$

Define the random variables X_n, $n \geq 0$ by

$$X_0 = 0$$

$$X_{n+1} = \alpha X_n + Y_{n+1}.$$

Prove that

$$P\{X_n = 0\} = \alpha^n$$

$$P\{X_n > x\} = (1 - \alpha^n)e^{-x}, \qquad x > 0.$$

1.43. For a nonnegative random variable X, show that for $a > 0$,

$$P\{X \geq a\} \leq E[X^t]/a^t$$

REFERENCES

References 6 and 7 are elementary introductions to probability and its applications. Rigorous approaches to probability and stochastic processes, based on measure theory, are given in References 2, 4, 5, and 8. Example 1.3(C) is taken from Reference 1. Proposition 1.5.2 is due to Blackwell and Girshick (Reference 3).

1. N. Alon, J. Spencer, and P. Erdos, *The Probabilistic Method,* Wiley, New York, 1992.
2. P. Billingsley, *Probability and Measure,* 3rd ed., Wiley, New York, 1995.
3. D. Blackwell and M. A. Girshick, *Theory of Games and Statistical Decisions,* Wiley, New York, 1954.
4. L. Breiman, *Probability,* Addison-Wesley, Reading, MA, 1968.
5. R. Durrett, *Probability: Theory and Examples,* Brooks/Cole, California, 1991.
6. W. Feller, *An Introduction to Probability Theory and its Applications,* Vol. 1, Wiley, New York, 1957.
7. S. M. Ross, *A First Course in Probability,* 4th ed., Macmillan, New York, 1994.
8. D. Williams, *Probability with Martingales,* Cambridge University Press, Cambridge, England, 1991.

APPENDIX

The Strong Law of Large Numbers

If $X_1, X_2, \ldots$ is a sequence of independent and identically distributed random variables with mean μ, then

$$P\left\{\lim_{n\to\infty} \frac{X_1 + X_2 + \cdots + X_n}{n} = \mu\right\} = 1.$$

Although the theorem can be proven without this assumption, our proof of the strong law of large numbers will assume that the random variables X_i have a finite fourth moment. That is, we will suppose that $E[X_i^4] = K < \infty$.

Proof of the Strong Law of Large Numbers To begin, assume that μ, the mean of the X_i, is equal to 0. Let $S_n = \sum_{i=1}^{n} X_i$ and consider

$$E[S_n^4] = E[(X_1 + \cdots + X_n)(X_1 + \cdots + X_n)(X_1 + \cdots + X_n)(X_1 + \cdots + X_n)].$$

Expanding the right side of the above will result in terms of the form

$$X_i^4, \quad X_i^3X_j, \quad X_i^2X_j^2, \quad X_i^2X_jX_k, \qquad \text{and} \qquad X_iX_jX_kX_l$$

where i, j, k, l are all different. As all the X_i have mean 0, it follows by independence that

$$E[X_i^3X_j] = E[X_i^3]E[X_j] = 0$$
$$E[X_i^2X_jX_k] = E[X_i^2]E[X_j]E[X_k] = 0$$
$$E[X_iX_jX_kX_l] = 0.$$

Now, for a given pair i and j there will be $\binom{4}{2} = 6$ terms in the expansion that will equal $X_i^2X_j^2$. Hence, it follows upon expanding the above product and taking expectations terms by term that

$$E[S_n^4] = nE[X_i^4] + 6\binom{n}{2} E[X_i^2 X_j^2]$$

$$= nK + 3n(n-1)E[X_i^2]E[X_j^2]$$

where we have once again made use of the independence assumption. Now, since

$$0 \leq \mathrm{Var}(X_i^2) = E[X_i^4] - (E[X_i^2])^2,$$

we see that

$$(E[X_i^2])^2 \leq E[X_i^4] = K.$$

Therefore, from the preceding we have that

$$E[S_n^4] \leq nK + 3n(n-1)K$$

which implies that

$$E[S_n^4/n^4] \leq K/n^3 + 3K/n^2.$$

Therefore, it follows that

$$E\left[\sum_{n=1}^{\infty} S_n^4/n^4\right] = \sum_{n=1}^{\infty} E[S_n^4/n^4] < \infty. \qquad (*)$$

Now, for any $\varepsilon > 0$ it follows from the Markov inequality that

$$P\{S_n^4/n^4 > \varepsilon\} \leq E[S_n^4/n^4]/\varepsilon$$

and thus from (*)

$$\sum_{n=1}^{\infty} P\{S_n^4/n^4 > \varepsilon\} < \infty$$

which implies by the Borel-Cantelli lemma that with probability 1, $S_n^4/n^4 > \varepsilon$ for only finitely many n. As this is true for any $\varepsilon > 0$, we can thus conclude that with probability 1,

$$\lim_{n\to\infty} S_n^4/n^4 = 0.$$

But if $S_n^4/n^4 = (S_n/n)^4$ goes to 0 then so must S_n/n; and so we have proven that, with probability 1,

$$S_n/n \to 0 \qquad \text{as} \qquad n \to \infty.$$

When μ, the mean of the X_i, is not equal to 0 we can apply the preceding argument to the random variables $X_i - \mu$ to obtain that, with probability 1,

$$\lim_{n\to\infty} \sum_{i=1}^{n} (X_i - \mu)/n = 0,$$

or equivalently,

$$\lim_{n\to\infty} \sum_{i=1}^{n} X_i/n = \mu$$

which proves the result.

CHAPTER 2

The Poisson Process

2.1 The Poisson Process

A stochastic process $\{N(t), t \geq 0\}$ is said to be a *counting process* if $N(t)$ represents the total number of 'events' that have occurred up to time t. Hence, a counting process $N(t)$ must satisfy:

(i) $N(t) \geq 0$.

(ii) $N(t)$ is integer valued.

(iii) If $s < t$, then $N(s) \leq N(t)$.

(iv) For $s < t$, $N(t) - N(s)$ equals the number of events that have occurred in the interval $(s, t]$.

A counting process is said to possess *independent increments* if the numbers of events that occur in disjoint time intervals are independent. For example, this means that the number of events that have occurred by time t (that is, $N(t)$) must be independent of the number of events occurring between times t and $t + s$ (that is, $N(t + s) - N(t)$).

A counting process is said to possess *stationary increments* if the distribution of the number of events that occur in any interval of time depends only on the length of the time interval. In other words, the process has stationary increments if the number of events in the interval $(t_1 + s, t_2 + s]$ (that is, $N(t_2 + s) - N(t_1 + s)$) has the same distribution as the number of events in the interval $(t_1, t_2]$ (that is, $N(t_2) - N(t_1)$) for all $t_1 < t_2$, and $s > 0$.

One of the most important types of counting processes is the Poisson process, which is defined as follows.

Definition 2.1.1

The counting process $\{N(t), t \geq 0\}$ is said to be a *Poisson process having rate* λ, $\lambda > 0$, if:

(i) $N(0) = 0$.

(ii) The process has independent increments.

(iii) The number of events in any interval of length t is Poisson distributed with mean λt. That is, for all $s, t \geq 0$,

$$P\{N(t+s) - N(s) = n\} = e^{-\lambda t}\frac{(\lambda t)^n}{n!}, \qquad n = 0, 1, \ldots.$$

Note that it follows from condition (iii) that a Poisson process has stationary increments and also that

$$E[N(t)] = \lambda t,$$

which explains why λ is called the rate of the process.

In order to determine if an arbitrary counting process is actually a Poisson process, we must show that conditions (i), (ii), and (iii) are satisfied. Condition (i), which simply states that the counting of events begins at time $t = 0$, and condition (ii) can usually be directly verified from our knowledge of the process. However, it is not at all clear how we would determine that condition (iii) is satisfied, and for this reason an equivalent definition of a Poisson process would be useful.

As a prelude to giving a second definition of a Poisson process, we shall define the concept of a function f being $o(h)$.

Definition

The function f is said to be $o(h)$ if

$$\lim_{h \to 0} \frac{f(h)}{h} = 0.$$

We are now in a position to give an alternative definition of a Poisson process.

Definition 2.1.2

The counting process $\{N(t), t \geq 0\}$ is said to be a Poisson process with rate λ, $\lambda > 0$, if:

(i) $N(0) = 0$.
(ii) The process has stationary and independent increments.
(iii) $P\{N(h) = 1\} = \lambda h + o(h)$.
(iv) $P\{N(h) \geq 2\} = o(h)$.

THEOREM 2.1.1

Definitions 2.1.1 and 2.1.2 are equivalent.

Proof We first show that Definition 2.1.2 implies Definition 2.1.1. To do this let

$$P_n(t) = P\{N(t) = n\}.$$

We derive a differential equation for $P_0(t)$ in the following manner:

$$\begin{aligned} P_0(t+h) &= P\{N(t+h) = 0\} \\ &= P\{N(t) = 0, N(t+h) - N(t) = 0\} \\ &= P\{N(t) = 0\}P\{N(t+h) - N(t) = 0\} \\ &= P_0(t)[1 - \lambda h + o(h)], \end{aligned}$$

where the final two equations follow from Assumption (ii) and the fact that (iii) and (iv) imply that $P\{N(h) = 0\} = 1 - \lambda h + o(h)$. Hence,

$$\frac{P_0(t+h) - P_0(t)}{h} = -\lambda P_0(t) + \frac{o(h)}{h}.$$

Letting $h \to 0$ yields

$$P_0'(t) = -\lambda P_0(t)$$

or

$$\frac{P_0'(t)}{P_0(t)} = -\lambda,$$

which implies, by integration,

$$\log P_0(t) = -\lambda t + c$$

or

$$P_0(t) = Ke^{-\lambda t}.$$

Since $P_0(0) = P\{N(0) = 0\} = 1$, we arrive at

$$P_0(t) = e^{-\lambda t}. \tag{2.1.1}$$

Similarly, for $n \geq 1$,

$$\begin{aligned} P_n(t+h) &= P\{N(t+h) = n\} \\ &= P\{N(t) = n, N(t+h) - N(t) = 0\} \\ &\quad + P\{N(t) = n-1, N(t+h) - N(t) = 1\} \\ &\quad + P\{N(t+h) = n, N(t+h) - N(t) \geq 2\}. \end{aligned}$$

However, by (iv), the last term in the above is $o(h)$; hence, by using (ii), we obtain

$$\begin{aligned} P_n(t+h) &= P_n(t)P_0(h) + P_{n-1}(t)P_1(h) + o(h) \\ &= (1-\lambda h)P_n(t) + \lambda h P_{n-1}(t) + o(h). \end{aligned}$$

Thus,

$$\frac{P_n(t+h) - P_n(t)}{h} = -\lambda P_n(t) + \lambda P_{n-1}(t) + \frac{o(h)}{h}.$$

Letting $h \to 0$,

$$P_n'(t) = -\lambda P_n(t) + \lambda P_{n-1}(t),$$

or, equivalently,

$$e^{\lambda t}[P_n'(t) + \lambda P_n(t)] = \lambda e^{\lambda t} P_{n-1}(t).$$

Hence,

(2.1.2) $$\frac{d}{dt}(e^{\lambda t}P_n(t)) = \lambda e^{\lambda t} P_{n-1}(t).$$

Now by (2.1.1) we have when $n = 1$

$$\frac{d}{dt}(e^{\lambda t}P_1(t)) = \lambda$$

or

$$P_1(t) = (\lambda t + c)e^{-\lambda t},$$

which, since $P_1(0) = 0$, yields

$$P_1(t) = \lambda t e^{-\lambda t}.$$

To show that $P_n(t) = e^{-\lambda t}(\lambda t)^n/n!$, we use mathematical induction and hence first assume it for $n - 1$. Then by (2.1.2),

$$\frac{d}{dt}(e^{\lambda t}P_n(t)) = \frac{\lambda(\lambda t)^{n-1}}{(n-1)!}$$

implying that

$$e^{\lambda t}P_n(t) = \frac{(\lambda t)^n}{n!} + c,$$

or, since $P_n(0) = P\{N(0) = n\} = 0$,

$$P_n(t) = e^{-\lambda t}\frac{(\lambda t)^n}{n!}.$$

Thus Definition 2.1.2 implies Definition 2.1.1. We will leave it for the reader to prove the reverse.

Remark The result that $N(t)$ has a Poisson distribution is a consequence of the Poisson approximation to the binomial distribution. To see this subdivide the interval $[0, t]$ into k equal parts where k is very large (Figure 2.1.1). First we note that the probability of having 2 or more events in any subinterval goes to 0 as $k \to \infty$. This follows from

$$\begin{aligned}
&P\{2 \text{ or more events in any subinterval}\} \\
&\qquad \leq \sum_{i=1}^{k} P\{2 \text{ or more events in the } i\text{th subinterval}\} \\
&\qquad = ko(t/k) \\
&\qquad = t\frac{o(t/k)}{t/k} \\
&\qquad \to 0 \qquad \text{as } k \to \infty.
\end{aligned}$$

Hence, $N(t)$ will (with a probability going to 1) just equal the number of subintervals in which an event occurs. However, by stationary and independent increments this number will have a binomial distribution with parameters k and $p = \lambda t/k + o(t/k)$. Hence by the Poisson approximation to the binomial we see by letting k approach ∞ that $N(t)$ will have a Poisson distribution with mean equal to

$$\lim_{k\to\infty} k\left[\lambda\frac{t}{k} + o\left(\frac{t}{k}\right)\right] = \lambda t + \lim_{k\to\infty}\left[t\frac{o(t/k)}{t/k}\right] = \lambda t.$$

$0 \quad \frac{t}{k} \quad \frac{2t}{k} \quad \cdots \quad t = \frac{kt}{k}$

Figure 2.1.1

2.2 Interarrival and Waiting Time Distributions

Consider a Poisson process, and let X_1 denote the time of the first event. Further, for $n \geq 1$, let X_n denote the time between the $(n-1)$st and the nth event. The sequence $\{X_n, n \geq 1\}$ is called the *sequence of interarrival times.*

We shall now determine the distribution of the X_n. To do so we first note that the event $\{X_1 > t\}$ takes place if, and only if, no events of the Poisson process occur in the interval $[0, t]$, and thus

$$P\{X_1 > t\} = P\{N(t) = 0\} = e^{-\lambda t}.$$

Hence, X_1 has an exponential distribution with mean $1/\lambda$. To obtain the distribution of X_2 condition on X_1. This gives

$$\begin{aligned} P\{X_2 > t | X_1 = s\} &= P\{0 \text{ events in } (s, s+t] | X_1 = s\} \\ &= P\{0 \text{ events in } (s, s+t]\} \quad \text{(by independent increments)} \\ &= e^{-\lambda t} \quad \text{(by stationary increments).} \end{aligned}$$

Therefore, from the above we conclude that X_2 is also an exponential random variable with mean $1/\lambda$, and furthermore, that X_2 is independent of X_1. Repeating the same argument yields the following.

PROPOSITION 2.2.1

$X_n, n = 1, 2, \ldots$ are independent identically distributed exponential random variables having mean $1/\lambda$.

Remark The proposition should not surprise us. The assumption of stationary and independent increments is equivalent to asserting that, at any point in time, the process *probabilistically* restarts itself. That is, the process from any point on is independent of all that has previously occurred (by independent increments), and also has the same distribution as the original process (by stationary increments). In other words, the process has no *memory,* and hence exponential interarrival times are to be expected.

Another quantity of interest is S_n, the arrival time of the nth event, also called the *waiting time* until the nth event. Since

$$S_n = \sum_{i=1}^{n} X_i, \qquad n \geq 1,$$

it is easy to show, using moment generating functions, that Proposition 2.2.1 implies that S_n has a gamma distribution with parameters n and λ. That is, its probability density is

$$f(t) = \lambda e^{-\lambda t} \frac{(\lambda t)^{n-1}}{(n-1)!}, \qquad t \geq 0.$$

The above could also have been derived by noting that the nth event occurs prior or at time t if, and only if, the number of events occurring by time t is at least n. That is,

$$N(t) \geq n \Leftrightarrow S_n \leq t.$$

Hence,

$$\begin{aligned} P\{S_n \leq t\} &= P\{N(t) \geq n\} \\ &= \sum_{j=n}^{\infty} e^{-\lambda t} \frac{(\lambda t)^j}{j!}, \end{aligned}$$

which upon differentiation yields that the density function of S_n is

$$\begin{aligned} f(t) &= -\sum_{j=n}^{\infty} \lambda e^{-\lambda t} \frac{(\lambda t)^j}{j!} + \sum_{j=n}^{\infty} \lambda e^{-\lambda t} \frac{(\lambda t)^{j-1}}{(j-1)!} \\ &= \lambda e^{-\lambda t} \frac{(\lambda t)^{n-1}}{(n-1)!} \end{aligned}$$

Remark Another way of obtaining the density of S_n is to use the independent increment assumption as follows:

$$\begin{aligned} P\{t < S_n < t + dt\} &= P\{N(t) = n - 1, 1 \text{ event in } (t, t + dt)\} + o(dt) \\ &= P\{N(t) = n - 1\} P\{1 \text{ event in } (t, t + dt)\} + o(dt) \\ &= \frac{e^{-\lambda t}(\lambda t)^{n-1}}{(n-1)!} \lambda dt + o(dt) \end{aligned}$$

which yields, upon dividing by $d(t)$ and then letting it approach 0, that

$$f_{S_n}(t) = \frac{\lambda e^{-\lambda t}(\lambda t)^{n-1}}{(n-1)!}$$

Proposition 2.2.1 also gives us another way of defining a Poisson process. For suppose that we start out with a sequence $\{X_n, n \geq 1\}$ of independent identically distributed exponential random variables each having mean $1/\lambda$. Now let us define a counting process by saying that the nth event of this process occurs at time S_n, where

$$S_n \equiv X_1 + X_2 + \cdots + X_n.$$

The resultant counting process $\{N(t), t \geq 0\}$ will be Poisson with rate λ.

2.3 Conditional Distribution of the Arrival Times

Suppose we are told that exactly one event of a Poisson process has taken place by time t, and we are asked to determine the distribution of the time at which the event occurred. Since a Poisson process possesses stationary and independent increments, it seems reasonable that each interval in $[0, t]$ of equal length should have the same probability of containing the event. In other words, the time of the event should be uniformly distributed over $[0, t]$. This is easily checked since, for $s \leq t$,

$$\begin{aligned} P\{X_1 < s | N(t) = 1\} &= \frac{P\{X_1 < s, N(t) = 1\}}{P\{N(t) = 1\}} \\ &= \frac{P\{1 \text{ event in } [0, s), 0 \text{ events in } [s, t)\}}{P\{N(t) = 1\}} \\ &= \frac{P\{1 \text{ event in } [0, s)\} P\{0 \text{ events in } [s, t)\}}{P\{N(t) = 1\}} \\ &= \frac{\lambda s e^{-\lambda s} e^{-\lambda(t-s)}}{\lambda t e^{-\lambda t}} \\ &= \frac{s}{t}. \end{aligned}$$

This result may be generalized, but before doing so we need to introduce the concept of order statistics.

Let $Y_1, Y_2, \ldots, Y_n$ be n random variables. We say that $Y_{(1)}, Y_{(2)}, \ldots, Y_{(n)}$ are the order statistics corresponding to $Y_1, Y_2, \ldots, Y_n$ if $Y_{(k)}$ is the kth smallest value among $Y_1, \ldots, Y_n$, $k = 1, 2, \ldots, n$. If the Y_i's are independent identically distributed continuous random variables with probability density f, then the joint density of the order statistics $Y_{(1)}, Y_{(2)}, \ldots, Y_{(n)}$ is given by

$$f(y_1, y_2, \ldots, y_n) = n! \prod_{i=1}^{n} f(y_i), \qquad y_1 < y_2 < \cdots < y_n.$$

The above follows since (i) $(Y_{(1)}, Y_{(2)}, \ldots, Y_{(n)})$ will equal $(y_1, y_2, \ldots, y_n)$ if $(Y_1, Y_2, \ldots, Y_n)$ is equal to any of the $n!$ permutations of $(y_1, y_2, \ldots, y_n)$, and (ii) the probability density that $(Y_1, Y_2, \ldots, Y_n)$ is equal to $y_{i_1}, y_{i_2}, \ldots, y_{i_n}$ is $f(y_{i_1})f(y_{i_2}) \cdots f(y_{i_n}) = \prod_1^n f(y_i)$ when $(y_{i_1}, y_{i_2}, \ldots, y_{i_n})$ is a permutation of $(y_1, y_2, \ldots, y_n)$.

If the Y_i, $i = 1, \ldots, n$, are uniformly distributed over $(0, t)$, then it follows from the above that the joint density function of the order statistics $Y_{(1)}, Y_{(2)}, \ldots, Y_{(n)}$ is

$$f(y_1, y_2, \ldots, y_n) = \frac{n!}{t^n}, \qquad 0 < y_1 < y_2 < \cdots < y_n < t.$$

We are now ready for the following useful theorem.

THEOREM 2.3.1

Given that $N(t) = n$, the n arrival times $S_1, \ldots, S_n$ have the same distribution as the order statistics corresponding to n independent random variables uniformly distributed on the interval $(0, t)$.

Proof We shall compute the conditional density function of $S_1, \ldots, S_n$ given that $N(t) = n$. So let $0 < t_1 < t_2 < \cdots < t_{n+1} = t$ and let h_i be small enough so that $t_i + h_i < t_{i+1}$, $i = 1, \ldots, n$. Now,

$$\begin{aligned} &P\{t_i \le S_i \le t_i + h_i, i = 1, 2, \ldots, n \mid N(t) = n\} \\ &\quad = \frac{P\{\text{exactly 1 event in } [t_i, t_i + h_i], i = 1, \ldots, n, \text{ no events elsewhere in } [0, t]\}}{P\{N(t) = n\}} \\ &\quad = \frac{\lambda h_1 e^{-\lambda h_1} \cdots \lambda h_n e^{-\lambda h_n} e^{-\lambda(t - h_1 - h_2 - \cdots - h_n)}}{e^{-\lambda t}(\lambda t)^n / n!} \\ &\quad = \frac{n!}{t^n} h_1 \cdot h_2 \cdot \cdots \cdot h_n. \end{aligned}$$

Hence,

$$\frac{P\{t_i \le S_i \le t_i + h_i, i = 1, 2, \ldots, n \mid N(t) = n\}}{h_1 \cdot h_2 \cdot \cdots \cdot h_n} = \frac{n!}{t^n},$$

and by letting the $h_i \to 0$, we obtain that the conditional density of $S_1, \ldots, S_n$ given that $N(t) = n$ is

$$f(t_1, \ldots, t_n) = \frac{n!}{t^n}, \qquad 0 < t_1 < \cdots < t_n,$$

which completes the proof.

Remark Intuitively, we usually say that under the condition that n events have occurred in $(0, t)$, the times $S_1, \ldots, S_n$ at which events occur, considered as unordered random variables, are distributed independently and uniformly in the interval $(0, t)$.

EXAMPLE 2.3(A) Suppose that travelers arrive at a train depot in accordance with a Poisson process with rate λ. If the train departs at time t, let us compute the expected sum of the waiting times of travelers arriving in $(0, t)$. That is, we want $E[\sum_{i=1}^{N(t)} (t - S_i)]$, where S_i is the arrival time of the ith traveler. Conditioning on $N(t)$ yields

$$E\left[\sum_{i=1}^{N(t)} (t - S_i) | N(t) = n\right] = E\left[\sum_{i=1}^{n} (t - S_i) | N(t) = n\right]$$

$$= nt - E\left[\sum_{i=1}^{n} S_i | N(t) = n\right].$$

Now if we let $U_1, \ldots, U_n$ denote a set of n independent uniform $(0, t)$ random variables, then

$$E\left[\sum_{i=1}^{n} S_i | N(t) = n\right] = E\left[\sum_{i=1}^{n} U_{(i)}\right] \qquad \text{(by Theorem 2.3.1)}$$

$$= E\left[\sum_{i=1}^{n} U_i\right] \qquad \left(\text{since } \sum_1^n U_{(i)} = \sum_1^n U_i\right)$$

$$= \frac{nt}{2}.$$

Hence,

$$E\left[\sum_1^{N(t)} (t - S_i) | N(t) = n\right] = nt - \frac{nt}{2} = \frac{nt}{2}$$

and

$$E\left[\sum_1^{N(t)} (t - S_i)\right] = \frac{t}{2} E[N(t)] = \frac{\lambda t^2}{2}.$$

As an important application of Theorem 2.3.1 suppose that each event of a Poisson process with rate λ is classified as being either a type-I or type-II event, and suppose that the probability of an event being classified as type-I depends on the time at which it occurs. Specifically, suppose that if an event occurs at time s, then, independently of all else, it is classified as being a type-I event with probability $P(s)$ and a type-II event with probability $1 - P(s)$. By using Theorem 2.3.1 we can prove the following proposition.

PROPOSITION 2.3.2

If $N_i(t)$ represents the number of type-i events that occur by time t, $i = 1, 2$, then $N_1(t)$ and $N_2(t)$ are independent Poisson random variables having respective means λtp and $\lambda t(1 - p)$, where

$$p = \frac{1}{t}\int_0^t P(s)\, ds.$$

Proof We compute the joint distribution of $N_1(t)$ and $N_2(t)$ by conditioning on $N(t)$:

$$\begin{aligned}
&P\{N_1(t) = n, N_2(t) = m\} \\
&\quad = \sum_{k=0}^{\infty} P\{N_1(t) = n, N_2(t) = m \mid N(t) = k\} P\{N(t) = k\} \\
&\quad = P\{N_1(t) = n, N_2(t) = m \mid N(t) = n + m\} P\{N(t) = n + m\}.
\end{aligned}$$

Now consider an arbitrary event that occurred in the interval $[0, t]$. If it had occurred at time s, then the probability that it would be a type-I event would be $P(s)$. Hence, since by Theorem 2.3.1 this event will have occurred at some time uniformly distributed on $(0, t)$, it follows that the probability that it will be a type-I event is

$$p = \frac{1}{t}\int_0^t P(s)\, ds$$

independently of the other events. Hence, $P\{N_1(t) = n, N_2(t) = m \mid N(t) = n + m\}$ will just equal the probability of n successes and m failures in $n + m$ independent trials when p is the probability of success on each trial. That is,

$$P\{N_1(t) = n, N_2(t) = m \mid N(t) = n + m\} = \binom{n+m}{n} p^n (1 - p)^m.$$

Consequently,

$$\begin{aligned}
P\{N_1(t) = n, N_2(t) = m\} &= \frac{(n+m)!}{n!\,m!} p^n (1-p)^m e^{-\lambda t} \frac{(\lambda t)^{n+m}}{(n+m)!} \\
&= e^{-\lambda tp} \frac{(\lambda tp)^n}{n!} e^{-\lambda t(1-p)} \frac{(\lambda t(1-p))^m}{m!},
\end{aligned}$$

which completes the proof.

The importance of the above proposition is illustrated by the following example.

EXAMPLE 2.3(B) ***The Infinite Server Poisson Queue.*** Suppose that customers arrive at a service station in accordance with a Poisson process with rate λ. Upon arrival the customer is immediately served by one of an infinite number of possible servers, and the service times are assumed to be independent with a common distribution G.

To compute the joint distribution of the number of customers that have completed their service and the number that are in service at t, call an entering customer a type-I customer if it completes its service by time t and a type-II customer if it does not complete its service by time t. Now, if the customer enters at time s, $s \le t$, then it will be a type-I customer if its service time is less than $t - s$, and since the service time distribution is G, the probability of this will be $G(t - s)$. Hence,

$$P(s) = G(t - s), \qquad s \le t,$$

and thus from Proposition 2.3.2 we obtain that the distribution of $N_1(t)$—the number of customers that have completed service by time t—is Poisson with mean

$$E[N_1(t)] = \lambda \int_0^t G(t - s)\, ds = \lambda \int_0^t G(y)\, dy.$$

Similarly $N_2(t)$, the number of customers being served at time t, is Poisson distributed with mean

$$E[N_2(t)] = \lambda \int_0^t \overline{G}(y)\, dy.$$

Further $N_1(t)$ and $N_2(t)$ are independent.

The following example further illustrates the use of Theorem 2.3.1.

EXAMPLE 2.3(C) Suppose that a device is subject to shocks that occur in accordance with a Poisson process having rate λ. The ith shock gives rise to a damage D_i. The D_i, $i \ge 1$, are assumed to be independent and identically distributed and also to be independent of $\{N(t), t \ge 0\}$, where $N(t)$ denotes the number of shocks in $[0, t]$. The damage due to a shock is assumed to decrease exponentially in time. That is, if a shock has an initial damage D, then a time t later its damage is $De^{-\alpha t}$.

If we suppose that the damages are additive, then $D(t)$, the damage at t, can be expressed as

$$D(t) = \sum_{i=1}^{N(t)} D_i e^{-\alpha(t - S_i)},$$

where S_i represents the arrival time of the ith shock. We can determine $E[D(t)]$ as follows:

$$\begin{aligned}
E[D(t)|N(t) = n] &= E\left[\sum_{i=1}^{N(t)} D_i e^{-\alpha(t-S_i)} \Big| N(t) = n\right] \\
&= E\left[\sum_{i=1}^{n} D_i e^{-\alpha(t-S_i)} \Big| N(t) = n\right] \\
&= \sum_{i=1}^{n} E[D_i e^{-\alpha(t-S_i)} | N(t) = n] \\
&= \sum_{i=1}^{n} E[D_i | N(t) = n] E[e^{-\alpha(t-S_i)} | N(t) = n] \\
&= E[D] \sum_{i=1}^{n} E[e^{-\alpha(t-S_i)} | N(t) = n] \\
&= E[D] E\left[\sum_{i=1}^{n} e^{-\alpha(t-S_i)} \Big| N(t) = n\right] \\
&= E[D] e^{-\alpha t} E\left[\sum_{i=1}^{n} e^{\alpha S_i} \Big| N(t) = n\right].
\end{aligned}$$

Now, letting $U_1, \ldots, U_n$ be independent and identically distributed uniform $[0, t]$ random variables, then, by Theorem 2.3.1,

$$\begin{aligned}
E\left[\sum_{i=1}^{n} e^{\alpha S_i} \Big| N(t) = n\right] &= E\left[\sum_{i=1}^{n} e^{\alpha U_{(i)}}\right] \\
&= E\left[\sum_{i=1}^{n} e^{\alpha U_i}\right] \\
&= \frac{n}{t} \int_0^t e^{\alpha x}\, dx \\
&= \frac{n}{\alpha t}(e^{\alpha t} - 1).
\end{aligned}$$

Hence,

$$E[D(t)|N(t)] = \frac{N(t)}{\alpha t}(1 - e^{-\alpha t}) E[D]$$

and, taking expectations,

$$E[D(t)] = \frac{\lambda E[D]}{\alpha}(1 - e^{-\alpha t}).$$

Remark Another approach to obtaining $E[D(t)]$ is to break up the interval $(0, t)$ into nonoverlapping intervals of length h and then add the contribution at time t of shocks originating in these intervals. More specifically, let h be given and define X_i as the sum of the damages at time t of all shocks arriving in the interval $I_i \equiv (ih, (i+1)h)$, $i = 0, 1, \ldots, [t/h]$, where $[a]$ denotes the largest integer less than or equal to a. Then we have the representation

$$D(t) = \sum_{i=0}^{[t/h]} X_i,$$

and so

$$E[D(t)] = \sum_{i=0}^{[t/h]} E[X_i].$$

To compute $E[X_i]$ condition on whether or not a shock arrives in the interval I_i. This yields

$$E[D(t)] = \sum_{i=0}^{[t/h]} (\lambda h E[De^{-\alpha(t-L_i)}] + o(h)),$$

where L_i is the arrival time of the shock in the interval I_i. Hence,

$$E[D(t)] = \lambda E[D] E\left[\sum_{i=0}^{[t/h]} he^{-\alpha(t-L_i)}\right] + \left[\frac{t}{h}\right] o(h). \tag{2.3.1}$$

But, since $L_i \in I_i$, it follows upon letting $h \to 0$ that

$$\sum_{i=0}^{[t/h]} he^{-\alpha(t-L_i)} \to \int_0^t e^{-\alpha(t-y)}\, dy = \frac{1-e^{-\alpha t}}{\alpha}$$

and thus from (2.3.1) upon letting $h \to 0$

$$E[D(t)] = \frac{\lambda E[D]}{\alpha}(1 - e^{-\alpha t}).$$

It is worth noting that the above is a more rigorous version of the following argument: Since a shock occurs in the interval $(y, y + dy)$ with probability $\lambda\, dy$ and since its damage at time t will equal $e^{-\alpha(t-y)}$ times its initial damage, it follows that the expected damage at t from shocks originating in $(y, y + dy)$ is

$$\lambda\, dy E[D] e^{-\alpha(t-y)},$$

and so

$$E[D(t)] = \lambda E[D] \int_0^t e^{-\alpha(t-y)}\, dy$$

$$= \frac{\lambda E[D]}{\alpha}(1 - e^{-\alpha t}).$$

2.3.1 The $M/G/1$ Busy Period

Consider the queueing system, known as $M/G/1$, in which customers arrive in accordance with a Poisson process with rate λ. Upon arrival they either enter service if the server is free or else they join the queue. The successive service times are independent and identically distributed according to G, and are also independent of the arrival process. When an arrival finds the server free, we say that a busy period begins. It ends when there are no longer any customers in the system. We would like to compute the distribution of the length of a busy period.

Suppose that a busy period has just begun at some time, which we shall designate as time 0. Let S_k denote the time until k additional customers have arrived. (Thus, for instance, S_k has a gamma distribution with parameters k, λ.) Also let $Y_1, Y_2, \ldots$ denote the sequence of service times. Now the busy period will last a time t and will consist of n services if, and only if,

(i) $S_k \le Y_1 + \cdots + Y_k,\ k = 1, \ldots, n - 1.$
(ii) $Y_1 + \cdots + Y_n = t.$
(iii) There are $n - 1$ arrivals in $(0, t)$.

Equation (i) is necessary for, if $S_k > Y_1 + \cdots + Y_k$, then the kth arrival after the initial customer will find the system empty of customers and thus the busy period would have ended prior to $k + 1$ (and thus prior to n) services. The reasoning behind (ii) and (iii) is straightforward and left to the reader.

Hence, reasoning heuristically (by treating densities as if they were probabilities) we see from the above that

(2.3.2) $P\{\text{busy period is of length } t \text{ and consists of } n \text{ services}\}$

$$= P\{Y_1 + \cdots + Y_n = t, n - 1 \text{ arrivals in } (0, t), S_k \le Y_1 + \cdots + Y_k,\ k = 1, \ldots, n - 1\}$$

$$= P\{S_k \le Y_1 + \cdots + Y_k, k = 1, \ldots, n - 1 \mid n - 1 \text{ arrivals in } (0, t),\ Y_1 + \cdots + Y_n = t\} \times P\{n - 1 \text{ arrivals in } (0, t), Y_1 + \cdots + Y_n = t\}.$$

Now the arrival process is independent of the service times and thus

$$(2.3.3)\qquad P\{n-1 \text{ arrivals in } (0,t), Y_1+\cdots+Y_n=t\} = e^{-\lambda t}\frac{(\lambda t)^{n-1}}{(n-1)!}\,dG_n(t),$$

where G_n is the n-fold convolution of G with itself. In addition, we have from Theorem 2.3.1 that, given $n-1$ arrivals in $(0, t)$, the ordered arrival times are distributed as the ordered values of a set of $n-1$ independent uniform $(0, t)$ random variables. Hence using this fact along with (2.3.3) and (2.3.2) yields

$$(2.3.4)\qquad P\{\text{busy period is of length } t \text{ and consists of } n \text{ services}\}$$
$$= e^{-\lambda t}\frac{(\lambda t)^{n-1}}{(n-1)!}\,dG_n(t) \times P\{\tau_k \le Y_1+\cdots+Y_k, k=1,\ldots,n-1 \mid Y_1+\cdots+Y_n=t\},$$

where $\tau_1, \ldots, \tau_{n-1}$ are independent of $\{Y_1, \ldots, Y_n\}$ and represent the ordered values of a set of $n-1$ uniform $(0, t)$ random variables.

To compute the remaining probability in (2.3.4) we need some lemmas. Lemma 2.3.3 is elementary and its proof is left as an exercise.

Lemma 2.3.3

Let $Y_1, Y_2, \ldots, Y_n$ be independent and identically distributed nonnegative random variables. Then

$$E[Y_1+\cdots+Y_k \mid Y_1+\cdots+Y_n=y] = \frac{k}{n}y, \qquad k=1,\ldots,n.$$

Lemma 2.3.4

Let $\tau_1, \ldots, \tau_n$ denote the ordered values from a set of n independent uniform $(0, t)$ random variables. Let $Y_1, Y_2, \ldots$ be independent and identically distributed nonnegative random variables that are also independent of $\{\tau_1, \ldots, \tau_n\}$. Then

$$(2.3.5)\qquad P\{Y_1+\cdots+Y_k<\tau_k, k=1,\ldots,n \mid Y_1+\cdots+Y_n=y\} = \begin{cases} 1-y/t & 0<y<t \\ 0 & \text{otherwise.} \end{cases}$$

Proof The proof is by induction on n. When $n = 1$ we must compute $P\{Y_1 < \tau_1 | Y_1 = y\}$ when τ_1 is uniform $(0, t)$. But

$$P\{Y_1 < \tau_1 | Y_1 = y\} = P\{y < \tau_1\} = 1 - y/t, \qquad 0 < y < t.$$

So assume the lemma when n is replaced by $n - 1$ and now consider the n case. Since the result is obvious for $y \geq t$, suppose that $y < t$. To make use of the induction hypothesis we will compute the left-hand side of (2.3.5) by conditioning on the values of $Y_1 + \cdots + Y_{n-1}$ and τ_n and then using the fact that conditional on $\tau_n = u$, $\tau_1, \ldots, \tau_{n-1}$ are distributed as the order statistics from a set of $n - 1$ uniform $(0, u)$ random variables (see Problem 2.18). Doing so, we have for $s < y$

(2.3.6)

$$P\{Y_1 + \cdots + Y_k < \tau_k, k = 1, \ldots, n | Y_1 + \cdots + Y_{n-1} = s, \tau_n = u, Y_1 + \cdots + Y_n = y\}$$

$$= \begin{cases} P\{Y_1 + \cdots + Y_k < \tau_k^*, k = 1, \ldots, n-1 | Y_1 + \cdots + Y_{n-1} = s\} & y < u \\ 0 \quad \text{if } y \geq u, \end{cases}$$

where $\tau_1^*, \ldots, \tau_{n-1}^*$ are the ordered values from a set of $n - 1$ independent uniform $(0, u)$ random variables. Hence by the induction hypothesis we see that the right-hand side of (2.3.6) is equal to

$$\text{R.H.S.} = \begin{cases} 1 - s/u & y < u \\ 0 & \text{otherwise.} \end{cases}$$

Hence, for $y < u$,

$$P\{Y_1 + \cdots + Y_k < \tau_k, k = 1, \ldots, n | Y_1 + \cdots + Y_{n-1}, \tau_n = u, Y_1 + \cdots + Y_n = y\}$$
$$= 1 - \frac{Y_1 + \cdots + Y_{n-1}}{\tau_n}$$

and thus, for $y < u$,

$$P\{Y_1 + \cdots + Y_k < \tau_k, k = 1, \ldots, n | \tau_n = u, Y_1 + \cdots + Y_n = y\}$$
$$= E\left[1 - \frac{Y_1 + \cdots + Y_{n-1}}{\tau_n} \,\middle|\, \tau_n = u, Y_1 + \cdots + Y_n = y\right]$$
$$= 1 - \frac{1}{u} E[Y_1 + \cdots + Y_{n-1} | Y_1 + \cdots + Y_n = y]$$
$$= 1 - \frac{n-1}{n} \frac{y}{u},$$

where we have made use of Lemma 2.3.3 in the above. Taking expectations once more yields

$$\text{(2.3.7)} \qquad P\{Y_1 + \cdots + Y_k < \tau_k, k = 1, \ldots, n \mid Y_1 + \cdots + Y_n = y\}$$
$$= E\left[1 - \frac{n-1}{n}\frac{y}{\tau_n} \,\middle|\, y < \tau_n\right] P\{y < \tau_n\}$$
$$= P\{y < \tau_n\} - \frac{n-1}{n} y E\left[\frac{1}{\tau_n} \,\middle|\, y < \tau_n\right] P\{y < \tau_n\}.$$

Now the distribution of τ_n is given by

$$P\{\tau_n < x\} = P\left\{\max_{1\le i\le n} U_i < x\right\}$$
$$= P\{U_i < x, i = 1, \ldots, n\}$$
$$= (x/t)^n, \qquad 0 < x < t,$$

where U_i, $i = 1, \ldots, n$ are independent uniform $(0, t)$ random variables. Hence its density is given by

$$f_{\tau_n}(x) = \frac{n}{t}\left(\frac{x}{t}\right)^{n-1}, \qquad 0 < x < t$$

and thus

$$\text{(2.3.8)} \qquad E\left[\frac{1}{\tau_n} \,\middle|\, \tau_n > y\right] P\{\tau_n > y\} = \int_y^t \frac{1}{x}\frac{n}{t}\left(\frac{x}{t}\right)^{n-1} dx$$
$$= \frac{n}{n-1}\left(\frac{t^{n-1} - y^{n-1}}{t^n}\right).$$

Thus from (2.3.7) and (2.3.8), when $y < t$,

$$P\{Y_1 + \cdots + Y_k < \tau_k, k = 1, \ldots, n \mid Y_1 + \cdots + Y_n = y\}$$
$$= 1 - \left(\frac{y}{t}\right)^n - \frac{y(t^{n-1} - y^{n-1})}{t^n}$$
$$= 1 - \frac{y}{t}.$$

and the proof is complete.

∎

We need one additional lemma before going back to the busy period problem.

Lemma 2.3.5

Let $\tau_1, \ldots, \tau_{n-1}$ denote the ordered values from a set of $n - 1$ independent uniform $(0, t)$ random variables; and let $Y_1, Y_2, \ldots$ be independent and identically distributed nonnegative random variables that are also independent of $\{\tau_1, \ldots, \tau_{n-1}\}$. Then

$$P\{Y_1 + \cdots + Y_k < \tau_k, k = 1, \ldots, n-1 \mid Y_1 + \cdots + Y_n = t\} = 1/n.$$

Proof To compute the above probability we will make use of Lemma 2.3.4 by conditioning on $Y_1 + \cdots + Y_{n-1}$. That is, by Lemma 2.3.4,

$$\begin{aligned} &P\{Y_1 + \cdots + Y_k < \tau_k, k = 1, \ldots, n-1 \mid Y_1 + \cdots + Y_{n-1} = y, Y_1 + \cdots + Y_n = t\} \\ &\quad = P\{Y_1 + \cdots + Y_k < \tau_k, k = 1, \ldots, n-1 \mid Y_1 + \cdots + Y_{n-1} = y\} \\ &\quad = \begin{cases} 1 - y/t & 0 < y < t \\ 0 & \text{otherwise.} \end{cases} \end{aligned}$$

Hence, as $Y_1 + \cdots + Y_{n-1} \le Y_1 + \cdots + Y_n$, we have that

$$\begin{aligned} &P\{Y_1 + \cdots + Y_k < \tau_k, k = 1, \ldots, n-1 \mid Y_1 + \cdots + Y_n = t\} \\ &\quad = E\left[1 - \frac{Y_1 + \cdots + Y_{n-1}}{t} \,\middle|\, Y_1 + \cdots + Y_n = t\right] \\ &\quad = 1 - \frac{n-1}{n} \qquad \text{(by Lemma 2.3.3),} \end{aligned}$$

which proves the result.

Returning to the joint distribution of the length of a busy period and the number of customers served, we must, from (2.3.4), compute

$$P\{\tau_k \le Y_1 + \cdots + Y_k, k = 1, \ldots, n-1 \mid Y_1 + \cdots + Y_n = t\}.$$

Now since $t - U$ will also be a uniform $(0, t)$ random variable whenever U is, it follows that $\tau_1, \ldots, \tau_{n-1}$ has the same joint distribution as $t - \tau_{n-1}, \ldots, t - \tau_1$. Hence, upon replacing τ_k by $t - \tau_{n-k}$ throughout, $1 \le k \le n-1$, we obtain

$$\begin{aligned} &P\{\tau_k \le Y_1 + \cdots + Y_k, k = 1, \ldots, n-1 \mid Y_1 + \cdots + Y_n = t\} \\ &\quad = P\{t - \tau_{n-k} \le Y_1 + \cdots + Y_k, k = 1, \ldots, n-1 \mid Y_1 + \cdots + Y_n = t\} \\ &\quad = P\{t - \tau_{n-k} \le t - (Y_{k+1} + \cdots + Y_n), k = 1, \ldots, n-1 \mid Y_1 + \cdots + Y_n = t\} \\ &\quad = P\{\tau_{n-k} \ge Y_{k+1} + \cdots + Y_n, k = 1, \ldots, n-1 \mid Y_1 + \cdots + Y_n = t\} \\ &\quad = P\{\tau_{n-k} \ge Y_{n-k} + \cdots + Y_1, k = 1, \ldots, n-1 \mid Y_1 + \cdots + Y_n = t\}, \end{aligned}$$

where the last equality follows since $Y_1, \ldots, Y_n$ has the same joint distribution as $Y_n, \ldots, Y_1$ and so any probability statement about the Y_i's remains valid if Y_1 is replaced by Y_n, Y_2 by $Y_{n-1}, \ldots, Y_k$ by $Y_{n-k+1}, \ldots, Y_n$ by Y_1. Hence, we see that

$$\begin{aligned} P\{\tau_k \le Y_1 + \cdots + Y_k, k = 1, \ldots, n-1 \mid Y_1 + \cdots + Y_n = t\} \\ = P\{\tau_k \ge Y_1 + \cdots + Y_k, k = 1, \ldots, n-1 \mid Y_1 + \cdots + Y_n = t\} \\ = 1/n \qquad \text{(from Lemma 2.3.5)}. \end{aligned}$$

Hence, from (2.3.4), if we let

$$B(t, n) = P\{\text{busy period is of length} \le t, n \text{ customers served in a busy period}\},$$

then

$$\frac{d}{dt} B(t, n) = e^{-\lambda t} \frac{(\lambda t)^{n-1}}{n!} dG_n(t)$$

or

$$B(t, n) = \int_0^t e^{-\lambda t} \frac{(\lambda t)^{n-1}}{n!} dG_n(t).$$

The distribution of the length of a busy period, call it $B(t) = \sum_{n=1}^{\infty} B(t, n)$, is thus given by

$$B(t) = \sum_{n=1}^{\infty} \int_0^t e^{-\lambda t} \frac{(\lambda t)^{n-1}}{n!} dG_n(t).$$

2.4 Nonhomogeneous Poisson Process

In this section we generalize the Poisson process by allowing the arrival rate at time t to be a function of t.

Definition 2.4.1

The counting process $\{N(t), t \ge 0\}$ is said to be a nonstationary or nonhomogeneous Poisson process with intensity function $\lambda(t)$, $t \ge 0$ if:

(i) $N(0) = 0$.
(ii) $\{N(t), t \ge 0\}$ has independent increments.

(iii) $P\{N(t+h) - N(t) \geq 2\} = o(h)$.
(iv) $P\{N(t+h) - N(t) = 1\} = \lambda(t)h + o(h)$.

If we let

$$m(t) = \int_0^t \lambda(s)\, ds,$$

then it can be shown that

$$P\{N(t+s) - N(t) = n\} = \exp\{-(m(t+s) - m(t))\}[m(t+s) - m(t)]^n/n!, \qquad n \geq 0. \tag{2.4.1}$$

That is, $N(t+s) - N(t)$ is Poisson distributed with mean $m(t+s) - m(t)$.

The proof of (2.4.1) follows along the lines of the proof of Theorem 2.1.1 with a slight modification: Fix t and define

$$P_n(s) = P\{N(t+s) - N(t) = n\}.$$

Then,

$$\begin{aligned} P_0(s+h) &= P\{N(t+s+h) - N(t) = 0\} \\ &= P\{0 \text{ events in } (t, t+s), 0 \text{ events in } (t+s, t+s+h)\} \\ &= P\{0 \text{ events in } (t, t+s)\}P\{0 \text{ events in } (t+s, t+s+h)\} \\ &= P_0(s)[1 - \lambda(t+s)h + o(h)], \end{aligned}$$

where the next-to-last equality follows from Axiom (ii) and the last from Axioms (iii) and (iv). Hence,

$$\frac{P_0(s+h) - P_0(s)}{h} = -\lambda(t+s)P_0(s) + \frac{o(h)}{h}.$$

Letting $h \to 0$ yields

$$P_0'(s) = -\lambda(t+s)P_0(s)$$

or

$$\log P_0(s) = -\int_0^s \lambda(t+u)\, du$$

or

$$P_0(s) = e^{-[m(t+s)-m(t)]}.$$

The remainder of the verification of (2.4.1) follows similarly and is left as an exercise.

The importance of the nonhomogeneous Poisson process resides in the fact that we no longer require stationary increments, and so we allow for the possibility that events may be more likely to occur at certain times than at other times.

When the intensity function $\lambda(t)$ is bounded, we can think of the nonhomogeneous process as being a random sample from a homogeneous Poisson process. Specifically, let λ be such that

$$\lambda(t) \le \lambda \qquad \text{for all } t \ge 0$$

and consider a Poisson process with rate λ. Now if we suppose that an event of the Poisson process that occurs at time t is counted with probability $\lambda(t)/\lambda$, then the process of counted events is a nonhomogeneous Poisson process with intensity function $\lambda(t)$. This last statement easily follows from Definition 2.4.1. For instance (i), (ii), and (iii) follow since they are also true for the homogeneous Poisson process. Axiom (iv) follows since

$$\begin{aligned} P\{\text{one counted event in } (t, t+h)\} &= P\{\text{one event in } (t, t+h)\}\frac{\lambda(t)}{\lambda} + o(h) \\ &= \lambda h \frac{\lambda(t)}{\lambda} + o(h) \\ &= \lambda(t)h + o(h). \end{aligned}$$

The interpretation of a nonhomogeneous Poisson process as a sampling from a homogeneous one also gives us another way of understanding Proposition 2.3.2 (or, equivalently, it gives us another way of proving that $N(t)$ of Proposition 2.3.2 is Poisson distributed).

Example 2.4(a) ***Record Values.*** Let $X_1, X_2, \ldots$ denote a sequence of independent and identically distributed nonnegative continuous random variables whose hazard rate function is given by $\lambda(t)$. (That is, $\lambda(t) = f(t)/\overline{F}(t)$, where f and F are respectively the density and distribution function of X.) We say that a *record* occurs at time n if $X_n > \max(X_1, \ldots, X_{n-1})$, where $X_0 \equiv 0$. If a record occurs at time n, then X_n is called a *record value*. Let $N(t)$ denote the number of record values less than or equal to t. That is, $N(t)$ is a counting process of events where an event is said to occur at time x if x is a record value.

We claim that $\{N(t), t \geq 0\}$ will be a nonhomogeneous Poisson process with intensity function $\lambda(t)$. To verify this claim note that there will be a record value between t and $t + h$ if, and only if, the first X_i whose value is greater than t lies between t and $t + h$. But we have (conditioning on which X_i this is, say, $i = n$), by the definition of a hazard rate function,

$$P\{X_n \in (t, t + h) \mid X_n > t\} = \lambda(t)h + o(h),$$

which proves the claim.

EXAMPLE 2.4(B) ***The Output Process of an Infinite Server Poisson Queue (M/G/∞).*** It turns out that the output process of the $M/G/\infty$ queue—that is, of the infinite server queue having Poisson arrivals and general service distribution G—is a nonhomogeneous Poisson process having intensity function $\lambda(t) = \lambda G(t)$. To prove this we shall first argue that

(1) the number of departures in $(s, s + t)$ is Poisson distributed with mean $\lambda \int_s^{s+t} G(y)\, dy$, and

(2) the numbers of departures in disjoint time intervals are independent.

To prove statement (1), call an arrival type I if it departs in the interval $(s, s + t)$. Then an arrival at y will be type I with probability

$$P(y) = \begin{cases} G(s + t - y) - G(s - y) & \text{if } y < s \\ G(s + t - y) & \text{if } s < y < s + t \\ 0 & \text{if } y > s + t. \end{cases}$$

Hence, from Proposition 2.3.2 the number of such departures will be Poisson distributed with mean

$$\begin{aligned} \lambda \int_0^\infty P(y)\, dy &= \lambda \int_0^s (G(s + t - y) - G(s - y))\, dy \\ &\quad + \lambda \int_s^{s+t} G(s + t - y)\, dy \\ &= \lambda \int_s^{s+t} G(y)\, dy. \end{aligned}$$

To prove statement (2) let I_1 and I_2 denote disjoint time intervals and call an arrival type I if it departs in I_1, call it type II if it departs in I_2, and call it type III otherwise. Again, from Proposition 2.3.2, or, more precisely, from its generalization to three types of customers, it follows that the number of departures in I_1 and I_2 (that is, the number of type-I and type-II arrivals) are independent Poisson random variables.

Using statements (1) and (2) it is a simple matter to verify all the axiomatic requirements for the departure process to be a nonhomogeneous Poisson process (it is much like showing that Definition 2.1.1 of a Poisson process implies Definition 2.1.2).

Since $\lambda(t) \to \lambda$ as $t \to \infty$, it is interesting to note that the limiting output process after time t (as $t \to \infty$) is a Poisson process with rate λ.

2.5 Compound Poisson Random Variables and Processes

Let $X_1, X_2, \ldots$ be a sequence of independent and identically distributed random variables having distribution function F, and suppose that this sequence is independent of N, a Poisson random variable with mean λ. The random variable

$$W = \sum_{i=1}^{N} X_i$$

is said to be a *compound Poisson* random variable with Poisson parameter λ and component distribution F.

The moment generating function of W is obtained by conditioning on N. This gives

$$\begin{aligned}
E[e^{tW}] &= \sum_{n=0}^{\infty} E[e^{tW} \mid N = n] P\{N = n\} \\
&= \sum_{n=0}^{\infty} E[e^{t(X_1 + \cdots + X_n)} \mid N = n] e^{-\lambda t} (\lambda t)^n / n! \\
&= \sum_{n=0}^{\infty} E[e^{t(X_1 + \cdots + X_n)}] e^{-\lambda t} (\lambda t)^n / n! \qquad (2.5.1) \\
&= \sum_{n=0}^{\infty} E[e^{tX_1}]^n e^{-\lambda t} (\lambda t)^n / n! \qquad (2.5.2)
\end{aligned}$$

where (2.5.1) follows from the independence of $\{X_1, X_2, \ldots\}$ and N, and (2.5.2) follows from the independence of the X_i. Hence, letting

$$\phi_X(t) = E[e^{tX_i}]$$

denote the moment generating function of the X_i, we have from (2.5.2) that

$$\begin{aligned}
E[e^{tW}] &= \sum_{n=0}^{\infty} [\phi_X(t)]^n e^{-\lambda t} (\lambda t)^n / n! \\
&= \exp\{\lambda t (\phi_X(t) - 1)\}. \qquad (2.5.3)
\end{aligned}$$

It is easily shown, either by differentiating (2.5.3) or by directly using a conditioning argument, that

$$E[W] = \lambda E[X]$$
$$\mathrm{Var}(W) = \lambda E[X^2]$$

where X has distribution F.

Example 2.5(a) Aside from the way in which they are defined, compound Poisson random variables often arise in the following manner. Suppose that events are occurring in accordance with a Poisson process having rate (say) α, and that whenever an event occurs a certain contribution results. Specifically, suppose that an event occurring at time s will, independent of the past, result in a contribution whose value is a random variable with distribution F_s. Let W denote the sum of the contributions up to time t—that is,

$$W = \sum_{i=1}^{N(t)} X_i$$

where $N(t)$ is the number of events occurring by time t, and X_i is the contribution made when event i occurs. Then, even though the X_i are neither independent nor identically distributed it follows that W is a compound Poisson random variable with parameters

$$\lambda = \alpha t \qquad \text{and} \qquad F(x) = \frac{1}{t}\int_0^t F_s(x)\, ds.$$

This can be shown by calculating the distribution of W by first conditioning on $N(t)$, and then using the result that, given $N(t)$, the unordered set of $N(t)$ event times are independent uniform $(0, t)$ random variables (see Section 2.3).

When F is a discrete probability distribution function there is an interesting representation of W as a linear combination of independent Poisson random variables. Suppose that the X_i are discrete random variables such that

$$P\{X_i = j\} = p_j, j = 1, \ldots, k, \sum_{j=1}^{k} p_j = 1.$$

If we let N_j denote the number of the X_i's that are equal to j, $j = 1, \ldots, k$, then we can express W as

$$W = \sum_j jN_j \tag{2.5.4}$$

where, using the results of Example 1.5(I), the N_j are independent Poisson random variables with respective means λp_j, $j = 1, \ldots, k$. As a check, let us use the representation (2.5.4) to compute the mean and variance of W.

$$E[W] = \sum_j jE[N_j] = \sum_j j\lambda p_j = \lambda E[X]$$

$$\mathrm{Var}(W) = \sum_j j^2 \mathrm{Var}(N_j) = \sum_j j^2 \lambda p_j = \lambda E[X^2]$$

which check with our previous results.

2.5.1 A Compound Poisson Identity

As before, let $W = \sum_{i=1}^{N} X_i$ be a compound Poisson random variable with N being Poisson with mean λ and the X_i having distribution F. We now present a useful identity concerning W.

PROPOSITION 2.5.2

Let X be a random variable having distribution F that is independent of W. Then, for any function $h(x)$

$$E[Wh(W)] = \lambda E[Xh(W + X)].$$

Proof

$$\begin{aligned} E[Wh(W)] &= \sum_{n=0}^{\infty} E[Wh(W) \mid N = n] e^{-\lambda} \frac{\lambda^n}{n!} \\ &= \sum_{n=0}^{\infty} e^{-\lambda} \frac{\lambda^n}{n!} E\left[\sum_{i=1}^{n} X_i h\left(\sum_{j=1}^{n} X_j\right)\right] \\ &= \sum_{n=0}^{\infty} e^{-\lambda} \frac{\lambda^n}{n!} \sum_{i=1}^{n} E\left[X_i h\left(\sum_{j=1}^{n} X_j\right)\right] \\ &= \sum_{n=0}^{\infty} e^{-\lambda} \frac{\lambda^n}{n!} nE\left[X_n h\left(\sum_{j=1}^{n} X_j\right)\right] \end{aligned}$$

where the preceding equation follows because all of the random variables $X_i h\left(\sum_{j=1}^{n} X_j\right)$ have the same distribution. Hence, from the above we obtain, upon conditioning on X_n, that

$$
\begin{aligned}
E[Wh(W)] &= \sum_{n=1}^{\infty} e^{-\lambda} \frac{\lambda^n}{(n-1)!} \int E\left[X_n h\left(\sum_{j=1}^{n} X_j\right) \middle| X_n = x \right] dF(x) \\
&= \lambda \sum_{n=1}^{\infty} e^{-\lambda} \frac{\lambda^{n-1}}{(n-1)!} \int x E\left[h\left(\sum_{j=1}^{n-1} X_j + x\right)\right] dF(x) \\
&= \lambda \int x \sum_{m=0}^{\infty} e^{-\lambda} \frac{\lambda^m}{m!} E\left[h\left(\sum_{j=1}^{m} X_j + x\right)\right] dF(x) \\
&= \lambda \int x \sum_{m=0}^{\infty} E[h(W+x) \mid N=m] P\{N=m\}\, dF(x) \\
&= \lambda \int x E[h(W+x)]\, dF(x) \\
&= \lambda \int E[Xh(W+X) \mid X=x]\, dF(x) \\
&= \lambda E[Xh(W+X)].
\end{aligned}
$$

Proposition 2.5.2 gives an easy way of computing the moments of W.

Corollary 2.5.3

If X has distribution F, then for any positive integer n

$$E[W^n] = \lambda \sum_{j=0}^{n-1} \binom{n-1}{j} E[W^j] E[X^{n-j}].$$

Proof Let $h(x) = x^{n-1}$ and apply Proposition 2.5.2 to obtain

$$
\begin{aligned}
E[W^n] &= \lambda E[X(W+X)^{n-1}] \\
&= \lambda E\left[X \sum_{j=0}^{n-1} \binom{n-1}{j} W^j X^{n-1-j} \right] \\
&= \lambda \sum_{j=0}^{n-1} \binom{n-1}{j} E[W^j] E[X^{n-j}]
\end{aligned}
$$

Thus, by starting with $n = 1$ and successively increasing the value of n, we see that

$$
\begin{aligned}
E[W] &= \lambda E[X] \\
E[W^2] &= \lambda(E[X^2] + E[W]E[X]) \\
&= \lambda E[X^2] + \lambda^2(E[X])^2 \\
E[W^3] &= \lambda(E[X^3] + 2E[W]E[X^2] + E[W^2]E[X]) \\
&= \lambda E[X^3] + 3\lambda^2 E[X]E[X^2] + \lambda^3(E[X])^3
\end{aligned}
$$

and so on.

We will now show that Proposition 2.5.2 yields an elegant recursive formula for the probability mass function of W when the X_i are positive integer valued random variables. Suppose this is the situation, and let

$$\alpha_j = P\{X_i = j\}, \qquad j \geq 1$$

and

$$P_j = P\{W = j\}, \qquad j \geq 0.$$

The successive values of P_n can be obtained by making use of the following.

Corollary 2.5.4

$$
\begin{aligned}
P_0 &= e^{-\lambda} \\
P_n &= \frac{\lambda}{n} \sum_{j=1}^{n} j\alpha_j P_{n-j}, \qquad n \geq 1
\end{aligned}
$$

Proof That $P_0 = e^{-\lambda}$ is immediate, so take $n > 0$. Let

$$h(x) = \begin{cases} 0 & \text{if } x \neq n \\ 1/n & \text{if } x = n. \end{cases}$$

Since $Wh(W) = I\{W = n\}$, which is defined to equal 1 if $W = n$ and 0 otherwise, we obtain upon applying Proposition 2.5.2 that

$$
\begin{aligned}
P\{W = n\} &= \lambda E[Xh(W + X)] \\
&= \lambda \sum_j E[Xh(W + X) \mid X = j]\alpha_j \\
&= \lambda \sum_j jE[h(W + j)]\alpha_j \\
&= \lambda \sum_j j \frac{1}{n} P\{W + j = n\}\alpha_j.
\end{aligned}
$$

Remark When the X_i are identically equal to 1, the preceding recursion reduces to the well-known identity for Poisson probabilities

$$P\{N = 0\} = e^{-\lambda}$$

$$P\{N = n\} = \frac{\lambda}{n} P\{N = n - 1\}, \qquad n \geq 1.$$

EXAMPLE 2.5(B) Let W be a compound Poisson random variable with Poisson parameter $\lambda = 4$ and with

$$P\{X_i = i\} = 1/4, \qquad i = 1, 2, 3, 4.$$

To determine $P\{W = 5\}$, we use the recursion of Corollary 2.5.4 as follows:

$$P_0 = e^{-\lambda} = e^{-4}$$

$$P_1 = \lambda\alpha_1 P_0 = e^{-4}$$

$$P_2 = \frac{\lambda}{2}\{\alpha_1 P_1 + 2\alpha_2 P_0\} = \frac{3}{2} e^{-4}$$

$$P_3 = \frac{\lambda}{3}\{\alpha_1 P_2 + 2\alpha_2 P_1 + 3\alpha_3 P_0\} = \frac{13}{6} e^{-4}$$

$$P_4 = \frac{\lambda}{4}\{\alpha_1 P_3 + 2\alpha_2 P_2 + 3\alpha_3 P_1 + 4\alpha_4 P_0\} = \frac{73}{24} e^{-4}$$

$$P_5 = \frac{\lambda}{5}\{\alpha_1 P_4 + 2\alpha_2 P_3 + 3\alpha_3 P_2 + 4\alpha_4 P_1 + 5\alpha_5 P_0\} = \frac{501}{120} e^{-4}.$$

2.5.2 Compound Poisson Processes

A stochastic process $\{X(t), t \geq 0\}$ is said to be a *compound Poisson process* if it can be represented, for $t \geq 0$, by

$$X(t) = \sum_{i=1}^{N(t)} X_i$$

where $\{N(t), t \geq 0\}$ is a Poisson process, and $\{X_i, i = 1, 2, \ldots\}$ is a family of independent and identically distributed random variables that is independent of the process $\{N(t), t \geq 0\}$. Thus, if $\{X(t), t \geq 0\}$ is a compound Poisson process then $X(t)$ is a compound Poisson random variable.

As an example of a compound Poisson process, suppose that customers arrive at a store at a Poisson rate λ. Suppose, also, that the amounts of money spent by each customer form a set of independent and identically distributed random variables that is independent of the arrival process. If $X(t)$ denotes the total amount spent in the store by all customers arriving by time t, then $\{X(t), t \geq 0\}$ is a compound Poisson process.

2.6 Conditional Poisson Processes

Let Λ be a positive random variable having distribution G and let $\{N(t), t \geq 0\}$ be a counting process such that, given that $\Lambda = \lambda$, $\{N(t), t \geq 0\}$ is a Poisson process having rate λ. Thus, for instance,

$$P\{N(t+s) - N(s) = n\} = \int_0^\infty e^{-\lambda t} \frac{(\lambda t)^n}{n!} \, dG(\lambda).$$

The process $\{N(t), t \geq 0\}$ is called a *conditional Poisson process* since, conditional on the event that $\Lambda = \lambda$, it is a Poisson process with rate λ. It should be noted, however, that $\{N(t), t \geq 0\}$ is *not* a Poisson process. For instance, whereas it does have stationary increments, it does not have independent ones. (Why not?)

Let us compute the conditional distribution of Λ given that $N(t) = n$. For $d\lambda$ small,

$$\begin{aligned} &P\{\Lambda \in (\lambda, \lambda + d\lambda) \mid N(t) = n\} \\ &= \frac{P\{N(t) = n \mid \Lambda \in (\lambda, \lambda + d\lambda)\} P\{\Lambda \in (\lambda, \lambda + d\lambda)\}}{P\{N(t) = n\}} \\ &= \frac{e^{-\lambda t} \dfrac{(\lambda t)^n}{n!} \, dG(\lambda)}{\displaystyle\int_0^\infty e^{-\lambda t} \frac{(\lambda t)^n}{n!} \, dG(\lambda)} \end{aligned}$$

and so the conditional distribution of Λ, given that $N(t) = n$, is given by

$$P\{\Lambda \leq x \mid N(t) = n\} = \frac{\int_0^x e^{-\lambda t} (\lambda t)^n \, dG(\lambda)}{\int_0^\infty e^{-\lambda t} (\lambda t)^n \, dG(\lambda)}.$$

Example 2.6(a) Suppose that, depending on factors not at present understood, the average rate at which seismic shocks occur in a certain region over a given season is either λ_1 or λ_2. Suppose also that it is λ_1 for 100 p percent of the seasons and λ_2 the remaining time. A simple model for such a situation would be to suppose that $\{N(t), 0 \leq t < \infty\}$ is a conditional Poisson process such that Λ is either λ_1 or λ_2 with respective probabilities p and $1 - p$. Given n shocks in the first t time units of a season, then the probability it is a λ_1 season is

$$P\{\Lambda = \lambda_1 \mid N(t) = n\} = \frac{pe^{-\lambda_1 t}(\lambda_1 t)^n}{pe^{-\lambda_1 t}(\lambda_1 t)^n + e^{-\lambda_2 t}(\lambda_2 t)^n(1-p)}.$$

Also, by conditioning on whether $\Lambda = \lambda_1$ or $\Lambda = \lambda_2$, we see that the time from t until the next shock, given $N(t) = n$, has the distribution

$$P\{\text{time from } t \text{ until next shock is} \le x \mid N(t) = n\}$$
$$= \frac{p(1 - e^{-\lambda_1 x})e^{-\lambda_1 t}(\lambda_1 t)^n + (1 - e^{-\lambda_2 x})e^{-\lambda_2 t}(\lambda_2 t)^n(1-p)}{pe^{-\lambda_1 t}(\lambda_1 t)^n + e^{-\lambda_2 t}(\lambda_2 t)^n(1-p)}.$$

PROBLEMS

2.1. Show that Definition 2.1.1 of a Poisson process implies Definition 2.1.2.

2.2. For another approach to proving that Definition 2.1.2 implies Definition 2.1.1:

(a) Prove, using Definition 2.1.2, that

$$P_0(t + s) = P_0(t)P_0(s).$$

(b) Use (a) to infer that the interarrival times $X_1, X_2, \ldots$ are independent exponential random variables with rate λ.

(c) Use (b) to show that $N(t)$ is Poisson distributed with mean λt.

2.3. For a Poisson process show, for $s < t$, that

$$P\{N(s) = k \mid N(t) = n\} = \binom{n}{k}\left(\frac{s}{t}\right)^k\left(1 - \frac{s}{t}\right)^{n-k}, \qquad k = 0, 1, \ldots, n.$$

2.4. Let $\{N(t), t \ge 0\}$ be a Poisson process with rate λ. Calculate $E[N(t) \cdot N(t + s)]$.

2.5. Suppose that $\{N_1(t), t \ge 0\}$ and $\{N_2(t), t \ge 0\}$ are independent Poisson processes with rates λ_1 and λ_2. Show that $\{N_1(t) + N_2(t), t \ge 0\}$ is a Poisson process with rate $\lambda_1 + \lambda_2$. Also, show that the probability that the first event of the combined process comes from $\{N_1(t), t \ge 0\}$ is $\lambda_1/(\lambda_1 + \lambda_2)$, independently of the time of the event.

2.6. A machine needs two types of components in order to function. We have a stockpile of n type-1 components and m type-2 components.

Type-i components last for an exponential time with rate μ_i before failing. Compute the mean length of time the machine is operative if a failed component is replaced by one of the same type from the stockpile; that is, compute $E[\min(\sum_1^n X_i, \sum_1^m Y_i)]$, where the $X_i(Y_i)$ are exponential with rate $\mu_1(\mu_2)$.

2.7. Compute the joint distribution of S_1, S_2, S_3.

2.8. Generating a Poisson Random Variable. Let $U_1, U_2, \ldots$ be independent uniform (0, 1) random variables.

(a) If $X_i = (-\log U_i)/\lambda$, show that X_i is exponentially distributed with rate λ.

(b) Use part (a) to show that N is Poisson distributed with mean λ when N is defined to equal that value of n such that

$$\prod_{i=1}^{n} U_i \geq e^{-\lambda} > \prod_{i=1}^{n+1} U_i,$$

where $\prod_{i=1}^{0} U_i \equiv 1$. Compare with Problem 1.21 of Chapter 1.

2.9. Suppose that events occur according to a Poisson process with rate λ. Each time an event occurs we must decide whether or not to stop, with our objective being to stop at the last event to occur prior to some specified time T. That is, if an event occurs at time t, $0 \leq t \leq T$ and we decide to stop, then we lose if there are any events in the interval $(t, T]$, and win otherwise. If we do not stop when an event occurs, and no additional events occur by time T, then we also lose. Consider the strategy that stops at the first event that occurs after some specified time s, $0 \leq s \leq T$.

(a) If the preceding strategy is employed, what is the probability of winning?

(b) What value of s maximizes the probability of winning?

(c) Show that the probability of winning under the optimal strategy is $1/e$.

2.10. Buses arrive at a certain stop according to a Poisson process with rate λ. If you take the bus from that stop then it takes a time R, measured from the time at which you enter the bus, to arrive home. If you walk from the bus stop then it takes a time W to arrive home. Suppose that your policy when arriving at the bus stop is to wait up to a time s, and if a bus has not yet arrived by that time then you walk home.

(a) Compute the expected time from when you arrive at the bus stop until you reach home.

(b) Show that if $W < 1/\lambda + R$ then the expected time of part (a) is minimized by letting $s = 0$; if $W > 1/\lambda + R$ then it is minimized by letting $s = \infty$ (that is, you continue to wait for the bus); and when $W = 1/\lambda + R$ all values of s give the same expected time.

(c) Give an intuitive explanation of why we need only consider the cases $s = 0$ and $s = \infty$ when minimizing the expected time.

2.11. Cars pass a certain street location according to a Poisson process with rate λ. A person wanting to cross the street at that location waits until she can see that no cars will come by in the next T time units. Find the expected time that the person waits before starting to cross. (Note, for instance, that if no cars will be passing in the first T time units then the waiting time is 0.)

2.12. Events, occurring according to a Poisson process with rate λ, are registered by a counter. However, each time an event is registered the counter becomes inoperative for the next b units of time and does not register any new events that might occur during that interval. Let $R(t)$ denote the number of events that occur by time t and are registered.

(a) Find the probability that the first k events are all registered.

(b) For $t \geq (n - 1)b$, find $P\{R(t) \geq n\}$.

2.13. Suppose that shocks occur according to a Poisson process with rate λ, and suppose that each shock, independently, causes the system to fail with probability p. Let N denote the number of shocks that it takes for the system to fail and let T denote the time of failure. Find $P\{N = n \mid T = t\}$.

2.14. Consider an elevator that starts in the basement and travels upward. Let N_i denote the number of people that get in the elevator at floor i. Assume the N_i are independent and that N_i is Poisson with mean λ_i. Each person entering at i will, independent of everything else, get off at j with probability P_{ij}. $\sum_{j>i} P_{ij} = 1$. Let O_j = number of people getting off the elevator at floor j.

(a) Compute $E[O_j]$.

(b) What is the distribution of O_j?

(c) What is the joint distribution of O_j and O_k?

2.15. Consider an r-sided coin and suppose that on each flip of the coin exactly one of the sides appears: side i with probability P_i, $\sum_1^r P_i = 1$. For given numbers $n_1, \ldots, n_r$, let N_i denote the number of flips required until side i has appeared for the n_i time, $i = 1, \ldots, r$, and let

$$N = \min_{i=1,\ldots,r} N_i.$$

Thus N is the number of flips required until side i has appeared n_i times for some $i = 1, \ldots, r$.

(a) What is the distribution of N_i?

(b) Are the N_i independent?

Now suppose that the flips are performed at random times generated by a Poisson process with rate $\lambda = 1$. Let T_i denote the time until side i has appeared for the n_i time, $i = 1, \ldots, r$ and let

$$T = \min_{i=1,\ldots,r} T_i.$$

(c) What is the distribution of T_i?

(d) Are the T_i independent?

(e) Derive an expression for $E[T]$.

(f) Use (e) to derive an expression for $E[N]$.

2.16. The number of trials to be performed is a Poisson random variable with mean λ. Each trial has n possible outcomes and, independent of everything else, results in outcome number i with probability P_i, $\sum_1^n P_i = 1$. Let X_j denote the number of outcomes that occur exactly j times, $j = 0, 1, \ldots$. Compute $E[X_j]$, $\text{Var}(X_j)$.

2.17. Let $X_1, X_2, \ldots, X_n$ be independent continuous random variables with common density function f. Let $X_{(i)}$ denote the ith smallest of $X_1, \ldots, X_n$.

(a) Note that in order for $X_{(i)}$ to equal x, exactly $i - 1$ of the X's must be less than x, one must equal x, and the other $n - i$ must all be greater than x. Using this fact argue that the density function of $X_{(i)}$ is given by

$$f_{X_{(i)}}(x) = \frac{n!}{(i-1)!(n-i)!}(F(x))^{i-1}(\bar{F}(x))^{n-i}f(x).$$

(b) $X_{(i)}$ will be less than x if, and only if, how many of the X's are less than x?

(c) Use (b) to obtain an expression for $P\{X_{(i)} \le x\}$.

(d) Using (a) and (c) establish the identity

$$\sum_{k=i}^{n} \binom{n}{k} y^k (1-y)^{n-k} = \int_0^y \frac{n!}{(i-1)!(n-i)!} x^{i-1}(1-x)^{n-i}\, dx$$

for $0 \le y \le 1$.

(e) Let S_i denote the time of the ith event of the Poisson process $\{N(t), t \geq 0\}$. Find

$$E[S_i \mid N(t) = n] = \begin{cases} & i \leq n \\ & i > n. \end{cases}$$

2.18. Let $U_{(1)}, \ldots, U_{(n)}$ denote the order statistics of a set of n uniform (0, 1) random variables. Show that given $U_{(n)} = y$, $U_{(1)}, \ldots, U_{(n-1)}$ are distributed as the order statistics of a set of $n - 1$ uniform $(0, y)$ random variables.

2.19. Busloads of customers arrive at an infinite server queue at a Poisson rate λ. Let G denote the service distribution. A bus contains j customers with probability $\alpha_j, j = 1, \ldots$. Let $X(t)$ denote the number of customers that have been served by time t.

(a) $E[X(t)] = ?$

(b) Is $X(t)$ Poisson distributed?

2.20. Suppose that each event of a Poisson process with rate λ is classified as being either of type $1, 2, \ldots, k$. If the event occurs at s, then, independently of all else, it is classified as type i with probability $P_i(s)$, $i = 1, \ldots, k$, $\sum_1^k P_i(s) = 1$. Let $N_i(t)$ denote the number of type i arrivals in $[0, t]$. Show that the $N_i(t)$, $i = 1, \ldots, k$ are independent and $N_i(t)$ is Poisson distributed with mean $\lambda \int_0^t P_i(s)\, ds$.

2.21. Individuals enter the system in accordance with a Poisson process having rate λ. Each arrival independently makes its way through the states of the system. Let $\alpha_i(s)$ denote the probability that an individual is in state i a time s after it arrived. Let $N_i(t)$ denote the number of individuals in state i at time t. Show that the $N_i(t)$, $i \geq 1$, are independent and $N_i(t)$ is Poisson with mean equal to

$$\lambda E[\text{amount of time an individual is in state } i \text{ during its first } t \text{ units in the system}].$$

2.22. Suppose cars enter a one-way infinite highway at a Poisson rate λ. The ith car to enter chooses a velocity V_i and travels at this velocity. Assume that the V_i's are independent positive random variables having a common distribution F. Derive the distribution of the number of cars that are located in the interval (a, b) at time t. Assume that no time is lost when one car overtakes another car.

2.23. For the model of Example 2.3(C), find

(a) $\text{Var}[D(t)]$.

(b) $\text{Cov}[D(t), D(t + s)]$.

2.24. Suppose that cars enter a one-way highway of length L in accordance with a Poisson process with rate λ. Each car travels at a constant speed that is randomly determined, independently from car to car, from the distribution F. When a faster car encounters a slower one, it passes it with no loss of time. Suppose that a car enters the highway at time t. Show that as $t \to \infty$ the speed of the car that minimizes the expected number of encounters with other cars, where we say an encounter occurs when a car is either passed by or passes another car, is the median of the distribution G.

2.25. Suppose that events occur in accordance with a Poisson process with rate λ, and that an event occurring at time s, independent of the past, contributes a random amount having distribution F_s, $s \geq 0$. Show that W, the sum of all contributions by time t, is a compound Poisson random variable. That is, show that W has the same distribution as $\sum_{i=1}^{N} X_i$, where the X_i are independent and identically distributed random variables and are independent of N, a Poisson random variable. Identify the distribution of the X_i and the mean of N.

2.26. Compute the conditional distribution of $S_1, S_2, \ldots, S_n$ given that $S_n = t$.

2.27. Compute the moment generating function of $D(t)$ in Example 2.3(C).

2.28. Prove Lemma 2.3.3.

2.29. Complete the proof that for a nonhomogeneous Poisson process $N(t + s) - N(t)$ is Poisson with mean $m(t + s) - m(t)$.

2.30. Let $T_1, T_2, \ldots$ denote the interarrival times of events of a nonhomogeneous Poisson process having intensity function $\lambda(t)$.

(a) Are the T_i independent?

(b) Are the T_i identically distributed?

(c) Find the distribution of T_1.

(d) Find the distribution of T_2.

2.31. Consider a nonhomogeneous Poisson process $\{N(t), t \geq 0\}$, where $\lambda(t) > 0$ for all t. Let

$$N^*(t) = N(m^{-1}(t)).$$

Show that $\{N^*(t), t \geq 0\}$ is a Poisson process with rate $\lambda = 1$.

2.32. **(a)** Let $\{N(t), t \geq 0\}$ be a nonhomogeneous Poisson process with mean value function $m(t)$. Given $N(t) = n$, show that the unordered set of arrival times has the same distribution as n independent and identically distributed random variables having distribution function

$$F(x) = \begin{cases} \dfrac{m(x)}{m(t)} & x \leq t \\ 1 & x > t. \end{cases}$$

(b) Suppose that workers incur accidents in accordance with a nonhomogeneous Poisson process with mean value function $m(t)$. Suppose further that each injured person is out of work for a random amount of time having distribution F. Let $X(t)$ be the number of workers who are out of work at time t. Compute $E[X(t)]$ and $\text{Var}(X(t))$.

2.33. A two-dimensional Poisson process is a process of events in the plane such that (i) for any region of area A, the number of events in A is Poisson distributed with mean λA, and (ii) the numbers of events in nonoverlapping regions are independent. Consider a fixed point, and let X denote the distance from that point to its nearest event, where distance is measured in the usual Euclidean manner. Show that:

(a) $P\{X > t\} = e^{-\lambda \pi t^2}$.

(b) $E[X] = 1/(2\sqrt{\lambda})$.

Let R_i, $i \geq 1$ denote the distance from an arbitrary point to the ith closest event to it. Show that, with $R_0 = 0$,

(c) $\pi R_i^2 - \pi R_{i-1}^2$, $i \geq 1$ are independent exponential random variables, each with rate λ.

2.34. Repeat Problem 2.25 when the events occur according to a nonhomogeneous Poisson process with intensity function $\lambda(t)$, $t \geq 0$.

2.35. Let $\{N(t), t \geq 0\}$ be a nonhomogeneous Poisson process with intensity function $\lambda(t)$, $t \geq 0$. However, suppose one starts observing the process at a random time τ having distribution function F. Let $N^*(t) = N(\tau + t) - N(\tau)$ denote the number of events that occur in the first t time units of observation.

(a) Does the process $\{N^*(t), t \geq 0\}$ possess independent increments?

(b) Repeat (a) when $\{N(t), t \geq 0\}$ is a Poisson process.

2.36. Let C denote the number of customers served in an $M/G/1$ busy period. Find

(a) $E[C]$.

(b) $\text{Var}(C)$.

2.37. Let $\{X(t), t \geq 0\}$ be a compound Poisson process with $X(t) = \sum_{i=1}^{N(t)} X_i$, and suppose that the X_i can only assume a finite set of possible values. Argue that, for t large, the distribution of $X(t)$ is approximately normal.

2.38. Let $\{X(t), t \geq 0\}$ be a compound Poisson process with $X(t) = \sum_{i=1}^{N(t)} X_i$, and suppose that $\lambda = 1$ and $P\{X_i = j\} = j/10$, $j = 1, 2, 3, 4$. Calculate $P\{X(4) = 20\}$.

2.39. Compute $\mathrm{Cov}(X(s), X(t))$ for a compound Poisson process.

2.40. Give an example of a counting process $\{N(t), t \geq 0\}$ that is not a Poisson process but which has the property that conditional on $N(t) = n$ the first n event times are distributed as the order statistics from a set of n independent uniform $(0, t)$ random variables.

2.41. For a conditional Poisson process:

(a) Explain why a conditional Poisson process has stationary but not independent increments.

(b) Compute the conditional distribution of Λ given $\{N(s), 0 \leq s \leq t\}$, the history of the process up to time t, and show that it depends on the history only through $N(t)$. Explain why this is true.

(c) Compute the conditional distribution of the time of the first event after t given that $N(t) = n$.

(d) Compute

$$\lim_{h \to 0} \frac{P\{N(h) \geq 1\}}{h}.$$

(e) Let $X_1, X_2, \ldots$ denote the interarrival times. Are they independent? Are they identically distributed?

2.42. Consider a conditional Poisson process where the distribution of Λ is the gamma distribution with parameters m and α: that is, the density is given by

$$g(\lambda) = \alpha e^{-\lambda\alpha}(\lambda\alpha)^{m-1}/(m-1)!, \qquad 0 < \lambda < \infty.$$

(a) Show that

$$P\{N(t) = n\} = \binom{m+n-1}{n}\left(\frac{\alpha}{\alpha+t}\right)^m\left(\frac{t}{\alpha+t}\right)^n, \qquad n \geq 0.$$

(b) Show that the conditional distribution of Λ given $N(t) = n$ is again gamma with parameters $m + n, \alpha + t$.

(c) What is

$$\lim_{h \to 0} P\{N(t + h) - N(t) = 1 \mid N(t) = n\}/h?$$

REFERENCES

Reference 3 provides an alternate and mathematically easier treatment of the Poisson process. Corollary 2.5.4 was originally obtained in Reference 1 by using a generating function argument. Another approach to the Poisson process is given in Reference 2.

1. R. M. Adelson, "Compound Poisson Distributions," Operations Research Quarterly, 17, 73–75, (1966).
2. E. Cinlar, *Introduction to Stochastic Processes*, Prentice-Hall, Englewood Cliffs, NJ, 1975.
3. S. M. Ross, *Introduction to Probability Models*, 5th ed., Academic Press, Orlando, FL, 1993.

CHAPTER 3

Renewal Theory

3.1 Introduction and Preliminaries

In the previous chapter we saw that the interarrival times for the Poisson process are independent and identically distributed exponential random variables. A natural generalization is to consider a counting process for which the interarrival times are independent and identically distributed with an arbitrary distribution. Such a counting process is called a *renewal process.*

Formally, let $\{X_n, n = 1, 2, \ldots\}$ be a sequence of nonnegative independent random variables with a common distribution F, and to avoid trivialities suppose that $F(0) = P\{X_n = 0\} < 1$. We shall interpret X_n as the time between the $(n - 1)$st and nth event. Let

$$\mu = E[X_n] = \int_0^\infty x\, dF(x)$$

denote the mean time between successive events and note that from the assumptions that $X_n \geq 0$ and $F(0) < 1$, it follows that $0 < \mu \leq \infty$. Letting

$$S_0 = 0, \qquad S_n = \sum_{i=1}^{n} X_i, \qquad n \geq 1,$$

it follows that S_n is the time of the nth event. As the number of events by time t will equal the largest value of n for which the nth event occurs before or at time t, we have that $N(t)$, the number of events by time t, is given by

$$N(t) = \sup\{n: \quad S_n \leq t\}. \tag{3.1.1}$$

Definition 3.1.1

The counting process $\{N(t), t \geq 0\}$ is called a *renewal* process.

We will use the terms *events* and *renewals* interchangeably, and so we say that the nth renewal occurs at time S_n. Since the interarrival times are independent and identically distributed, it follows that at each renewal the process probabilistically starts over.

The first question we shall attempt to answer is whether an infinite number of renewals can occur in a finite time. To show that it cannot, we note, by the strong law of large numbers, that with probability 1,

$$\frac{S_n}{n} \to \mu \qquad \text{as } n \to \infty.$$

But since $\mu > 0$, this means that S_n must be going to infinity as n goes to infinity. Thus, S_n can be less than or equal to t for at most a finite number of values of n. Hence, by (3.1.1), $N(t)$ must be finite, and we can write

$$N(t) = \max\{n: S_n \le t\}.$$

3.2 Distribution of $N(t)$

The distribution of $N(t)$ can be obtained, at least in theory, by first noting the important relationship that *the number of renewals by time t is greater than or equal to n if, and only if, the nth renewal occurs before or at time t. That is,*

$$N(t) \ge n \Leftrightarrow S_n \le t. \tag{3.2.1}$$

From (3.2.1) we obtain

$$\begin{aligned} P\{N(t) = n\} &= P\{N(t) \ge n\} - P\{N(t) \ge n + 1\} \\ &= P\{S_n \le t\} - P\{S_{n+1} \le t\}. \end{aligned} \tag{3.2.2}$$

Now since the random variables $X_i, i \ge 1$, are independent and have a common distribution F, it follows that $S_n = \sum_{i=1}^{n} X_i$ is distributed as F_n, the n-fold convolution of F with itself. Therefore, from (3.2.2) we obtain

$$P\{N(t) = n\} = F_n(t) - F_{n+1}(t).$$

Let

$$m(t) = E[N(t)].$$

$m(t)$ is called the *renewal function*, and much of renewal theory is concerned with determining its properties. The relationship between $m(t)$ and F is given by the following proposition.

PROPOSITION 3.2.1

$$m(t) = \sum_{n=1}^{\infty} F_n(t) \tag{3.2.3}$$

Proof

$$N(t) = \sum_{n=1}^{\infty} I_n,$$

where

$$I_n = \begin{cases} 1 & \text{if the } n\text{th renewal occurred in } [0, t] \\ 0 & \text{otherwise.} \end{cases}$$

Hence,

$$\begin{aligned} E[N(t)] &= E\left[\sum_{n=1}^{\infty} I_n\right] \\ &= \sum_{n=1}^{\infty} E[I_n] \\ &= \sum_{n=1}^{\infty} P\{I_n = 1\} \\ &= \sum_{n=1}^{\infty} P\{S_n \le t\} \\ &= \sum_{n=1}^{\infty} F_n(t), \end{aligned}$$

where the interchange of expectation and summation is justified by the nonnegativity of the I_n.

The next proposition shows that $N(t)$ has finite expectation.

PROPOSITION 3.2.2

$$m(t) < \infty \qquad \text{for all } 0 \le t < \infty.$$

Proof Since $P\{X_n = 0\} < 1$, it follows by the continuity property of probabilities that there exists an $\alpha > 0$ such that $P\{X_n \ge \alpha\} > 0$. Now define a related renewal process

$\{\overline{X}_n, n \geq 1\}$ by

$$\overline{X}_n = \begin{cases} 0 & \text{if } X_n < \alpha \\ \alpha & \text{if } X_n \geq \alpha, \end{cases}$$

and let $\overline{N}(t) = \sup\{n: \overline{X}_1 + \cdots + \overline{X}_n \leq t\}$. Then it is easy to see that for the related process, renewals can only take place at times $t = n\alpha$, $n = 0, 1, 2, \ldots$, and also the number of renewals at each of these times are independent geometric random variables with mean

$$\frac{1}{P\{X_n \geq \alpha\}}.$$

Thus,

$$E[\overline{N}(t)] \leq \frac{t/\alpha + 1}{P\{X_n \geq \alpha\}} < \infty$$

and the result follows since $\overline{X}_n \leq X_n$ implies that $\overline{N}(t) \geq N(t)$.

Remark The above proof also shows that $E[N^r(t)] < \infty$ for all $t \geq 0$, $r \geq 0$.

3.3 Some Limit Theorems

If we let $N(\infty) = \lim_{t\to\infty} N(t)$ denote the total number of renewals that occurs, then it is easy to see that

$$N(\infty) = \infty \qquad \text{with probability 1.}$$

This follows since the only way in which $N(\infty)$—the total number of renewals that occurs—can be finite, is for one of the interarrival times to be infinite. Therefore,

$$\begin{aligned} P\{N(\infty) < \infty\} &= P\{X_n = \infty \text{ for some } n\} \\ &= P\left\{\bigcup_{n=1}^{\infty} \{X_n = \infty\}\right\} \\ &\leq \sum_{n=1}^{\infty} P\{X_n = \infty\} \\ &= 0. \end{aligned}$$

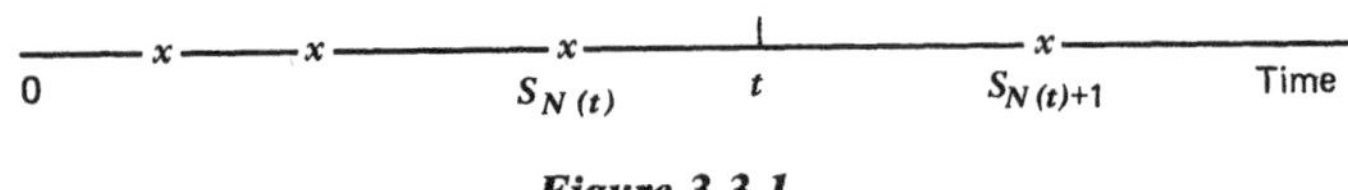

Figure 3.3.1

Thus $N(t)$ goes to infinity as t goes to infinity. However, it would be nice to know the rate at which $N(t)$ goes to infinity. That is, we would like to be able to say something about $\lim_{t\to\infty} N(t)/t$.

As a prelude to determining the rate at which $N(t)$ grows, let us first consider the random variable $S_{N(t)}$. In words, just what does this random variable represent? Proceeding inductively, suppose, for instance, that $N(t) = 3$. Then $S_{N(t)} = S_3$ represents the time of the third event. Since there are only three events that have occurred by time t, S_3 also represents the time of the last event prior to (or at) time t. This is, in fact, what $S_{N(t)}$ represents—namely, the time of the last renewal *prior to or at* time t. Similar reasoning leads to the conclusion that $S_{N(t)+1}$ represents the time of the first renewal *after* time t (see Figure 3.3.1).

We may now prove the following.

PROPOSITION 3.3.1

With probability 1,

$$\frac{N(t)}{t} \to \frac{1}{\mu} \qquad \text{as } t \to \infty.$$

Proof Since $S_{N(t)} \le t < S_{N(t)+1}$, we see that

$$\frac{S_{N(t)}}{N(t)} \le \frac{t}{N(t)} < \frac{S_{N(t)+1}}{N(t)}. \tag{3.3.1}$$

However, since $S_{N(t)}/N(t)$ is the average of the first $N(t)$ interarrival times, it follows by the strong law of large numbers that $S_{N(t)}/N(t) \to \mu$ as $N(t) \to \infty$. But since $N(t) \to \infty$ when $t \to \infty$, we obtain

$$\frac{S_{N(t)}}{N(t)} \to \mu \qquad \text{as } t \to \infty.$$

Furthermore, writing

$$\frac{S_{N(t)+1}}{N(t)} = \left[\frac{S_{N(t)+1}}{N(t)+1}\right]\left[\frac{N(t)+1}{N(t)}\right],$$

we have, by the same reasoning,

$$\frac{S_{N(t)+1}}{N(t)} \to \mu \qquad \text{as } t \to \infty.$$

The result now follows by (3.3.1) since $t/N(t)$ is between two numbers, each of which converges to μ as $t \to \infty$.

EXAMPLE 3.3(A) A container contains an infinite collection of coins. Each coin has its own probability of landing heads, and these probabilities are the values of independent random variables that are uniformly distributed over (0, 1). Suppose we are to flip coins sequentially, at any time either flipping a new coin or one that had previously been used. If our objective is to maximize the long-run proportion of flips that lands on heads, how should we proceed?

Solution. We will exhibit a strategy that results in the long-run proportion of heads being equal to 1. To begin, let $N(n)$ denote the number of tails in the first n flips, and so the long-run proportion of heads, call it P_h, is given by

$$P_h = \lim_{n\to\infty} \frac{n - N(n)}{n} = 1 - \lim_{n\to\infty} \frac{N(n)}{n}.$$

Consider the strategy that initially chooses a coin and continues to flip it until it comes up tails. At this point that coin is discarded (never to be used again) and a new one is chosen. The process is then repeated. To determine P_h for this strategy, note that the times at which a flipped coin lands on tails constitute renewals. Hence, by Proposition 3.3.1

$$\lim_{n\to\infty} \frac{N(n)}{n} = 1/E[\text{number of flips between successive tails}].$$

But, given its probability p of landing heads, the number of flips of a coin until it lands tails is geometric with mean $1/(1 - p)$. Hence, conditioning gives

$$E[\text{number of flips between successive tails}] = \int_0^1 \frac{1}{1-p}\,dp = \infty$$

implying that, with probability 1, $\lim_{n\to\infty} \frac{N(n)}{n} = 0$.

Thus Proposition 3.3.1 states that with probability 1, the long-run rate at which renewals occur will equal $1/\mu$. For this reason $1/\mu$ is called the *rate* of the renewal process.

We show that the expected average rate of renewals $m(t)/t$ also converges to $1/\mu$. However, before presenting the proof, we will find it useful to digress to stopping times and Wald's equation.

3.3.1 Wald's Equation

Let $X_1, X_2, \ldots$ denote a sequence of independent random variables. We have the following definition.

Definition

An integer-valued random variable N is said to be a *stopping time* for the sequence $X_1, X_2, \ldots$ if the event $\{N = n\}$ is independent of $X_{n+1}, X_{n+2}, \ldots$ for all $n = 1, 2, \ldots$.

Intuitively, we observe the X_n's in sequential order and N denotes the number observed before stopping. If $N = n$, then we have stopped after observing $X_1, \ldots, X_n$ and before observing $X_{n+1}, X_{n+2}, \ldots$.

EXAMPLE 3.3(B) Let $X_n, n = 1, 2, \ldots$, be independent and such that

$$P\{X_n = 0\} = P\{X_n = 1\} = \tfrac{1}{2}, \qquad n = 1, 2, \ldots.$$

If we let

$$N = \min\{n\colon\ X_1 + \cdots + X_n = 10\},$$

then N is a stopping time. We may regard N as being the stopping time of an experiment that successively flips a fair coin and then stops when the number of heads reaches 10.

EXAMPLE 3.3(C) Let $X_n, n = 1, 2, \ldots$, be independent and such that

$$P\{X_n = -1\} = P\{X_n = 1\} = \tfrac{1}{2}.$$

Then

$$N = \min\{n\colon\ X_1 + \cdots + X_n = 1\}$$

is a stopping time. It can be regarded as the stopping time for a gambler who on each play is equally likely to either win or lose 1 unit and who decides to stop the first time he is ahead. (It will be shown in the next chapter that N is finite with probability 1.)

THEOREM 3.3.2 (Wald's Equation).

If $X_1, X_2, \ldots$ are independent and identically distributed random variables having finite expectations, and if N is a stopping time for $X_1, X_2, \ldots$ such that $E[N] < \infty$, then

$$E\left[\sum_1^N X_n\right] = E[N]E[X].$$

Proof Letting

$$I_n = \begin{cases} 1 & \text{if } N \geq n \\ 0 & \text{if } N < n, \end{cases}$$

we have that

$$\sum_{n=1}^N X_n = \sum_{n=1}^\infty X_n I_n.$$

Hence,

$$E\left[\sum_{n=1}^N X_n\right] = E\left[\sum_{n=1}^\infty X_n I_n\right] = \sum_{n=1}^\infty E[X_n I_n]. \tag{3.3.2}$$

However, $I_n = 1$ if, and only if, we have not stopped after successively observing $X_1, \ldots, X_{n-1}$. Therefore, I_n is determined by $X_1, \ldots, X_{n-1}$ and is thus independent of X_n. From (3.3.2), we thus obtain

$$\begin{aligned} E\left[\sum_{n=1}^N X_n\right] &= \sum_{n=1}^\infty E[X_n]E[I_n] \\ &= E[X]\sum_{n=1}^\infty E[I_n] \\ &= E[X]\sum_{n=1}^\infty P\{N \geq n\} \\ &= E[X]E[N]. \end{aligned}$$

Remark In Equation (3.3.2) we interchanged expectation and summation without justification. To justify this interchange, replace X_i by its absolute value throughout. In this case, the interchange is justified as all terms are nonnegative. However, this implies that the original interchange is allowable by Lebesgue's dominated convergence theorem.

For Example 3.3(B), Wald's equation implies

$$E[X_1 + \cdots + X_N] = \tfrac{1}{2}EN.$$

However, $X_1 + \cdots + X_N = 10$ by definition of N, and so $E[N] = 20$.

An application of the conclusion of Wald's equation to Example 3.3(C) would yield $E[X_1 + \cdots + X_N] = E[N]E[X]$. However, $X_1 + \cdots + X_N = 1$ and $E[X] = 0$, and so we would arrive at a contradiction. Thus Wald's equation is not applicable, which yields the conclusion that $E[N] = \infty$.

3.3.2 Back to Renewal Theory

Let $X_1, X_2, \ldots$ denote the interarrival times of a renewal process and let us stop at the first renewal after t—that is, at the $N(t) + 1$ renewal. To verify that $N(t) + 1$ is indeed a stopping time for the sequence of X_i, note that

$$\begin{aligned} N(t) + 1 = n &\Leftrightarrow N(t) = n - 1 \\ &\Leftrightarrow X_1 + \cdots + X_{n-1} \le t, \qquad X_1 + \cdots + X_n > t. \end{aligned}$$

Thus the event $\{N(t) + 1 = n\}$ depends only on $X_1, \ldots, X_n$ and is thus independent of $X_{n+1}, \ldots$; hence $N(t) + 1$ is a stopping time. From Wald's equation we obtain that, when $E[X] < \infty$,

$$E[X_1 + \cdots + X_{N(t)+1}] = E[X]E[N(t) + 1],$$

or, equivalently, we have the following.

Corollary 3.3.3

If $\mu < \infty$, then

$$E[S_{N(t)+1}] = \mu[m(t) + 1]. \tag{3.3.3}$$

We are now in position to prove the following.

THEOREM 3.3.4 (The Elementary Renewal Theorem).

$$\frac{m(t)}{t} \to \frac{1}{\mu} \qquad \text{as } t \to \infty \qquad \left(\textit{where } \frac{1}{\infty} \equiv 0\right).$$

Proof Suppose first that $\mu < \infty$. Now (see Figure 3.3.1)

$$S_{N(t)+1} > t.$$

Taking expectations and using Corollary 3.3.3 gives

$$\mu(m(t) + 1) > t,$$

implying that

$$\liminf_{t\to\infty} \frac{m(t)}{t} \geq \frac{1}{\mu}. \tag{3.3.4}$$

To go the other way, we fix a constant M, and define a new renewal process $\{\overline{X}_n, n = 1, 2, \ldots\}$ by letting

$$\overline{X}_n = \begin{cases} X_n & \text{if } X_n \leq M, \\ M & \text{if } X_n > M. \end{cases} \qquad n = 1, 2, \ldots$$

Let $\overline{S}_n = \sum_1^n \overline{X}_i$, and $\overline{N}(t) = \sup\{n: \overline{S}_n \leq t\}$. Since the interarrival times for this truncated renewal process are bounded by M, we obtain

$$\overline{S}_{N(t)+1} \leq t + M.$$

Hence by Corollary 3.3.3,

$$(\overline{m}(t) + 1)\mu_M \leq t + M,$$

where $\mu_M = E[\overline{X}_n]$. Thus

$$\limsup_{t\to\infty} \frac{\overline{m}(t)}{t} \leq \frac{1}{\mu_M}.$$

Now, since $\overline{S}_n \leq S_n$, it follows that $\overline{N}(t) \geq N(t)$ and $\overline{m}(t) \geq m(t)$, thus

$$\limsup_{t\to\infty} \frac{m(t)}{t} \leq \frac{1}{\mu_M}. \tag{3.3.5}$$

Letting $M \to \infty$ yields

$$\limsup_{t\to\infty} \frac{m(t)}{t} \leq \frac{1}{\mu}, \tag{3.3.6}$$

and the result follows from (3.3.4) and (3.3.6).

When $\mu = \infty$, we again consider the truncated process; since $\mu_M \to \infty$ as $M \to \infty$, the result follows from (3.3.5).

Remark At first glance it might seem that the elementary renewal theorem should be a simple consequence of Proposition 3.3.1. That is, since the average renewal rate will, with probability 1, converge to $1/\mu$, should this not imply that the expected average renewal rate also converges to $1/\mu$? We must, however, be careful; consider the following example.

Let U be a random variable that is uniformly distributed on (0, 1). Define the random variables Y_n, $n \geq 1$, by

$$Y_n = \begin{cases} 0 & \text{if } U > 1/n \\ n & \text{if } U \leq 1/n. \end{cases}$$

Now, since, with probability 1, U will be greater than 0, it follows that Y_n will equal 0 for all sufficiently large n. That is, Y_n will equal 0 for all n large enough so that $1/n < U$. Hence, with probability 1,

$$Y_n \to 0 \qquad \text{as } n \to \infty.$$

However,

$$E[Y_n] = nP\left\{U \leq \frac{1}{n}\right\} = n\frac{1}{n} = 1.$$

Therefore, even though the sequence of random variables Y_n converges to 0, the expected values of the Y_n are all identically 1.

We will end this section by showing that $N(t)$ is asymptotically, as $t \to \infty$, normally distributed. To prove this result we make use both of the central limit theorem (to show that S_n is asymptotically normal) and the relationship

$$N(t) < n \Leftrightarrow S_n > t. \tag{3.3.7}$$

THEOREM 3.3.5

Let μ and σ^2, assumed finite, represent the mean and variance of an interarrival time. Then

$$P\left\{\frac{N(t) - t/\mu}{\sigma\sqrt{t/\mu^3}} < y\right\} \to \frac{1}{\sqrt{2\pi}}\int_{-\infty}^{y} e^{-x^2/2}\,dx \qquad \text{as } t \to \infty.$$

Proof Let $r_t = t/\mu + y\sigma\sqrt{t/\mu^3}$. Then

$$P\left\{\frac{N(t) - t/\mu}{\sigma\sqrt{t/\mu^3}} < y\right\} = P\{N(t) < r_t\}$$

$$= P\{S_{r_t} > t\} \qquad \text{(by (3.3.7))}$$

$$= P\left\{\frac{S_{r_t} - r_t\mu}{\sigma\sqrt{r_t}} > \frac{t - r_t\mu}{\sigma\sqrt{r_t}}\right\}$$

$$= P\left\{\frac{S_{r_t} - r_t\mu}{\sigma\sqrt{r_t}} > -y\left(1 + \frac{y\sigma}{\sqrt{t\mu}}\right)^{-1/2}\right\}.$$

Now, by the central limit theorem, $(S_{r_t} - r_t\mu)/\sigma\sqrt{r_t}$ converges to a normal random variable having mean 0 and variance 1 as t (and thus r_t) approaches ∞. Also, since

$$-y\left(1 + \frac{y\sigma}{\sqrt{t\mu}}\right)^{-1/2} \to -y \qquad \text{as } t \to \infty,$$

we see that

$$P\left\{\frac{N(t) - t/\mu}{\sigma\sqrt{t/\mu^3}} < y\right\} \to \frac{1}{\sqrt{2\pi}}\int_{-y}^{\infty} e^{-x^2/2}\,dx$$

and since

$$\int_{-y}^{\infty} e^{-x^2/2}\,dx = \int_{-\infty}^{y} e^{-x^2/2}\,dx,$$

the result follows.

Remarks

(i) There is a slight difficulty in the above argument since r_t should be an integer for us to use the relationship (3.3.7). It is not, however, too difficult to make the above argument rigorous.

(ii) Theorem 3.3.5 states that $N(t)$ is asymptotically normal with mean t/μ and variance $t\sigma^2/\mu^3$.

3.4 The Key Renewal Theorem and Applications

A nonnegative random variable X is said to be *lattice* if there exists $d \geq 0$ such that $\sum_{n=0}^{\infty} P\{X = nd\} = 1$. That is, X is lattice if it only takes on integral multiples of some nonnegative number d. The largest d having this property is said to be the period of X. If X is lattice and F is the distribution function of X, then we say that F is lattice.

We shall state without proof the following theorem.

THEOREM 3.4.1 (Blackwell's Theorem).

(i) *If F is not lattice, then*

$$m(t + a) - m(t) \to a/\mu \qquad \text{as } t \to \infty$$

for all $a \geq 0$.

(ii) *If F is lattice with period d, then*

$$E[\text{number of renewals at } nd] \to d/\mu \qquad \text{as } n \to \infty.$$

Thus Blackwell's theorem states that if F is not lattice, then the expected number of renewals in an interval of length a, far from the origin, is approximately a/μ. This is quite intuitive, for as we go further away from the origin it would seem that the initial effects wear away and thus

$$g(a) \equiv \lim_{t \to \infty} [m(t + a) - m(t)] \tag{3.4.1}$$

should exist. However, if the above limit does in fact exist, then, as a simple consequence of the elementary renewal theorem, it must equal a/μ. To see this note first that

$$\begin{aligned} g(a + b) &= \lim_{t \to \infty} [m(t + a + b) - m(t)] \\ &= \lim_{t \to \infty} [m(t + a + b) - m(t + a) + m(t + a) - m(t)] \\ &= g(b) + g(a). \end{aligned}$$

However, the only (increasing) solution of $g(a + b) = g(a) + g(b)$ is

$$g(a) = ca, \qquad a > 0$$

for some constant c. To show that $c = 1/\mu$ define

$$\begin{aligned} x_1 &= m(1) - m(0) \\ x_2 &= m(2) - m(1) \\ &\vdots \\ x_n &= m(n) - m(n-1) \\ &\vdots \end{aligned}$$

Then

$$\lim_{n\to\infty} x_n = c$$

implying that

$$\lim_{n\to\infty} \frac{x_1 + \cdots + x_n}{n} = c$$

or

$$\lim_{n\to\infty} \frac{m(n)}{n} = c.$$

Hence, by the elementary renewal theorem, $c = 1/\mu$.

When F is lattice with period d, then the limit in (3.4.1) cannot exist. For now renewals can only occur at integral multiples of d and thus the expected number of renewals in an interval far from the origin would clearly depend not on the intervals' length per se but rather on how many points of the form nd, $n \geq 0$, it contains. Thus in the lattice case the relevant limit is that of the expected number of renewals at nd and, again, if $\lim_{n\to\infty} E$[number of renewals at nd] exists, then by the elementary renewal theorem it must equal d/μ. If interarrivals are always positive, then part (ii) of Blackwell's theorem states that, in the lattice case,

$$\lim_{n\to\infty} P\{\text{renewal at } nd\} = \frac{d}{\mu}.$$

Let h be a function defined on $[0, \infty]$. For any $a > 0$ let $\underline{m}_n(a)$ be the supremum, and $\overline{m}_n(a)$ the infinum of $h(t)$ over the interval $(n - 1)a \leq t \leq na$. We say that h is *directly Riemann integrable* if $\sum_{n=1}^{\infty} \overline{m}_n(a)$ and

$\sum_{n=1}^{\infty} \underline{m}_n(a)$ are finite for all $a > 0$ and

$$\lim_{a \to 0} a \sum_{n=1}^{\infty} \overline{m}_n(a) = \lim_{a \to 0} a \sum_{n=1}^{\infty} \underline{m}_n(a).$$

A sufficient condition for h to be directly Riemann integrable is that:

(i) $h(t) \geq 0$ for all $t \geq 0$,
(ii) $h(t)$ is nonincreasing,
(iii) $\int_0^{\infty} h(t)dt < \infty$.

The following theorem, known as the *key renewal theorem*, will be stated without proof.

THEOREM 3.4.2 (The Key Renewal Theorem).

If F is not lattice, and if h(t) is directly Riemann integrable, then

$$\lim_{t \to \infty} \int_0^t h(t-x)\, dm(x) = \frac{1}{\mu} \int_0^t h(t)\, dt,$$

where

$$m(x) = \sum_{n=1}^{\infty} F_n(x) \qquad \text{and} \qquad \mu = \int_0^{\infty} \overline{F}(t)\, dt.$$

To obtain a feel for the key renewal theorem start with Blackwell's theorem and reason as follows: By Blackwell's theorem, we have that

$$\lim_{t \to \infty} \frac{m(t+a) - m(t)}{a} = \frac{1}{\mu}$$

and, hence,

$$\lim_{a \to 0} \lim_{t \to \infty} \frac{m(t+a) - m(t)}{a} = \frac{1}{\mu}.$$

Now, assuming that we can justify interchanging the limits, we obtain

$$\lim_{t \to \infty} \frac{dm(t)}{dt} = \frac{1}{\mu}.$$

The key renewal theorem is a formalization of the above.

Blackwell's theorem and the key renewal theorem can be shown to be equivalent. Problem 3.12 asks the reader to deduce Blackwell from the key renewal theorem; and the reverse can be proven by approximating a directly Riemann integrable function with step functions. In Section 9.3 a probabilistic proof of Blackwell's theorem is presented when F is continuous and has a failure rate function bounded away from 0 and ∞.

The key renewal theorem is a very important and useful result. It is used when one wants to compute the limiting value of $g(t)$, some probability or expectation at time t. The technique we shall employ for its use is to derive an equation for $g(t)$ by first conditioning on the time of the last renewal prior to (or at) t. This, as we will see, will yield an equation of the form

$$g(t) = h(t) + \int_0^t h(t-x)\, dm(x).$$

We start with a lemma that gives the distribution of $S_{N(t)}$, the time of the last renewal prior to (or at) time t.

Lemma 3.4.3

$$P\{S_{N(t)} \leq s\} = \overline{F}(t) + \int_0^s \overline{F}(t-y)\, dm(y), \qquad t \geq s \geq 0.$$

Proof

$$\begin{aligned}
P\{S_{N(t)} \leq s\} &= \sum_{n=0}^{\infty} P\{S_n \leq s, S_{n+1} > t\} \\
&= \overline{F}(t) + \sum_{n=1}^{\infty} P\{S_n \leq s, S_{n+1} > t\} \\
&= \overline{F}(t) + \sum_{n=1}^{\infty} \int_0^{\infty} P\{S_n \leq s, S_{n+1} > t \mid S_n = y\}\, dF_n(y) \\
&= \overline{F}(t) + \sum_{n=1}^{\infty} \int_0^s \overline{F}(t-y)\, dF_n(y) \\
&= \overline{F}(t) + \int_0^s \overline{F}(t-y)\, d\left(\sum_{n=1}^{\infty} F_n(y)\right) \\
&= \overline{F}(t) + \int_0^s \overline{F}(t-y)\, dm(y),
\end{aligned}$$

where the interchange of integral and summation is justified since all terms are nonnegative.

Remarks

(1) It follows from Lemma 3.4.3 that

$$P\{S_{N(t)} = 0\} = \bar{F}(t),$$
$$dF_{S_{N(t)}}(y) = \bar{F}(t-y)\,dm(y), \qquad 0 < y < \infty.$$

(2) To obtain an intuitive feel for the above, suppose that F is continuous with density f. Then $m(y) = \sum_{n=1}^{\infty} F_n(y)$, and so, for $y > 0$,

$$\begin{aligned} dm(y) &= \sum_{n=1}^{\infty} f_n(y)\,dy \\ &= \sum_{n=1}^{\infty} P\{n\text{th renewal occurs in } (y, y+dy)\} \\ &= P\{\text{renewal occurs in } (y, y+dy)\}. \end{aligned}$$

So, the probability density of $S_{N(t)}$ is

$$\begin{aligned} f_{S_{N(t)}}(y)\,dy &= P\{\text{renewal in } (y, y+dy), \text{ next interarrival} > t-y\} \\ &= dm(y)\bar{F}(t-y). \end{aligned}$$

We now present some examples of the utility of the key renewal theorem. Once again the technique we employ will be to condition on $S_{N(t)}$.

3.4.1 Alternating Renewal Processes

Consider a system that can be in one of two states: *on* or *off*. Initially it is on and it remains on for a time Z_1; it then goes off and remains off for a time Y_1; it then goes on for a time Z_2; then off for a time Y_2; then on, and so forth.

We suppose that the random vectors (Z_n, Y_n), $n \geq 1$, are independent and identically distributed. Hence, both the sequence of random variables $\{Z_n\}$ and the sequence $\{Y_n\}$ are independent and identically distributed; but we allow Z_n and Y_n to be dependent. In other words, each time the process goes on everything starts over again, but when it goes off we allow the length of the off time to depend on the previous on time.

Let H be the distribution of Z_n, G the distribution of Y_n, and F the distribution of $Z_n + Y_n$, $n \geq 1$. Furthermore, let

$$P(t) = P\{\text{system is on at time } t\}.$$

THEOREM 3.4.4

If $E[Z_n + Y_n] < \infty$ and F is nonlattice, then

$$\lim_{t\to\infty} P(t) = \frac{E[Z_n]}{E[Z_n] + E[Y_n]}.$$

Proof Say that a renewal takes place each time the system goes on. Conditioning on the time of the last renewal prior to (or at) t yields

$$P(t) = P\{\text{on at } t | S_{N(t)} = 0\}P\{S_{N(t)} = 0\} + \int_0^\infty P\{\text{on at } t | S_{N(t)} = y\}\, dF_{S_{N(t)}}(y).$$

Now

$$P\{\text{on at } t | S_{N(t)} = 0\} = P\{Z_1 > t | Z_1 + Y_1 > t\} = \overline{H}(t)/\overline{F}(t),$$

and, for $y < t$,

$$P\{\text{on at } t | S_{N(t)} = y\} = P\{Z > t - y | Z + Y > t - y\} = \overline{H}(t-y)/\overline{F}(t-y).$$

Hence, using Lemma 3.4.3,

$$P(t) = \overline{H}(t) + \int_0^t \overline{H}(t-y)\, dm(y),$$

where $m(y) = \sum_{n=1}^\infty F_n(y)$. Now $\overline{H}(t)$ is clearly nonnegative, nonincreasing, and $\int_0^\infty \overline{H}(t)\, dt = E[Z] < \infty$. Since this last statement also implies that $\overline{H}(t) \to 0$ as $t \to \infty$, we have, upon application of the key renewal theorem,

$$P(t) \to \frac{\int_0^\infty \overline{H}(t)\, dt}{\mu_F} = \frac{E[Z_n]}{E[Z_n] + E[Y_n]}.$$

If we let $Q(t) = P\{\text{off at } t\} = 1 - P(t)$, then

$$Q(t) \to \frac{E[Y]}{E[Z] + E[Y]}.$$

We note that the fact that the system was initially on makes no difference in the limit.

Theorem 3.4.4 is quite important because many systems can be modelled by an alternating renewal process. For instance, consider a renewal process and let $Y(t)$ denote the time from t until the next renewal and let $A(t)$ be the time from t since the last renewal. That is,

$$Y(t) = S_{N(t)+1} - t,$$
$$A(t) = t - S_{N(t)}.$$

$Y(t)$ is called the *excess* or *residual life* at t, and $A(t)$ is called the *age* at t. If we imagine that the renewal process is generated by putting an item in use and then replacing it upon failure, then $A(t)$ would represent the age of the item in use at time t and $Y(t)$ its remaining life.

Suppose we want to derive $P\{A(t) \le x\}$. To do so let an on–off cycle correspond to a renewal and say that the system is "on" at time t if the age at t is less than or equal to x. In other words, the system is "on" the first x units of a renewal interval and "off" the remaining time. Then, if the renewal distribution is not lattice, we have by Theorem 3.4.4 that

$$\begin{aligned}\lim_{t\to\infty} P\{A(t) \le x\} &= E[\min(X, x)]/E[X] \\ &= \int_0^\infty P\{\min(X, x) > y\}\, dy/E[X] \\ &= \int_0^x \overline{F}(y)\, dy/\mu.\end{aligned}$$

Similarly to obtain the limiting value of $P\{Y(t) \le x\}$, say that the system is "off" the last x units of a renewal cycle and "on" otherwise. Thus the off time in a cycle is $\min(x, X)$, and so

$$\begin{aligned}\lim_{t\to\infty} P\{Y(t) \le x\} &= \lim_{t\to\infty} P\{\text{off at } t\} \\ &= E[\min(x, X)]/E[X] \\ &= \int_0^x \overline{F}(y)\, dy/\mu.\end{aligned}$$

Thus summing up we have proven the following.

PROPOSITION 3.4.5

If the interarrival distribution is nonlattice and $\mu < \infty$, then

$$\lim_{t\to\infty} P\{Y(t) \le x\} = \lim_{t\to\infty} P\{A(t) \le x\} = \int_0^x \overline{F}(y)\, dy/\mu.$$

Remark To understand why the limiting distribution of excess and age are identical, consider the process after it has been in operation for a long time; for instance, suppose it started at $t = -\infty$. Then if we look backwards in time, the time between successive events will still be independent and have distribution F. Hence, looking backwards we see an identically distributed renewal process. But looking backwards the excess life at t is exactly the age at t of the original process. We will find this technique of looking backward in time to be quite valuable in our studies of Markov chains in Chapters 4 and 5. (See Problem 3.14 for another way of relating the distributions of excess and age.)

Another random variable of interest is $X_{N(t)+1} = S_{N(t)+1} - S_{N(t)}$, or, equivalently,

$$X_{N(t)+1} = A(t) + Y(t).$$

Thus $X_{N(t)+1}$ represents the length of the renewal interval that contains the point t. In Problem 3.3 we prove that

$$P\{X_{N(t)+1} > x\} \geq \bar{F}(x).$$

That is, for any x it is more likely that the length of the interval containing the point t is greater than x than it is that an ordinary renewal interval is greater than x. This result, which at first glance may seem surprising, is known as the *inspection* paradox.

We will now use alternating renewal process theory to obtain the limiting distribution of $X_{N(t)+1}$. Again let an on–off cycle correspond to a renewal interval, and say that the on time in the cycle is the total cycle time if that time is greater than x and is zero otherwise. That is, the system is either totally on during a cycle (if the renewal interval is greater than x) or totally off otherwise. Then

$$\begin{aligned} P\{X_{N(t)+1} > x\} &= P\{\text{length of renewal interval containing } t > x\} \\ &= P\{\text{on at time } t\}. \end{aligned}$$

Thus by Theorem 3.4.4, provided F is not lattice, we obtain

$$\begin{aligned} \lim_{t\to\infty} P\{X_{N(t)+1} > x\} &= \frac{E[\text{on time in cycle}]}{\mu} \\ &= E[X \mid X > x]\bar{F}(x)/\mu \\ &= \int_x^\infty y\, dF(y)/\mu, \end{aligned}$$

or, equivalently,

$$\lim_{t\to\infty} P\{X_{N(t)+1} \le x\} = \int_0^x y\, dF(y)/\mu. \tag{3.4.2}$$

Remark To better understand the inspection paradox, reason as follows: Since the line is covered by renewal intervals, is it not more likely that a larger interval—as opposed to a shorter one—covers the point t? In fact, in the limit (as $t \to \infty$) it is exactly true that an interval of length y is y times more likely to cover t than one of length 1. For if this were the case, then the density of the interval containing the point t, call it g, would be $g(y) = y\, dF(y)/c$ (since $dF(y)$ is the probability that an arbitrary interval is of length y and y/c the conditional probability that it contains the point). But by (3.4.2) we see that this is indeed the limiting density.

For another illustration of the varied uses of alternating renewal processes consider the following example.

Example 3.4(a) ***An Inventory Example.*** Suppose that customers arrive at a store, which sells a single type of commodity, in accordance with a renewal process having nonlattice interarrival distribution F. The amounts desired by the customers are assumed to be independent with a common distribution G. The store uses the following (s, S) ordering policy: If the inventory level after serving a customer is below s, then an order is placed to bring it up to S. Otherwise no order is placed. Thus if the inventory level after serving a customer is x, then the amount ordered is

$$\begin{array}{ll} S - x & \text{if } x < s, \\ 0 & \text{if } x \ge s. \end{array}$$

The order is assumed to be instantaneously filled.

Let $X(t)$ denote the inventory level at time t, and suppose we desire $\lim_{t\to\infty} P\{X(t) \ge x\}$. If $X(0) = S$, then if we say that the system is "on" whenever the inventory level is at least x and "off" otherwise, the above is just an alternating renewal process. Hence, from Theorem 3.4.4,

$$\lim_{t\to\infty} P\{X(t) \ge x\} = \frac{E[\text{amount of time the inventory} \ge x \text{ in a cycle}]}{E[\text{time of a cycle}]}.$$

Now if we let $Y_1, Y_2, \ldots$ denote the successive customer demands and let

$$N_x = \min\{n: \quad Y_1 + \cdots + Y_n > S - x\}, \tag{3.4.3}$$

then it is the N_x customer in the cycle that causes the inventory level to fall below x, and it is the N_s customer that ends the cycle. Hence if X_i, $i \geq 1$, denote the interarrival times of customers, then

$$\text{amount of "on" time in cycle} = \sum_{i=1}^{N_x} X_i,$$

$$\text{time of cycle} = \sum_{i=1}^{N_s} X_i.$$

Assuming that the interarrival times are independent of the successive demands, we thus have upon taking expectations

$$\lim_{t\to\infty} P\{X(t) \geq x\} = \frac{E\left[\sum_{i=1}^{N_x} X_i\right]}{E\left[\sum_{i=1}^{N_s} X_i\right]} = \frac{E[N_x]}{E[N_s]}. \tag{3.4.4}$$

However, as the $Y_i, i \geq 1$, are independent and identically distributed, it follows from (3.4.3) that we can interpret $N_x - 1$ as the number of renewals by time $S - x$ of a renewal process with interarrival time Y_i, $i \geq 1$. Hence,

$$E[N_x] = m_G(S - x) + 1,$$
$$E[N_s] = m_G(S - s) + 1,$$

where G is the customer demand distribution and

$$m_G(t) = \sum_{n=1}^{\infty} G_n(t).$$

Hence, from (3.4.4), we arrive at

$$\lim_{t\to\infty} P\{X(t) \geq x\} = \frac{1 + m_G(S - x)}{1 + m_G(S - s)}, \qquad s \leq x \leq S.$$

3.4.2 Limiting Mean Excess and the Expansion of $m(t)$

Let us start by computing the mean excess of a nonlattice renewal process. Conditioning on $S_{N(t)}$ yields (by Lemma 3.4.3)

$$E[Y(t)] = E[Y(t) | S_{N(t)} = 0]\bar{F}(t) + \int_0^t E[Y(t) | S_{N(t)} = y]\bar{F}(t - y)\, dm(y).$$

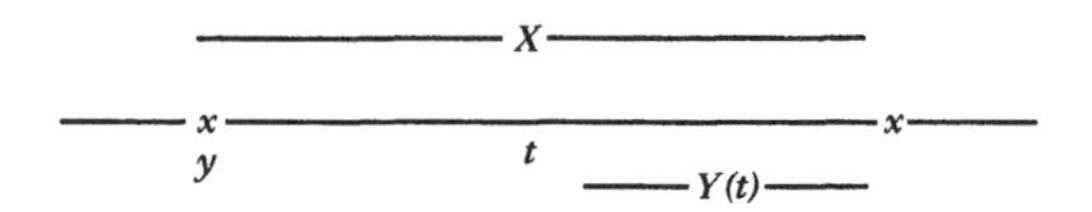

Figure 3.4.1. **$S_{N(t)} = y$; x = renewal.**

Now,

$$E[Y(t)|S_{N(t)} = 0] = E[X - t|X > t],$$
$$E[Y(t)|S_{N(t)} = y] = E[X - (t - y)|X > t - y].$$

where the above follows since $S_{N(t)} = y$ means that there is a renewal at y and the next interarrival time—call it X—is greater than $t - y$ (see Figure 3.4.1). Hence,

$$E[Y(t)] = E[X - t|X > t]\bar{F}(t) + \int_0^t E[X - (t - y)|X > t - y]\bar{F}(t - y)\,dm(y).$$

Now it can be shown that the function $h(t) = E[X - t|X > t]\bar{F}(t)$ is directly Riemann integrable provided $E[X^2] < \infty$, and so by the key renewal theorem

$$\begin{aligned} E[Y(t)] &\to \int_0^\infty E[X - t|X > t]\bar{F}(t)\,dt/\mu \\ &= \int_0^\infty \int_t^\infty (x - t)\,dF(x)\,dt/\mu \\ &= \int_0^\infty \int_0^x (x - t)\,dt\,dF(x)/\mu \qquad \text{(by interchange of order of integration)} \\ &= \int_0^\infty x^2\,dF(x)/2\mu \\ &= E[X^2]/2\mu. \end{aligned}$$

Thus we have proven the following.

PROPOSITION 3.4.6

If the interarrival distribution is nonlattice and $E[X^2] < \infty$, then

$$\lim_{t\to\infty} E[Y(t)] = E[X^2]/2\mu.$$

Now $S_{N(t)+1}$, the time of the first renewal after t, can be expressed as

$$S_{N(t)+1} = t + Y(t).$$

Taking expectations and using Corollary 3.3.3, we have

$$\mu(m(t) + 1) = t + E[Y(t)]$$

or

$$m(t) - \frac{t}{\mu} = \frac{E[Y(t)]}{\mu} - 1.$$

Hence, from Proposition 3.4.6 we obtain the following.

Corollary 3.4.7

If $E[X^2] < \infty$ and F is nonlattice, then

$$m(t) - \frac{t}{\mu} \to \frac{E[X^2]}{2\mu^2} - 1 \qquad \text{as } t \to \infty.$$

3.4.3 Age-Dependent Branching Processes

Suppose that an organism at the end of its lifetime produces a random number of offspring in accordance with the probability distribution $\{P_j, j = 0, 1, 2, \ldots\}$. Assume further that all offspring act independently of each other and produce their own offspring in accordance with the same probability distribution $\{P_j\}$. Finally, let us assume that the lifetimes of the organisms are independent random variables with some common distribution F.

Let $X(t)$ denote the number of organisms alive at t. The stochastic process $\{X(t), t \geq 0\}$ is called an *age-dependent branching process*. We shall concern ourselves with determining the asymptotic form of $M(t) = E[X(t)]$, when $m = \sum_{j=0}^{\infty} jP_j > 1$.

THEOREM 3.4.8

If $X_0 = 1$, $m > 1$, and F is not lattice, then

$$e^{-\alpha t}M(t) \to \frac{m - 1}{m^2\alpha \int_0^\infty xe^{-\alpha x}\, dF(x)} \qquad \text{as } t \to \infty,$$

where α is the unique positive number such that

$$\int_0^\infty e^{-\alpha x}\, dF(x) = \frac{1}{m}.$$

Proof By conditioning on T_1, the lifetime of the initial organism, we obtain

$$M(t) = \int_0^\infty E[X(t)|T_1 = s]\,dF(s).$$

However,

$$E[X(t)|T_1 = s] = \begin{cases} 1 & \text{if } s > t \\ m \cdot M(t-s) & \text{if } s \le t. \end{cases} \tag{3.4.5}$$

To see why (3.4.5) is true, suppose that $T_1 = s$, $s \le t$, and suppose further that the organism has j offspring. Then the number of organisms alive at t may be written as $Y_1 + \cdots + Y_j$, where Y_i is the number of descendants (including himself) of the ith offspring that are alive at t. Clearly, $Y_1, \ldots, Y_j$ are independent with the same distribution as $X(t - s)$. Thus, $E(Y_1 + \cdots + Y_j) = jM(t - s)$; and (3.4.5) follows by taking the expectation (with respect to j) of $jM(t - s)$.

Thus, from the above we obtain

$$M(t) = \bar{F}(t) + m\int_0^t M(t-s)\,dF(s). \tag{3.4.6}$$

Now, let α denote the unique positive number such that

$$\int_0^\infty e^{-\alpha y}\,dF(y) = \frac{1}{m}$$

and define the distribution G by

$$G(s) = m\int_0^s e^{-\alpha y}\,dF(y), \qquad 0 \le s < \infty.$$

Upon multiplying both sides of (3.4.6) by $e^{-\alpha t}$ and using the fact that $dG(s) = me^{-\alpha s}\,dF(s)$, we obtain

$$e^{-\alpha t}M(t) = e^{-\alpha t}\bar{F}(t) + \int_0^t e^{-\alpha(t-s)}M(t-s)\,dG(s). \tag{3.4.7}$$

Letting $f(t) = e^{-\alpha t}M(t)$ and $h(t) = e^{-\alpha t}\bar{F}(t)$, we have from (3.4.7), using convolution notation,

$$\begin{aligned} f &= h + f * G \\ &= h + G * f \\ &= h + G * (h + G * f) \\ &= h + G * h + G_2 * f \\ &= h + G * h + G_2 * (h + G * f) \\ &= h + G * h + G_2 * h + G_3 * f \\ &\;\;\vdots \\ &= h + G * h + G_2 * h + \cdots + G_n * h + G_{n+1} * f. \end{aligned}$$

Now since G is the distribution of some nonnegative random variable, it follows that $G_n(t) \to 0$ as $n \to \infty$ (why?), and so letting $n \to \infty$ in the above yields

$$f = h + h * \sum_{i=1}^{\infty} G_i$$
$$= h + h * m_G$$

or

$$f(t) = h(t) + \int_0^t h(t-s)\, dm_G(s).$$

Now it can be shown that $h(t)$ is directly Riemann integrable, and thus by the key renewal theorem

$$f(t) \to \frac{\int_0^\infty h(t)\, dt}{\mu_G} = \frac{\int_0^\infty e^{-\alpha t}\overline{F}(t)\, dt}{\int_0^\infty x\, dG(x)}. \tag{3.4.8}$$

Now

$$\begin{aligned} \int_0^\infty e^{-\alpha t}\overline{F}(t)\, dt &= \int_0^\infty e^{-\alpha t}\int_t^\infty dF(x)\, dt \\ &= \int_0^\infty \int_0^x e^{-\alpha t}\, dt\, dF(x) \\ &= \frac{1}{\alpha}\int_0^\infty (1 - e^{-\alpha x})\, dF(x) \\ &= \frac{1}{\alpha}\left(1 - \frac{1}{m}\right) \qquad \text{(by the definition of } \alpha). \end{aligned} \tag{3.4.9}$$

Also,

$$\int_0^\infty x\, dG(x) = m\int_0^\infty xe^{-\alpha x}\, dF(x). \tag{3.4.10}$$

Thus from (3.4.8), (3.4.9), and (3.4.10) we obtain

$$e^{-\alpha t}M(t) \to \frac{m-1}{m^2\alpha \int_0^\infty xe^{-\alpha x}\, dF(x)} \qquad \text{as } t \to \infty.$$

3.5 Delayed Renewal Processes

We often consider a counting process for which the first interarrival time has a different distribution from the remaining ones. For instance, we might start observing a renewal process at some time $t > 0$. If a renewal does not occur at t, then the distribution of the time we must wait until the first observed renewal will not be the same as the remaining interarrival distributions.

Formally, let $\{X_n, n = 1, 2, \ldots\}$ be a sequence of independent nonnegative random variables with X_1 having distribution G, and X_n having distribution F, $n > 1$. Let $S_0 = 0$, $S_n = \sum_1^n X_i$, $n \geq 1$, and define

$$N_D(t) = \sup\{n: \quad S_n \leq t\}.$$

Definition

The stochastic process $\{N_D(t), t \geq 0\}$ is called a *general or a delayed renewal process.*

When $G = F$, we have, of course, an ordinary renewal process. As in the ordinary case, we have

$$\begin{aligned} P\{N_D(t) = n\} &= P\{S_n \leq t\} - P\{S_{n+1} \leq t\} \\ &= G * F_{n-1}(t) - G * F_n(t). \end{aligned}$$

Let

$$m_D(t) = E[N_D(t)].$$

Then it is easy to show that

$$m_D(t) = \sum_{n=1}^{\infty} G * F_{n-1}(t) \tag{3.5.1}$$

and by taking transforms of (3.5.1), we obtain

$$\tilde{m}_D(s) = \frac{\tilde{G}(s)}{1 - \tilde{F}(s)}. \tag{3.5.2}$$

By using the corresponding result for the ordinary renewal process, it is easy to prove similar limit theorems for the delayed process. We leave the proof of the following proposition for the reader.

Let $\mu = \int_0^\infty x \, dF(x)$.

PROPOSITION 3.5.1

(i) With probability 1,

$$\frac{N_D(t)}{t} \to \frac{1}{\mu} \qquad \text{as } t \to \infty.$$

(ii)
$$\frac{m_D(t)}{t} \to \frac{1}{\mu} \qquad \text{as } t \to \infty.$$

(iii) If F is not lattice, then

$$m_D(t+a) - m_D(t) \to \frac{a}{\mu} \qquad \text{as } t \to \infty.$$

(iv) If F and G are lattice with period d, then

$$E[\text{number of renewals at } nd] \to \frac{d}{\mu} \qquad \text{as } n \to \infty.$$

(v) If F is not lattice, $\mu < \infty$, and h directly Riemann integrable, then

$$\int_0^\infty h(t-x)\, dm_D(x) \to \int_0^\infty h(t)\, dt/\mu.$$

Example 3.5(a) Suppose that a sequence of independent and identically distributed discrete random variables $X_1, X_2, \ldots$ is observed, and suppose that we keep track of the number of times that a given subsequence of outcomes, or *pattern*, occurs. That is, suppose the pattern is $x_1, x_2, \ldots, x_k$ and say that it occurs at time n if $X_n = x_k$, $X_{n-1} = x_{k-1}, \ldots, X_{n-k+1} = x_1$. For instance, if the pattern is 0, 1, 0, 1 and the sequence is $(X_1, X_2, \ldots) = (1, 0, 1, 0, 1, 0, 1, 1, 1, 0, 1, 0, 1, \ldots)$, then the pattern occurred at times 5, 7, 13. If we let $N(n)$ denote the number of times the pattern occurs by time n, then $\{N(n), n \geq 1\}$ is a delayed renewal process. The distribution until the first renewal is the distribution of the time until the pattern first occurs; whereas the subsequent interarrival distribution is the distribution of time between replications of the pattern.

Suppose we want to determine the rate at which the pattern occurs. By the strong law for delayed renewal processes (part (i) of Theorem 3.5.1) this will equal the reciprocal of the mean time between patterns. But by Blackwell's theorem (part (iv) of Theorem 3.5.1) this is just the limiting probability of a renewal at time n. That is,

$$(E[\text{time between patterns}])^{-1} = \lim_{n\to\infty} P\{\text{pattern at time } n\}$$
$$= \prod_{i=1}^{k} P\{X = x_i\}.$$

Hence the rate at which the pattern occurs is $\prod_{i=1}^{k} P\{X = x_i\}$ and the mean time between patterns is $(\prod_1^k P\{X = x_i\})^{-1}$.

For instance, if each random variable is 1 with probability p and 0 with probability q then the mean time between patterns of 0, 1, 0, 1 is $p^{-2}q^{-2}$. Suppose now that we are interested in the expected time that the pattern 0, 1, 0, 1 first occurs. Since the expected time to go from 0, 1, 0, 1 to 0, 1, 0, 1 is $p^{-2}q^{-2}$, it follows that starting with 0, 1 the expected number of additional outcomes to obtain 0, 1, 0, 1 is $p^{-2}q^{-2}$. But since in order for the pattern 0, 1, 0, 1 to occur we must first obtain 0, 1 it follows that

$$E[\text{time to } 0, 1, 0, 1] = E[\text{time to } 0, 1] + p^{-2}q^{-2}.$$

By using the same logic on the pattern 0, 1 we see that the expected time between occurrences of this pattern is $1/(pq)$; and as this is equal to the expected time of its first occurrence we obtain that

$$E[\text{time to } 0, 1, 0, 1] = p^{-2}q^{-2} + p^{-1}q^{-1}.$$

The above argument can be used to compute the expected number of outcomes needed for any specified pattern to appear. For instance, if a coin with probability p of coming up heads is successively flipped then

$$\begin{aligned} E[\text{time until HTHHTHH}] &= E[\text{time until HTHH}] + p^{-5}q^{-2} \\ &= E[\text{time until H}] + p^{-3}q^{-1} + p^{-5}q^{-2} \\ &= p^{-1} + p^{-3}q^{-1} + p^{-5}q^{-2}. \end{aligned}$$

Also, by the same reasoning,

$$\begin{aligned} &E[\text{time until } k \text{ consecutive heads appear}] \\ &\quad = (1/p)^k + E[\text{time until } k-1 \text{ consecutive heads appear}] \\ &\quad = \sum_{i=1}^{k} (1/p)^i. \end{aligned}$$

Suppose now that we want to compute the probability that a given pattern, say pattern A, occurs before a second pattern, say pattern B. For instance, consider independent flips of a coin that lands on heads with probability p, and suppose we are interested in the probability that A = HTHT occurs before B = THTT. To obtain this probability we will find it useful to first consider the expected additional time after a given one of these patterns occur until the other one does. Let $N_{B|A}$ denote respectively the number of additional flips needed for B to appear starting with A, and similarly for $N_{A|B}$. Also let N_A denote the number of flips until A occurs. Then

$$\begin{aligned} E[N_{B|A}] &= E[\text{additional number to THTT starting with HTHT}] \\ &= E[\text{additional number to THTT starting with THT}]. \end{aligned}$$

But since

$$E[N_{\text{THTT}}] = E[N_{\text{THT}}] + E[N_{\text{THTT}|\text{THT}}],$$

we see that

$$E[N_{B|A}] = E[N_{\text{THTT}}] - E[N_{\text{THT}}].$$

But,

$$\begin{aligned} E[N_{\text{THTT}}] &= E[N_T] + q^{-3}p^{-1} = q^{-1} + q^{-3}p^{-1} \\ E[N_{\text{THT}}] &= E[N_T] + q^{-2}p^{-1} = q^{-1} + q^{-2}p^{-1} \end{aligned}$$

and so,

$$E[N_{B|A}] = q^{-3}p^{-1} - q^{-2}p^{-1}.$$

Also,

$$E[N_{A|B}] = E[N_A] = p^{-2}q^{-2} + p^{-1}q^{-1}.$$

To compute $P_A = P\{A \text{ before } B\}$ let $M = \text{Min}(N_A, N_B)$. Then

$$\begin{aligned} E[N_A] &= E[M] + E[N_A - M] \\ &= E[M] + E[N_A - M|B \text{ before } A](1 - P_A) \\ &= E[M] + E[N_{A|B}](1 - P_A). \end{aligned}$$

Similarly,

$$E[N_B] = E[M] + E[N_{B|A}]P_A.$$

Solving these equations yields

$$\begin{aligned} P_A &= \frac{E[N_B] + E[N_{A|B}] - E[N_A]}{E[N_{B|A}] + E[N_{A|B}]} \\ E[M] &= E[N_B] - E[N_{B|A}]P_A. \end{aligned}$$

For instance, suppose that $p = 1/2$. Then as

$$E[N_{A|B}] = E[N_A] = 2^4 + 2^2 = 20$$
$$E[N_B] = 2 + 2^4 = 18, E[N_{B|A}] = 2^4 - 2^3 = 8,$$

we obtain that

$$P_A = \frac{18 + 20 - 20}{8 + 20} = 9/14, E[M] = 18 - \frac{8 \cdot 9}{14} = 90/7.$$

Therefore, the expected number of flips until the pattern A appears is 20, the expected number until the pattern B appears is 18, the expected number until either appears is 90/7, and the probability that pattern A appears first is 9/14 (which is somewhat counterintuitive since $E[N_A] > E[N_B]$).

Example 3.5(b) A system consists of n independent components, each of which acts like an exponential alternating renewal process. More specifically, component $i, i = 1, \ldots, n$, is up for an exponential time with mean λ_i and then goes down, in which state it remains for an exponential time with mean μ_i, before going back up and starting anew.

Suppose that the system is said to be functional at any time if at least one component is up at that time (such a system is called *parallel*). If we let $N(t)$ denote the number of times the system becomes nonfunctional (that is, breaks down) in $[0, t]$, then $\{N(t), t \geq 0)$ is a delayed renewal process.

Suppose we want to compute the mean time between system breakdowns. To do so let us first look at the probability of a breakdown in $(t, t + h)$ for large t and small h. Now one way for a breakdown to occur in $(t, t + h)$ is to have exactly 1 component up at time t and all others down, and then have that component fail. Since all other possibilities taken together clearly have probability $o(h)$, we see that

$$\lim_{t \to \infty} P\{\text{breakdown in } (t, t + h)\} = \sum_{i=1}^{n} \left\{ \frac{\lambda_i}{\lambda_i + \mu_i} \prod_{j \neq i} \frac{\mu_j}{\lambda_j + \mu_j} \right\} \frac{1}{\lambda_i} h + o(h).$$

But by Blackwell's theorem the above is just h times the reciprocal of the mean time between breakdowns, and so upon letting $h \to 0$ we obtain

$$E[\text{time between breakdowns}] = \left(\prod_{j=1}^{n} \frac{\mu_j}{\lambda_j + \mu_j} \sum_{i=1}^{n} \frac{1}{\mu_i} \right)^{-1}.$$

As the expected length of a breakdown period is $(\sum_{i=1}^{n} 1/\mu_i)^{-1}$, we can compute the average length of an up (or functional) period from

$$E[\text{length of up period}] = E[\text{time between breakdowns}] - \left(\sum_1^n \frac{1}{\mu_i}\right)^{-1}$$

$$= \frac{1 - \prod_{j=1}^{n} \frac{\mu_j}{\lambda_j + \mu_j}}{\prod_{j=1}^{n} \frac{\mu_j}{\lambda_j + \mu_j} \sum_{i=1}^{n} \frac{1}{\mu_i}}.$$

As a check of the above, note that the system may be regarded as a delayed alternating renewal process whose limiting probability of being down is

$$\lim_{t\to\infty} P\{\text{system is down at } t\} = \prod_{j=1}^{n} \frac{\mu_j}{\lambda_j + \mu_j}.$$

We can now verify that the above is indeed equal to the expected length of a down period divided by the expected time length of a cycle (or time between breakdowns).

Example 3.5(c) Consider two coins, and suppose that each time coin i is flipped it lands on tails with some unknown probability p_i, $i = 1, 2$. Our objective is to continually flip among these coins so as to make the long-run proportion of tails equal to $\min(p_1, p_2)$. The following strategy, having a very small memory requirement, will accomplish this objective. Start by flipping coin 1 until a tail occurs, at which point switch to coin 2 and flip it until a tail occurs. Say that cycle 1 ends at this point. Now flip coin 1 until two tails in a row occur, and then switch and do the same with coin 2. Say that cycle 2 ends at this point. In general, when cycle n ends, return to coin 1 and flip it until $n + 1$ tails in a row occur and then flip coin 2 until this occurs, which ends cycle $n + 1$.

To show that the preceding policy meets our objective, let $p = \max(p_1, p_2)$ and $\alpha p = \min(p_1, p_2)$, where $\alpha < 1$ (if $\alpha = 1$ then all policies meet the objective). Call the coin with tail probability p the bad coin and the one with probability αp the good one. Let B_m denote the number of flips in cycle m that use the bad coin and let G_m be the number that use the good coin. We will need the following lemma.

Lemma

For any $\varepsilon > 0$,

$$P\{B_m \geq \varepsilon G_m \text{ for infinitely many } m\} = 0.$$

Proof We will show that

$$\sum_{m=1}^{\infty} P\{B_m \geq \varepsilon G_m\} < \infty$$

which, by the Borel-Cantelli lemma (see Section 1.1), will establish the result. Now,

$$\begin{aligned} P\{G_m \leq B_m/\varepsilon\} &= E[P\{G_m \leq B_m/\varepsilon \mid B_m\}] \\ &= E\left[\sum_{i=1}^{B_m/\varepsilon} P\{G_m = i \mid B_m\}\right] \\ &\leq E\left[\sum_{i=1}^{B_m/\varepsilon} (\alpha p)^m\right] \end{aligned}$$

where the preceding inequality follows from the fact that $G_m = i$ implies that $i \geq m$ and that cycle m good flips numbered $i - m + 1$ to i must all be tails. Thus, we see that

$$P\{G_m \leq B_m/\varepsilon\} \leq \varepsilon^{-1}(\alpha p)^m E[B_m].$$

But, from Example 3.3(A),

$$E[B_m] = \sum_{i=1}^{m} (1/p)^i = \frac{(1/p)^m - 1}{1 - p}.$$

Therefore,

$$\sum_{m=1}^{\infty} P\{B_m \geq \varepsilon G_m\} \leq \frac{1}{\varepsilon(1-p)} \sum_{m=1}^{\infty} \alpha^m < \infty$$

which proves the lemma. ∎

Thus, using the lemma, we see that in all but a finite number of cycles the proportion $B/(B + G)$ of flips that use the bad coin will be less than $\frac{\varepsilon}{1+\varepsilon} < \varepsilon$. Hence, supposing that the first coin used in each cycle is the good coin it follows, with probability 1, that the long-run proportion of flips that use the bad coin will be less than or equal to ε. Since ε is arbitrary, this implies, from the continuity property of probabilities (Proposition 1.1.1), that the long-run proportion of flips that use the bad coin is 0. As a result, it follows that, with probability 1, the long-run proportion of flips that land tails is equal to the long-run proportion of good coin flips that land tails, and this, by the strong law of large numbers, is αp. (Whereas the preceding argument supposed that the good coin is used first in each cycle, this is not necessary and a similar argument can be given to show that the result holds without this assumption.)

In the same way we proved the result in the case of an ordinary renewal process, it follows that the distribution of the time of the last renewal before (or at t) t is given by

$$P\{S_{N(t)} \le s\} = \bar{G}(t) + \int_0^s \bar{F}(t-y)\, dm_D(y). \tag{3.5.3}$$

When $\mu < \infty$, the distribution function

$$F_e(x) = \int_0^x \bar{F}(y)\, dy/\mu, \qquad x \ge 0,$$

is called the *equilibrium distribution* of F. Its Laplace transform is given by

$$\begin{aligned} \tilde{F}_e(s) &= \int_0^\infty e^{-sx}\, dF_e(x) \\ &= \int_0^\infty e^{-sx} \int_x^\infty dF(y)\, dx/\mu \\ &= \int_0^\infty \int_0^y e^{-sx}\, dx\, dF(y)/\mu \\ &= \frac{1}{s\mu} \int_0^\infty (1 - e^{-sy})\, dF(y) \\ &= \frac{1 - \tilde{F}(s)}{\mu s}. \end{aligned} \tag{3.5.4}$$

The delayed renewal process with $G = F_e$ is called the *equilibrium renewal* process and is extremely important. For suppose that we start observing a renewal process at time t. Then the process we observe is a delayed renewal process whose initial distribution is the distribution of $Y(t)$. Thus, for t large, it follows from Proposition 3.4.5 that the observed process is the equilibrium renewal process. The stationarity of this process is proven in the next theorem.

Let $Y_D(t)$ denote the excess at t for a delayed renewal process.

THEOREM 3.5.2

For the equilibrium renewal process:

(i) $m_D(t) = t/\mu$;
(ii) $P\{Y_D(t) \le x\} = F_e(x)$ *for all* $t \ge 0$;
(iii) $\{N_D(t), t \ge 0\}$ *has stationary increments.*

Proof (i) From (3.5.2) and (3.5.4), we have that

$$\tilde{m}_D(s) = \frac{1}{\mu s}.$$

However, simple calculus shows that $1/\mu s$ is the Laplace transform of the function $h(t) = t/\mu$, and thus by the uniqueness of transforms, we obtain

$$m_D(t) = t/\mu.$$

(ii) For a delayed renewal process, upon conditioning on $S_{N(t)}$ we obtain, using (3.5.3),

$$P\{Y_D(t) > x\} = P\{Y_D(t) > x | S_{N(t)} = 0\}\overline{G}(t) + \int_0^t P\{Y_D(t) > x | S_{N(t)} = s\}\overline{F}(t-s)\, dm_D(s).$$

Now

$$P\{Y_D(t) > x | S_{N(t)} = 0\} = P\{X_1 > t + x | X_1 > t\} = \frac{\overline{G}(t+x)}{\overline{G}(t)}$$

$$P\{Y_D(t) > x | S_{N(t)} = s\} = P\{X > t + x - s | X > t - s\} = \frac{\overline{F}(t+x-s)}{\overline{F}(t-s)}.$$

Hence,

$$P\{Y_D(t) > x\} = \overline{G}(t+x) + \int_0^t \overline{F}(t+x-s)\, dm_D(s).$$

Now, letting $G = F_e$ and using part (i) yields

$$\begin{aligned} P\{Y_D(t) > x\} &= \overline{F}_e(t+x) + \int_0^t \overline{F}(t+x-s)\, ds/\mu \\ &= \overline{F}_e(t+x) + \int_x^{t+x} \overline{F}(y)\, dy/\mu \\ &= \overline{F}_e(x). \end{aligned}$$

(iii) To prove (iii) we note that $N_D(t+s) - N_D(s)$ may be interpreted as the number of renewals in time t of a delayed renewal process, where the initial distribution is the distribution of $Y_D(s)$. The result then follows from (ii).

3.6 Renewal Reward Processes

A large number of probability models are special cases of the following model. Consider a renewal process $\{N(t), t \geq 0\}$ having interarrival times X_n, $n \geq 1$ with distribution F, and suppose that each time a renewal occurs we receive a reward. We denote by R_n the reward earned at the time of the nth renewal. We shall assume that the R_n, $n \geq 1$, are independent and identically distributed.

However, we do allow for the possibility that R_n may (and usually will) depend on X_n, the length of the nth renewal interval, and so we assume that the pairs (X_n, R_n), $n \geq 1$, are independent and identically distributed. If we let

$$R(t) = \sum_{n=1}^{N(t)} R_n,$$

then $R(t)$ represents the total reward earned by time t. Let

$$E[R] = E[R_n], \qquad E[X] = E[X_n].$$

THEOREM 3.6.1

If $E[R] < \infty$ and $E[X] < \infty$, then

(i) with probability 1,

$$\frac{R(t)}{t} \to \frac{E[R]}{E[X]} \quad \text{as } t \to \infty,$$

(ii)

$$\frac{E[R(t)]}{t} \to \frac{E[R]}{E[X]} \quad \text{as } t \to \infty.$$

Proof To prove (i) write

$$\frac{R(t)}{t} = \frac{\sum_{n=1}^{N(t)} R_n}{t}$$

$$= \left(\frac{\sum_{n=1}^{N(t)} R_n}{N(t)}\right)\left(\frac{N(t)}{t}\right).$$

By the strong law of large numbers we obtain that

$$\frac{\sum_{n=1}^{N(t)} R_n}{N(t)} \to E[R] \quad \text{as } t \to \infty,$$

and by the strong law for renewal processes

$$\frac{N(t)}{t} \to \frac{1}{E[X]} \quad \text{as } t \to \infty.$$

Thus (i) is proven.

To prove (ii) we first note that since $N(t) + 1$ is a stopping time for the sequence $X_1, X_2, \ldots$, it is also a stopping time for $R_1, R_2, \ldots$. (Why?) Thus, by Wald's equation,

$$E\left[\sum_{i=1}^{N(t)} R_i\right] = E\left[\sum_{i=1}^{N(t)+1} R_i\right] - E[R_{N(t)+1}]$$
$$= (m(t) + 1)E[R] - E[R_{N(t)+1}]$$

and so

$$\frac{E[R(t)]}{t} = \frac{m(t) + 1}{t} E[R] - \frac{E[R_{N(t)+1}]}{t},$$

and the result will follow from the elementary renewal theorem if we can show that $E[R_{N(t)+1}]/t \to 0$ as $t \to \infty$. So, towards this end, let $g(t) = E[R_{N(t)+1}]$. Then conditioning on $S_{N(t)}$ yields

$$g(t) = E[R_{N(t)+1} | S_{N(t)} = 0]\bar{F}(t)$$
$$+ \int_0^t E[R_{N(t)+1} | S_{N(t)} = s]\bar{F}(t - s)\, dm(s).$$

However,

$$E[R_{N(t)+1} | S_{N(t)} = 0] = E[R_1 | X_1 > t],$$
$$E[R_{N(t)+1} | S_{N(t)} = s] = E[R_n | X_n > t - s],$$

and so

$$g(t) = E[R_1 | X_1 > t]\bar{F}(t) + \int_0^t E[R_n | X_n > t - s]\bar{F}(t - s)\, dm(s). \tag{3.6.1}$$

Now, let

$$h(t) = E[R_1 | X_1 > t]\bar{F}(t) = \int_t^\infty E[R_1 | X_1 = x]\, dF(x),$$

and note that since

$$E|R_1| = \int_0^\infty E[|R_1| \,| X_1 = x]\, dF(x) < \infty,$$

it follows that

$$h(t) \to 0 \text{ as } t \to \infty \quad \text{and} \quad h(t) \le E|R_1| \text{ for all } t,$$

and thus we can choose T so that $|h(t)| < \varepsilon$ whenever $t \ge T$. Hence, from (3.6.1),

$$\frac{|g(t)|}{t} \le \frac{|h(t)|}{t} + \int_0^{t-T} \frac{|h(t - x)|\, dm(x)}{t} + \int_{t-T}^t \frac{|h(t - x)|\, dm(x)}{t}$$
$$\le \frac{\varepsilon}{t} + \frac{\varepsilon m(t - T)}{t} + E|R_1| \frac{m(t) - m(t - T)}{t}$$
$$\to \frac{\varepsilon}{EX} \quad \text{as } t \to \infty$$

by the elementary renewal theorem. Since ε is arbitrary, it follows that $g(t)/t \to 0$, and the result follows.

Remarks If we say that a cycle is completed every time a renewal occurs, then the theorem states that the (expected) long-run average return is just the expected return earned during a cycle, divided by the expected time of a cycle.

In the proof of the theorem it is tempting to say that $E[R_{N(t)+1}] = E[R_1]$ and thus $1/tE[R_{N(t)+1}]$ trivially converges to zero. However, $R_{N(t)+1}$ is related to $X_{N(t)+1}$, and $X_{N(t)+1}$ is the length of the renewal interval containing the point t. Since larger renewal intervals have a greater chance of containing t, it (heuristically) follows that $X_{N(t)+1}$ tends to be larger than an ordinary renewal interval (see Problem 3.3), and thus the distribution of $R_{N(t)+1}$ is not that of R_1.

Also, up to now we have assumed that the reward is earned all at once at the end of the renewal cycle. However, this is not essential, and Theorem 3.6.1 remains true if the reward is earned gradually during the renewal cycle. To see this, let $R(t)$ denote the reward earned by t, and suppose first that all returns are nonnegative. Then

$$\frac{\sum_{n=1}^{N(t)} R_n}{t} \le \frac{R(t)}{t} \le \frac{\sum_{n=1}^{N(t)} R_n}{t} + \frac{R_{N(t)+1}}{t}$$

and (ii) of Theorem 3.6.1 follows since

$$\frac{E[R_{N(t)+1}]}{t} \to 0.$$

Part (i) of Theorem 3.6.1 follows by noting that both $\sum_{n=1}^{N(t)} R_n/t$ and $\sum_{n=1}^{N(t)+1} R_n/t$ converge to $E[R]/E[X]$ by the argument given in the proof. A similar argument holds when the returns are nonpositive, and the general case follows by breaking up the returns into their positive and negative parts and applying the above argument separately to each.

Example 3.6(a) ***Alternating Renewal Process.*** For an alternating renewal process (see Section 3.4.1) suppose that we earn at a rate of one per unit time when the system is on (and thus the reward for a cycle equals the on time of that cycle). Then the total reward earned by t is just the total on time in $[0, t]$, and thus by Theorem 3.6.1, with probability 1,

$$\text{amount of on time in } \frac{[0, t]}{t} \to \frac{E[X]}{E[X] + E[Y]},$$

where X is an on time and Y an off time in a cycle. Thus by Theorem 3.4.4 when the cycle distribution is nonlattice the limiting probability of the system being on is equal to the long-run proportion of time it is on.

Example 3.6(b) ***Average Age and Excess.*** Let $A(t)$ denote the age at t of a renewal process, and suppose we are interested in computing

$$\lim_{t\to\infty} \int_0^t A(s)\, ds/t.$$

To do so assume that we are being paid money at any time at a rate equal to the age of the renewal process at that time. That is, at time s we are being paid at a rate $A(s)$, and so $\int_0^t A(s)\, ds$ represents our total earnings by time t. As everything starts over again when a renewal occurs, it follows that, with probability 1,

$$\frac{\int_0^t A(s)\, ds}{t} \to \frac{E[\text{reward during a renewal cycle}]}{E[\text{time of a renewal cycle}]}.$$

Now since the age of the renewal process a time s into a renewal cycle is just s, we have

$$\text{reward during a renewal cycle} = \int_0^X s\, ds = \frac{X^2}{2},$$

where X is the time of the renewal cycle. Hence, with probability 1,

$$\lim_{t\to\infty} \frac{\int_0^t A(s)\, ds}{t} = \frac{E[X^2]}{2E[X]}.$$

Similarly if $Y(t)$ denotes the excess at t, we can compute the average excess by supposing that we are earning rewards at a rate equal to the excess at that time. Then the average value of the excess will, by Theorem 3.6.1, be given by

$$\begin{aligned}\lim_{t\to\infty} \int_0^t Y(s)\, ds/t &= \frac{E[\text{reward during a renewal cycle}]}{E[X]} \\ &= \frac{E\left[\int_0^X (X-t)\, dt\right]}{E[X]} \\ &= \frac{E[X^2]}{2E[X]}.\end{aligned}$$

Thus the average values of the age and excess are equal. (Why was this to be expected?)

The quantity $X_{N(t)+1} = S_{N(t)+1} - S_{N(t)}$ represents the length of the renewal interval containing the point t. Since it may also be expressed by

$$X_{N(t)+1} = A(t) + Y(t),$$

we see that its average value is given by

$$\lim_{t\to\infty} \int_0^t X_{N(s)+1}\, ds/t = \frac{E[X^2]}{E[X]}.$$

Since

$$\frac{E[X^2]}{E[X]} \geq E[X]$$

(with equality only when $\text{Var}(X) = 0$) we see that the average value of $X_{N(t)+1}$ is greater than $E[X]$. (Why is this not surprising?)

Example 3.6(c) Suppose that travelers arrive at a train depot in accordance with a renewal process having a mean interarrival time μ. Whenever there are N travelers waiting in the depot, a train leaves. If the depot incurs a cost at the rate of nc dollars per unit time whenever there are n travelers waiting and an additional cost of K each time a train is dispatched, what is the average cost per unit time incurred by the depot?

If we say that a cycle is completed whenever a train leaves, then the above is a renewal reward process. The expected length of a cycle is the expected time required for N travelers to arrive, and, since the mean interarrival time is μ, this equals

$$E[\text{length of cycle}] = N\mu.$$

If we let X_n denote the time between the nth and $(n + 1)$st arrival in a cycle, then the expected cost of a cycle may be expressed as

$$\begin{aligned} E[\text{cost of a cycle}] &= E[cX_1 + 2cX_2 + \cdots + (N-1)cX_{N-1}] + K \\ &= \frac{c\mu N(N-1)}{2} + K \end{aligned}$$

Hence the average cost incurred is

$$\frac{c(N-1)}{2} + \frac{K}{N\mu}.$$

3.6.1 A Queueing Application

Suppose that customers arrive at a single-server service station in accordance with a nonlattice renewal process. Upon arrival, a customer is immediately served if the server is idle, and he or she waits in line if the server is busy. The service times of customers are assumed to be independent and identically distributed, and are also assumed independent of the arrival stream.

Let $X_1, X_2, \ldots$ denote the interarrival times between customers; and let $Y_1, Y_2, \ldots$ denote the service times of successive customers. We shall assume that

$$E[Y_i] < E[X_i] < \infty. \tag{3.6.2}$$

Suppose that the first customer arrives at time 0 and let $n(t)$ denote the number of customers in the system at time t. Define

$$L = \lim_{t\to\infty} \int_0^t n(s)\, ds/t.$$

To show that L exists and is constant, with probability 1, imagine that a reward is being earned at time s at rate $n(s)$. If we let a cycle correspond to the start of a busy period (that is, a new cycle begins each time an arrival finds the system empty), then it is easy to see that the process restarts itself each cycle. As L represents the long-run average reward, it follows from Theorem 3.6.1 that

$$L = \frac{E[\text{reward during a cycle}]}{E[\text{time of a cycle}]} = \frac{E\left[\int_0^T n(s)\, ds\right]}{E[T]}. \tag{3.6.3}$$

Also, let W_i denote the amount of time the ith customer spends in the system and define

$$W = \lim_{n\to\infty} \frac{W_1 + \cdots + W_n}{n}.$$

To argue that W exists with probability 1, imagine that we receive a reward W_i on day i. Since the queueing process begins anew after each cycle, it follows that if we let N denote the number of customers served in a cycle, then W is the average reward per unit time of a renewal process in which the cycle time is N and the cycle reward is $W_1 + \cdots + W_N$, and, hence,

$$W = \frac{E[\text{reward during a cycle}]}{E[\text{time of a cycle}]} = \frac{E\left[\sum_{i=1}^{N} W_i\right]}{E[N]}. \tag{3.6.4}$$

We should remark that it can be shown (see Proposition 7.1.1 of Chapter 7) that (3.6.2) implies that $E[N] < \infty$.

The following theorem is quite important in queueing theory.

THEOREM 3.6.2

Let $\lambda = 1/E[X_i]$ denote the arrival rate. Then

$$L = \lambda W.$$

Proof We start with the relationship between T, the length of a cycle, and N, the number of customers served in that cycle. If n customers are served in a cycle, then the next cycle begins when the $(n + 1)$st customer arrives; hence,

$$T = \sum_{i=1}^{N} X_i.$$

Now it is easy to see that N is a stopping time for the sequence $X_1, X_2, \ldots$ since

$$N = n \leftrightarrow X_1 + \cdots + X_k < Y_1 + \cdots + Y_k, \qquad k = 1, \ldots, n - 1$$
$$\text{and} \quad X_1 + \cdots + X_n > Y_1 + \cdots + Y_n$$

and thus $\{N = n\}$ is independent of $X_{n+1}, X_{n+2}, \ldots$. Hence by Wald's equation

$$E[T] = E[N]E[X] = E[N]/\lambda$$

and so by (3.6.3) and (3.6.4)

$$L = \lambda W \frac{E\left[\int_0^T n(s)\, ds\right]}{E\left[\sum_{i=1}^{N} W_i\right]}. \tag{3.6.5}$$

But by imagining that each customer pays at a rate of 1 per unit time while in the system (and so the total amount paid by the ith arrival is just W_i), we see

$$\int_0^T n(s)\,ds = \sum_{i=1}^{N} W_i = \text{total paid during a cycle},$$

and so the result follows from (3.6.5). ∎

Remarks

(1) The proof of Theorem 3.6.2 does not depend on the particular queueing model we have assumed. The proof goes through without change for any queueing system that contains times at which the process probabilistically restarts itself and where the mean time between such cycles is finite. For example, if in our model we suppose that there are k available servers, then it can be shown that a sufficient condition for the mean cycle time to be finite is that

$$E[Y_i] < kE[X_i] \quad \text{and} \quad P\{Y_i < X_i\} > 0.$$

(2) Theorem 3.6.2 states that the

$$\text{(time) average number in "the system"} = \lambda \cdot (\text{average time a customer spends in "the system"}).$$

By replacing "the system" by "the queue" the same proof shows that the

$$\text{average number in the queue} = \lambda \cdot (\text{average time a customer spends in the queue}),$$

or, by replacing "the system" by "service" we have that the

$$\text{average number in service} = \lambda E[Y].$$

3.7 Regenerative Processes

Consider a stochastic process $\{X(t), t \geq 0\}$ with state space $\{0, 1, 2, \ldots\}$ having the property that there exist time points at which the process (probabilistically) restarts itself. That is, suppose that with probability 1, there exists a time S_1 such that the continuation of the process beyond S_1 is a probabilistic replica of the whole process starting at 0. Note that this property implies the existence of further times $S_2, S_3, \ldots$ having the same property as S_1. Such a stochastic process is known as a *regenerative process*.

From the above, it follows that $\{S_1, S_2, \ldots,\}$ constitute the event times of a renewal process. We say that a cycle is completed every time a renewal occurs. Let $N(t) = \max\{n: S_n \le t\}$ denote the number of cycles by time t.

The proof of the following important theorem is a further indication of the power of the key renewal theorem.

THEOREM 3.7.1

If F, the distribution of a cycle, has a density over some interval, and if $E[S_1] < \infty$, then

$$P_j \equiv \lim_{t\to\infty} P\{X(t) = j\} = \frac{E[\text{amount of time in state } j \text{ during a cycle}]}{E[\text{time of a cycle}]}.$$

Proof Let $P(t) = P\{X(t) = j\}$. Conditioning on the time of the last cycle before t yields

$$P(t) = P\{X(t) = j \mid S_{N(t)} = 0\}\bar{F}(t) + \int_0^t P\{X(t) = j \mid S_{N(t)} = s\}\bar{F}(t-s)\,dm(s).$$

Now,

$$P\{X(t) = j \mid S_{N(t)} = 0\} = P\{X(t) = j \mid S_1 > t\},$$
$$P\{X(t) = j \mid S_{N(t)} = s\} = P\{X(t-s) = j \mid S_1 > t-s\},$$

and thus

$$P(t) = P\{X(t) = j, S_1 > t\} + \int_0^t P\{X(t-s) = j, S_1 > t-s\}\,dm(s).$$

Hence, as it can be shown that $h(t) \equiv P\{X(t) = j, S_1 > t\}$ is directly Riemann integrable, we have by the key renewal theorem that

$$P(t) \to \int_0^\infty P\{X(t) = j, S_1 > t\}\,dt / E[S_1].$$

Now, letting

$$I(t) = \begin{cases} 1 & \text{if } X(t) = j, S_1 > t \\ 0 & \text{otherwise,} \end{cases}$$

then $\int_0^\infty I(t)\,dt$ represents the amount of time in the first cycle that $X(t) = j$. Since

$$E\left[\int_0^\infty I(t)\,dt\right] = \int_0^\infty E[I(t)]\,dt$$
$$= \int_0^\infty P\{X(t) = j, S_1 > t\}\,dt,$$

the result follows.

Example 3.7(a) ***Queueing Models with Renewal Arrivals.*** Most queueing processes in which customers arrive in accordance to a renewal process (such as those in Section 3.6) are regenerative processes with cycles beginning each time an arrival finds the system empty. Thus for instance in the single-server queueing model with renewal arrivals, $X(t)$, the number in the system at time t, constitutes a regenerative process provided the initial customer arrives at $t = 0$ (if not then it is a *delayed* regenerative process and Theorem 3.7.1 remains valid).

From the theory of renewal reward processes it follows that P_j also equals the *long-run proportion* of time that $X(t) = j$. In fact, we have the following.

PROPOSITION 3.7.2

For a regenerative process with $E[S_1] < \infty$, with probability 1,

$$\lim_{t\to\infty} \frac{[\text{amount of time in } j \text{ during } (0, t)]}{t} = \frac{E[\text{time in } j \text{ during a cycle}]}{E[\text{time of a cycle}]}.$$

Proof Suppose that a reward is earned at *rate* 1 whenever the process is in state j. This generates a renewal reward process and the proposition follows directly from Theorem 3.6.1.

3.7.1 The Symmetric Random Walk and the Arc Sine Laws

Let $Y_1, Y_2, \ldots$ be independent and identically distributed with

$$P\{Y_i = 1\} = P\{Y_i = -1\} = \tfrac{1}{2},$$

and define

$$Z_0 = 0, \qquad Z_n = \sum_{i=1}^{n} Y_i.$$

The process $\{Z_n, n \geq 0\}$ is called the symmetric random walk process.

If we now define X_n by

$$X_n = \begin{cases} 0 & \text{if } Z_n = 0 \\ 1 & \text{if } Z_n > 0 \\ -1 & \text{if } Z_n < 0, \end{cases}$$

then $\{X_n, n \geq 1\}$ is a regenerative process that regenerates whenever X_n takes value 0. To obtain some of the properties of this regenerative process, we will first study the symmetric random walk $\{Z_n, n \geq 0\}$.

Let

$$u_n = P\{Z_{2n} = 0\}$$
$$= \binom{2n}{n} (\tfrac{1}{2})^{2n}$$

and note that

$$u_n = \frac{2n-1}{2n} u_{n-1}. \tag{3.7.1}$$

Now let us recall from the results of Example 1.5(E) of Chapter 1 (the ballot problem example) the expression for the probability that the first visit to 0 in the symmetric random walk occurs at time $2n$. Namely,

$$P\{Z_1 \neq 0, Z_2 \neq 0, \ldots, Z_{2n-1} \neq 0, Z_{2n} = 0\} = \frac{\binom{2n}{n} (\tfrac{1}{2})^{2n}}{2n-1} \tag{3.7.2}$$
$$= \frac{u_n}{2n-1}.$$

We will need the following lemma, which states that u_n—the probability that the symmetric random walk is at 0 at time $2n$—is also equal to the probability that the random walk does not hit 0 by time $2n$.

Lemma 3.7.3

$$P\{Z_1 \neq 0, Z_2 \neq 0, \ldots, Z_{2n} \neq 0\} = u_n.$$

Proof From (3.7.2) we see that

$$P\{Z_1 \neq 0, \ldots, Z_{2n} \neq 0\} = 1 - \sum_{k=1}^{n} \frac{u_k}{2k-1}.$$

Hence we must show that

$$u_n = 1 - \sum_{k=1}^{n} \frac{u_k}{2k-1}, \tag{3.7.3}$$

which we will do by induction on n. When $n = 1$, the above identity holds since $u_1 = \frac{1}{2}$. So assume (3.7.3) for $n - 1$. Now

$$\begin{aligned} 1 - \sum_{k=1}^{n} \frac{u_k}{2k-1} &= 1 - \sum_{k=1}^{n-1} \frac{u_k}{2k-1} - \frac{u_n}{2n-1} \\ &= u_{n-1} - \frac{u_n}{2n-1} \qquad \text{(by the induction hypothesis)} \\ &= u_n \qquad \text{(by (3.7.1))}. \end{aligned}$$

Thus the proof is complete.

Since

$$u_n = \binom{2n}{n} \left(\tfrac{1}{2}\right)^{2n},$$

it follows upon using an approximation due to Stirling—which states that $n! \sim n^{n+1/2}e^{-n}\sqrt{2\pi}$—that

$$u_n \sim \frac{(2n)^{2n+1/2}e^{-2n}\sqrt{2\pi}}{n^{2n+1}e^{-2n}(2\pi)2^{2n}} = \frac{1}{\sqrt{n\pi}},$$

and so $u_n \to 0$ as $n \to \infty$. Thus from Lemma 3.7.3 we see that, with probability 1, the symmetric random walk will return to the origin.

The next proposition gives the distribution of the time of the last visit to 0 up to and including time $2n$.

PROPOSITION 3.7.4

For $k = 0, 1, \ldots, n$,

$$P\{Z_{2k} = 0, Z_{2k+1} \neq 0, Z_{2k+2} \neq 0, \ldots, Z_{2n} \neq 0\} = u_k u_{n-k}.$$

Proof

$$\begin{aligned} P\{Z_{2k} = 0, Z_{2k+1} \neq 0, \ldots, Z_{2n} \neq 0\} &= P\{Z_{2k} = 0\}P\{Z_{2k+1} \neq 0, \ldots, Z_{2n} \neq 0 \mid Z_{2k} = 0\} \\ &= u_k u_{n-k}. \end{aligned}$$

where we have used Lemma 3.7.3 to evaluate the second term on the right in the above.

We are now ready for our major result, which is that if we plot the symmetric random walk (starting with $Z_0 = 0$) by connecting Z_k and Z_{k+1} by a straight line (see Figure 3.7.1), then the probability that up to time $2n$ the process has been positive for $2k$ time units and negative for $2n - 2k$ time units is the same as the probability given in Proposition 3.7.4. (For the sample path presented in Figure 3.7.1, of the first eight time units the random walk was positive for six and negative for two.)

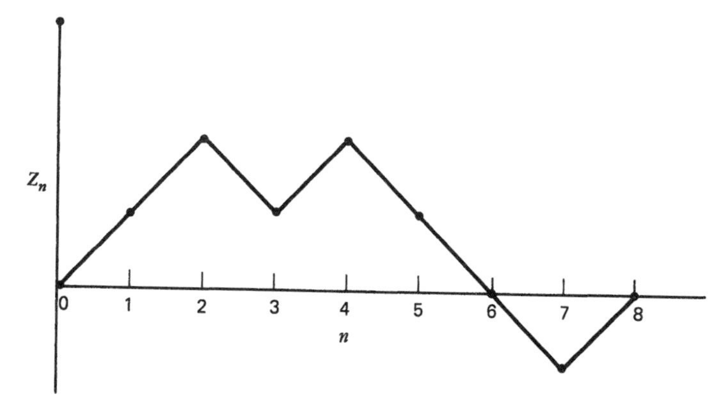

Figure 3.7.1. A sample path for the random walk.

THEOREM 3.7.5

Let $E_{k,n}$ denote the event that by time $2n$ the symmetric random walk will be positive for $2k$ time units and negative for $2n - 2k$ time units, and let $b_{k,n} = P(E_{k,n})$. Then

$$b_{k,n} = u_k u_{n-k}. \tag{3.7.4}$$

Proof The proof is by induction on n. Since

$$b_{0,1} = b_{1,1} = \tfrac{1}{2}, \qquad u_0 = 1, \qquad u_1 = \tfrac{1}{2},$$

it follows that (3.7.4) is valid when $n = 1$. So assume that $b_{k,m} = u_k u_{m-k}$ for all values of m such that $m < n$. To prove (3.7.4) we first consider the case where $k = n$. Then, conditioning on T, the time of the first return to 0, yields

$$b_{n,n} = \sum_{r=1}^{n} P\{E_{n,n} | T = 2r\} P\{T = 2r\} + P\{E_{n,n} | T > 2n\} P\{T > 2n\}.$$

Now given that $T = 2r$, it is equally likely that the random walk has always been positive or always negative in $(0, 2r)$ and it is at 0 at $2r$. Hence,

$$P\{E_{n,n} | T = 2r\} = b_{n-r,n-r}/2, \qquad P\{E_{n,n} | T > 2n\} = \tfrac{1}{2},$$

and so

$$\begin{aligned} b_{n,n} &= \tfrac{1}{2} \sum_{r=1}^{n} b_{n-r,n-r} P\{T = 2r\} + \tfrac{1}{2} P\{T > 2n\} \\ &= \tfrac{1}{2} \sum_{r=1}^{n} u_{n-r} P\{T = 2r\} + \tfrac{1}{2} P\{T > 2n\}, \end{aligned}$$

where the last equality, $b_{n-r,n-r} = u_{n-r} u_0$, follows from the induction hypothesis. Now,

$$\begin{aligned} \sum_{r=1}^{n} u_{n-r} P\{T = 2r\} &= \sum_{r=1}^{n} P\{Z_{2n-2r} = 0\} P\{T = 2r\} \\ &= \sum_{r=1}^{n} P\{Z_{2n} = 0 | T = 2r\} P\{T = 2r\} \\ &= u_n, \end{aligned}$$

and so

$$\begin{aligned} b_{n,n} &= \tfrac{1}{2} u_n + \tfrac{1}{2} P\{T > 2n\} \\ &= \tfrac{1}{2} u_n + \tfrac{1}{2} u_n \qquad \text{(by Lemma 3.7.3)} \\ &= u_n. \end{aligned}$$

Hence (3.7.4) is valid for $k = n$ and, in fact, by symmetry, also for $k = 0$. The proof that (3.7.4) is valid for $0 < k < n$ follows in a similar fashion. Again, conditioning on T yields

$$b_{k,n} = \sum_{r=1}^{n} P\{E_{k,n} | T = 2r\} P\{T = 2r\}.$$

Now given that $T = 2r$, it is equally likely that the random walk has either always been positive or always negative in $(0, 2r)$. Hence in order for $E_{k,n}$ to occur, the continuation from time $2r$ to $2n$ would need $2k - 2r$ positive units in the former case and $2k$ in the latter. Hence,

$$\begin{aligned} b_{k,n} &= \tfrac{1}{2} \sum_{r=1}^{n} b_{k-r,n-r} P\{T = 2r\} + \tfrac{1}{2} \sum_{r=1}^{n} b_{k,n-r} P\{T = 2r\} \\ &= \tfrac{1}{2} u_{n-k} \sum_{r=1}^{n} u_{k-r} P\{T = 2r\} + \tfrac{1}{2} u_k \sum_{r=1}^{n} u_{n-r-k} P\{T = 2r\}, \end{aligned}$$

where the last equality follows from the induction hypothesis. As

$$\sum_{r=1}^{n} u_{k-r} P\{T = 2r\} = u_k,$$

$$\sum_{r=1}^{n} u_{n-r-k} P\{T = 2r\} = u_{n-k},$$

we see that

$$b_{k,n} = u_k u_{n-k},$$

which completes the proof.

The probability distribution given by Theorem 3.7.5, namely,

$$P\{X = 2k\} = u_k u_{n-k},$$

is called the discrete arc sine distribution. We call it such since for large k and n we have, by Stirling's approximation, that

$$u_k u_{n-k} \sim \frac{1}{\pi \sqrt{k(n-k)}}.$$

Hence for any x, $0 < x < 1$, we see that the probability that the proportion of time in $(0, 2n)$ that the symmetric random is positive is less than x, is given by

(3.7.5)
$$\begin{aligned}\sum_{k=0}^{nx} u_k u_{n-k} &\approx \frac{1}{\pi}\int_0^{nx} \frac{1}{\sqrt{y(n-y)}}\,dy \\ &= \frac{1}{\pi}\int_0^{x} \frac{1}{\sqrt{w(1-w)}}\,dw \qquad \left(\text{by } w = \frac{y}{n}\right) \\ &= \frac{2}{\pi}\text{ arc sine } \sqrt{x}.\end{aligned}$$

Thus we see from the above that for n large the proportion of time that the symmetric random walk is positive in the first $2n$ time units has approximately the arc sine distribution given by (3.7.5). Thus, for instance, the probability that it is positive less than one-half of the time is $(2/\pi)$ arc sine $\sqrt{\frac{1}{2}} = \frac{1}{2}$.

One interesting consequence of the above is that it tells us that the proportion of time the symmetric random walk is positive is not converging to the constant value $\frac{1}{2}$ (for if it were then the limiting distribution rather than being arc sine would be the distribution of the constant random variable). Hence if we consider the regenerative process $\{X_n\}$, which keeps track of the sign of the symmetric random walk, it follows that the proportion of time that X_n equals 1 does not converge to some constant. On the other hand, it is clear by symmetry and the fact that $u_n \to 0$ and $n \to \infty$ that

$$P\{X_n = 1\} = P\{Z_n > 0\} \to \tfrac{1}{2} \qquad \text{as } n \to \infty.$$

The reason why the above (the fact that the limiting probability that a regenerative process is in some state does not equal the long-run proportion of time it spends in that state) is not a contradiction, is that the expected cycle time is infinite. That is, the mean time between visits of the symmetric random walk to state 0 is such that

$$E[T] = \infty.$$

The above must be true, for otherwise we would have a contradiction. It also follows directly from Lemma 3.7.3 since

$$\begin{aligned}E[T] &\geq \sum_{n=0}^{\infty} P\{T > 2n\} \\ &= \sum_{n=0}^{\infty} u_n \qquad \text{(from Lemma 3.7.3)}\end{aligned}$$

and as $u_n \sim (\sqrt{n\pi})^{-1}$ we see that

$$E[T] = \infty.$$

Remark It follows from Proposition 3.7.4 and Stirling's approximation that, for $0 < x < 1$ and n large,

$$
\begin{aligned}
P\{\text{no zeroes between } 2nx \text{ and } 2n\} &= 1 - \sum_{k=nx}^{n} u_k u_{n-k} \\
&= \sum_{k=0}^{nx-1} u_k u_{n-k} \\
&\approx \frac{2}{\pi} \text{arc sine } \sqrt{x},
\end{aligned}
$$

where the final approximation follows from (3.7.5).

3.8 Stationary Point Processes

A counting process $\{N(t), t \geq 0\}$ that possesses stationary increments is called a *stationary point process*. We note from Theorem 3.5.2 that the equilibrium renewal process is one example of a stationary point process.

THEOREM 3.8.1

For any stationary point process, excluding the trivial one with $P\{N(t) = 0\} = 1$ for all $t \geq 0$,

$$
\lim_{t \to 0} \frac{P\{N(t) > 0\}}{t} = \lambda > 0, \tag{3.8.1}
$$

where $\lambda = \infty$ is not excluded.

Proof Let $f(t) = P\{N(t) > 0\}$ and note that $f(t)$ is nonnegative and nondecreasing. Also

$$
\begin{aligned}
f(s + t) &= P\{N(s + t) - N(s) > 0 \quad \text{or} \quad N(s) > 0\} \\
&\leq P\{N(s + t) - N(s) > 0\} + P\{N(s) > 0\} \\
&= f(t) + f(s).
\end{aligned}
$$

Hence

$$
f(t) \leq 2f(t/2)
$$

and by induction

$$
f(t) \leq nf(t/n) \qquad \text{for all } n = 1, 2, \ldots.
$$

Thus, letting a be such that $f(a) > 0$, we have

$$\frac{f(a)}{a} \le \frac{f(a/n)}{a/n} \qquad \text{all } n = 1, 2, \ldots. \tag{3.8.2}$$

Now define $\lambda = \limsup_{t \to 0} f(t)/t$. By (3.8.2) we obtain

$$\lambda \ge \frac{f(a)}{a} > 0.$$

To show that $\lambda = \lim_{t\to 0} f(t)/t$, we consider two cases. First, suppose that $\lambda < \infty$. In this case, fix $\varepsilon > 0$, and let $s > 0$ be such that $f(s)/s > \lambda - \varepsilon$. Now, for any $t \in (0, s)$ there is an integer n such that

$$\frac{s}{n} \le t \le \frac{s}{n-1}.$$

From the monotonicity of $f(t)$ and from (3.8.2), we obtain that for all t in this interval

$$\frac{f(t)}{t} \ge \frac{f(s/n)}{s/(n-1)} = \frac{n-1}{n}\frac{f(s/n)}{s/n} \ge \frac{n-1}{n}\frac{f(s)}{s}. \tag{3.8.3}$$

Hence,

$$\frac{f(t)}{t} > \frac{n-1}{n}(\lambda - \varepsilon).$$

Since ε is arbitrary, and since $n \to \infty$ as $t \to 0$, it follows that $\lim_{t\to 0} f(t)/t = \lambda$.

Now assume $\lambda = \infty$. In this case, fix any large $A > 0$ and choose s such that $f(s)/s > A$. Then, from (3.8.3), it follows that for all $t \in (0, s)$

$$\frac{f(t)}{t} \ge \frac{n-1}{n}\frac{f(s)}{s} > \frac{n-1}{n}A,$$

which implies $\lim_{t\to 0} f(t)/t = \infty$, and the proof is complete.

EXAMPLE 3.8(A) For the equilibrium renewal process

$$P\{N(t) > 0\} = F_e(t)$$
$$= \int_0^t \bar{F}(y)\,dy/\mu.$$

Hence, using L'hospital's rule,

$$\lambda = \lim_{t\to 0} \frac{P\{N(t) > 0\}}{t} = \lim_{t\to 0} \frac{\bar{F}(t)}{\mu} = \frac{1}{\mu}.$$

Thus, for the equilibrium renewal process, λ is the rate of the renewal process.

For any stationary point process $\{N(t), t \geq 0\}$, we have that

$$E[N(t+s)] = E[N(t+s) - N(s)] + E[N(s)]$$
$$= E[N(t)] + E[N(s)]$$

implying that, for some constant c,

$$E[N(t)] = ct.$$

What is the relationship between c and λ? (In the case of the equilibrium renewal process, it follows from Example 3.8(A) and Theorem 3.5.2 that $\lambda = c = 1/\mu$.) In general we note that since

$$c = \sum_{n=1}^{\infty} \frac{nP\{N(t) = n\}}{t}$$
$$\geq \sum_{n=1}^{\infty} \frac{P\{N(t) = n\}}{t}$$
$$= \frac{P\{N(t) > 0\}}{t},$$

it follows that $c \geq \lambda$. In order to determine when $c = \lambda$, we need the following concept. A stationary point process is said to be *regular* or *orderly* if

$$P\{N(t) \geq 2\} = o(t). \tag{3.8.4}$$

It should be noted that, for a stationary point process, (3.8.4) implies that the probability that two or more events will occur simultaneously at any point is 0. To see this, divide the interval [0, 1] into n equal parts. The probability of a simultaneous occurrence of events is less than the probability of two or more events in any of the intervals

$$\left(\frac{j}{n}, \frac{j+1}{n}\right), \qquad j = 0, 1, \ldots, n-1,$$

and thus this probability is bounded by $nP\{N(1/n) \geq 2\}$, which by (3.8.4) goes to zero as $n \to \infty$. When $c < \infty$, the converse of this is also true. That is, any stationary point process for which c is finite and for which there is zero probability of simultaneous events, is necessarily regular. The proof in this direction is more complicated and will not be given.

We will end this section by proving that $c = \lambda$ for a regular stationary point process. This is known as *Korolyook's theorem*.

KOROLYOOK'S THEOREM

For a regular stationary point process, c, the mean number of events per unit time and the intensity λ, *defined by* (3.8.1), *are equal. The case* $\lambda = c = \infty$ *is not excluded.*

Proof Let us define the following notation:

A_k for the event $\{N(1) > k\}$;

B_{nj} for the event $\left\{N\left(\frac{j+1}{n}\right) - N\left(\frac{j}{n}\right) \geq 2\right\}$;

$$B_n = \bigcup_{j=0}^{n-1} B_{nj};$$

C_{nkj} for the event $\left\{N\left(\frac{j+1}{n}\right) - N\left(\frac{j}{n}\right) \geq 1, N(1) - N\left(\frac{j+1}{n}\right) = k\right\}$.

Let $\varepsilon > 0$ and a positive integer m be given. From the assumed regularity of the process, it follows that

$$P(B_{nj}) < \frac{\varepsilon}{n(m+1)}, \qquad j = 0, 1, \ldots, n-1$$

for all sufficiently large n. Hence,

$$P(B_n) \leq \sum_{j=0}^{n-1} P(B_{nj}) \leq \frac{\varepsilon}{m+1}.$$

Therefore,

$$\begin{aligned} P(A_k) &= P(A_k\overline{B}_n) + P(A_kB_n) \\ &\leq P(A_k\overline{B}_n) + \frac{\varepsilon}{m+1}, \end{aligned} \tag{3.8.5}$$

where $\overline{B}_n$ is the complement of B_n. However, a little thought reveals that

$$A_k\overline{B}_n = \bigcup_{j=0}^{n-1} C_{nkj}\overline{B}_n,$$

and hence

$$P(A_k\overline{B}_n) \leq \sum_{j=0}^{n-1} P(C_{nkj}),$$

which together with (3.8.5) implies that

$$(3.8.6) \quad \sum_{k=0}^{m} P(A_k) \le \sum_{j=0}^{n-1} \sum_{k=0}^{m} P(C_{nkj}) + \varepsilon$$

$$= \sum_{j=0}^{n-1} P\left\{N\left(\frac{j+1}{n}\right) - N\left(\frac{j}{n}\right) \ge 1, N(1) - N\left(\frac{j+1}{n}\right) \le m\right\} + \varepsilon$$

$$\le \sum_{j=0}^{n-1} P\left\{N\left(\frac{j+1}{n}\right) - N\left(\frac{j}{n}\right) \ge 1\right\} + \varepsilon$$

$$= nP\{N(1/n) \ge 1\} + \varepsilon$$

$$\le \lambda + 2\varepsilon$$

for all n sufficiently large. Now since (3.8.6) is true for all m, it follows that

$$\sum_{k=0}^{\infty} P(A_k) \le \lambda + 2\varepsilon.$$

Hence,

$$c = E[N(1)] = \sum_{k=0}^{\infty} P\{N(1) > k\} = \sum_{k=0}^{\infty} P(A_k) \le \lambda + 2\varepsilon$$

and the result is obtained as ε is arbitrary and it is already known that $c \ge \lambda$.

PROBLEMS

3.1. Is it true that:

(a) $N(t) < n$ if and only if $S_n > t$?

(b) $N(t) \le n$ if and only if $S_n \ge t$?

(c) $N(t) > n$ if and only if $S_n < t$?

3.2. In defining a renewal process we suppose that $F(\infty)$, the probability that an interarrival time is finite, equals 1. If $F(\infty) < 1$, then after each renewal there is a positive probability $1 - F(\infty)$ that there will be no further renewals. Argue that when $F(\infty) < 1$ the total number of renewals, call it $N(\infty)$, is such that $1 + N(\infty)$ has a geometric distribution with mean $1/(1 - F(\infty))$.

3.3. Express in words what the random variable $X_{N(t)+1}$ represents. (*Hint:* It is the length of which renewal interval?) Show that

$$P\{X_{N(t)+1} \ge x\} \ge \bar{F}(x).$$

Compute the above exactly when $F(x) = 1 - e^{-\lambda x}$.

3.4. Prove the renewal equation

$$m(t) = F(t) + \int_0^t m(t - x)\, dF(x).$$

3.5. Prove that the renewal function $m(t)$, $0 \le t < \infty$ uniquely determines the interarrival distribution F.

3.6. Let $\{N(t), t \ge 0\}$ be a renewal process and suppose that for all n and t, conditional on the event that $N(t) = n$, the event times $S_1, \ldots, S_n$ are distributed as the order statistics of a set of independent uniform $(0, t)$ random variables. Show that $\{N(t), t \ge 0\}$ is a Poisson process. (*Hint:* Consider $E[N(s) \mid N(t)]$ and then use the result of Problem 3.5.)

3.7. If F is the uniform (0, 1) distribution function show that

$$m(t) = e^t - 1, \qquad 0 \le t \le 1.$$

Now argue that the expected number of uniform (0, 1) random variables that need to be added until their sum exceeds 1 has mean e.

3.8. The random variables $X_1, \ldots, X_n$ are said to be exchangeable if $X_{i_1}, \ldots, X_{i_n}$ has the same joint distribution as $X_1, \ldots, X_n$ whenever $i_1, i_2, \ldots, i_n$ is a permutation of $1, 2, \ldots, n$. That is, they are exchangeable if the joint distribution function $P\{X_1 \le x_1, X_2 \le x_2, \ldots, X_n \le x_n\}$ is a symmetric function of $(x_1, x_2, \ldots, x_n)$. Let $X_1, X_2, \ldots$ denote the interarrival times of a renewal process.

(a) Argue that conditional on $N(t) = n$, $X_1, \ldots, X_n$ are exchangeable. Would $X_1, \ldots, X_n, X_{n+1}$ be exchangeable (conditional on $N(t) = n$)?

(b) Use (a) to prove that for $n > 0$

$$E\left[\frac{X_1 + \cdots + X_{N(t)}}{N(t)} \,\middle|\, N(t) = n\right] = E[X_1 \mid N(t) = n].$$

(c) Prove that

$$E\left[\frac{X_1 + \cdots + X_{N(t)}}{N(t)} \,\middle|\, N(t) > 0\right] = E[X_1 \mid X_1 < t].$$

3.9. Consider a single-server bank in which potential customers arrive at a Poisson rate λ. However, an arrival only enters the bank if the server is free when he or she arrives. Let G denote the service distribution.

(a) At what rate do customers enter the bank?

(b) What fraction of potential customers enter the bank?

(c) What fraction of time is the server busy?

3.10. Let $X_1, X_2, \ldots$ be independent and identically distributed with $E[X_i] < \infty$. Also let $N_1, N_2, \ldots$ be independent and identically distributed stopping times for the sequence $X_1, X_2, \ldots$ with $E[N_i] < \infty$. Observe the X_i sequentially, stopping at N_1. Now start sampling the remaining X_i—acting as if the sample was just beginning with X_{N_1+1}—stopping after an additional N_2. (Thus, for instance, $X_1 + \cdots + X_{N_1}$ has the same distribution as $X_{N_1+1} + \cdots + X_{N_1+N_2}$). Now start sampling the remaining X_i—again acting as if the sample was just beginning—and stop after an additional N_3, and so on.

(a) Let

$$S_1 = \sum_{i=1}^{N_1} X_i, \qquad S_2 = \sum_{i=N_1+1}^{N_1+N_2} X_i, \ldots, \qquad S_m = \sum_{i=N_1+\cdots+N_{m-1}+1}^{N_1+\cdots+N_m} X_i.$$

Use the strong law of large numbers to compute

$$\lim_{m\to\infty} \left(\frac{S_1 + \cdots + S_m}{N_1 + \cdots + N_m} \right).$$

(b) Writing

$$\frac{S_1 + \cdots + S_m}{N_1 + \cdots + N_m} = \frac{S_1 + \cdots + S_m}{m} \frac{m}{N_1 + \cdots + N_m},$$

derive another expression for the limit in part (a).

(c) Equate the two expressions to obtain Wald's equation.

3.11. Consider a miner trapped in a room that contains three doors. Door 1 leads her to freedom after two-days' travel; door 2 returns her to her room after four-days' journey; and door 3 returns her to her room after eight-days' journey. Suppose at all times she is equally to choose any of the three doors, and let T denote the time it takes the miner to become free.

(a) Define a sequence of independent and identically distributed random variables $X_1, X_2, \ldots$ and a stopping time N such that

$$T = \sum_{i=1}^{N} X_i.$$

Note: You may have to imagine that the miner continues to randomly choose doors even after she reaches safety.

(b) Use Wald's equation to find $E[T]$.

(c) Compute $E[\sum_{i=1}^{N} X_i | N = n]$ and note that it is not equal to $E[\sum_{i=1}^{n} X_i]$.

(d) Use part (c) for a second derivation of $E[T]$.

3.12. Show how Blackwell's theorem follows from the key renewal theorem.

3.13. A process is in one of n states, $1, 2, \ldots, n$. Initially it is in state 1, where it remains for an amount of time having distribution F_1. After leaving state 1 it goes to state 2, where it remains for a time having distribution F_2. When it leaves 2 it goes to state 3, and so on. From state n it returns to 1 and starts over. Find

$$\lim_{t\to\infty} P\{\text{process is in state } i \text{ at time } t\}.$$

Assume that H, the distribution of time between entrances to state 1, is nonlattice and has finite mean.

3.14. Let $A(t)$ and $Y(t)$ denote the age and excess at t of a renewal process. Fill in the missing terms:

(a) $A(t) > x \leftrightarrow 0$ events in the interval _____?

(b) $Y(t) > x \leftrightarrow 0$ events in the interval _____?

(c) $P\{Y(t) > x\} = P\{A(\quad) > \qquad\}$.

(d) Compute the joint distribution of $A(t)$ and $Y(t)$ for a Poisson process.

3.15. Let $A(t)$ and $Y(t)$ denote respectively the age and excess at t. Find:

(a) $P\{Y(t) > x | A(t) = s\}$.

(b) $P\{Y(t) > x | A(t + x/2) = s\}$.

(c) $P\{Y(t) > x | A(t + x) > s\}$ for a Poisson process.

(d) $P\{Y(t) > x, A(t) > y\}$.

(e) If $\mu < \infty$, show that, with probability 1, $A(t)/t \to 0$ as $t \to \infty$.

3.16. Consider a renewal process whose interarrival distribution is the gamma distribution with parameters (n, λ). Use Proposition 3.4.6 to show that

$$\lim_{t\to\infty} E[Y(t)] = \frac{n+1}{2\lambda}.$$

Now explain how this could have been obtained without any computations.

3.17. An equation of the form

$$g(t) = h(t) + \int_0^t g(t - x)\, dF(x)$$

is called a renewal-type equation. In convolution notation the above states that

$$g = h + g * F.$$

Either iterate the above or use Laplace transforms to show that a renewal-type equation has the solution

$$g(t) = h(t) + \int_0^t h(t - x)\, dm(x),$$

where $m(x) = \sum_{n=1}^{\infty} F_n(x)$. If h is directly Riemann integrable and F nonlattice with finite mean, one can then apply the key renewal theorem to obtain

$$\lim_{t\to\infty} g(t) = \frac{\int_0^\infty h(t)\, dt}{\int_0^\infty \overline{F}(t)\, dt}.$$

Renewal-type equations for $g(t)$ are obtained by conditioning on the time at which the process probabilistically starts over. Obtain a renewal-type equation for:

(a) $P(t)$, the probability an alternating renewal process is on at time t;

(b) $g(t) = E[A(t)]$, the expected age of a renewal process at t.

Apply the key renewal theorem to obtain the limiting values in (a) and (b).

3.18. In Problem 3.9 suppose that potential customers arrive in accordance with a renewal process having interarrival distribution F. Would the number of events by time t constitute a (possibly delayed) renewal process if an event corresponds to a customer:

(a) entering the bank?

(b) leaving the bank?

What if F were exponential?

3.19. Prove Equation (3.5.3).

3.20. Consider successive flips of a fair coin.

(a) Compute the mean number of flips until the pattern HHTHHTT appears.

(b) Which pattern requires a larger expected time to occur: HHTT or HTHT?

3.21. On each bet a gambler, independently of the past, either wins or loses 1 unit with respective probabilities p and $1 - p$. Suppose the gambler's strategy is to quit playing the first time she wins k consecutive bets. At the moment she quits

(a) find her expected winnings.

(b) find the expected number of bets that she has won.

3.22. Consider successive flips of a coin having probability p of landing heads. Find the expected number of flips until the following sequences appear:

(a) A = HHTTHH.

(b) B = HTHTT.

Suppose now that $p = 1/2$.

(c) Find $P\{A \text{ occurs before } B\}$.

(d) Find the expected number of flips until either A or B occurs.

3.23. A coin having probability p of landing heads is flipped k times. Additional flips of the coin are then made until the pattern of the first k is repeated (possibly by using some of the first k flips). Show that the expected number of additional flips after the initial k is 2^k.

3.24. Draw cards one at a time, with replacement, from a standard deck of playing cards. Find the expected number of draws until four successive cards of the same suit appear.

3.25. Consider a delayed renewal process $\{N_D(t), t \geq 0\}$ whose first interarrival has distribution G and the others have distribution F. Let $m_D(t) = E[N_D(t)]$.

(a) Prove that

$$m_D(t) = G(t) + \int_0^t m(t - x)\, dG(x),$$

where $m(t) = \sum_{n=1}^{\infty} F_n(t)$.

(b) Let $A_D(t)$ denote the age at time t. Show that if F is nonlattice with $\int x^2\, dF(x) < \infty$ and $t\overline{G}(t) \to 0$ as $t \to \infty$, then

$$E[A_D(t)] \to \frac{\int_0^\infty x^2\, dF(x)}{2\int_0^\infty x\, dF(x)}.$$

(c) Show that if G has a finite mean, then $t\overline{G}(t) \to 0$ as $t \to \infty$.

3.26. Prove Blackwell's theorem for renewal reward processes. That is, assuming that the cycle distribution is not lattice, show that, as $t \to \infty$,

$$E[\text{reward in } (t, t+a)] \to a \frac{E[\text{reward in cycle}]}{E[\text{time of cycle}]}.$$

Assume that any relevant function is directly Riemann integrable.

3.27. For a renewal reward process show that

$$\lim_{t\to\infty} E[R_{N(t)+1}] = \frac{E[R_1 X_1]}{E[X_1]}.$$

Assume the distribution of X_i is nonlattice and that any relevant function is directly Riemann integrable. When the cycle reward is defined to equal the cycle length, the above yields

$$\lim_{t\to\infty} E[X_{N(t)+1}] = \frac{E[X^2]}{E[X]},$$

which is always greater than $E[X]$ except when X is constant with probability 1. (Why?)

3.28. In Example 3.6(C) suppose that the renewal process of arrivals is a Poisson process with mean μ. Let N^* denote that value of N that minimizes the long-run average cost if a train leaves whenever there are N travelers waiting. Another type of policy is to have a train depart every T time units. Compute the long-run average cost under this policy and let T^* denote the value of T that minimizes it. Show that the policy that departs whenever N^* travelers are waiting leads to a smaller average cost than the one that departs every T^* time units.

3.29. The life of a car is a random variable with distribution F. An individual has a policy of trading in his car either when it fails or reaches the age of A. Let $R(A)$ denote the resale value of an A-year-old car. There is no resale value of a failed car. Let C_1 denote the cost of a new car and suppose that an additional cost C_2 is incurred whenever the car fails.

(a) Say that a cycle begins each time a new car is purchased. Compute the long-run average cost per unit time.

(b) Say that a cycle begins each time a car in use fails. Compute the long-run average cost per unit time.

Note: In both (a) and (b) you are expected to compute the ratio of the expected cost incurred in a cycle to the expected time of a cycle. The answer should, of course, be the same in both parts.

3.30. Suppose in Example 3.3(A) that a coin's probability of landing heads is a beta random variable with parameters n and m; that is, the probability density is

$$f(p) = Cp^{n-1}(1 - p)^{m-1}, \qquad 0 \le p \le 1.$$

Consider the policy that flips each newly chosen coin until m consecutive flips land tails, then discards that coin and does the same with a new one. For this policy show that, with probability 1, the long-run proportion of flips that land heads is 1.

3.31. A system consisting of four components is said to work whenever both at least one of components 1 and 2 work and at least one of components 3 and 4 work. Suppose that component i alternates between working and being failed in accordance with a nonlattice alternating renewal process with distributions F_i and G_i, $i = 1, 2, 3, 4$. If these alternating renewal processes are independent, find $\lim_{t\to\infty} P\{\text{system is working at time } t\}$.

3.32 Consider a single-server queueing system having Poisson arrivals at rate λ and service distribution G with mean μ_G. Suppose that $\lambda\mu_G < 1$.

(a) Find P_0, the proportion of time the system is empty.

(b) Say the system is busy whenever it is nonempty (and so the server is busy). Compute the expected length of a busy period.

(c) Use part (b) and Wald's equation to compute the expected number of customers served in a busy period.

3.33. For the queueing system of Section 3.6.1 define $V(t)$, the work in the system at time t, as the sum of the remaining service times of all customers in the system at t. Let

$$V = \lim_{t\to\infty} \int_0^t V(s)\, ds/t.$$

Also, let D_i denote the amount of time the ith customer spends waiting in queue and define

$$W_Q = \lim_{n\to\infty} (D_1 + \cdots + D_n)/n.$$

(a) Argue that V and W_Q exist and are constant with probability 1.

(b) Prove the identity

$$V = \lambda E[Y]W_Q + \lambda E[Y^2]/2,$$

where $1/\lambda$ is the mean interarrival time and Y has the distribution of a service time.

3.34. In a k server queueing model with renewal arrivals show by counterexample that the condition $E[Y] < kE[X]$, where Y is a service time and X an interarrival time, is not sufficient for a cycle time to be necessarily finite. (*Hint:* Give an example where the system is never empty after the initial arrival.)

3.35. Packages arrive at a mailing depot in accordance with a Poisson process having rate λ. Trucks, picking up all waiting packages, arrive in accordance to a renewal process with nonlattice interarrival distribution F. Let $X(t)$ denote the number of packages waiting to be picked up at time t.

(a) What type of stochastic process is $\{X(t), t \geq 0\}$?

(b) Find an expression for

$$\lim_{t\to\infty} P\{X(t) = i\}, \qquad i \geq 0.$$

3.36. Consider a regenerative process satisfying the conditions of Theorem 3.7.1. Suppose that a reward at rate $r(j)$ is earned whenever the process is in state j. If the expected reward during a cycle is finite, show that the long-run average reward per unit time is, with probability 1, given by

$$\lim_{t\to\infty} \int_0^t \frac{r(X(s))\,ds}{t} = \sum_j P_j r(j),$$

where P_j is the limiting probability that $X(t)$ equals j.

References

References 1, 6, 7, and 11 present renewal theory at roughly the same mathematical level as the present text. A simpler, more intuitive approach is given in Reference 8. Reference 10 has many interesting applications.

Reference 9 provides an illuminating review paper in renewal theory. For a proof of the key renewal theorem under the most stringent conditions, the reader should see Volume 2 of Feller (Reference 4). Theorem 3.6.2 is known in the queueing literature as "Little's Formula." Our approach to the arc sine law is similar to that given in Volume 1 of Feller (Reference 4); the interested reader should also see Reference 3. Examples 3.3(A) and 3.5(C) are from Reference 5.

1. D. R. Cox, *Renewal Theory*, Methuen, London, 1962.
2. H. Cramer and M. Leadbetter, *Stationary and Related Stochastic Processes*, Wiley, New York, 1966.

3. B. DeFinetti, *Theory of Probability*, Vol. 1, Wiley, New York, 1970.
4. W. Feller, *An Introduction to Probability Theory and Its Applications,* Vols. I and II, Wiley, New York, 1957 and 1966.
5. S. Herschkorn, E. Pekoz, and S. M. Ross, "Policies Without Memory for the Infinite-Armed Bernoulli Bandit Under the Average Reward Criterion," *Probability in the Engineering and Informational Sciences,* Vol. 10, 1, 1996.
6. D. Heyman and M. Sobel, *Stochastic Models in Operations Research, Volume I*, McGraw-Hill, New York, 1982.
7. S. Karlin and H. Taylor, *A First Course in Stochastic Processes*, 2nd ed., Academic Press, Orlando, FL, 1975.
8. S. M. Ross, *Introduction to Probability Models*, 5th ed., Academic Press, Orlando, FL, 1993.
9. W. Smith, "Renewal Theory and its Ramifications," *Journal of the Royal Statistical Society,* Series B, 20, (1958), pp. 243–302.
10. H. C. Tijms, *Stochastic Models, An Algorithmic Approach*, Wiley, New York, 1994.
11. R. Wolff, *Stochastic Modeling and the Theory of Queues*, Prentice-Hall, NJ, 1989.

CHAPTER 4

Markov Chains

4.1 Introduction and Examples

Consider a stochastic process $\{X_n, n = 0, 1, 2, \ldots\}$ that takes on a finite or countable number of possible values. Unless otherwise mentioned, this set of possible values of the process will be denoted by the set of nonnegative integers $\{0, 1, 2, \ldots\}$. If $X_n = i$, then the process is said to be in state i at time n. We suppose that whenever the process is in state i, there is a fixed probability P_{ij} that it will next be in state j. That is, we suppose that

$$(4.1.1) \qquad P\{X_{n+1} = j \mid X_n = i, X_{n-1} = i_{n-1}, \ldots, X_1 = i_1, X_0 = i_0\} = P_{ij}$$

for all states $i_0, i_1, \ldots, i_{n-1}, i, j$ and all $n \geq 0$. Such a stochastic process is known as a *Markov chain*. Equation (4.1.1) may be interpreted as stating that, for a Markov chain, the conditional distribution of any future state X_{n+1}, given the past states $X_0, X_1, \ldots, X_{n-1}$ and the present state X_n, is independent of the past states and depends only on the present state. This is called the *Markovian* property. The value P_{ij} represents the probability that the process will, when in state i, next make a transition into state j. Since probabilities are nonnegative and since the process must make a transition into some state, we have that

$$P_{ij} \geq 0, \qquad i, j \geq 0; \qquad \sum_{j=0}^{\infty} P_{ij} = 1, \qquad i = 0, 1, \ldots.$$

Let P denote the matrix of one-step transition probabilities P_{ij}, so that

$$P = \begin{Vmatrix} P_{00} & P_{01} & P_{02} & \cdots \\ P_{10} & P_{11} & P_{12} & \cdots \\ \vdots & & & \\ P_{i0} & P_{i1} & P_{i2} & \cdots \\ \vdots & \vdots & & \vdots \end{Vmatrix}.$$

Example 4.1(a) ***The M/G/1 Queue.*** Suppose that customers arrive at a service center in accordance with a Poisson process with rate λ. There is a single server and those arrivals finding the server free go immediately into service; all others wait in line until their service turn. The service times of successive customers are assumed to be independent random variables having a common distribution G; and they are also assumed to be independent of the arrival process.

The above system is called the $M/G/1$ queueing system. The letter M stands for the fact that the interarrival distribution of customers is exponential, G for the service distribution; the number 1 indicates that there is a single server.

If we let $X(t)$ denote the number of customers in the system at t, then $\{X(t), t \geq 0\}$ would not possess the Markovian property that the conditional distribution of the future depends only on the present and not on the past. For if we knew the number in the system at time t, then, to predict future behavior, whereas we would not care how much time had elapsed since the last arrival (since the arrival process is memoryless), we would care how long the person in service had already been there (since the service distribution G is arbitrary and therefore not memoryless).

As a means of getting around the above dilemma let us only look at the system at moments when customers depart. That is, let X_n denote the number of customers left behind by the nth departure, $n \geq 1$. Also, let Y_n denote the number of customers arriving during the service period of the $(n + 1)$st customer.

When $X_n > 0$, the nth departure leaves behind X_n customers—of which one enters service and the other $X_n - 1$ wait in line. Hence, at the next departure the system will contain the $X_n - 1$ customers that were in line in addition to any arrivals during the service time of the $(n + 1)$st customer. Since a similar argument holds when $X_n = 0$, we see that

$$X_{n+1} = \begin{cases} X_n - 1 + Y_n & \text{if } X_n > 0 \\ Y_n & \text{if } X_n = 0. \end{cases} \tag{4.1.2}$$

Since $Y_n, n \geq 1$, represent the number of arrivals in nonoverlapping service intervals, it follows, the arrival process being a Poisson process, that they are independent and

$$P\{Y_n = j\} = \int_0^\infty e^{-\lambda x} \frac{(\lambda x)^j}{j!} dG(x), \qquad j = 0, 1, \ldots. \tag{4.1.3}$$

From (4.1.2) and (4.1.3) it follows that $\{X_n, n = 1, 2, \ldots\}$ is a Markov chain with transition probabilities given by

$$P_{0j} = \int_0^\infty e^{-\lambda x} \frac{(\lambda x)^j}{j!} dG(x), \qquad j \geq 0,$$

$$P_{ij} = \int_0^\infty e^{-\lambda x} \frac{(\lambda x)^{j-i+1}}{(j-i+1)!} dG(x), \qquad j \geq i - 1,\ i \geq 1,$$

$$P_{ij} = 0 \qquad \text{otherwise.}$$

EXAMPLE 4.1(B) ***The G/M/1 Queue.*** Suppose that customers arrive at a single-server service center in accordance with an arbitrary renewal process having interarrival distribution G. Suppose further that the service distribution is exponential with rate μ.

If we let X_n denote the number of customers in the system as seen by the nth arrival, it is easy to see that the process $\{X_n, n \geq 1\}$ is a Markov chain. To compute the transition probabilities P_{ij} for this Markov chain, let us first note that, as long as there are customers to be served, the number of services in any length of time t is a Poisson random variable with mean μt. This is true since the time between successive services is exponential and, as we know, this implies that the number of services thus constitutes a Poisson process. Therefore,

$$P_{i,i+1-j} = \int_0^\infty e^{-\mu t} \frac{(\mu t)^j}{j!} \, dG(t), \qquad j = 0, 1, \ldots, i,$$

which follows since if an arrival finds i in the system, then the next arrival will find $i + 1$ minus the number served, and the probability that j will be served is easily seen (by conditioning on the time between the successive arrivals) to equal the right-hand side of the above.

The formula for P_{i0} is little different (it is the probability that *at least* $i + 1$ Poisson events occur in a random length of time having distribution G) and thus is given by

$$P_{i0} = \int_0^\infty \sum_{k=i+1}^\infty e^{-\mu t} \frac{(\mu t)^k}{k!} \, dG(t), \qquad i \geq 0.$$

Remark The reader should note that in the previous two examples we were able to discover an *embedded* Markov chain by looking at the process only at certain time points, and by choosing these time points so as to exploit the lack of memory of the exponential distribution. This is often a fruitful approach for processes in which the exponential distribution is present.

EXAMPLE 4.1(C) ***Sums of Independent, Identically Distributed Random Variables. The General Random Walk.*** Let X_i, $i \geq 1$, be independent and identically distributed with

$$P\{X_i = j\} = a_j, \qquad j = 0, \pm 1, \ldots.$$

If we let

$$S_0 = 0 \quad \text{and} \quad S_n = \sum_{i=1}^n X_i,$$

then $\{S_n, n \geq 0\}$ is a Markov chain for which

$$P_{ij} = a_{j-i}.$$

$\{S_n, n \geq 0\}$ is called the general random walk and will be studied in Chapter 7.

Example 4.1(d) ***The Absolute Value of the Simple Random Walk.*** The random walk $\{S_n, n \geq 1\}$, where $S_n = \sum_1^n X_i$, is said to be a *simple random walk* if for some p, $0 < p < 1$,

$$P\{X_i = 1\} = p,$$
$$P\{X_i = -1\} = q \equiv 1 - p.$$

Thus in the simple random walk the process always either goes up one step (with probability p) or down one step (with probability q).

Now consider $|S_n|$, the absolute value of the simple random walk. The process $\{|S_n|, n \geq 1\}$ measures at each time unit the absolute distance of the simple random walk from the origin. Somewhat surprisingly $\{|S_n|\}$ is itself a Markov chain. To prove this we will first show that if $|S_n| = i$, then no matter what its previous values the probability that S_n equals i (as opposed to $-i$) is $p^i/(p^i + q^i)$.

PROPOSITION 4.1.1

If $\{S_n, n \geq 1\}$ is a simple random walk, then

$$P\{S_n = i \,|\, |S_n| = i, |S_{n-1}| = i_{n-1}, \ldots, |S_1| = i_1\} = \frac{p^i}{p^i + q^i}.$$

Proof If we let $i_0 = 0$ and define

$$j = \max\{k: 0 \leq k \leq n: i_k = 0\},$$

then, since we know the actual value of S_j, it is clear that

$$P\{S_n = i \,|\, |S_n| = i, |S_{n-1}| = i_{n-1}, \ldots, |S_1| = i_1\}$$
$$= P\{S_n = i \,|\, |S_n| = i, \ldots, |S_{j+1}| = i_{j+1}, |S_j| = 0\}.$$

Now there are two possible values of the sequence $S_{j+1}, \ldots, S_n$ for which $|S_{j+1}| = i_{j+1}, \ldots, |S_n| = i$. The first of which results in $S_n = i$ and has probability

$$p^{\frac{n-j}{2} + \frac{i}{2}} q^{\frac{n-j}{2} - \frac{i}{2}},$$

and the second results in $S_n = -i$ and has probability

$$p^{\frac{n-j}{2} - \frac{i}{2}} q^{\frac{n-j}{2} + \frac{i}{2}}.$$

Hence,

$$P\{S_n = i \mid |S_n| = i, \ldots, |S_1| = i_1\} = \frac{p^{\frac{n-j}{2}+\frac{i}{2}} q^{\frac{n-j}{2}-\frac{i}{2}}}{p^{\frac{n-j}{2}+\frac{i}{2}} q^{\frac{n-j}{2}-\frac{i}{2}} + p^{\frac{n-j}{2}-\frac{i}{2}} q^{\frac{n-j}{2}+\frac{i}{2}}}$$

$$= \frac{p^i}{p^i + q^i}$$

and the proposition is proven.

From Proposition 4.1.1 it follows upon conditioning on whether $S_n = +i$ or $-i$ that

$$P\{|S_{n+1}| = i + 1 \mid |S_n| = i, |S_{n-1}|, \ldots, |S_1|\}$$

$$= P\{S_{n+1} = i + 1 \mid S_n = i\} \frac{p^i}{p^i + q^i}$$

$$+ P\{S_{n+1} = -(i+1) \mid S_n = -i\} \frac{q^i}{p^i + q^i} = \frac{p^{i+1} + q^{i+1}}{p^i + q^i}.$$

Hence, $\{|S_n|, n \geq 1\}$ is a Markov chain with transition probabilities

$$P_{i,i+1} = \frac{p^{i+1} + q^{i+1}}{p^i + q^i} = 1 - P_{i,i-1}, \qquad i > 0,$$

$$P_{01} = 1.$$

4.2 Chapman–Kolmogorov Equations and Classification of States

We have already defined the one-step transition probabilities P_{ij}. We now define the n-step transition probabilities P_{ij}^n to be the probability that a process in state i will be in state j after n additional transitions. That is,

$$P_{ij}^n = P\{X_{n+m} = j \mid X_m = i\}, \qquad n \geq 0, \quad i, j \geq 0.$$

Of course $P_{ij}^1 = P_{ij}$. The *Chapman–Kolmogorov equations* provide a method for computing these n-step transition probabilities. These equations are

$$P_{ij}^{n+m} = \sum_{k=0}^{\infty} P_{ik}^n P_{kj}^m \qquad \text{for all } n, m \geq 0, \quad \text{all } i, j, \tag{4.2.1}$$

and are established by observing that

$$\begin{aligned} P_{ij}^{n+m} &= P\{X_{n+m} = j \mid X_0 = i\} \\ &= \sum_{k=0}^{\infty} P\{X_{n+m} = j, X_n = k \mid X_0 = i\} \\ &= \sum_{k=0}^{\infty} P\{X_{n+m} = j \mid X_n = k, X_0 = i\} P\{X_n = k \mid X_0 = i\} \\ &= \sum_{k=0}^{\infty} P_{kj}^m P_{ik}^n. \end{aligned}$$

If we let $P^{(n)}$ denote the matrix of n-step transition probabilities P_{ij}^n, then Equation (4.2.1) asserts that

$$P^{(n+m)} = P^{(n)} \cdot P^{(m)},$$

where the dot represents matrix multiplication. Hence,

$$P^{(n)} = P \cdot P^{(n-1)} = P \cdot P \cdot P^{(n-2)} = \cdots = P^n,$$

and thus $P^{(n)}$ may be calculated by multiplying the matrix P by itself n times.

State j is said to be accessible from state i if for some $n \geq 0$, $P_{ij}^n > 0$. Two states i and j accessible to each other are said to *communicate*, and we write $i \leftrightarrow j$.

PROPOSITION 4.2.1

Communication is an equivalence relation. That is:

(i) $i \leftrightarrow i$;
(ii) if $i \leftrightarrow j$, then $j \leftrightarrow i$;
(iii) if $i \leftrightarrow j$ and $j \leftrightarrow k$, then $i \leftrightarrow k$.

Proof The first two parts follow trivially from the definition of communication. To prove (iii), suppose that $i \leftrightarrow j$ and $j \leftrightarrow k$; then there exists m, n such that $P_{ij}^m > 0$, $P_{jk}^n > 0$. Hence,

$$P_{ik}^{m+n} = \sum_{r=0}^{\infty} P_{ir}^m P_{rk}^n \geq P_{ij}^m P_{jk}^n > 0.$$

Similarly, we may show there exists an s for which $P_{ki}^s > 0$.

Two states that communicate are said to be in the same *class*; and by Proposition 4.2.1, any two classes are either disjoint or identical. We say that the Markov chain is *irreducible* if there is only one class—that is, if all states communicate with each other.

State i is said to have period d if $P_{ii}^n = 0$ whenever n is not divisible by d and d is the greatest integer with this property. (If $P_{ii}^n = 0$ for all $n > 0$, then define the period of i to be infinite.) A state with period 1 is said to be *aperiodic*. Let $d(i)$ denote the period of i. We now show that periodicity is a class property.

PROPOSITION 4.2.2

If $i \leftrightarrow j$, then $d(i) = d(j)$.

Proof Let m and n be such that $P_{ij}^m P_{ji}^n > 0$, and suppose that $P_{ii}^s > 0$. Then

$$P_{jj}^{n+m} \geq P_{ji}^n P_{ij}^m > 0$$

$$P_{jj}^{n+s+m} \geq P_{ji}^n P_{ii}^s P_{ij}^m > 0,$$

where the second inequality follows, for instance, since the left-hand side represents the probability that starting in j the chain will be back in j after $n + s + m$ transitions, whereas the right-hand side is the probability of the same event subject to the further restriction that the chain is in i both after n and $n + s$ transitions. Hence, $d(j)$ divides both $n + m$ and $n + s + m$; thus $n + s + m - (n + m) = s$, whenever $P_{ii}^s > 0$. Therefore, $d(j)$ divides $d(i)$. A similar argument yields that $d(i)$ divides $d(j)$, thus $d(i) = d(j)$.

For any states i and j define f_{ij}^n to be the probability that, starting in i, the first transition into j occurs at time n. Formally,

$$f_{ij}^0 = 0,$$

$$f_{ij}^n = P\{X_n = j, X_k \neq j, k = 1, \ldots, n-1 \mid X_0 = i\}.$$

Let

$$f_{ij} = \sum_{n=1}^{\infty} f_{ij}^n.$$

Then f_{ij} denotes the probability of ever making a transition into state j, given that the process starts in i. (Note that for $i \neq j$, f_{ij} is positive if, and only if, j is accessible from i.) State j is said to be *recurrent* if $f_{jj} = 1$, and *transient* otherwise.

PROPOSITION 4.2.3

State j is recurrent if, and only if,

$$\sum_{n=1}^{\infty} P_{jj}^n = \infty.$$

Proof State j is recurrent if, with probability 1, a process starting at j will eventually return. However, by the Markovian property it follows that the process probabilistically restarts itself upon returning to j. Hence, with probability 1, it will return again to j. Repeating this argument, we see that, with probability 1, the number of visits to j will be infinite and will thus have infinite expectation. On the other hand, suppose j is transient. Then each time the process returns to j there is a positive probability $1 - f_{jj}$ that it will never again return; hence the number of visits is geometric with finite mean $1/(1 - f_{jj})$.

By the above argument we see that state j is recurrent if, and only if,

$$E[\text{number of visits to } j | X_0 = j] = \infty.$$

But, letting

$$I_n = \begin{cases} 1 & \text{if } X_n = j \\ 0 & \text{otherwise,} \end{cases}$$

it follows that $\sum_0^\infty I_n$ denotes the number of visits to j. Since

$$E\left[\sum_{n=0}^{\infty} I_n | X_0 = j\right] = \sum_{n=0}^{\infty} E\left[I_n | X_0 = j\right] = \sum_{n=0}^{\infty} P_{jj}^n,$$

the result follows.

The argument leading to the above proposition is doubly important for it also shows that a transient state will only be visited a finite number of times (hence the name transient). This leads to the conclusion that in a finite-state Markov chain not all states can be transient. To see this, suppose the states are $0, 1, \ldots, M$ and suppose that they are all transient. Then after a finite amount of time (say after time T_0) state 0 will never be visited, and after a time (say T_1) state 1 will never be visited, and after a time (say T_2) state 2 will never be visited, and so on. Thus, after a finite time $T = \max\{T_0, T_1, \ldots, T_M\}$ no states will be visited. But as the process must be in some state after time T, we arrive at a contradiction, which shows that at least one of the states must be recurrent.

We will use Proposition 4.2.3 to prove that recurrence, like periodicity, is a class property.

Corollary 4.2.4

If i is recurrent and $i \leftrightarrow j$, then j is recurrent.

Proof Let m and n be such that $P^n_{ij} > 0, P^m_{ji} > 0$. Now for any $s \geq 0$

$$P^{m+n+s}_{jj} \geq P^m_{ji} P^s_{ii} P^n_{ij}$$

and thus

$$\sum_s P^{m+n+s}_{jj} \geq P^m_{ji} P^n_{ij} \sum_s P^s_{ii} = \infty,$$

and the result follows from Proposition 4.2.3.

Example 4.2(A) ***The Simple Random Walk.*** The Markov chain whose state space is the set of all integers and has transition probabilities

$$P_{i,i+1} = p = 1 - P_{i,i-1}, \qquad i = 0, \pm 1, \ldots,$$

where $0 < p < 1$, is called the simple random walk. One interpretation of this process is that it represents the wanderings of a drunken man as he walks along a straight line. Another is that it represents the winnings of a gambler who on each play of the game either wins or loses one dollar.

Since all states clearly communicate it follows from Corollary 4.2.4 that they are either all transient or all recurrent. So let us consider state 0 and attempt to determine if $\sum_{n=1}^{\infty} P^n_{00}$ is finite or infinite.

Since it is impossible to be even (using the gambling model interpretation) after an odd number of plays, we must, of course, have that

$$P^{2n+1}_{00} = 0, \qquad n = 1, 2, \ldots.$$

On the other hand, the gambler would be even after $2n$ trials if, and only if, he won n of these and lost n of these. As each play of the game results in a win with probability p and a loss with probability $1 - p$, the desired probability is thus the binomial probability

$$P^{2n}_{00} = \binom{2n}{n} p^n (1-p)^n = \frac{(2n)!}{n!n!} (p(1-p))^n, \qquad n = 1, 2, 3, \ldots.$$

By using an approximation, due to Stirling, which asserts that

$$n! \sim n^{n+1/2} e^{-n} \sqrt{2\pi},$$

where we say that $a_n \sim b_n$ when $\lim_{n\to\infty}(a_n/b_n) = 1$, we obtain

$$P_{00}^{2n} \sim \frac{(4p(1-p))^n}{\sqrt{\pi n}}.$$

Now it is easy to verify that if $a_n \sim b_n$, then $\sum_n a_n < \infty$, if, and only if, $\sum_n b_n < \infty$. Hence $\sum_{n=1}^{\infty} P_{00}^n$ will converge if, and only if,

$$\sum_{n=1}^{\infty} \frac{(4p(1-p))^n}{\sqrt{\pi n}}$$

does. However, $4p(1-p) \leq 1$ with equality holding if, and only if, $p = \frac{1}{2}$. Hence, $\sum_{n=1}^{\infty} P_{00}^n = \infty$ if, and only if, $p = \frac{1}{2}$. Thus, the chain is recurrent when $p = \frac{1}{2}$ and transient if $p \neq \frac{1}{2}$.

When $p = \frac{1}{2}$, the above process is called a *symmetric random walk*. We could also look at symmetric random walks in more than one dimension. For instance, in the two-dimensional symmetric random walk the process would, at each transition, either take one step to the left, right, up, or down, each having probability $\frac{1}{4}$. Similarly, in three dimensions the process would, with probability $\frac{1}{6}$, make a transition to any of the six adjacent points. By using the same method as in the one-dimensional random walk it can be shown that the two-dimensional symmetric random walk is recurrent, but all higher-dimensional random walks are transient.

Corollary 4.2.5

If $i \leftrightarrow j$ and j is recurrent, then $f_{ij} = 1$.

Proof Suppose $X_0 = i$, and let n be such that $P_{ij}^n > 0$. Say that we miss opportunity 1 if $X_n \neq j$. If we miss opportunity 1, then let T_1 denote the next time we enter i (T_1 is finite with probability 1 by Corollary 4.2.4). Say that we miss opportunity 2 if $X_{T_1+n} \neq j$. If opportunity 2 is missed, let T_2 denote the next time we enter i and say that we miss opportunity 3 if $X_{T_2+n} \neq j$, and so on. It is easy to see that the opportunity number of the first success is a geometric random variable with mean $1/P_{ij}^n$, and is thus finite with probability 1. The result follows since i being recurrent implies that the number of potential opportunities is infinite.

Let $N_j(t)$ denote the number of transitions into j by time t. If j is recurrent and $X_0 = j$, then as the process probabilistically starts over upon transitions into j, it follows that $\{N_j(t), t \geq 0\}$ is a renewal process with interarrival distribution $\{f_{jj}^n, n \geq 1\}$. If $X_0 = i$, $i \leftrightarrow j$, and j is recurrent, then $\{N_j(t), t \geq 0\}$ is a delayed renewal process with initial interarrival distribution $\{f_{ij}^n, n \geq 1\}$.

4.3 Limit Theorems

It is easy to show that if state j is transient, then

$$\sum_{n=1}^{\infty} P_{ij}^n < \infty \qquad \text{for all } i,$$

meaning that, starting in i, the expected number of transitions into state j is finite. As a consequence it follows that for j transient $P_{ij}^n \to 0$ as $n \to \infty$.

Let μ_{jj} denote the expected number of transitions needed to return to state j. That is,

$$\mu_{jj} = \begin{cases} \infty & \text{if } j \text{ is transient} \\ \sum_{n=1}^{\infty} n f_{jj}^n & \text{if } j \text{ is recurrent.} \end{cases}$$

By interpreting transitions into state j as being renewals, we obtain the following theorem from Propositions 3.3.1, 3.3.4, and 3.4.1 of Chapter 3.

THEOREM 4.3.1

If i and j communicate, then:

(i) $P\left\{\lim_{t\to\infty} N_j(t)/t = 1/\mu_{jj} \mid X_0 = i\right\} = 1.$

(ii) $\lim_{n\to\infty} \sum_{k=1}^{n} P_{ij}^k/n = 1/\mu_{jj}.$

(iii) *If j is aperiodic, then* $\lim_{n\to\infty} P_{ij}^n = 1/\mu_{jj}$.

(iv) *If j has period d, then* $\lim_{n\to\infty} P_{jj}^{nd} = d/\mu_{jj}$.

If state j is recurrent, then we say that it is *positive* recurrent if $\mu_{jj} < \infty$ and *null* recurrent if $\mu_{jj} = \infty$. If we let

$$\pi_j = \lim_{n\to\infty} P_{jj}^{nd(j)},$$

it follows that a recurrent state j is positive recurrent if $\pi_j > 0$ and null recurrent if $\pi_j = 0$. The proof of the following proposition is left as an exercise.

PROPOSITION 4.3.2

Positive (null) recurrence is a class property.

A positive recurrent, aperiodic state is called *ergodic*. Before presenting a theorem that shows how to obtain the limiting probabilities in the ergodic case, we need the following definition.

Definition

A probability distribution $\{P_j, j \geq 0\}$ is said to be *stationary* for the Markov chain if

$$P_j = \sum_{i=0}^{\infty} P_i P_{ij}, \qquad j \geq 0.$$

If the probability distribution of X_0—say $P_j = P\{X_0 = j\}$, $j \geq 0$—is a stationary distribution, then

$$P\{X_1 = j\} = \sum_{i=0}^{\infty} P\{X_1 = j \mid X_0 = i\} P\{X_0 = i\}$$

$$= \sum_{i=0}^{\infty} P_i P_{ij} = P_j$$

and, by induction,

$$P\{X_n = j\} = \sum_{i=0}^{\infty} P\{X_n = j \mid X_{n-1} = i\} P\{X_{n-1} = i\} \tag{4.3.1}$$

$$= \sum_{i=0}^{\infty} P_{ij} P_i = P_j.$$

Hence, if the initial probability distribution is the stationary distribution, then X_n will have the same distribution for all n. In fact, as $\{X_n, n \geq 0\}$ is a Markov chain, it easily follows from this that for each $m \geq 0$, $X_n, X_{n+1}, \ldots, X_{n+m}$ will have the same joint distribution for each n; in other words, $\{X_n, n \geq 0\}$ will be a stationary process.

THEOREM 4.3.3

An irreducible aperiodic Markov chain belongs to one of the following two classes:

(i) *Either the states are all transient or all null recurrent; in this case, $P_{ij}^n \to 0$ as $n \to \infty$ for all i, j and there exists no stationary distribution.*

(ii) *Or else, all states are positive recurrent, that is,*

$$\pi_j = \lim_{n\to\infty} P_{ij}^n > 0.$$

In this case, $\{\pi_j, j = 0, 1, 2, \ldots\}$ is a stationary distribution and there exists no other stationary distribution.

Proof We will first prove (ii). To begin, note that

$$\sum_{j=0}^{M} P_{ij}^n \le \sum_{j=0}^{\infty} P_{ij}^n = 1 \qquad \text{for all } M.$$

Letting $n \to \infty$ yields

$$\sum_{j=0}^{M} \pi_j \le 1 \qquad \text{for all } M,$$

implying that

$$\sum_{j=0}^{\infty} \pi_j \le 1.$$

Now

$$P_{ij}^{n+1} = \sum_{k=0}^{\infty} P_{ik}^n P_{kj} \ge \sum_{k=0}^{M} P_{ik}^n P_{kj} \qquad \text{for all } M.$$

Letting $n \to \infty$ yields

$$\pi_j \ge \sum_{k=0}^{M} \pi_k P_{kj} \qquad \text{for all } M,$$

implying that

$$\pi_j \ge \sum_{k=0}^{\infty} \pi_k P_{kj}, \qquad j \ge 0.$$

To show that the above is actually an equality, suppose that the inequality is strict for some j. Then upon adding these inequalities we obtain

$$\sum_{j=0}^{\infty} \pi_j > \sum_{j=0}^{\infty} \sum_{k=0}^{\infty} \pi_k P_{kj} = \sum_{k=0}^{\infty} \pi_k \sum_{j=0}^{\infty} P_{kj} = \sum_{k=0}^{\infty} \pi_k,$$

which is a contradiction. Therefore,

$$\pi_j = \sum_{k=0}^{\infty} \pi_k P_{kj}, \qquad j = 0, 1, 2, \ldots.$$

Putting $P_j = \pi_j / \sum_0^{\infty} \pi_k$, we see that $\{P_j, j = 0, 1, 2, \ldots\}$ is a stationary distribution, and hence at least one stationary distribution exists. Now let $\{P_j, j = 0, 1, 2, \ldots\}$ be any stationary distribution. Then if $\{P_j, j = 0, 1, 2, \ldots\}$ is the probability distribution of X_0, then by (4.3.1)

$$\begin{aligned} P_j &= P\{X_n = j\} \\ &= \sum_{i=0}^{\infty} P\{X_n = j \mid X_0 = i\} P\{X_0 = i\} \qquad (4.3.2) \\ &= \sum_{i=0}^{\infty} P_{ij}^n P_i. \end{aligned}$$

From (4.3.2) we see that

$$P_j \geq \sum_{i=0}^{M} P_{ij}^n P_i \qquad \text{for all } M.$$

Letting n and then M approach ∞ yields

$$P_j \geq \sum_{i=0}^{\infty} \pi_j P_i = \pi_j. \qquad (4.3.3)$$

To go the other way and show that $P_j \leq \pi_j$, use (4.3.2) and the fact that $P_{ij}^n \leq 1$ to obtain

$$P_j \leq \sum_{i=0}^{M} P_{ij}^n P_i + \sum_{i=M+1}^{\infty} P_i \qquad \text{for all } M,$$

and letting $n \to \infty$ gives

$$P_j \leq \sum_{i=0}^{M} \pi_j P_i + \sum_{i=M+1}^{\infty} P_i \qquad \text{for all } M.$$

Since $\sum_0^\infty P_i = 1$, we obtain upon letting $M \to \infty$ that

$$(4.3.4) \qquad P_j \le \sum_{i=0}^{\infty} \pi_j P_i = \pi_j.$$

If the states are transient or null recurrent and $\{P_j, j = 0, 1, 2, \ldots\}$ is a stationary distribution, then Equations (4.3.2) hold and $P_{ij}^n \to 0$, which is clearly impossible. Thus, for case (i), no stationary distribution exists and the proof is complete.

Remarks

(1) When the situation is as described in part (ii) of Theorem 4.3.3 we say that the Markov chain is *ergodic*.

(2) It is quite intuitive that if the process is started with the limiting probabilities, then the resultant Markov chain is stationary. For in this case the Markov chain at time 0 is equivalent to an independent Markov chain with the same P matrix at time ∞. Hence the original chain at time t is equivalent to the second one at time $\infty + t = \infty$, and is therefore stationary.

(3) In the irreducible, positive recurrent, *periodic* case we still have that the $\pi_j, j \ge 0$, are the unique nonnegative solution of

$$\pi_j = \sum_i \pi_i P_{ij},$$

$$\sum_j \pi_j = 1.$$

But now π_j must be interpreted as the long-run proportion of time that the Markov chain is in state j (see Problem 4.17). Thus, $\pi_j = 1/\mu_{jj}$, whereas the limiting probability of going from j to j in $nd(j)$ steps is, by (iv) of Theorem 4.3.1, given by

$$\lim_{n\to\infty} P_{jj}^{nd} = \frac{d}{\mu_{jj}} = d\pi_j,$$

where d is the period of the Markov chain.

EXAMPLE 4.3(A) ***Limiting Probabilities for the Embedded M/G/1 Queue.*** Consider the embedded Markov chain of the $M/G/1$ system as in Example 4.1(A) and let

$$a_j = \int_0^\infty e^{-\lambda x} \frac{(\lambda x)^j}{j!}\, dG(x).$$

That is, a_j is the probability of j arrivals during a service period. The transition probabilities for this chain are

$$P_{0j} = a_j,$$
$$P_{ij} = a_{j-i+1}, \qquad i > 0, \quad j \geq i - 1,$$
$$P_{ij} = 0, \qquad j < i - 1.$$

Let $\rho = \sum_j j a_j$. Since ρ equals the mean number of arrivals during a service period, it follows, upon conditioning on the length of that period, that

$$\rho = \lambda E[S],$$

where S is a service time having distribution G.

We shall now show that the Markov chain is positive recurrent when $\rho < 1$ by solving the system of equations

$$\pi_j = \sum_i \pi_i P_{ij}.$$

These equations take the form

$$\pi_j = \pi_0 a_j + \sum_{i=1}^{j+1} \pi_i a_{j-i+1}, \qquad j \geq 0. \tag{4.3.5}$$

To solve, we introduce the generating functions

$$\pi(s) = \sum_{j=0}^\infty \pi_j s^j, \qquad A(s) = \sum_{j=0}^\infty a_j s^j.$$

Multiplying both sides of (4.3.5) by s^j and summing over j yields

$$\begin{aligned}\pi(s) &= \pi_0 A(s) + \sum_{j=0}^\infty \sum_{i=1}^{j+1} \pi_i a_{j-i+1} s^j \\ &= \pi_0 A(s) + s^{-1} \sum_{i=1}^\infty \pi_i s^i \sum_{j=i-1}^\infty a_{j-i+1} s^{j-i+1} \\ &= \pi_0 A(s) + (\pi(s) - \pi_0) A(s)/s,\end{aligned}$$

or

$$\pi(s) = \frac{(s-1)\pi_0 A(s)}{s - A(s)}.$$

To compute π_0 we let $s \to 1$ in the above. As

$$\lim_{s \to 1} A(s) = \sum_{i=0}^{\infty} a_i = 1,$$

this gives

$$\lim_{s \to 1} \pi(s) = \pi_0 \lim_{s \to 1} \frac{s-1}{s - A(s)}$$
$$= \pi_0 (1 - A'(1))^{-1},$$

where the last equality follows from L'hospital's rule. Now

$$A'(1) = \sum_{i=0}^{\infty} i a_i = \rho,$$

and thus

$$\lim_{s \to 1} \pi(s) = \frac{\pi_0}{1 - \rho}.$$

However, since $\lim_{s \to 1} \pi(s) = \sum_{i=0}^{\infty} \pi_i$, this implies that $\sum_{i=0}^{\infty} \pi_i = \pi_0/(1 - \rho)$; thus stationary probabilities exist if and only if $\rho < 1$, and in this case,

$$\pi_0 = 1 - \rho = 1 - \lambda E[S].$$

Hence, when $\rho < 1$, or, equivalently, when $E[S] < 1/\lambda$,

$$\pi(s) = \frac{(1 - \lambda E[S])(s-1)A(s)}{s - A(s)}.$$

EXAMPLE 4.3(B) ***Limiting Probabilities for the Embedded G/M/1 Queue.*** Consider the embedded Markov chain for the $G/M/1$ queueing system as presented in Example 4.1(B). The limiting probabilities π_k, $k = 0, 1, \ldots$ can be obtained as the unique solution of

$$\pi_k = \sum_i \pi_i P_{ik}, \qquad k \geq 0,$$

$$\sum_k \pi_k = 1,$$

which in this case reduce to

$$(4.3.6)\qquad \begin{aligned} \pi_k &= \sum_{i=k-1}^{\infty} \pi_i \int_0^\infty e^{-\mu t} \frac{(\mu t)^{i+1-k}}{(i+1-k)!}\, dG(t), \qquad k \ge 1, \\ \sum_0^\infty \pi_k &= 1. \end{aligned}$$

(We've not included the equation $\pi_0 = \sum_i \pi_i P_{i0}$ since one of the equations is always redundant.)

To solve the above let us try a solution of the form $\pi_k = c\beta^k$. Substitution into (4.3.6) leads to

$$(4.3.7)\qquad \begin{aligned} c\beta^k &= c \sum_{i=k-1}^{\infty} \beta^i \int_0^\infty e^{-\mu t} \frac{(\mu t)^{i+1-k}}{(i+1-k)!}\, dG(t) \\ &= c \int_0^\infty e^{-\mu t} \beta^{k-1} \sum_{i=k-1}^{\infty} \frac{(\beta\mu t)^{i+1-k}}{(i+1-k)!}\, dG(t) \\ &= c \int_0^\infty e^{-\mu t} \beta^{k-1} e^{\beta\mu t}\, dG(t) \end{aligned}$$

or

$$(4.3.8)\qquad \beta = \int_0^\infty e^{-\mu t(1-\beta)}\, dG(t).$$

The constant c can be obtained from $\sum_k \pi_k = 1$, which implies that

$$c = 1 - \beta.$$

Since the π_k are the *unique* solution to (4.3.6) and $\pi_k = (1-\beta)\beta^k$ satisfies, it follows that

$$\pi_k = (1-\beta)\beta^k, \qquad k = 0, 1, \ldots,$$

where β is the solution of Equation (4.3.8). (It can be shown that if the mean of G is greater than the mean service time $1/\mu$, then there is a unique value of β satisfying (4.3.8) that is between 0 and 1.) The exact value of β can usually only be obtained by numerical methods.

Example 4.3(c) ***The Age of a Renewal Process.*** Initially an item is put into use, and when it fails it is replaced at the beginning of the next time period by a new item. Suppose that the lives of the items are independent and each will fail in its ith period of use with probability P_i, $i \ge 1$, where the distribution $\{P_i\}$ is aperiodic and $\sum_i iP_i < \infty$. Let X_n denote the age of the item in use at time

n—that is, the number of periods (including the nth) it has been in use. Then if we let

$$\lambda(i) = \frac{P_i}{\sum_{j=i}^{\infty} P_j}$$

denote the probability that an i unit old item fails, then $\{X_n, n \geq 0\}$ is a Markov chain with transition probabilities given by

$$P_{i,1} = \lambda(i) = 1 - P_{i,i+1}, \qquad i \geq 1.$$

Hence the limiting probabilities are such that

$$\pi_1 = \sum_i \pi_i \lambda(i), \tag{4.3.9}$$

$$\pi_{i+1} = \pi_i(1 - \lambda(i)), \qquad i \geq 1. \tag{4.3.10}$$

Iterating (4.3.10) yields

$$\begin{aligned}
\pi_{i+1} &= \pi_i(1 - \lambda(i)) \\
&= \pi_{i-1}(1 - \lambda(i))(1 - \lambda(i-1)) \\
&= \pi_1(1 - \lambda(1))(1 - \lambda(2)) \cdots (1 - \lambda(i)) \\
&= \pi_1 \sum_{j=i+1}^{\infty} P_j \\
&= \pi_1 P\{X \geq i + 1\},
\end{aligned}$$

where X is the life of an item. Using $\sum_1^\infty \pi_i = 1$ yields

$$1 = \pi_1 \sum_{i=1}^{\infty} P\{X \geq i\}$$

or

$$\pi_1 = 1/E[X]$$

and

$$\pi_i = P\{X \geq i\}/E[X], \qquad i \geq 1, \tag{4.3.11}$$

which is easily seen to satisfy (4.3.9).

It is worth noting that (4.3.11) is as expected since the limiting distribution of age in the nonlattice case is the equilibrium distribution (see Section 3.4 of Chapter 3) whose density is $\overline{F}(x)/E[X]$.

Our next two examples illustrate how the stationary probabilities can sometimes be determined not by algebraically solving the stationary equations but by reasoning directly that when the initial state is chosen according to a certain set of probabilities then the resulting chain is stationary.

EXAMPLE 4.3(D) Suppose that during each time period, every member of a population independently dies with probability p, and also that the number of new members that join the population in each time period is a Poisson random variable with mean λ. If we let X_n denote the number of members of the population at the beginning of period n, then it is easy to see that $\{X_n, n = 1, \ldots\}$ is a Markov chain.

To find the stationary probabilities of this chain, suppose that X_0 is distributed as a Poisson random variable with parameter α. Since each of these X_0 individuals will independently be alive at the beginning of the next period with probability $1 - p$, it follows that the number of them that are still in the population at time 1 is a Poisson random variable with mean $\alpha(1 - p)$. As the number of new members that join the population by time 1 is an independent Poisson random variable with mean λ, it thus follows that X_1 is a Poisson random variable with mean $\alpha(1 - p) + \lambda$. Hence, if

$$\alpha = \alpha(1 - p) + \lambda$$

then the chain would be stationary. Hence, by the uniqueness of the stationary distribution, we can conclude that the stationary distribution is Poisson with mean λ/p. That is,

$$\pi_j = e^{-\lambda/p}(\lambda/p)^j/j!, \qquad j = 0, 1, \ldots.$$

EXAMPLE 4.3(E) ***The Gibbs Sampler.*** Let $p(x_1, \ldots, x_n)$ be the joint probability mass function of the random vector $X_1, \ldots, X_n$. In cases where it is difficult to directly generate the values of such a random vector, but where it is relatively easy to generate, for each i, a random variable having the conditional distribution of X_i given all of the other $X_j, j \neq i$, we can generate a random vector whose probability mass function is approximately $p(x_1, \ldots, x_n)$ by using the Gibbs sampler. It works as follows.

Let $\boldsymbol{X}^0 = (x_1^0, \ldots, x_n^0)$ be any vector for which $p(x_1^0, \ldots, x_n^0) > 0$.

Then generate a random variable whose distribution is the conditional distribution of X_1 given that $X_j = x_j^0, j = 2, \ldots, n$, and call its value x_1^1.

Then generate a random variable whose distribution is the conditional distribution of X_2 given that $X_1 = x_1^1$, $X_j = x_j^0, j = 3, \ldots, n$, and call its value x_2^1.

Then continue in this fashion until you have generated a random variable whose distribution is the conditional distribution of X_n given that $X_j = x_j^1, j = 1, \ldots, n - 1$, and call its value x_n^1.

Let $\boldsymbol{X}^1 = (x_1^1, \ldots, x_n^1)$, and repeat the process, this time starting with $\boldsymbol{X}^1$ in place of $\boldsymbol{X}^0$, to obtain the new vector $\boldsymbol{X}^2$, and so on. It is easy to see that the sequence of vectors $\boldsymbol{X}^j$, $j \geq 0$ is a Markov chain and the claim is that its stationary probabilities are given by $p(x_1, \ldots, x_n)$.

To verify the claim, suppose that X^0 has probability mass function $p(x_1, \ldots, x_n)$. Then it is easy to see that at any point in this algorithm the vector $x_1^j, \ldots, x_{i-1}^j, x_i^{j-1} \ldots, x_n^{j-1}$ will be the value of a random variable with mass function $p(x_1, \ldots, x_n)$. For instance, letting X_i^j be the random variable that takes on the value denoted by x_i^j then

$$\begin{aligned}
P\{X_1^1 &= x_1, X_j^0 = x_j, j = 2, \ldots, n\} \\
&= P\{X_1^1 = x_1 | X_j^0 = x_j, j = 2, \ldots, n\} P\{X_j^0 = x_j, j = 2, \ldots, n\} \\
&= P\{X_1 = x_1 | X_j = x_j, j = 2, \ldots, n\} P\{X_j = x_j, j = 2, \ldots, n\} \\
&= p(x_1, \ldots, x_n).
\end{aligned}$$

Thus $p(x_1, \ldots, x_n)$ is a stationary probability distribution and so, provided that the Markov chain is irreducible and aperiodic, we can conclude that it is the limiting probability vector for the Gibbs sampler. It also follows from the preceding that $p(x_1, \ldots, x_n)$ would be the limiting probability vector even if the Gibbs sampler were not systematic in first changing the value of X_1, then X_2, and so on. Indeed, even if the component whose value was to be changed was always randomly determined, then $p(x_1, \ldots, x_n)$ would remain a stationary distribution, and would thus be the limiting probability mass function provided that the resulting chain is aperiodic and irreducible.

Now, consider an irreducible, positive recurrent Markov chain with stationary probabilities, π_j, $j \geq 0$—that is, π_j is the long-run proportion of transitions that are into state j. Consider a given state of this chain, say state 0, and let N denote the number of transitions between successive visits to state 0. Since visits to state 0 constitute renewals it follows from Theorem 3.3.5 that the number of visits by time n is, for large n, approximately normally distributed with mean $n/E[N] = n\pi_0$ and variance $n\text{Var}(N)/(E[N])^3 = n\text{Var}(N)\pi_0^3$. It remains to determine $\text{Var}(N) = E[N^2] - 1/\pi_0^2$.

To derive an expression for $E[N^2]$, let us first determine the average number of transitions until the Markov chain next enters state 0. That is, let T_n denote the number of transitions from time n onward until the Markov chain enters state 0, and consider $\lim_{n\to\infty} \dfrac{T_1 + T_2 + \cdots + T_n}{n}$. By imagining that we receive a reward at any time that is equal to the number of transitions from that time onward until the next visit to state 0, we obtain a renewal reward process in which a new cycle begins each time a transition into state 0 occurs, and for which the average reward per unit time is

$$\text{Long-Run Average Reward Per Unit Time} = \lim_{n\to\infty} \frac{T_1 + T_2 + \cdots + T_n}{n}.$$

By the theory of renewal reward processes this long-run average reward per unit time is equal to the expected reward earned during a cycle divided by the expected time of a cycle. But if N is the number of transitions between successive visits into state 0, then

$$E[\text{Reward Earned during a Cycle}] = E[N + N - 1 + \cdots + 1]$$
$$E[\text{Time of a Cycle}] = E[N].$$

Thus,

$$\text{(4.3.12)} \qquad \text{Average Reward Per Unit Time} = \frac{E[N(N+1)/2]}{E[N]} = \frac{E[N^2] + E[N]}{2E[N]}.$$

However, since the average reward is just the average number of transitions it takes the Markov chain to make a transition into state 0, and since the proportion of time that the chain is in state i is π_i, it follows that

$$\text{(4.3.13)} \qquad \text{Average Reward Per Unit Time} = \sum_i \pi_i \mu_{i0},$$

where μ_{i0} denotes the mean number of transitions until the chain enters state 0 given that it is presently in state i. By equating the two expressions for the average reward given by Equations (4.3.12) and (4.3.13), and using that $E[N] = 1/\pi_0$, we obtain that

$$\pi_0 E[N^2] + 1 = 2 \sum_i \pi_i \mu_{i0}$$

or

$$E[N^2] = \frac{2}{\pi_0} \sum_i \pi_i \mu_{i0} - \frac{1}{\pi_0}$$

$$\text{(4.3.14)} \qquad = \frac{2}{\pi_0} \sum_{i \neq 0} \pi_i \mu_{i0} + \frac{1}{\pi_0}$$

where the final equation used that $\mu_{00} = E[N] = 1/\pi_0$. The values μ_{i0} can be obtained by solving the following set of linear equations (obtained by conditioning on the next state visited).

$$\mu_{i0} = 1 + \sum_{j \neq 0} P_{ij} \mu_{j0}, \qquad i \geq 0.$$

EXAMPLE 4.3(F) Consider the two-state Markov chain with

$$P_{00} = \alpha = 1 - P_{01}$$
$$P_{10} = \beta = 1 - P_{11}.$$

For this chain,

$$\pi_0 = \frac{\beta}{1 - \alpha + \beta}, \qquad \pi_1 = \frac{1 - \alpha}{1 - \alpha + \beta}$$
$$\mu_{00} = 1/\pi_0, \qquad \mu_{10} = 1/\beta.$$

Hence, from (4.3.14),

$$E[N^2] = 2\pi_1\mu_{10}/\pi_0 + 1/\pi_0 = 2(1 - \alpha)/\beta^2 + (1 - \alpha + \beta)/\beta$$

and so,

$$\begin{aligned}\mathrm{Var}(N) &= 2(1 - \alpha)/\beta^2 + (1 - \alpha + \beta)/\beta - (1 - \alpha + \beta)^2/\beta^2 \\ &= (1 - \beta + \alpha\beta - \alpha^2)/\beta^2.\end{aligned}$$

Hence, for n large, the number of transitions into state 0 by time n is approximately normal with mean $n\pi_0 = n\beta/(1 - \alpha + \beta)$ and variance $n\pi_0^3\mathrm{Var}(N) = n\beta(1 - \beta + \alpha\beta - \alpha^2)/(1 - \alpha + \beta)^3$. For instance, if $\alpha = \beta = 1/2$, then the number of visits to state 0 by time n is for large n approximately normal with mean $n/2$ and variance $n/4$.

4.4 Transitions among Classes, the Gambler's Ruin Problem, and Mean Times in Transient States

We begin this section by showing that a recurrent class is a *closed* class in the sense that once entered it is never left.

PROPOSITION 4.4.1

Let R be a recurrent class of states. If $i \in R$, $j \notin R$, then $P_{ij} = 0$.

Proof Suppose $P_{ij} > 0$. Then, as i and j do not communicate (since $j \notin R$), $P^n_{ji} = 0$ for all n. Hence if the process starts in state i, there is a positive probability of at least P_{ij} that the process will never return to i. This contradicts the fact that i is recurrent, and so $P_{ij} = 0$.

Let j be a given recurrent state and let T denote the set of all transient states. For $i \in T$, we are often interested in computing f_{ij}, the probability of ever entering j given that the process starts in i. The following proposition, by conditioning on the state after the initial transition, yields a set of equations satisfied by the f_{ij}.

PROPOSITION 4.4.2

If j is recurrent, then the set of probabilities $\{f_{ij}, i \in T\}$ satisfies

$$f_{ij} = \sum_{k \in T} P_{ik} f_{kj} + \sum_{k \in R} P_{ik}, \qquad i \in T,$$

where R denotes the set of states communicating with j.

Proof

$$\begin{aligned} f_{ij} &= P\{N_j(\infty) > 0 | X_0 = i\} \\ &= \sum_{\text{all } k} P\{N_j(\infty) > 0 | X_0 = i, X_1 = k\} P\{X_1 = k | X_0 = i\} \\ &= \sum_{k \in T} f_{kj} P_{ik} + \sum_{k \in R} f_{kj} P_{ik} + \sum_{\substack{k \notin R \\ k \notin T}} f_{kj} P_{ik} \\ &= \sum_{k \in T} f_{kj} P_{ik} + \sum_{k \in R} P_{ik}, \end{aligned}$$

where we have used Corollary 4.2.5 in asserting that $f_{kj} = 1$ for $k \in R$ and Proposition 4.4.1 in asserting that $f_{kj} = 0$ for $k \notin T$, $k \notin R$.

EXAMPLE 4.4(A) ***The Gambler's Ruin Problem.*** Consider a gambler who at each play of the game has probability p of winning 1 unit and probability $q = 1 - p$ of losing 1 unit. Assuming successive plays of the game are independent, what is the probability that, starting with i units, the gambler's fortune will reach N before reaching 0.

If we let X_n denote the player's fortune at time n, then the process $\{X_n, n = 0, 1, 2, \ldots\}$ is a Markov chain with transition probabilities

$$\begin{aligned} P_{00} &= P_{NN} = 1, \\ P_{i,i+1} &= p = 1 - P_{i,i-1}, \qquad i = 1, 2, \ldots, N - 1. \end{aligned}$$

This Markov chain has three classes, namely, $\{0\}$, $\{1, 2, \ldots, N - 1\}$, and $\{N\}$, the first and third class being recurrent and the second transient. Since each transient state is only visited finitely often, it follows that, after some finite amount of time, the gambler will either attain her goal of N or go broke.

Let $f_i \equiv f_{iN}$ denote the probability that, starting with i, $0 \le i \le N$, the gambler's fortune will eventually reach N. By conditioning on the outcome of the initial play of the game (or, equivalently, by using Proposition 4.4.2), we obtain

$$f_i = pf_{i+1} + qf_{i-1}, \qquad i = 1, 2, \ldots, N-1,$$

or, equivalently, since $p + q = 1$,

$$f_{i+1} - f_i = \frac{q}{p}(f_i - f_{i-1}), \qquad i = 1, 2, \ldots, N-1.$$

Since $f_0 = 0$, we see from the above that

$$f_2 - f_1 = \frac{q}{p}(f_1 - f_0) = \frac{q}{p} f_1$$

$$f_3 - f_2 = \frac{q}{p}(f_2 - f_1) = \left(\frac{q}{p}\right)^2 f_1$$

$$\vdots$$

$$f_i - f_{i-1} = \frac{q}{p}(f_{i-1} - f_{i-2}) = \left(\frac{q}{p}\right)^{i-1} f_1$$

$$\vdots$$

$$f_N - f_{N-1} = \left(\frac{q}{p}\right)(f_{N-1} - f_{N-2}) = \left(\frac{q}{p}\right)^{N-1} f_1.$$

Adding the first $i - 1$ of these equations yields

$$f_i - f_1 = f_1 \left[\left(\frac{q}{p}\right) + \left(\frac{q}{p}\right)^2 + \cdots + \left(\frac{q}{p}\right)^{i-1} \right]$$

or

$$f_i = \begin{cases} \dfrac{1 - (q/p)^i}{1 - (q/p)} f_1 & \text{if } \dfrac{q}{p} \neq 1 \\[2ex] i f_1 & \text{if } \dfrac{q}{p} = 1. \end{cases}$$

Using $f_N = 1$ yields

$$f_i = \begin{cases} \dfrac{1 - (q/p)^i}{1 - (q/p)^N} & \text{if } p \neq \frac{1}{2} \\[2ex] \dfrac{i}{N} & \text{if } p = \frac{1}{2}. \end{cases}$$

It is interesting to note that as $N \to \infty$

$$f_i \to \begin{cases} 1 - (q/p)^i & \text{if } p > \frac{1}{2} \\ 0 & \text{if } p \leq \frac{1}{2}. \end{cases}$$

Hence, from the continuity property of probabilities, it follows that if $p > \frac{1}{2}$, there is a positive probability that the gambler's fortune will converge to infinity; whereas if $p \leq \frac{1}{2}$, then, with probability 1, the gambler will eventually go broke when playing against an infinitely rich adversary.

Suppose now that we want to determine the expected number of bets that the gambler, starting at i, makes before reaching either 0 or n. Whereas we could call this quantity m_i and derive a set of linear equations for m_i, $i = 1, \ldots, n - 1$, by conditioning on the outcome of the initial gamble, we will obtain a more elegant solution by using Wald's equation along with the preceding.

Imagining that the gambler continues to play after reaching either 0 or n, let X_j be her winnings on the jth bet, $j \geq 1$. Also, let B denote the number of bets until the gambler's fortune reaches either 0 or n. That is,

$$B = \text{Min}\left\{m: \sum_{j=1}^{m} X_j = -i \quad \text{or} \quad \sum_{j=1}^{m} X_j = n - i\right\}.$$

Since the X_j are independent random variables with mean $E[X_j] = 1(p) - 1(1 - p) = 2p - 1$, and N is a stopping time for the X_j, it follows by Wald's equation that

$$E\left[\sum_{j=1}^{B} X_j\right] = (2p - 1)E[B].$$

But, if we let $\alpha = [1 - (q/p)^i]/[1 - (q/p)^n]$ be the probability that n is reached before 0, then

$$\sum_{j=1}^{B} X_j = \begin{cases} n - i & \text{with probability } \alpha \\ -i & \text{with probability } 1 - \alpha. \end{cases}$$

Hence, we obtain that

$$(2p - 1)E[B] = n\alpha - i$$

or

$$E[B] = \frac{1}{2p - 1}\left\{\frac{n[1 - (q/p)^i]}{1 - (q/p)^n} - i\right\}.$$

Consider now a finite state Markov chain and suppose that the states are numbered so that $T = \{1, 2, \ldots, t\}$ denotes the set of transient states. Let

$$Q = \begin{bmatrix} P_{11} & P_{12} & \cdots & P_{1t} \\ P_{i1} & P_{i2} & \cdots & P_{it} \\ P_{t1} & P_{t2} & \cdots & P_{tt} \end{bmatrix}$$

and note that since $\boldsymbol{Q}$ specifies only the transition probabilities from transient states into transient states, some of its row sums are less than 1 (for otherwise, T would be a closed class of states).

For transient states i and j, let m_{ij} denote the expected total number of time periods spent in state j given that the chain starts in state i. Conditioning on the initial transition yields:

(4.4.1)
$$\begin{aligned} m_{ij} &= \delta(i,j) + \sum_k P_{ik} m_{kj} \\ &= \delta(i,j) + \sum_{k=1}^{t} P_{ik} m_{kj} \end{aligned}$$

where $\delta(i, j)$ is equal to 1 when $i = j$ and is 0 otherwise, and where the final equality follows from the fact that $m_{kj} = 0$ when k is a recurrent state.

Let $\boldsymbol{M}$ denote the matrix of values m_{ij}, $i, j = 1, \ldots, t$, that is,

$$M = \begin{bmatrix} m_{11} & m_{12} & \cdots & m_{1t} \\ m_{i1} & m_{i2} & \cdots & m_{it} \\ m_{t1} & m_{t2} & \cdots & m_{tt} \end{bmatrix}$$

In matrix notation, (4.4.1) can be written as

$$\boldsymbol{M} = \boldsymbol{I} + \boldsymbol{QM}$$

where $\boldsymbol{I}$ is the identity matrix of size t. As the preceding equation is equivalent to

$$(\boldsymbol{I} - \boldsymbol{Q})\boldsymbol{M} = \boldsymbol{I}$$

we obtain, upon multiplying both sides by $(\boldsymbol{I} - \boldsymbol{Q})^{-1}$, that

$$\boldsymbol{M} = (\boldsymbol{I} - \boldsymbol{Q})^{-1}.$$

That is, the quantities m_{ij}, $i \varepsilon T$, $j \varepsilon T$, can be obtained by inverting the matrix $\boldsymbol{I} - \boldsymbol{Q}$. (The existence of the inverse is easily established.)

For $i \varepsilon T$, $j \varepsilon T$, the quantity f_{ij}, equal to the probability of ever making a transition into state j given that the chain starts in state i, is easily determined from $\boldsymbol{M}$. To determine the relationship, we start by deriving an expression for m_{ij} by conditioning on whether state j is ever entered.

$$
\begin{aligned}
m_{ij} &= E[\text{number of transitions into state } j \mid \text{start in } i] \\
&= m_{jj} f_{ij}
\end{aligned}
$$

where m_{jj} is the expected number of time periods spent in state j given that it is eventually entered from state i. Thus, we see that

$$f_{ij} = m_{ij}/m_{jj}.$$

Example 4.4(b) Consider the gambler's ruin problem with $p = .4$ and $n = 6$. Starting in state 3, determine:

(a) the expected amount of time spent in state 3.
(b) the expected number of visits to state 2.
(c) the probability of ever visiting state 4.

Solution. The matrix $\boldsymbol{Q}$, which specifies P_{ij}, $i, j \varepsilon \{1, 2, 3, 4, 5\}$ is as follows:

$$
\boldsymbol{Q} = \begin{array}{c} \\ 1 \\ 2 \\ 3 \\ 4 \\ 5 \end{array}
\begin{array}{c} \begin{array}{ccccc} 1 & 2 & 3 & 4 & 5 \end{array} \\
\begin{bmatrix}
0 & .4 & 0 & 0 & 0 \\
.6 & 0 & .4 & 0 & 0 \\
0 & .6 & 0 & .4 & 0 \\
0 & 0 & .6 & 0 & .4 \\
0 & 0 & 0 & .6 & 0
\end{bmatrix} \end{array}
$$

Inverting $(\boldsymbol{I} - \boldsymbol{Q})$ gives that

$$
\boldsymbol{M} = (\boldsymbol{I} - \boldsymbol{Q})^{-1} = \begin{bmatrix}
1.5865 & 0.9774 & 0.5714 & 0.3008 & 0.1203 \\
1.4662 & 2.4436 & 1.4286 & 0.7519 & 0.3008 \\
1.2857 & 2.1429 & 2.7143 & 1.4286 & 0.5714 \\
1.0150 & 1.6917 & 2.1429 & 2.4436 & 0.9774 \\
0.6090 & 1.0150 & 1.2857 & 1.4662 & 1.5865
\end{bmatrix}
$$

Hence,

$$m_{3,3} = 2.7143, \quad m_{3,2} = 2.1429$$

$$f_{3,4} = m_{3,4}/m_{4,4} = 1.4286/2.4436 = .5846.$$

As a check, note that $f_{3,4}$ is just the probability, starting with 3, of visiting 4 before 0, and thus

$$f_{3,4} = \frac{1 - (.6/.4)^3}{1 - (.6/.4)^4} = 38/65 = .5846.$$

4.5 Branching Processes

Consider a population consisting of individuals able to produce offspring of the same kind. Suppose that each individual will, by the end of its lifetime, have produced j new offspring with probability P_j, $j \geq 0$, independently of the number produced by any other individual. The number of individuals initially present, denoted by X_0, is called the size of the zeroth generation. All offspring of the zeroth generation constitute the first generation and their number is denoted by X_1. In general, let X_n denote the size of the nth generation. The Markov chain $\{X_n, n \geq 0\}$ is called a *branching process*.

Suppose that $X_0 = 1$. We can calculate the mean of X_n by noting that

$$X_n = \sum_{i=1}^{X_{n-1}} Z_i,$$

where Z_i represents the number of offspring of the ith individual of the $(n-1)$st generation. Conditioning on X_{n-1} yields

$$\begin{aligned} E[X_n] &= E[E[X_n|X_{n-1}]] \\ &= \mu E[X_{n-1}] \\ &= \mu^2 E[X_{n-2}] \\ &= \mu^n, \end{aligned}$$

where μ is the mean number of offspring per individual.

Let π_0 denote the probability that, starting with a single individual, the population ever dies out.

An equation determining π_0 may be derived by conditioning on the number of offspring of the initial individual, as follows:

$$\begin{aligned} \pi_0 &= P\{\text{population dies out}\} \\ &= \sum_{j=0}^{\infty} P\{\text{population dies out} | X_1 = j\} P_j. \end{aligned}$$

Now, given that $X_1 = j$, the population will eventually die out if, and only if, each of the j families started by the members of the first generation eventually die out. Since each family is assumed to act independently, and since the probability that any particular family dies out is just π_0, this yields

$$\pi_0 = \sum_{j=0}^{\infty} \pi_0^j P_j. \tag{4.5.1}$$

In fact we can prove the following.

THEOREM 4.5.1

Suppose that $P_0 > 0$ and $P_0 + P_1 < 1$. Then (i) *π_0 is the smallest positive number satisfying*

$$\pi_0 = \sum_{j=0}^{\infty} \pi_0^j P_j.$$

(ii) $\pi_0 = 1$ *if, and only if,* $\mu \le 1$.

Proof To show that π_0 is the smallest solution of (4.5.1), let $\pi \ge 0$ satisfy (4.5.1). We'll first show by induction that $\pi \ge P\{X_n = 0\}$ for all n. Now

$$\pi = \sum_j \pi^j P_j \ge \pi^0 P_0 = P_0 = P\{X_1 = 0\}$$

and assume that $\pi \ge P\{X_n = 0\}$. Then

$$\begin{aligned} P\{X_{n+1} = 0\} &= \sum_j P\{X_{n+1} = 0 \mid X_1 = j\} P_j \\ &= \sum_j (P\{X_n = 0\})^j P_j \\ &\le \sum_j \pi^j P_j \qquad \text{(by the induction hypothesis)} \\ &= \pi. \end{aligned}$$

Hence,

$$\pi \ge P\{X_n = 0\} \qquad \text{for all } n,$$

and letting $n \to \infty$,

$$\pi \ge \lim_n P\{X_n = 0\} = P\{\text{population dies out}\} = \pi_0.$$

To prove (ii) define the generating function

$$\phi(s) = \sum_{j=0}^{\infty} s^j P_j.$$

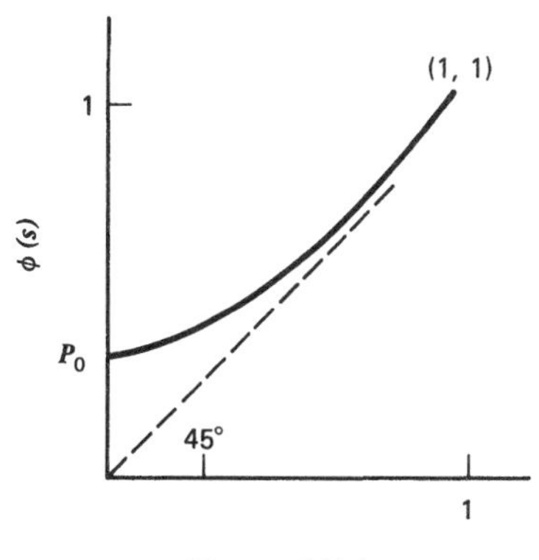

Figure 4.5.1

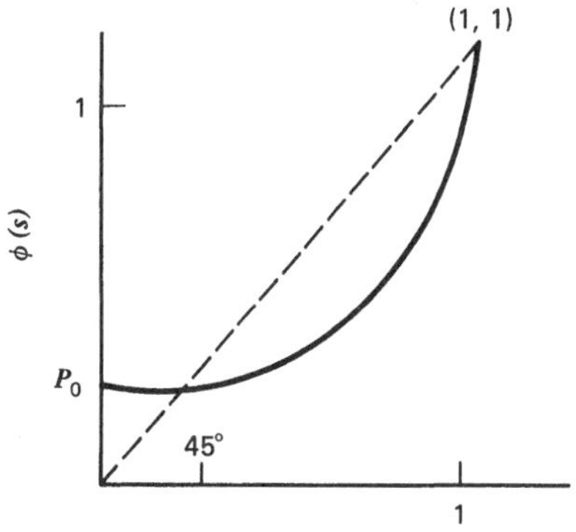

Figure 4.5.2

Since $P_0 + P_1 < 1$, it follows that

$$\phi''(s) = \sum_{j=0}^{\infty} j(j-1)s^{j-2}P_j > 0$$

for all $s \in (0, 1)$. Hence, $\phi(s)$ is a strictly convex function in the open interval (0, 1). We now distinguish two cases (Figures 4.5.1 and 4.5.2). In Figure 4.5.1 $\phi(s) > s$ for all $s \in (0, 1)$, and in Figure 4.5.2, $\phi(s) = s$ for some $s \in (0, 1)$. It is geometrically clear that Figure 4.5.1 represents the appropriate picture when $\phi'(1) \leq 1$, and Figure 4.5.2 is appropriate when $\phi'(1) > 1$. Thus, since $\phi(\pi_0) = \pi_0$, $\pi_0 = 1$ if, and only if, $\phi'(1) \leq 1$. The result follows, since $\phi'(1) = \sum_1^\infty jP_j = \mu$.

4.6 Applications of Markov Chains

4.6.1 A Markov Chain Model of Algorithmic Efficiency

Certain algorithms in operations research and computer science act in the following manner: the objective is to determine the best of a set of N ordered elements. The algorithm starts with one of the elements and then successively moves to a better element until it reaches the best. (The most important example is probably the simplex algorithm of linear programming, which attempts to maximize a linear function subject to linear constraints and where an element corresponds to an extreme point of the feasibility region.) If one looks at the algorithm's efficiency from a "worse case" point of view, then examples can usually be constructed that require roughly $N - 1$ steps to reach the optimal element. In this section, we will present a simple probabilistic model for the number of necessary steps. Specifically, we consider a Markov chain that when transiting from any state is equally likely to enter any of the better ones.

Consider a Markov chain for which $P_{11} = 1$ and

$$P_{ij} = \frac{1}{i-1}, \qquad j = 1, \ldots, i-1, \quad i > 1,$$

and let T_i denote the number of transitions to go from state i to state 1. A recursive formula for $E[T_i]$ can be obtained by conditioning on the initial transition:

$$E[T_i] = 1 + \frac{1}{i-1}\sum_{j=1}^{i-1} E[T_j]. \tag{4.6.1}$$

Starting with $E[T_1] = 0$, we successively see that

$$E[T_2] = 1,$$
$$E[T_3] = 1 + \tfrac{1}{2},$$
$$E[T_4] = 1 + \tfrac{1}{3}(1 + 1 + \tfrac{1}{2}) = 1 + \tfrac{1}{2} + \tfrac{1}{3},$$

and it is not difficult to guess and then prove inductively that

$$E[T_i] = \sum_{j=1}^{i-1} \frac{1}{j}.$$

However, to obtain a more complete description of T_N, we will use the representation

$$T_N = \sum_{j=1}^{N-1} I_j,$$

where

$$I_j = \begin{cases} 1 & \text{if the process ever enters } j \\ 0 & \text{otherwise.} \end{cases}$$

The importance of the above representation stems from the following.

Lemma 4.6.1

$I_1, \ldots, I_{N-1}$ are independent and

$$P\{I_j = 1\} = 1/j, \qquad 1 \le j \le N-1.$$

Proof Given $I_{j+1}, \ldots, I_N$, let $n = \min\{i: i > j, I_i = 1\}$. Then

$$P\{I_j = 1 | I_{j+1}, \ldots, I_N\} = \frac{1/(n-1)}{j/(n-1)} = \frac{1}{j}.$$

PROPOSITION 4.6.2

(i) $E[T_N] = \sum_{j=1}^{N-1} \frac{1}{j}$.

(ii) $\mathrm{Var}(T_N) = \sum_{j=1}^{N-1} \frac{1}{j}\left(1 - \frac{1}{j}\right)$.

(iii) For N large, T_N has approximately a Poisson distribution with mean $\log N$.

Proof Parts (i) and (ii) follow from Lemma 4.6.1 and the representation $T_N = \sum_{j=1}^{N-1} I_j$. Since the sum of a large number of independent Bernoulli random variables, each having a small probability of being nonzero, is approximately Poisson distributed, part (iii) follows since

$$\int_1^N \frac{dx}{x} < \sum_1^{N-1} \frac{1}{j} < 1 + \int_1^{N-1} \frac{dx}{x}$$

or

$$\log N < \sum_1^{N-1} \frac{1}{j} < 1 + \log(N-1),$$

and so

$$\log N \approx \sum_{j=1}^{N-1} \frac{1}{j}.$$

4.6.2 An Application to Runs—A Markov Chain with a Continuous State Space

Consider a sequence of numbers $x_1, x_2, \ldots$. If we place vertical lines before x_1 and between x_j and x_{j+1} whenever $x_j > x_{j+1}$, then we say that the *runs* are the segments between pairs of lines. For example, the partial sequence 3, 5, 8, 2, 4, 3, 1 contains four runs as indicated:

$$|3, 5, 8|2, 4|3|1.$$

Thus each run is an increasing segment of the sequence.

Suppose now that $X_1, X_2, \ldots$ are independent and identically distributed uniform (0, 1) random variables, and suppose we are interested in the distribution of the length of the successive runs. For instance, if we let L_1 denote the length of the initial run, then as L_1 will be at least m if, and only if, the first m values are in increasing order, we see that

$$P\{L_1 \geq m\} = \frac{1}{m!}, \qquad m = 1, 2, \ldots.$$

The distribution of a later run can be easily obtained if we know x, the initial value in the run. For if a run starts with x, then

$$P\{L \geq m | x\} = \frac{(1-x)^{m-1}}{(m-1)!} \tag{4.6.2}$$

since in order for the run length to be at least m, the next $m - 1$ values must all be greater than x and they must be in increasing order.

To obtain the unconditional distribution of the length of a given run, let I_n denote the initial value of the nth run. Now it is easy to see that $\{I_n, n \geq 1\}$ is a Markov chain having a continuous state space. To compute $p(y|x)$, the probability density that the next run begins with the value y given that a run has just begun with initial value x, reason as follows:

$$P\{I_{n+1} \in (y, y + dy) | I_n = x\} = \sum_{m=1}^{\infty} P\{I_{n+1} \in (y, y + dy), L_n = m | I_n = x\},$$

where L_n is the length of the nth run. Now the run will be of length m and the next one will start with value y if:

(i) the next $m - 1$ values are in increasing order and are all greater than x;
(ii) the mth value must equal y;
(iii) the maximum of the first $m - 1$ values must exceed y.

Hence,

$$\begin{aligned}
P\{I_{n+1} &\in (y, y + dy), L_n = m | I_n = x\} \\
&= \frac{(1-x)^{m-1}}{(m-1)!} dy\, P\{\max(X_1, \ldots, X_{m-1}) > y | X_i > x, i = 1, \ldots, m-1\} \\
&= \begin{cases} \dfrac{(1-x)^{m-1}}{(m-1)!} dy & \text{if } y < x \\[2ex] \dfrac{(1-x)^{m-1}}{(m-1)!} dy \left[1 - \left(\dfrac{y-x}{1-x}\right)^{m-1}\right] & \text{if } y > x. \end{cases}
\end{aligned}$$

Summing over m yields

$$p(y|x) = \begin{cases} e^{1-x} & \text{if } y < x \\ e^{1-x} - e^{y-x} & \text{if } y > x. \end{cases}$$

That is, $\{I_n, n \geq 1\}$ is a Markov chain having the continuous state space (0, 1) and a transition probability density $p(y|x)$ given by the above.

To obtain the limiting distribution of I_n we will first hazard a guess and then verify our guess by using the analog of Theorem 4.3.3. Now I_1, being the initial value of the first run, is uniformly distributed over (0, 1). However, the later runs begin whenever a value smaller than the previous one occurs. So it seems plausible that the long-run proportion of such values that are less than y would equal the probability that a uniform (0, 1) random variable is less than y given that it is smaller than a second, and independent, uniform (0, 1) random variable. Since

$$P\{X_2 > y | X_2 < X_1\} = \frac{\frac{1}{2}(1-y)^2}{\frac{1}{2}} = (1-y)^2,$$

it seems plausible that $\pi(y)$, the limiting density of I_n, is given by

$$\pi(y) = 2(1-y), \qquad 0 < y < 1.$$

[A second heuristic argument for the above limiting density is as follows: Each X_i of value y will be the starting point of a new run if the value prior to it is greater than y. Hence it would seem that the rate at which runs occur, beginning with an initial value in $(y, y + dy)$, equals $\overline{F}(y)f(y)\,dy = (1-y)\,dy$, and so the fraction of all runs whose initial value is in $(y, y + dy)$ will be $(1-y)\,dy/\int_0^1 (1-y)\,dy = 2(1-y)\,dy$.] Since a theorem analogous to Theorem 4.3.3 can be proven in the case of a Markov chain with a continuous state space, to prove the above we need to verify that

$$\pi(y) = \int_0^1 \pi(x)p(y|x)\,dx.$$

In this case the above reduces to showing that

$$1 - y = \int_0^y (e^{1-x} - e^{y-x})(1-x)\,dx + \int_y^1 e^{1-x}(1-x)\,dx$$

or

$$1 - y = \int_0^1 e^{1-x}(1-x)\,dx - \int_0^y e^{y-x}(1-x)\,dx,$$

which is easily shown upon application of the identity

$$\int ze^z\,dz = ze^z - e^z.$$

Thus we have shown that the limiting density of I_n, the initial value of the nth run, is $\pi(x) = 2(1-x), 0 < x < 1$. Hence the limiting distribution of L_n, the length of the nth run, is given by

$$\lim_{n\to\infty} P\{L_n \geq m\} = \int_0^1 \frac{(1-x)^{m-1}}{(m-1)!} 2(1-x)\,dx \tag{4.6.3}$$

$$= \frac{2}{(m+1)(m-1)!}$$

To compute the average length of a run note from (4.6.2) that

$$E[L|I = x] = \sum_{m=1}^{\infty} \frac{(1-x)^{m-1}}{(m-1)!}$$

$$= e^{1-x}$$

and so

$$\lim_{n\to\infty} E[L_n] = \int_0^1 e^{1-x} 2(1-x)\,dx$$

$$= 2.$$

The above could also have been computed from (4.6.3) as follows:

$$\lim_{n\to\infty} E[L_n] = 2\sum_{m=1}^{\infty} \frac{1}{(m+1)(m-1)!}$$

yielding the interesting identity

$$1 = \sum_{m=1}^{\infty} \frac{1}{(m+1)(m-1)!}.$$

4.6.3 List Ordering Rules—Optimality of the Transposition Rule

Suppose that we are given a set of n elements $e_1, \ldots, e_n$ that are to be arranged in some order. At each unit of time a request is made to retrieve one of these elements—e_i being requested (independently of the past) with probability P_i, $P_i \geq 0$, $\sum_1^n P_i = 1$. The problem of interest is to determine the optimal ordering so as to minimize the long-run average position of the element requested. Clearly if the P_i were known, the optimal ordering would simply be to order the elements in decreasing order of the P_i's. In fact, even if the P_i's were unknown we could do as well asymptotically by ordering the elements at each unit of time in decreasing order of the number of previous requests for them. However, the problem becomes more interesting if we do not allow such memory storage as would be necessary for the above rule, but rather restrict ourselves to reordering rules in which the reordered permutation of elements at any time is only allowed to depend on the present ordering and the position of the element requested.

For a given reordering rule, the average position of the element requested can be obtained, at least in theory, by analyzing the Markov chain of $n!$ states where the state at any time is the ordering at that time. However, for such a large number of states, the analysis quickly becomes intractable and so we shall simplify the problem by assuming that the probabilities satisfy

$$P_1 = p, \qquad P_2 = \cdots = P_n = \frac{1-p}{n-1} = q.$$

For such probabilities, since all the elements 2 through n are identical in the sense that they have the same probability of being requested, we can obtain the average position of the element requested by analyzing the much simpler Markov chain of n states with the state being the position of element e_1. We will now show that for such probabilities and among a wide class of rules the transposition rule, which always moves the requested element one closer to the front of the line, is optimal.

Consider the following restricted class of rules that, when an element is requested and found in position i, move the element to position j_i and leave the relative positions of the other elements unchanged. In addition, we suppose that $j_i < i$ for $i > 1$, $j_1 = 1$, and $j_i \geq j_{i-1}$, $i = 2, \ldots, n$. The set $\{j_i, i = 1, \ldots, n\}$ characterizes a rule in this class.

For a given rule in the above class, let

$$K(i) = \max\{l : j_{i+l} \leq i\}.$$

In other words, for any i, an element in any of the positions $i, i + 1, \ldots, i + K(i)$ will, if requested, be moved to a position less than or equal to i.

For a specified rule R in the above class—say the one having $K(i) = k(i)$, $i = 1, \ldots, n$—let us denote the stationary probabilities when this rule is employed by

$$\pi_i = P_\infty\{e_1 \text{ is in position } i\}, \qquad i \geq 1.$$

In addition, let

$$\Pi_i = \sum_{j=i+1}^{n} \pi_j = P_\infty\{e_1 \text{ is in a position greater than } i\}, \qquad i \geq 0$$

with the notation P_∞ signifying that the above are limiting probabilities. Before writing down the steady-state equations, it may be worth noting the following:

(i) Any element moves toward the back of the list at most one position at a time.

(ii) If an element is in position i and neither it nor any of the elements in the following $k(i)$ positions are requested, it will remain in position i.

(iii) Any element in one of the positions $i, i + 1, \ldots, i + k(i)$ will be moved to a position $\leq i$ if requested.

The steady-state probabilities can now easily be seen to be

$$\Pi_i = \Pi_{i+k(i)} + (\Pi_i - \Pi_{i+k(i)})(1 - p) + (\Pi_{i-1} - \Pi_i)qk(i).$$

The above follows since element 1 will be in a position higher than i if it was either in a position higher than $i + k(i)$ in the previous period, or it was in a position less than $i + k(i)$ but greater than i and was not selected, or if it was in position i and if any of the elements in positions $i + 1, \ldots, i + k(i)$ were selected. The above equations are equivalent to

$$\text{(4.6.4)} \qquad \begin{gathered} \Pi_i = a_i\Pi_{i-1} + (1 - a_i)\Pi_{i+k(i)}, \qquad i = 1, \ldots, n - 1, \\ \Pi_0 = 1, \qquad \Pi_n = 0, \end{gathered}$$

where

$$a_i = \frac{qk(i)}{qk(i) + p}.$$

Now consider a special rule of the above class, namely, the transposition rule, which has $j_i = i - 1, i = 2, \ldots, n, j_1 = 1$. Let the corresponding Π_i be denoted by $\overline{\Pi}_i$ for the transposition rule. Then from Equation (4.6.4), we have, since $K(i) = 1$ for this rule,

$$\overline{\Pi}_i = \frac{q\overline{\Pi}_{i-1} + p\overline{\Pi}_{i+1}}{p + q},$$

or, equivalently,

$$\overline{\Pi}_{i+1} - \overline{\Pi}_i = \frac{q}{p}(\overline{\Pi}_i - \overline{\Pi}_{i-1})$$

implying

$$\begin{aligned}\overline{\Pi}_{i+j} - \overline{\Pi}_{i+j-1} &= \frac{q}{p}(\overline{\Pi}_{i+j-1} - \overline{\Pi}_{i+j-2}) \\ &\vdots \\ &= \left(\frac{q}{p}\right)^j (\overline{\Pi}_i - \overline{\Pi}_{i-1}).\end{aligned}$$

Summing the above equations from $j = 1, \ldots, r$ yields

$$\overline{\Pi}_{i+r} - \overline{\Pi}_i = (\overline{\Pi}_i - \overline{\Pi}_{i-1})\left[\frac{q}{p} + \cdots + \left(\frac{q}{p}\right)^r\right], \qquad i + r \leq n.$$

Letting $r = k(i)$, where $k(i)$ is the value of $K(i)$ for a given rule R in our class, we see that

$$\overline{\Pi}_{i+k(i)} - \overline{\Pi}_i = (\overline{\Pi}_i - \overline{\Pi}_{i-1})\left[\frac{q}{p} + \cdots + \left(\frac{q}{p}\right)^{k(i)}\right],$$

or, equivalently,

$$\begin{aligned}\overline{\Pi}_i &= b_i\overline{\Pi}_{i-1} + (1 - b_i)\overline{\Pi}_{i+k(i)}, \qquad i = 1, \ldots, n-1, \\ \overline{\Pi}_0 &= 1, \qquad \overline{\Pi}_n = 0,\end{aligned} \tag{4.6.5}$$

where

$$b_i = \frac{(q/p) + \cdots + (q/p)^{k(i)}}{1 + (q/p) + \cdots + (q/p)^{k(i)}}, \qquad i = 1, \ldots, n-1.$$

We may now prove the following.

PROPOSITION 4.6.3

If $p \geq 1/n$, then $\overline{\Pi}_i \leq \Pi_i$ for all i.
If $p \leq 1/n$, then $\overline{\Pi}_i \geq \Pi_i$ for all i.

Proof Consider the case $p \geq 1/n$, which is equivalent to $p \geq q$, and note that in this case

$$a_i = 1 - \frac{1}{1 + (k(i)/p)q} \geq 1 - \frac{1}{1 + q/p + \cdots + (q/p)^{k(i)}} = b_i.$$

Now define a Markov chain with states 0, 1, . . . , n and transition probabilities

$$P_{0,0} = P_{n,n} = 1,$$

$$P_{ij} = \begin{cases} c_i & \text{if } j = i - 1, \\ 1 - c_i & \text{if } j = i + k(i), \end{cases} \qquad i = 1, \ldots, n-1. \tag{4.6.6}$$

Let f_i denote the probability that this Markov chain ever enters state 0 given that it starts in state i. Then f_i satisfies

$$f_i = c_i f_{i-1} + (1 - c_i) f_{i+k(i)}, \qquad i = 1, \ldots, n-1,$$

$$f_0 = 1, \qquad f_n = 0.$$

Hence, as it can be shown that the above set of equations has a unique solution, it follows from (4.6.4) that if we take c_i equal to a_i for all i, then f_i will equal the Π_i of rule R, and from (4.6.5) if we let $c_i = b_i$, then f_i equals $\overline{\Pi}_i$. Now it is intuitively clear (and we defer a formal proof until Chapter 8) that the probability that the Markov chain defined by (4.6.6) will ever enter 0 is an increasing function of the vector $\underline{c} = (c_1, \ldots, c_{n-1})$. Hence, since $a_i \geq b_i$, $i = 1, \ldots, n$, we see that

$$\Pi_i \geq \overline{\Pi}_i \qquad \text{for all } i.$$

When $p \leq 1/n$, then $a_i \leq b_i$, $i = 1, \ldots, n-1$, and the above inequality is reversed.

THEOREM 4.6.4

Among the rules considered, the limiting expected position of the element requested is minimized by the transposition rule.

Proof Letting X denote the position of e_1, we have upon conditioning on whether or not e_1 is requested that the expected position of the requested element can be expressed as

$$E[\text{position}] = pE[X] + (1-p)\frac{E[1 + 2 + \cdots + n - X]}{n-1}$$

$$= \left(p - \frac{1-p}{n-1}\right) E[X] + \frac{(1-p)n(n+1)}{2(n-1)}.$$

Thus, if $p \geq 1/n$, the expected position is minimized by minimizing $E[X]$, and if $p \leq 1/n$, by maximizing $E[X]$. Since $E[X] = \sum_{i=0}^{n} P\{X > i\}$, the result follows from Proposition 4.6.3.

4.7 Time-Reversible Markov Chains

An irreducible positive recurrent Markov chain is stationary if the initial state is chosen according to the stationary probabilities. (In the case of an ergodic chain this is equivalent to imagining that the process begins at time $t = -\infty$.) We say that such a chain is in *steady state*.

Consider now a stationary Markov chain having transition probabilities P_{ij} and stationary probabilities π_i, and suppose that starting at some time we trace the sequence of states going backwards in time. That is, starting at time n consider the sequence of states $X_n, X_{n-1}, \ldots$. It turns out that this sequence of states is itself a Markov chain with transition probabilities P^*_{ij} defined by

$$\begin{aligned} P^*_{ij} &= P\{X_m = j \mid X_{m+1} = i\} \\ &= \frac{P\{X_{m+1} = i \mid X_m = j\} P\{X_m = j\}}{P\{X_{m+1} = i\}} \\ &= \frac{\pi_j P_{ji}}{\pi_i}. \end{aligned}$$

To prove that the reversed process is indeed a Markov chain we need to verify that

$$P\{X_m = j \mid X_{m+1} = i, X_{m+2}, X_{m+3}, \ldots\} = P\{X_m = j \mid X_{m+1} = i\}.$$

To see that the preceding is true, think of the present time as being time $m + 1$. Then, since X_n, $n \geq 1$ is a Markov chain it follows that given the present state X_{m+1} the past state X_m and the future states $X_{m+2}, X_{m+3}, \ldots$ are independent. But this is exactly what the preceding equation states.

Thus the reversed process is also a Markov chain with transition probabilities given by

$$P^*_{ij} = \frac{\pi_j P_{ji}}{\pi_i}.$$

If $P^*_{ij} = P_{ij}$ for all i, j, then the Markov chain is said to be *time reversible*. The condition for time reversibility, namely, that

$$\pi_i P_{ij} = \pi_j P_{ji} \qquad \text{for all } i, j, \tag{4.7.1}$$

can be interpreted as stating that, for all states i and j, the rate at which the process goes from i to j (namely, $\pi_i P_{ij}$) is equal to the rate at which it goes from j to i (namely, $\pi_j P_{ji}$). It should be noted that this is an obvious necessary condition for time reversibility since a transition from i to j going backward in time is equivalent to a transition from j to i going forward in time; that is, if $X_m = i$ and $X_{m-1} = j$, then a transition from i to j is observed if we are looking backward in time and one from j to i if we are looking forward in time.

If we can find nonnegative numbers, summing to 1, which satisfy (4.7.1), then it follows that the Markov chain is time reversible and the numbers represent the stationary probabilities. This is so since if

$$x_i P_{ij} = x_j P_{ji} \qquad \text{for all } i, j \quad \sum_i x_i = 1, \tag{4.7.2}$$

then summing over i yields

$$\sum_i x_i P_{ij} = x_j \sum_i P_{ji} = x_j, \qquad \sum_i x_i = 1.$$

Since the stationary probabilities π_i are the unique solution of the above, it follows that $x_i = \pi_i$ for all i.

EXAMPLE 4.7(A) ***An Ergodic Random Walk.*** We can argue, without any need for computations, that an ergodic chain with $P_{i,i+1} + P_{i,i-1} = 1$ is time reversible. This follows by noting that the number of transitions from i to $i + 1$ must at all times be within 1 of the number from $i + 1$ to i. This is so since between any two transitions from i to $i + 1$ there must be one from $i + 1$ to i (and conversely) since the only way to re-enter i from a higher state is by way of state $i + 1$. Hence it follows that the rate of transitions from i to $i + 1$ equals the rate from $i + 1$ to i, and so the process is time reversible.

EXAMPLE 4.7(B) ***The Metropolis Algorithm.*** Let $a_j, j = 1, \ldots, m$ be positive numbers, and let $A = \sum_{j=1}^{m} a_j$. Suppose that m is large and that A is difficult to compute, and suppose we ideally want to simulate the values of a sequence of independent random variables whose probabilities are $p_j = a_j/A$, $j = 1, \ldots, m$. One way of simulating a sequence of random variables whose distributions converge to $\{p_j, j = 1, \ldots, m\}$ is to find a Markov chain that is both easy to simulate and whose limiting probabilities are the p_j. The *Metropolis algorithm* provides an approach for accomplishing this task.

Let $\boldsymbol{Q}$ be any irreducible transition probability matrix on the integers $1, \ldots, n$ such that $q_{ij} = q_{ji}$ for all i and j. Now define a Markov chain $\{X_n, n \geq 0\}$ as follows. If $X_n = i$, then generate a random variable that is equal to j with probability q_{ij}, $i, j = 1, \ldots, m$. If this random variable takes on the value j, then set X_{n+1} equal to j with probability $\min\{1, a_j/a_i\}$, and set it equal to i otherwise.

That is, the transition probabilities of $\{X_n, n \geq 0\}$ are

$$P_{ij} = \begin{cases} q_{ij} \min(1, a_j/a_i) & \text{if } j \neq i \\ q_{ii} + \sum_{j \neq i} q_{ij}\{1 - \min(1, a_j/a_i)\} & \text{if } j = i. \end{cases}$$

We will now show that the limiting probabilities of this Markov chain are precisely the p_j.

To prove that the p_j are the limiting probabilities, we will first show that the chain is time reversible with stationary probabilities $p_j, j = 1, \ldots, m$ by showing that

$$p_i P_{ij} = p_j P_{ji}.$$

To verify the preceding we must show that

$$p_i q_{ij} \min(1, a_j/a_i) = p_j q_{ji} \min(1, a_i/a_j).$$

Now, $q_{ij} = q_{ji}$ and $a_j/a_i = p_j/p_i$ and so we must verify that

$$p_i \min(1, p_j/p_i) = p_j \min(1, p_i/p_j).$$

However this is immediate since both sides of the equation are equal to $\min(p_i, p_j)$. That these stationary probabilities are also limiting probabilities follows from the fact that since $\boldsymbol{Q}$ is an irreducible transition probability matrix, $\{X_n\}$ will also be irreducible, and as (except in the trivial case where $p_i \equiv 1/n$) $P_{ii} > 0$ for some i, it is also aperiodic.

By choosing a transition probability matrix $\boldsymbol{Q}$ that is easy to simulate—that is, for each i it is easy to generate the value of a random variable that is equal to j with probability $q_{ij}, j = 1, \ldots, n$—we can use the preceding to generate a Markov chain whose limiting probabilities are $a_j/A, j = 1, \ldots, n$. This can also be accomplished without computing A.

Consider a graph having a positive number w_{ij} associated with each edge (i, j), and suppose that a particle moves from vertex to vertex in the following manner: If the particle is presently at vertex i then it will next move to vertex j with probability

$$P_{ij} = w_{ij} \Big/ \sum_j w_{ij}$$

where w_{ij} is 0 if (i, j) is not an edge of the graph. The Markov chain describing the sequence of vertices visited by the particle is called a random walk on an edge weighted graph.

PROPOSITION 4.7.1

Consider a random walk on an edge weighted graph with a finite number of vertices. If this Markov chain is irreducible then it is, in steady state, time reversible with stationary probabilities given by

$$\pi_i = \frac{\sum_i w_{ij}}{\sum_j \sum_i w_{ij}}.$$

Proof The time reversibility equations

$$\pi_i P_{ij} = \pi_j P_{ji}$$

reduce to

$$\frac{\pi_i w_{ij}}{\sum_k w_{ik}} = \frac{\pi_j w_{ji}}{\sum_k w_{jk}}$$

or, equivalently, since $w_{ij} = w_{ji}$

$$\frac{\pi_i}{\sum_k w_{ik}} = \frac{\pi_j}{\sum_k w_{jk}}$$

implying that

$$\pi_i = c \sum_k w_{ik}$$

which, since $\sum \pi_i = 1$, proves the result.

EXAMPLE 4.7(c) Consider a star graph consisting of r rays, with each ray consisting of n vertices. (See Example 1.9(C) for the definition of a star graph.) Let leaf i denote the leaf on ray i. Assume that a particle moves along the vertices of the graph in the following manner. Whenever it is at the central vertex 0, it is then equally likely to move to any of its neighbors. Whenever it is on an internal (nonleaf) vertex of a ray, then it moves towards the leaf of that ray with probability p and towards 0 with probability $1 - p$. Whenever it is at a leaf, it moves to its neighbor vertex with probability 1. Starting at vertex 0, we are interested in finding the expected number of transitions that it takes to visit all the vertices and then return to 0.

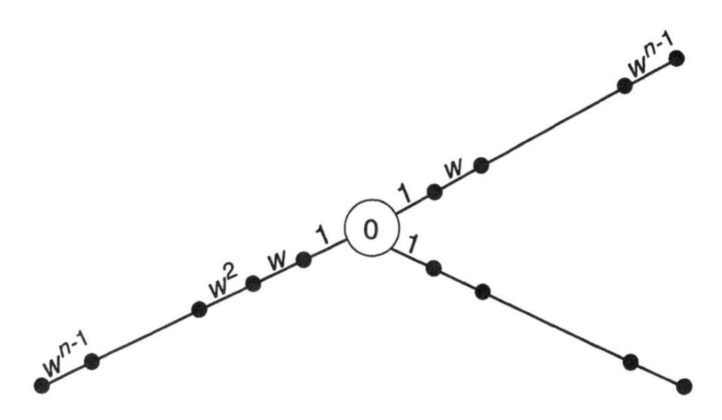

Figure 4.7.1. A star graph with weights: $w = p/(1 - p)$.

To begin, let us determine the expected number of transitions between returns to the central vertex 0. To evaluate this quantity, note the Markov chain of successive vertices visited is of the type considered in Proposition 4.7.1. To see this, attach a weight equal to 1 with each edge connected to 0, and a weight equal to w^i on an edge connecting the ith and $(i + 1)$st vertex (from 0) of a ray, where $w = p/(1 - p)$ (see Figure 4.7.1). Then, with these edge weights, the probability that a particle at a vertex i steps from 0 moves towards its leaf is $w^i/(w^i + w^{i-1}) = p$.

Since the total of the sum of the weights on the edges out of each of the vertices is

$$r + r\left[\sum_{i=1}^{n-1} (w^{i-1} + w^i) + w^{n-1}\right] = \frac{2r(1 - w^n)}{1 - w},$$

and the sum of the weights on the edges out of vertex 0 is r, we see from Proposition 4.7.1 that

$$\pi_0 = \frac{1 - w}{2(1 - w^n)}.$$

Therefore, μ_{00}, the expected number of steps between returns to vertex 0, is

$$\mu_{00} = 1/\pi_0 = \frac{2(1 - w^n)}{1 - w}.$$

Now, say that a new cycle begins whenever the particle returns to vertex 0, and let X_j be the number of transitions in the jth cycle, $j \geq 1$. Also, fix i and let N denote the number of cycles that it takes for the particle to visit leaf i and then return to 0. With these definitions, $\sum_{j=1}^{N} X_j$ is equal to the number of steps it takes to visit leaf i and then return to 0. As N is clearly a stopping time for the X_j, we obtain from Wald's equation that

$$E\left[\sum_{j=1}^{N} X_j\right] = \mu_{00} E[N] = \frac{2(1 - w^n)}{1 - w} E[N].$$

To determine $E[N]$, the expected number of cycles needed to reach leaf i, note that each cycle will independently reach leaf i with probability $\frac{1 - 1/w}{r[1 - (1/w)^n]}$ where $1/r$ is the probability that the transition from 0 is onto ray i, and $\frac{1 - 1/w}{1 - (1/w)^n}$ is the (gambler's ruin) probability that a particle on the first vertex of ray i will reach the leaf of that ray (that is, increase by $n - 1$) before returning to 0. Therefore N, the number of cycles needed to reach leaf i, is a geometric random variable with mean $r[1 - (1/w)^n]/(1 - 1/w)$, and so

$$E\left[\sum_{j=1}^{N} X_j\right] = \frac{2r(1 - w^n)[1 - (1/w)^n]}{(1 - w)(1 - 1/w)} = \frac{2r[2 - w^n - (1/w)^n]}{2 - w - 1/w}.$$

Now, let T denote the number of transitions that it takes to visit all the vertices of the graph and then return to vertex 0. To determine $E[T]$ we will use the representation

$$T = T_1 + T_2 + \cdots + T_r,$$

where T_1 is the time to visit the leaf 1 and then return to 0; T_2 is the additional time from T_1 until both the leafs 1 and 2 have been visited and the process returned to vertex 0; and, in general, T_i is the additional time from T_{i-1} until all of the leafs $1, \ldots, i$ have been visited and the process returned to 0. Note that if leaf i is not the last of the leafs $1, \ldots, i$ to be visited, then T_i will equal 0, and if it is the last of these leafs to be visited, then T_i will have the same distribution as the time until a specified leaf is first visited and the process then returned to 0. Hence, upon conditioning on whether leaf i is the last of leafs $1, \ldots, i$ to be visited (and the probability of this event is clearly $1/i$), we obtain from the preceding that

$$E[T] = \frac{2r[2 - w^n - 1/w^n]}{2 - w - 1/w} \sum_{i=1}^{r} 1/i.$$

If we try to solve Equations (4.7.2) for an arbitrary Markov chain, it will usually turn out that no solution exists. For example, from (4.7.2)

$$x_i P_{ij} = x_j P_{ji},$$
$$x_k P_{kj} = x_j P_{jk},$$

implying (if $P_{ij}P_{jk} > 0$) that

$$\frac{x_i}{x_k} = \frac{P_{ji}P_{kj}}{P_{ij}P_{jk}},$$

which need not in general equal P_{ki}/P_{ik}. Thus we see that a necessary condition for time reversibility is that

$$P_{ik}P_{kj}P_{ji} = P_{ij}P_{jk}P_{ki} \qquad \text{for all } i, j, k, \tag{4.7.3}$$

which is equivalent to the statement that, starting in state i, the path $i \to k \to j \to i$ has the same probability as the reversed path $i \to j \to k \to i$. To understand the necessity of this, note that time reversibility implies that the rate at which a sequence of transitions from i to k to j to i occur must equal the rate of ones from i to j to k to i (why?), and so we must have

$$\pi_i P_{ik}P_{kj}P_{ji} = \pi_i P_{ij}P_{jk}P_{ki},$$

implying (4.7.3).

In fact we can show the following.

THEOREM 4.7.2

A stationary Markov chain is time reversible if, and only if, starting in state i, any path back to i has the same probability as the reversed path, for all i. That is, if

$$P_{i,i_1}P_{i_1,i_2}\cdots P_{i_k,i} = P_{i,i_k}P_{i_k,i_{k-1}}\cdots P_{i_1,i} \tag{4.7.4}$$

for all states $i, i_1, \ldots, i_k$.

Proof The proof of necessity is as indicated. To prove sufficiency fix states i and j and rewrite (4.7.4) as

$$P_{i,i_1}P_{i_1,i_2}\cdots P_{i_k,j}P_{ji} = P_{ij}P_{j,i_k}\cdots P_{i_1,i}.$$

Summing the above over all states $i_1, i_2, \ldots, i_k$ yields

$$P_{ij}^{k+1}P_{ji} = P_{ij}P_{ji}^{k+1}.$$

Hence

$$P_{ji}\frac{\sum_{k=1}^{n} P_{ij}^{k+1}}{n} = P_{ij}\frac{\sum_{k=1}^{n} P_{ji}^{k+1}}{n}.$$

Letting $n \to \infty$ now yields

$$P_{ji}\pi_j = P_{ij}\pi_i,$$

which establishes the result.

Example 4.7(d) ***A List Problem.*** Suppose we are given a set of n elements—numbered 1 through n—that are to be arranged in some ordered list. At each unit of time a request is made to retrieve one of these elements—element i being requested (independently of the past) with probability P_i. After being requested, the element is then put back, but not necessarily in the same position. In fact, let us suppose that the element requested is moved one closer to the front of the list; for instance, if the present list ordering is 1, 3, 4, 2, 5 and element 2 is requested, then the new ordering becomes 1, 3, 2, 4, 5.

For any given probability vector $\underline{P} = (P_1, \ldots, P_n)$, the above can be modeled as a Markov chain with $n!$ states with the state at any time being the list order at that time. By using Theorem 4.7.1 it is easy to show that this chain is time reversible. For instance, suppose $n = 3$. Consider the following path from state (1, 2, 3) to itself:

$$(1,2,3) \to (2,1,3) \to (2,3,1) \to (3,2,1) \to (3,1,2) \to (1,3,2) \to (1,2,3).$$

The products of the transition probabilities in the forward direction and in the reverse direction are both equal to $P_1^2P_2^2P_3^2$. Since a similar result holds in general, the Markov chain is time reversible.

In fact, time reversibility and the limiting probabilities can also be verified by noting that for any permutation $(i_1, i_2, \ldots, i_n)$, the probabilities given by

$$\pi(i_1, \ldots, i_n) = CP_{i_1}^n P_{i_2}^{n-1} \cdots P_{i_n}$$

satisfy Equation (4.7.1), where C is chosen so that

$$\sum_{(i_1,\ldots,i_n)} \pi(i_1, \ldots, i_n) = 1.$$

Hence, we have a second argument that the chain is reversible, and the stationary probabilities are as given above.

The concept of the reversed chain is useful even when the process is not time reversible. To illustrate this we start with the following theorem.

THEOREM 4.7.3

Consider an irreducible Markov chain with transition probabilities P_{ij}. If one can find nonnegative numbers π_i, $i \geq 0$, summing to unity, and a transition probability matrix $P^* = [P^*_{ij}]$ such that

$$\pi_i P_{ij} = \pi_j P^*_{ji}, \tag{4.7.5}$$

then the π_i, $i \geq 0$, are the stationary probabilities and P^*_{ij} are the transition probabilities of the reversed chain.

Proof Summing the above equality over all i yields

$$\sum_i \pi_i P_{ij} = \pi_j \sum_i P^*_{ji}$$
$$= \pi_j.$$

Hence, the π_i's are the stationary probabilities of the forward chain (and also of the reversed chain; why?). Since

$$P^*_{ji} = \frac{\pi_i P_{ij}}{\pi_j},$$

it follows that the P^*_{ij} are the transition probabilities of the reversed chain.

The importance of Theorem 4.7.2 is that we can sometimes guess at the nature of the reversed chain and then use the set of equations (4.7.5) to obtain both the stationary probabilities and the P^*_{ij}.

EXAMPLE 4.7(E) Let us reconsider Example 4.3(C), which deals with the age of a discrete time renewal process. That is, let X_n denote the age at time n of a renewal process whose interarrival times are all integers. Since the state of this Markov chain always increases by one until it hits a value chosen by the interarrival distribution and then drops to 1, it follows that the reverse process will always decrease by one until it hits state 1 at which time it jumps to a state chosen by the interarrival distribution. Hence it seems that the reversed process is just the *excess* or *residual life* process.

Thus letting P_i denote the probability that an interarrival is i, $i \geq 1$, it seems likely that

$$P^*_{1i} = P_i, \qquad P^*_{i,i-1} = 1, \qquad i > 1.$$

Since

$$P_{i1} = \frac{P_i}{\sum_{j=i}^{\infty} P_j} = 1 - P_{i,i+1}, \qquad i \geq 1,$$

for the reversed chain to be as given above we would need from (4.7.5) that

$$\frac{\pi_i P_i}{\sum_{j=i}^{\infty} P_j} = \pi_1 P_i$$

or

$$\pi_i = \pi_1 P\{X \geq i\},$$

where X is an interarrival time. Since $\sum_i \pi_i = 1$, the above would necessitate that

$$\begin{aligned} 1 &= \pi_1 \sum_{i=1} P\{X \geq i\} \\ &= \pi_1 E[X], \end{aligned}$$

and so for the reversed chain to be as conjectured we would need that

$$\pi_i = \frac{P\{X \geq i\}}{E[X]}. \tag{4.7.6}$$

To complete the proof that the reversed process is the excess and the limiting probabilities are given by (4.7.6), we need verify that

$$\pi_i P_{i,i+1} = \pi_{i+1} P^*_{i+1,i},$$

or, equivalently,

$$P\{X \geq i\} \left[1 - \frac{P_i}{P\{X \geq i\}}\right) = P\{X \geq i + 1\},$$

which is immediate.

Thus by looking at the reversed chain we are able to show that it is the excess renewal process and obtain at the same time the limiting distribution (of both excess and age). In fact, this example yields additional insight as to why the renewal excess and age have the same limiting distribution.

The technique of using the reversed chain to obtain limiting probabilities will be further exploited in Chapter 5, where we deal with Markov chains in continuous time.

4.8 Semi-Markov Processes

A semi-Markov process is one that changes states in accordance with a Markov chain but takes a random amount of time between changes. More specifically consider a stochastic process with states 0, 1, . . . , which is such that, whenever it enters state i, $i \geq 0$:

(i) The next state it will enter is state j with probability P_{ij}, $i, j \geq 0$.
(ii) Given that the next state to be entered is state j, the time until the transition from i to j occurs has distribution F_{ij}.

If we let $Z(t)$ denote the state at time t, then $\{Z(t), t \geq 0\}$ is called a *semi-Markov process.*

Thus a semi-Markov process does not possess the Markovian property that given the present state the future is independent of the past. For in predicting the future not only would we want to know the present state, but also the length of time that has been spent in that state. Of course, at the moment of transition, all we would need to know is the new state (and nothing about the past). A Markov chain is a semi-Markov process in which

$$F_{ij}(t) = \begin{cases} 0 & t < 1 \\ 1 & t \geq 1. \end{cases}$$

That is, all transition times of a Markov chain are identically 1.

Let H_i denote the distribution of time that the semi-Markov process spends in state i before making a transition. That is, by conditioning on the next state, we see

$$H_i(t) = \sum_j P_{ij} F_{ij}(t),$$

and let μ_i denote its mean. That is,

$$\mu_i = \int_0^\infty x \, dH_i(x)$$

If we let X_n denote the nth state visited, then $\{X_n, n \geq 0\}$ is a Markov chain with transition probabilities P_{ij}. It is called the *embedded* Markov chain of the semi-Markov process. We say that the semi-Markov process is *irreducible* if the embedded Markov chain is irreducible as well.

Let T_{ii} denote the time between successive transitions into state i and let $\mu_{ii} = E[T_{ii}]$. By using the theory of alternating renewal processes, it is a simple matter to derive an expression for the limiting probabilities of a semi-Markov process.

PROPOSITION 4.8.1

If the semi-Markov process is irreducible and if T_{ii} has a nonlattice distribution with finite mean, then

$$P_i \equiv \lim_{t \to \infty} P\{Z(t) = i \mid Z(0) = j\}$$

exists and is independent of the initial state. Furthermore,

$$P_i = \frac{\mu_i}{\mu_{ii}}.$$

Proof Say that a cycle begins whenever the process enters state i, and say that the process is "on" when in state i and "off" when not in i. Thus we have a (delayed when $Z(0) \neq i$) alternating renewal process whose on time has distribution H_i and whose cycle time is T_{ii}. Hence, the result follows from Proposition 3.4.4 of Chapter 3.

As a corollary we note that P_i is also equal to the long-run proportion of time that the process is in state i.

Corollary 4.8.2

If the semi-Markov process is irreducible and $\mu_{ii} < \infty$, then, with probability 1,

$$\frac{\mu_i}{\mu_{ii}} = \lim_{t \to \infty} \frac{\text{amount of time in } i \text{ during } [0, t]}{t}.$$

That is, μ_i / μ_{ii} equals the long-run proportion of time in state i.

Proof Follows from Proposition 3.7.2 of Chapter 3.

While Proposition 4.8.1 gives us an expression for the limiting probabilities, it is not, however, the way one actually computes the P_i. To do so suppose that the embedded Markov chain $\{X_n, n \ge 0\}$ is irreducible and positive recurrent, and let its stationary probabilities be $\pi_j, j \ge 0$. That is, the $\pi_j, j \ge 0$, is the unique solution of

$$\pi_j = \sum_i \pi_i P_{ij},$$

$$\sum_j \pi_j = 1,$$

and π_j has the interpretation of being the proportion of the X_n's that equals j. (If the Markov chain is aperiodic, then π_j is also equal to $\lim_{n\to\infty} P\{X_n = j\}$.) Now as π_j equals the proportion of transitions that are into state j, and μ_j is the mean time spent in state j per transition, it seems intuitive that the limiting probabilities should be proportional to $\pi_j \mu_j$. We now prove this.

THEOREM 4.8.3

Suppose the conditions of Proposition 4.8.1 and suppose further that the embedded Markov chain $\{X_n, n \ge 0\}$ is positive recurrent. Then

$$P_i = \frac{\pi_i \mu_i}{\sum_j \pi_j \mu_j}.$$

Proof Define the notation as follows:

$Y_i(j)$ = amount of time spent in state i during the jth visit to that state, $i, j \ge 0$.

$N_i(m)$ = number of visits to state i in the first m transitions of the semi-Markov process.

In terms of the above notation we see that the proportion of time in i during the first m transitions, call it $P_{i=m}$, is as follows:

$$P_{i=m} = \frac{\sum_{j=1}^{N_i(m)} Y_i(j)}{\sum_i \sum_{j=1}^{N_i(m)} Y_i(j)} \tag{4.8.1}$$

$$= \frac{\dfrac{N_i(m)}{m} \displaystyle\sum_{j=1}^{N_i(m)} \frac{Y_i(j)}{N_i(m)}}{\displaystyle\sum_i \frac{N_i(m)}{m} \sum_{j=1}^{N_i(m)} \frac{Y_i(j)}{N_i(m)}}.$$

Now since $N_i(m) \to \infty$ as $m \to \infty$, it follows from the strong law of large numbers that

$$\sum_{j=1}^{N_i(m)} \frac{Y_i(j)}{N_i(m)} \to \mu_i,$$

and, by the strong law for renewal processes, that

$$\frac{N_i(m)}{m} \to (E[\text{number of transitions between visits to } i])^{-1} = \pi_i.$$

Hence, letting $m \to \infty$ in (4.8.1) shows that

$$\lim_{m\to\infty} P_{i=m} = \frac{\pi_i \mu_i}{\sum_j \pi_j \mu_j},$$

and the proof is complete.

From Theorem 4.8.3 it follows that the limiting probabilities depend only on the transition probabilities P_{ij} and the mean times μ_i, $i, j \geq 0$.

Example 4.8(a) Consider a machine that can be in one of three states: *good condition*, *fair condition*, or *broken down*. Suppose that a machine in good condition will remain this way for a mean time μ_1 and will then go to either the fair condition or the broken condition with respective probabilities $\frac{3}{4}$ and $\frac{1}{4}$. A machine in the fair condition will remain that way for a mean time μ_2 and will then break down. A broken machine will be repaired, which takes a mean time μ_3, and when repaired will be in the good condition with probability $\frac{2}{3}$ and the fair condition with probability $\frac{1}{3}$. What proportion of time is the machine in each state?

Solution. Letting the states be 1, 2, 3, we have that the π_i satisfy

$$\begin{aligned}
\pi_1 + \pi_2 + \pi_3 &= 1,\\
\pi_1 &= \tfrac{2}{3}\pi_3,\\
\pi_2 &= \tfrac{3}{4}\pi_1 + \tfrac{1}{3}\pi_3,\\
\pi_3 &= \tfrac{1}{4}\pi_1 + \pi_2.
\end{aligned}$$

The solution is

$$\pi_1 = \tfrac{4}{15}, \qquad \pi_2 = \tfrac{1}{3}, \qquad \pi_3 = \tfrac{2}{5}.$$

Hence, P_i, the proportion of time the machine is in state i, is given by

$$P_1 = \frac{4\mu_1}{4\mu_1 + 5\mu_2 + 6\mu_3},$$

$$P_2 = \frac{5\mu_2}{4\mu_1 + 5\mu_2 + 6\mu_3},$$

$$P_3 = \frac{6\mu_3}{4\mu_1 + 5\mu_2 + 6\mu_3}.$$

The problem of determining the limiting distribution of a semi-Markov process is not completely solved by deriving the P_i. For we may ask for the limit, as $t \to \infty$, of being in state i at time t of making the next transition after time $t + x$, and of this next transition being into state j. To express this probability let

$$Y(t) = \text{time from } t \text{ until the next transition},$$

$$S(t) = \text{state entered at the first transition after } t.$$

To compute

$$\lim_{t\to\infty} P\{Z(t) = i, Y(t) > x, S(t) = j\},$$

we again use the theory of alternating renewal processes.

THEOREM 4.8.4

If the semi-Markov process is irreducible and not lattice, then

$$\lim_{t\to\infty} P\{Z(t) = i, Y(t) > x, S(t) = j \mid Z(0) = k\} = \frac{P_{ij}\int_x^\infty \overline{F}_{ij}(y)\,dy}{\mu_{ii}}. \tag{4.8.2}$$

Proof Say that a cycle begins each time the process enters state i and say that it is "on" if the state is i and it will remain i for at least the next x time units and the next state is j. Say it is "off" otherwise. Thus we have an alternating renewal process. Conditioning on whether the state after i is j or not, we see that

$$E[\text{"on" time in a cycle}] = P_{ij}E[(X_{ij} - x)^+],$$

where X_{ij} is a random variable having distribution F_{ij} and representing the time to make a transition from i to j, and $y^+ = \max(0, y)$. Hence

$$E[\text{"on" time in cycle}] = P_{ij}\int_0^\infty P\{X_{ij} - x > a\}\,da$$
$$= P_{ij}\int_0^\infty \overline{F}_{ij}(a + x)\,da$$
$$= P_{ij}\int_x^\infty \overline{F}_{ij}(y)\,dy.$$

As $E[\text{cycle time}] = \mu_{ii}$, the result follows from alternating renewal processes.

By the same technique (or by summing (4.8.2) over j) we can prove the following.

Corollary 4.8.5

If the semi-Markov process is irreducible and not lattice, then

$$\lim_{t\to\infty} P\{Z(t) = i, Y(t) > x \mid Z(0) = k\} = \int_x^\infty \overline{H}_i(y)\,dy/\mu_{ii}. \tag{4.8.3}$$

Remarks

(1) Of course the limiting probabilities in Theorem 4.8.4 and Corollary 4.8.5 also have interpretations as long-run proportions. For instance, the long-run proportion of time that the semi-Markov process is in state i and will spend the next x time units without a transition and will then go to state j is given by Theorem 4.8.4.

(2) Multiplying and dividing (4.8.3) by μ_i, and using $P_i = \mu_i/\mu_{ii}$, gives

$$\lim_{t\to\infty} P\{Z(t) = i, Y(t) > x\} = P_i\overline{H}_{i,e}(x),$$

where $H_{i,e}$ is the equilibrium distribution of H_i. Hence the limiting probability of being in state i is P_i, and, given that the state at t is i, the time until transition (as t approaches ∞) has the equilibrium distribution of H_i.

PROBLEMS

4.1. A store that stocks a certain commodity uses the following (s, S) ordering policy; if its supply at the beginning of a time period is x, then it orders

$$\begin{array}{ll} 0 & \text{if } x \geq s, \\ S - x & \text{if } x < s. \end{array}$$

The order is immediately filled. The daily demands are independent and equal j with probability α_j. All demands that cannot be immediately met are lost. Let X_n denote the inventory level at the end of the nth time period. Argue that $\{X_n, n \geq 1\}$ is a Markov chain and compute its transition probabilities.

4.2. For a Markov chain prove that

$$P\{X_n = j | X_{n_1} = i_1, \ldots, X_{n_k} = i_k\} = P\{X_n = j | X_{n_k} = i_k\}$$

whenever $n_1 < n_2 < \cdots < n_k < n$.

4.3. Prove that if the number of states is n, and if state j is accessible from state i, then it is accessible in n or fewer steps.

4.4. Show that

$$P_{ij}^n = \sum_{k=0}^{n} f_{ij}^k P_{jj}^{n-k}.$$

4.5. For states $i, j, k, k \neq j$, let

$$P_{ij/k}^n = P\{X_n = j, X_\ell \neq k, \ell = 1, \ldots, n-1 | X_0 = i\}.$$

(a) Explain in words what $P_{ij/k}^n$ represents.

(b) Prove that, for $i \neq j$, $P_{ij}^n = \sum_{k=0}^{n} P_{ii}^k P_{ij/i}^{n-k}$.

4.6. Show that the symmetric random walk is recurrent in two dimensions and transient in three dimensions.

4.7. For the symmetric random walk starting at 0:

(a) What is the expected time to return to 0?

(b) Let N_n denote the number of returns by time n. Show that

$$E[N_{2n}] = (2n + 1)\binom{2n}{n}\left(\frac{1}{2}\right)^{2n} - 1.$$

(c) Use (b) and Stirling's approximation to show that for n large $E[N_n]$ is proportional to $\sqrt{n}$.

4.8. Let $X_1, X_2, \ldots$ be independent random variables such that $P\{X_i = j\} = \alpha_j, j \geq 0$. Say that a record occurs at time n if $X_n > \max(X_1, \ldots, X_{n-1})$, where $X_0 = -\infty$, and if a record does occur at time n call X_n the record value. Let R_i denote the ith record value.

(a) Argue that $\{R_i, i \geq 1\}$ is a Markov chain and compute its transition probabilities.

(b) Let T_i denote the time between the ith and $(i + 1)$st record. Is $\{T_i, i \geq 1\}$ a Markov chain? What about $\{(R_i, T_i), i \geq 1\}$? Compute transition probabilities where appropriate.

(c) Let $S_n = \sum_{i=1}^{n} T_i, n \geq 1$. Argue that $\{S_n, n \geq 1\}$ is a Markov chain when the X_i are continuous and find its transition probabilities.

4.9. For a Markov chain $\{X_n, n \geq 0\}$, show that

$$P\{X_k = i_k | X_j = i_j, \text{for all } j \neq k\} = P\{X_k = i_k | X_{k-1} = i_{k-1}, X_{k+1} = i_{k+1}\}.$$

4.10. At the beginning of every time period, each of N individuals is in one of three possible conditions: infectious, infected but not infectious, or noninfected. If a noninfected individual becomes infected during a time period then he or she will be in an infectious condition during the following time period, and from then on will be in an infected (but not infectious) condition. During every time period each of the $\binom{N}{2}$ pairs of individuals are independently in contact with probability p. If a pair is in contact and one of the members of the pair is infectious and the other is noninfected then the noninfected person becomes infected (and is thus in the infectious condition at the beginning of the next period). Let X_n and Y_n denote the number of infectious and the number of noninfected individuals, respectively, at the beginning of time period n.

(a) If there are i infectious individuals at the beginning of a time period, what is the probability that a specified noninfected individual will become infected in that period?

(b) Is $\{X_n, n \geq 0\}$ a Markov chain? If so, give its transition probabilities.

(c) Is $\{Y_n, n \geq 0\}$ a Markov chain? If so, give its transition probabilities.

(d) Is $\{(X_n, Y_n), n \geq 0\}$ a Markov chain? If so, give its transition probabilities.

4.11. If $f_{ii} < 1$ and $f_{jj} < 1$, show that:

$$\textbf{(a)}\ \sum_{n=1}^{\infty} P_{ij}^n < \infty; \quad \textbf{(b)}\ f_{ij} = \frac{\sum_{n=1}^{\infty} P_{ij}^n}{1 + \sum_{n=1}^{\infty} P_{jj}^n}.$$

4.12. A transition probability matrix P is said to be doubly stochastic if

$$\sum_i P_{ij} = 1 \qquad \text{for all } j.$$

That is, the column sums all equal 1. If a doubly stochastic chain has n states and is ergodic, calculate its limiting probabilities.

4.13. Show that positive and null recurrence are class properties.

4.14. Show that in a finite Markov chain there are no null recurrent states and not all states can be transient.

4.15. In the $M/G/1$ system (Example 4.3(A)) suppose that $\rho < 1$ and thus the stationary probabilities exist. Compute $\pi'(s)$ and find, by taking the limit as $s \to 1$, $\sum_0^\infty i\pi_i$.

4.16. An individual possesses r umbrellas, which she employs in going from her home to office and vice versa. If she is at home (the office) at the beginning (end) of a day and it is raining, then she will take an umbrella with her to the office (home), provided there is one to be taken. If it is not raining, then she never takes an umbrella. Assume that, independent of the past, it rains at the beginning (end) of a day with probability p.

(a) Define a Markov chain with $r + 1$ states that will help us determine the proportion of time that our individual gets wet. (Note: She gets wet if it is raining and all umbrellas are at her other location.)

(b) Compute the limiting probabilities.

(c) What fraction of time does the individual become wet?

4.17. Consider a positive recurrent irreducible periodic Markov chain and let π_j denote the long-run proportion of time in state j, $j \geq 0$. Prove that π_j, $j \geq 0$, satisfy $\pi_j = \sum_i \pi_i P_{ij}$, $\sum_j \pi_j = 1$.

4.18. Jobs arrive at a processing center in accordance with a Poisson process with rate λ. However, the center has waiting space for only N jobs and so an arriving job finding N others waiting goes away. At most 1 job per day can be processed, and the processing of this job must start at the beginning of the day. Thus, if there are any jobs waiting for processing at the beginning of a day, then one of them is processed that day, and if no jobs are waiting at the beginning of a day then no jobs are processed that day. Let X_n denote the number of jobs at the center at the beginning of day n.

(a) Find the transition probabilities of the Markov chain $\{X_n, n \geq 0\}$.

(b) Is this chain ergodic? Explain.

(c) Write the equations for the stationary probabilities.

4.19. Let $\pi_j, j \geq 0$, be the stationary probabilities for a specified Markov chain.

(a) Complete the following statement: $\pi_i P_{ij}$ is the proportion of all transitions that . . . .

Let A denote a set of states and let A^c denote the remaining states.

(b) Finish the following statement: $\sum_{j \in A^c} \sum_{i \in A} \pi_i P_{ij}$ is the proportion of all transitions that . . . .

(c) Let $N_n(A, A^c)$ denote the number of the first n transitions that are from a state in A to one in A^c; similarly, let $N_n(A^c, A)$ denote the number that are from a state in A^c to one in A. Argue that

$$|N_n(A, A^c) - N_n(A^c, A)| \leq 1.$$

(d) Prove and interpret the following result:

$$\sum_{j \in A^c} \sum_{i \in A} \pi_i P_{ij} = \sum_{j \in A^c} \sum_{i \in A} \pi_j P_{ji}.$$

4.20. Consider a recurrent Markov chain starting in state 0. Let m_i denote the expected number of time periods it spends in state i before returning to 0. Use Wald's equation to show that

$$m_j = \sum_i m_i P_{ij}, \qquad j > 0$$

$$m_0 = 1.$$

Now give a second proof that assumes the chain is positive recurrent and relates the m_j to the stationary probabilities.

4.21. Consider a Markov chain with states 0, 1, 2, . . . and such that

$$P_{i,i+1} = p_i = 1 - P_{i,i-1},$$

where $p_0 = 1$. Find the necessary and sufficient condition on the p_i's for this chain to be positive recurrent, and compute the limiting probabilities in this case.

4.22. Compute the expected number of plays, starting in i, in the gambler's ruin problem, until the gambler reaches either 0 or N.

4.23. In the gambler's ruin problem show that

P\{she wins the next gamble|present fortune is i, she eventually reaches N\}

$$= \begin{cases} p[1 - (q/p)^{i+1}]/[1 - (q/p)^i] & \text{if } p \neq \frac{1}{2} \\ (i + 1)/2i & \text{if } p = \frac{1}{2}. \end{cases}$$

4.24. Let $T = \{1, \ldots, t\}$ denote the transient states of a Markov chain, and let $\boldsymbol{Q}$ be, as in Section 4.4, the matrix of transition probabilities from states in T to states in T. Let $m_{ij}(n)$ denote the expected amount of time spent in state j during the first n transitions given that the chain begins in state i, for i and j in T. Let $\boldsymbol{M}_n$ be the matrix whose element in row i, column j, is $m_{ij}(n)$.

(a) Show that $\boldsymbol{M}_n = \boldsymbol{I} + \boldsymbol{Q} + \boldsymbol{Q}^2 + \cdots + \boldsymbol{Q}^n$.

(b) Show that $\boldsymbol{M}_n - \boldsymbol{I} + \boldsymbol{Q}^{n+1} = \boldsymbol{Q}[\boldsymbol{I} + \boldsymbol{Q} + \boldsymbol{Q}^2 + \cdots + \boldsymbol{Q}^n]$.

(c) Show that $\boldsymbol{M}_n = (\boldsymbol{I} - \boldsymbol{Q})^{-1}(\boldsymbol{I} - \boldsymbol{Q}^{n+1})$.

4.25. Consider the gambler's ruin problem with $N = 6$ and $p = .7$. Starting in state 3, determine:

(a) the expected number of visits to state 5.

(b) the expected number of visits to state 1.

(c) the expected number of visits to state 5 in the first 7 transitions.

(d) the probability of ever visting state 1.

4.26. Consider the Markov chain with states $0, 1, \ldots, n$ and transition probabilities

$$P_{0,1} = 1 = P_{n,n-1}, \qquad P_{i,i+1} = p = 1 - P_{i,i-1}, \qquad 0 < i < n.$$

Let $\mu_{i,n}$ denote the mean time to go from state i to state n.

(a) Derive a set of linear equations for the $\mu_{i,n}$.

(b) Let m_i denote the mean time to go from state i to state $i + 1$. Derive a set of equations for the m_i, $i = 0, \ldots, n - 1$, and show how they can be solved recursively, first for $i = 0$, then $i = 1$, and so on.

(c) What is the relation between $\mu_{i,n}$ and the m_j?

Starting at state 0, say that an excursion ends when the chain either returns to 0 or reaches state n. Let X_j denote the number of transitions in the jth excursion (that is, the one that begins at the jth return to 0), $j \geq 1$.

(d) Find $E[X_j]$.
(*Hint:* Relate it to the mean time of a gambler's ruin problem.)

(e) Let N denote the first excursion that ends in state n, and find $E[N]$.

(f) Find $\mu_{0,n}$.

(g) Find $\mu_{i,n}$.

4.27. Consider a particle that moves along a set of $m + 1$ nodes, labeled 0, $1, \ldots, m$. At each move it either goes one step in the clockwise direction with probability p or one step in the counterclockwise direction with probability $1 - p$. It continues moving until all the nodes $1, 2, \ldots, m$ have been visited at least once. Starting at node 0, find the probability that node i is the last node visited, $i = 1, \ldots, m$.

4.28. In Problem 4.27, find the expected number of additional steps it takes to return to the initial position after all nodes have been visited.

4.29. Each day one of n possible elements is requested; the ith one with probability P_i, $i \geq 1$, $\sum_1^n P_i = 1$. These elements are at all times arranged in an ordered list that is revised as follows: the element selected is moved to the front of the list with the relative positions of all the other elements remaining unchanged. Define the state at any time to be the list ordering at that time.

(a) Argue that the above is a Markov chain.

(b) For any state $i_1, \ldots, i_n$ (which is a permutation of $1, 2, \ldots, n$) let $\pi(i_1, \ldots, i_n)$ denote the limiting probability. Argue that

$$\pi(i_1, \ldots, i_n) = P_{i_1} \frac{P_{i_2}}{1 - P_{i_1}} \cdots \frac{P_{i_{n-1}}}{1 - P_{i_1} - \cdots - P_{i_{n-2}}}.$$

4.30. Suppose that two independent sequences $X_1, X_2, \ldots$ and $Y_1, Y_2, \ldots$ are coming in from some laboratory and that they represent Bernoulli trials with unknown success probabilities P_1 and P_2. That is, $P\{X_i = 1\} = 1 - P\{X_i = 0\} = P_1$, $P\{Y_i = 1\} = 1 - P\{Y_i = 0\} = P_2$, and all random variables are independent. To decide whether $P_1 > P_2$ or $P_2 > P_1$, we use the following test. Choose some positive integer M and stop at N, the first value of n such that either

$$X_1 + \cdots + X_n - (Y_1 + \cdots + Y_n) = M$$

or

$$X_1 + \cdots + X_n - (Y_1 + \cdots + Y_n) = -M.$$

In the former case we then assert that $P_1 > P_2$, and in the latter that $P_2 > P_1$. Show that when $P_1 \geq P_2$, the probability of making an error (that is, of asserting that $P_2 > P_1$) is

$$P\{\text{error}\} = \frac{1}{1 + \lambda^M},$$

and, also, that the expected number of pairs observed is

$$E[N] = \frac{M(\lambda^M - 1)}{(P_1 - P_2)(\lambda^M + 1)},$$

where

$$\lambda = \frac{P_1(1 - P_2)}{P_2(1 - P_1)}.$$

(*Hint:* Relate this to the gambler's ruin problem.)

4.31. A spider hunting a fly moves between locations 1 and 2 according to a Markov chain with transition matrix $\begin{bmatrix} 0.7 & 0.3 \\ 0.3 & 0.7 \end{bmatrix}$ starting in location 1. The fly, unaware of the spider, starts in location 2 and moves according to a Markov chain with transition matrix $\begin{bmatrix} 0.4 & 0.6 \\ 0.6 & 0.4 \end{bmatrix}$. The spider catches the fly and the hunt ends whenever they meet in the same location.

Show that the progress of the hunt, except for knowing the location where it ends, can be described by a three-state Markov chain where one absorbing state represents hunt ended and the other two that the spider and fly are at different locations. Obtain the transition matrix for this chain.

(a) Find the probability that at time n the spider and fly are both at their initial locations.

(b) What is the average duration of the hunt?

4.32. Consider a simple random walk on the integer points in which at each step a particle moves one step in the positive direction with probability

p, one step in the negative direction with probability p, and remains in the same place with probability $q = 1 - 2p (0 < p < \frac{1}{2})$. Suppose an absorbing barrier is placed at the origin—that is, $P_{00} = 1$—and a reflecting barrier at N—that is, $P_{N,N-1} = 1$—and that the particle starts at n $(0 < n < N)$.

Show that the probability of absorption is 1, and find the mean number of steps.

4.33. Given that $\{X_n, n \geq 0\}$ is a branching process:

(a) Argue that either X_n converges to 0 or to infinity.

(b) Show that

$$\text{Var}(X_n | X_0 = 1) = \begin{cases} \sigma^2 \mu^{n-1} \dfrac{\mu^n - 1}{\mu - 1} & \text{if } \mu \neq 1 \\ n\sigma^2 & \text{if } \mu = 1, \end{cases}$$

where μ and σ^2 are the mean and variance of the number of offspring an individual has.

4.34. In a branching process the number of offspring per individual has a binomial distribution with parameters $2, p$. Starting with a single individual, calculate:

(a) the extinction probability;

(b) the probability that the population becomes extinct for the first time in the third generation.

Suppose that, instead of starting with a single individual, the initial population size Z_0 is a random variable that is Poisson distributed with mean λ. Show that, in this case, the extinction probability is given, for $p > \frac{1}{2}$, by

$$\exp\{\lambda(1 - 2p)/p^2\}.$$

4.35. Consider a branching process in which the number of offspring per individual has a Poisson distribution with mean λ, $\lambda > 1$. Let π_0 denote the probability that, starting with a single individual, the population eventually becomes extinct. Also, let a, $a < 1$, be such that

$$ae^{-a} = \lambda e^{-\lambda}.$$

(a) Show that $a = \lambda \pi_0$.

(b) Show that, conditional on eventual extinction, the branching process follows the same probability law as the branching process in which the number of offspring per individual is Poisson with mean a.

4.36. For the Markov chain model of Section 4.6.1, namely,

$$P_{ij} = \frac{1}{i-1}, \qquad j = 1, \ldots, i-1, \quad i > 1,$$

suppose that the initial state is $N \equiv \binom{n}{m}$, where $n > m$. Show that when n, m, and $n - m$ are large the number of steps to reach 1 from state N has approximately a Poisson distribution with mean

$$m\left[c \log \frac{c}{c-1} + \log(c-1)\right],$$

where $c = n/m$. (*Hint:* Use Stirling's approximation.)

4.37. For any infinite sequence $x_1, x_2, \ldots$ we say that a new long run begins each time the sequence changes direction. That is, if the sequence starts 5, 2, 4, 5, 6, 9, 3, 4, then there are three long runs—namely, (5, 2), (4, 5, 6, 9), and (3, 4). Let $X_1, X_2, \ldots$ be independent uniform (0, 1) random variables and let I_n denote the initial value of the nth long run. Argue that $\{I_n, n \geq 1\}$ is a Markov chain having a continuous state space with transition probability density given by

$$p(y|x) = e^{1-x} + e^x - e^{|y-x|} - 1.$$

4.38. Suppose in Example 4.7(B) that if the Markov chain is in state i and the random variable distributed according to $q_{i\cdot}$ takes on the value j, then the next state is set equal to j with probability $a_j/(a_j + a_i)$ and equal to i otherwise. Show that the limiting probabilities for this chain are $\pi_j = a_j / \sum_j a_j$.

4.39. Find the transition probabilities for the Markov chain of Example 4.3(D) and show that it is time reversible.

4.40. Let $\{X_n, n \geq 0\}$ be a Markov chain with stationary probabilities $\pi_j, j \geq 0$. Suppose that $X_0 = i$ and define $T = \text{Min}\{n: n > 0 \text{ and } X_n = i\}$. Let $Y_j = X_{T-j}, j = 0, 1, \ldots, T$. Argue that $\{Y_j, j = 0, \ldots, T\}$ is distributed as the states of the reverse Markov chain (with transition probabilities $P_{ij}^* = \pi_j P_{ji}/\pi_i$) starting in state 0 until it returns to 0.

4.41. A particle moves among n locations that are arranged in a circle (with the neighbors of location n being $n - 1$ and 1). At each step, it moves one position either in the clockwise position with probability p or in the counterclockwise position with probability $1 - p$.

(a) Find the transition probabilities of the reverse chain.

(b) Is the chain time reversible?

4.42. Consider the Markov chain with states $0, 1, \ldots, n$ and with transition probabilities

$$P_{0,1} = P_{n,n-1} = 1$$
$$P_{i,i+1} = p_i = 1 - P_{i,i-1}, \qquad i = 1, \ldots, n-1.$$

Show that this Markov chain is of the type considered in Proposition 4.7.1 and find its stationary probabilities.

4.43. Consider the list model presented in Example 4.7(D). Under the one-closer rule show, by using time reversibility, that the limiting probability that element j precedes element i—call it $P\{j \text{ precedes } i\}$—is such that

$$P\{j \text{ precedes } i\} > \frac{P_j}{P_j + P_i} \qquad \text{when } P_j > P_i.$$

4.44. Consider a time-reversible Markov chain with transition probabilities P_{ij} and limiting probabilities π_i; and now consider the same chain truncated to the states $0, 1, \ldots, M$. That is, for the truncated chain its transition probabilities $\overline{P}_{ij}$ are

$$\overline{P}_{ij} = \begin{cases} P_{ij} + \sum_{k>M} P_{ik}, & 0 \le i \le M, j = i \\ P_{ij}, & 0 \le i \ne j \le M \\ 0, & \text{otherwise.} \end{cases}$$

Show that the truncated chain is also time reversible and has limiting probabilities given by

$$\overline{\pi}_i = \frac{\pi_i}{\sum_{i=0}^{M} \pi_i}.$$

4.45. Show that a finite state, ergodic Markov chain such that $P_{ij} > 0$ for all $i \ne j$ is time reversible if, and only if,

$$P_{ij} P_{jk} P_{ki} = P_{ik} P_{kj} P_{ji} \qquad \text{for all } i, j, k.$$

4.46. Let $\{X_n, n \geq 1\}$ denote an irreducible Markov chain having a countable state space. Now consider a new stochastic process $\{Y_n, n \geq 0\}$ that only accepts values of the Markov chain that are between 0 and N. That is, we define Y_n to be the nth value of the Markov chain that is between 0 and N. For instance, if $N = 3$ and $X_1 = 1$, $X_2 = 3$, $X_3 = 5$, $X_4 = 6$, $X_5 = 2$, then $Y_1 = 1$, $Y_2 = 3$, $Y_3 = 2$.

(a) Is $\{Y_n, n \geq 0\}$ a Markov chain? Explain briefly.

(b) Let π_j denote the proportion of time that $\{X_n, n \geq 1\}$ is in state j. If $\pi_j > 0$ for all j, what proportion of time is $\{Y_n, n \geq 0\}$ in each of the states $0, 1, \ldots, N$?

(c) Suppose $\{X_n\}$ is null recurrent and let $\pi_i(N)$, $i = 0, 1, \ldots, N$ denote the long-run proportions for $\{Y_n, n \geq 0\}$. Show that

$$\pi_j(N) = \pi_i(N) E[\text{time the } X \text{ process spends in } j \text{ between returns to } i], \qquad j \neq i.$$

(d) Use (c) to argue that in a symmetric random walk the expected number of visits to state i before returning to the origin equals 1.

(e) If $\{X_n, n \geq 0\}$ is time reversible, show that $\{Y_n, n \geq 0\}$ is also.

4.47. M balls are initially distributed among m urns. At each stage one of the balls is selected at random, taken from whichever urn it is in, and placed, at random, in one of the other $m - 1$ urns. Consider the Markov chain whose state at any time is the vector $(n_1, \ldots, n_m)$, where n_i denotes the number of balls in urn i. Guess at the limiting probabilities for this Markov chain and then verify your guess and show at the same time that the Markov chain is time reversible.

4.48. For an ergodic semi-Markov process:

(a) Compute the rate at which the process makes a transition from i into j.

(b) Show that

$$\sum_i P_{ij}/\mu_{ii} = 1/\mu_{jj}.$$

(c) Show that the proportion of time that the process is in state i and headed for state j is $P_{ij}\eta_{ij}/\mu_{ii}$ where $\eta_{ij} = \int_0^\infty \overline{F}_{ij}(t)\,dt$.

(d) Show that the proportion of time that the state is i and will next be j within a time x is

$$\frac{P_{ij}\eta_{ij}}{\mu_{ii}} F^e_{i,j}(x),$$

where F^e_{ij} is the equilibrium distribution of F_{ij}.

4.49. For an ergodic semi-Markov process derive an expression, as $t \to \infty$, for the limiting conditional probability that the next state visited after t is state j, given $X(t) = i$.

4.50. A taxi alternates between three locations. When it reaches location 1 it is equally likely to go next to either 2 or 3. When it reaches 2 it will next go to 1 with probability $\frac{1}{3}$ and to 3 with probability $\frac{2}{3}$. From 3 it always goes to 1. The mean times between locations i and j are $t_{12} = 20$, $t_{13} = 30$, $t_{23} = 30$ ($t_{ij} = t_{ji}$).

(a) What is the (limiting) probability that the taxi's most recent stop was at location i, $i = 1, 2, 3$?

(b) What is the (limiting) probability that the taxi is heading for location 2?

(c) What fraction of time is the taxi traveling from 2 to 3? *Note:* Upon arrival at a location the taxi immediately departs.

REFERENCES

References 1, 2, 3, 4, 6, 9, 10, 11, and 12 give alternate treatments of Markov chains. Reference 5 is a standard text in branching processes. References 7 and 8 have nice treatments of time reversibility.

1. N. Bhat, *Elements of Applied Stochastic Processes*, Wiley, New York, 1984.
2. C. L. Chiang, *An Introduction to Stochastic Processes and Their Applications*, Krieger, 1980.
3. E. Cinlar, *Introduction to Stochastic Processes*, Prentice-Hall, Englewood Cliffs, NJ, 1975.
4. D. R. Cox and H. D. Miller, *The Theory of Stochastic Processes*, Methuen, London, 1965.
5. T. Harris, *The Theory of Branching Processes*, Springer-Verlag, Berlin, 1963.
6. S. Karlin and H. Taylor, *A First Course in Stochastic Processes*, 2nd ed., Academic Press, Orlando, FL, 1975.
7. J. Keilson, *Markov Chain Models—Rarity and Exponentiality*, Springer-Verlag, Berlin, 1979.
8. F. Kelly, *Reversibility and Stochastic Networks*, Wiley, Chichester, England, 1979.
9. J. Kemeny, L. Snell, and A. Knapp, *Denumerable Markov Chains*, Van Nostrand, Princeton, NJ, 1966.
10. S. Resnick, *Adventures in Stochastic Processes*, Birkhauser, Boston, MA, 1992.
11. S. M. Ross, *Introduction to Probability Models*, 5th ed., Academic Press, Orlando, FL, 1993.
12. H. C. Tijms, *Stochastic Models, An Algorithmic Approach*, Wiley, Chichester, England, 1994.

CHAPTER 5

Continuous-Time Markov Chains

5.1 Introduction

In this chapter we consider the continuous-time analogs of discrete-time Markov chains. As in the case of their discrete-time analogs, they are characterized by the Markovian property that, given the present state, the future is independent of the past.

In Section 5.2 we define continuous-time Markov chains and relate them to the discrete-time Markov chains of Chapter 4. In Section 5.3 we introduce an important class of continuous-time Markov chains known as birth and death processes. These processes can be used to model populations whose size changes at any time by a single unit. In Section 5.4 we derive two sets of differential equations—the forward and backward equation—that describe the probability laws for the system. The material in Section 5.5 is concerned with determining the limiting (or long-run) probabilities connected with a continuous-time Markov chain. In Section 5.6 we consider the topic of time reversibility. Among other things, we show that all birth and death processes are time reversible, and then illustrate the importance of this observation to queueing systems. Applications of time reversibility to stochastic population models are also presented in this section. In Section 5.7 we illustrate the importance of the reversed chain, even when the process is not time reversible, by using it to study queueing network models, to derive the Erlang loss formula, and to analyze the shared processor system. In Section 5.8 we show how to "uniformize" Markov chains—a technique useful for numerical computations.

5.2 Continuous-Time Markov Chains

Consider a continuous-time stochastic process $\{X(t), t \geq 0\}$ taking on values in the set of nonnegative integers. In analogy with the definition of a discrete-time Markov chain, given in Chapter 4, we say that the process $\{X(t), t \geq 0\}$ is a *continuous-time Markov chain* if for all $s, t \geq 0$, and nonnegative integers

$i, j, x(u), 0 \le u \le s,$

$$P\{X(t+s) = j \mid X(s) = i, X(u) = x(u), 0 \le u < s\}$$
$$= P\{X(t+s) = j \mid X(s) = i\}.$$

In other words, a continuous-time Markov chain is a stochastic process having the Markovian property that the conditional distribution of the future state at time $t + s$, given the present state at s and all past states depends only on the present state and is independent of the past. If, in addition, $P\{X(t+s) = j \mid X(s) = i\}$ is independent of s, then the continuous-time Markov chain is said to have stationary or homogeneous transition probabilities. All Markov chains we consider will be assumed to have stationary transition probabilities.

Suppose that a continuous-time Markov chain enters state i at some time, say time 0, and suppose that the process does not leave state i (that is, a transition does not occur) during the next s time units. What is the probability that the process will not leave state i during the following t time units? To answer this, note that as the process is in state i at time s, it follows, by the Markovian property, that the probability it remains in that state during the interval $[s, s + t]$ is just the (unconditional) probability that it stays in state i for at least t time units. That is, if we let τ_i denote the amount of time that the process stays in state i before making a transition into a different state, then

$$P\{\tau_i > s + t \mid \tau_i > s\} = P\{\tau_i > t\}$$

for all $s, t \ge 0$. Hence, the random variable τ_i is memoryless and must thus be exponentially distributed.

In fact, the above gives us a way of constructing a continuous-time Markov chain. Namely, it is a stochastic process having the properties that each time it enters state i:

(i) the amount of time it spends in that state before making a transition into a different state is exponentially distributed with rate, say, v_i; and

(ii) when the process leaves state i, it will next enter state j with some probability, call it P_{ij}, where $\sum_{j \ne i} P_{ij} = 1$.

A state i for which $v_i = \infty$ is called an *instantaneous* state since when entered it is instantaneously left. Whereas such states are theoretically possible, we shall assume throughout that $0 \le v_i < \infty$ for all i. (If $v_i = 0$, then state i is called *absorbing* since once entered it is never left.) Hence, for our purposes a continuous-time Markov chain is a stochastic process that moves from state to state in accordance with a (discrete-time) Markov chain, but is such that the amount of time it spends in each state, before proceeding to the next state, is exponentially distributed. In addition, the amount of time the process spends in state i, and the next state visited, must be independent random variables. For if the next state visited were dependent on τ_i, then information as to how long the process has already been in state i would be relevant to the prediction of the next state—and this would contradict the Markovian assumption.

A continuous-time Markov chain is said to be *regular* if, with probability 1, the number of transitions in any finite length of time is finite. An example of a nonregular Markov chain is the one having

$$P_{i,i+1} = 1, \qquad v_i = i^2.$$

It can be shown that this Markov chain—which always goes from state i to $i + 1$, spending an exponentially distributed amount of time with mean $1/i^2$ in state i—will, with positive probability, make an infinite number of transitions in any time interval of length t, $t > 0$. We shall, however, assume from now on that all Markov chains considered are regular (some sufficient conditions for regularity are given in the Problem section).

Let q_{ij} be defined by

$$q_{ij} = v_i P_{ij}, \qquad \text{all } i \neq j.$$

Since v_i is the rate at which the process leaves state i and P_{ij} is the probability that it then goes to j, it follows that q_{ij} is the rate when in state i that the process makes a transition into state j; and in fact we call q_{ij} the *transition rate* from i to j.

Let us denote by $P_{ij}(t)$ the probability that a Markov chain, presently in state i, will be in state j after an additional time t. That is,

$$P_{ij}(t) = P\{X(t + s) = j \mid X(s) = i\}.$$

5.3 Birth and Death Processes

A continuous-time Markov chain with states $0, 1, \ldots$ for which $q_{ij} = 0$ whenever $|i - j| > 1$ is called a *birth and death process*. Thus a birth and death process is a continuous-time Markov chain with states $0, 1, \ldots$ for which transitions from state i can only go to either state $i - 1$ or state $i + 1$. The state of the process is usually thought of as representing the size of some population, and when the state increases by 1 we say that a birth occurs, and when it decreases by 1 we say that a death occurs. Let λ_i and μ_i be given by

$$\lambda_i = q_{i,i+1},$$
$$\mu_i = q_{i,i-1}.$$

The values $\{\lambda_i, i \geq 0\}$ and $\{\mu_i, i \geq 1\}$ are called respectively the birth rates and the death rates. Since $\sum_j q_{ij} = v_i$, we see that

$$v_i = \lambda_i + \mu_i,$$

$$P_{i,i+1} = \frac{\lambda_i}{\lambda_i + \mu_i} = 1 - P_{i,i-1}.$$

Hence, we can think of a birth and death process by supposing that whenever there are i people in the system the time until the next birth is exponential with rate λ_i and is independent of the time until the next death, which is exponential with rate μ_i.

EXAMPLE 5.3(A) ***Two Birth and Death Processes.*** (i) *The M/M/s Queue.* Suppose that customers arrive at an s-server service station in accordance with a Poisson process having rate λ. That is, the times between successive arrivals are independent exponential random variables having mean $1/\lambda$. Each customer, upon arrival, goes directly into service if any of the servers are free, and if not, then the customer joins the queue (that is, he waits in line). When a server finishes serving a customer, the customer leaves the system, and the next customer in line, if there are any waiting, enters the service. The successive service times are assumed to be independent exponential random variables having mean $1/\mu$. If we let $X(t)$ denote the number in the system at time t, then $\{X(t), t \geq 0\}$ is a birth and death process with

$$\mu_n = \begin{cases} n\mu & 1 \leq n \leq s \\ s\mu & n > s, \end{cases}$$

$$\lambda_n = \lambda, \qquad n \geq 0.$$

(ii) *A Linear Growth Model with Immigration.* A model in which

$$\mu_n = n\mu, \qquad n \geq 1,$$

$$\lambda_n = n\lambda + \theta, \qquad n \geq 0,$$

is called a linear growth process with immigration. Such processes occur naturally in the study of biological reproduction and population growth. Each individual in the population is assumed to give birth at an exponential rate λ; in addition, there is an exponential rate of increase θ of the population due to an external source such as immigration. Hence, the total birth rate where there are n persons in the system is $n\lambda + \theta$. Deaths are assumed to occur at an exponential rate μ for each member of the population, and hence $\mu_n = n\mu$.

A birth and death process is said to be a *pure birth* process if $\mu_n = 0$ for all n (that is, if death is impossible). The simplest example of a pure birth process is the Poisson process, which has a constant birth rate $\lambda_n = \lambda$, $n \geq 0$.

A second example of a pure birth process results from a population in which each member acts independently and gives birth at an exponential rate λ. If we suppose that no one ever dies, then, if $X(t)$ represents the population size at time t, $\{X(t), t \geq 0\}$ is a pure birth process with

$$\lambda_n = n\lambda, \qquad n \geq 0.$$

This pure birth process is called a *Yule* process.

Consider a Yule process starting with a single individual at time 0, and let T_i, $i \geq 1$, denote the time between the $(i-1)$st and ith birth. That is, T_i is the time it takes for the population size to go from i to $i+1$. It easily follows from the definition of a Yule process that the T_i, $i \geq 1$, are independent and T_i is exponential with rate $i\lambda$. Now

$$\begin{aligned}
P\{T_1 \leq t\} &= 1 - e^{-\lambda t},\\
P\{T_1 + T_2 \leq t\} &= \int_0^t P\{T_1 + T_2 \leq t \mid T_1 = x\}\lambda e^{-\lambda x}\,dx\\
&= \int_0^t (1 - e^{-2\lambda(t-x)})\lambda e^{-\lambda x}\,dx\\
&= (1 - e^{-\lambda t})^2,\\
P\{T_1 + T_2 + T_3 \leq t\} &= \int_0^t P\{T_1 + T_2 + T_3 \leq t \mid T_1 + T_2 = x\}\,dF_{T_1+T_2}(x)\\
&= \int_0^t (1 - e^{-3\lambda(t-x)})2\lambda e^{-\lambda x}(1 - e^{-\lambda x})\,dx\\
&= (1 - e^{-\lambda t})^3,
\end{aligned}$$

and, in general, we can show by induction that

$$P\{T_1 + \cdots + T_j \leq t\} = (1 - e^{-\lambda t})^j.$$

Hence, as $P\{T_1 + \cdots + T_j \leq t\} = P\{X(t) \geq j + 1 \mid X(0) = 1\}$, we see that, for a Yule process,

$$\begin{aligned}
P_{1j}(t) &= (1 - e^{-\lambda t})^{j-1} - (1 - e^{-\lambda t})^j\\
&= e^{-\lambda t}(1 - e^{-\lambda t})^{j-1}, \qquad j \geq 1.
\end{aligned}$$

Thus, we see from the above that, starting with a single individual, the population size at time t will have a geometric distribution with mean $e^{\lambda t}$. Hence if the population starts with i individuals, it follows that its size at t will be the sum of i independent and identically distributed geometric random variables, and will thus have a negative binomial distribution. That is, for the Yule process,

$$P_{ij}(t) = \binom{j-1}{i-1} e^{-\lambda t i}(1 - e^{-\lambda t})^{j-i}, \qquad j \geq i \geq 1.$$

Another interesting result about the Yule process, starting with a single individual, concerns the conditional distribution of the times of birth given the population size at t. Since the ith birth occurs at time $S_i \equiv T_1 + \cdots + T_i$, let us compute the conditional joint distribution of $S_1, \ldots, S_n$ given that $X(t) = n + 1$. Reasoning heuristically and treating densities as if they were probabilities yields that for $0 \leq s_1 \leq s_2 \leq \cdots \leq s_n \leq t$

$$\begin{aligned}
&P\{S_1 = s_1, S_2 = s_2, \ldots, S_n = s_n \mid X(t) = n + 1\} \\
&\quad = \frac{P\{T_1 = s_1, T_2 = s_2 - s_1, \ldots, T_n = s_n - s_{n-1}, T_{n+1} > t - s_n\}}{P\{X(t) = n + 1\}} \\
&\quad = \frac{\lambda e^{-\lambda s_1} 2\lambda e^{-2\lambda(s_2 - s_1)} \cdots n\lambda e^{-n\lambda(s_n - s_{n-1})} e^{-(n+1)\lambda(t - s_n)}}{P\{X(t) = n + 1\}} \\
&\quad = C e^{-\lambda(t - s_1)} e^{-\lambda(t - s_2)} \cdots e^{-\lambda(t - s_n)},
\end{aligned}$$

where C is some constant that does not depend on $s_1, \ldots, s_n$. Hence we see that the conditional density of $S_1, \ldots, S_n$ given that $X(t) = n + 1$ is given by

$$(5.3.1) \qquad f(s_1, \ldots, s_n \mid n + 1) = n! \prod_{i=1}^{n} f(s_i), \qquad 0 \leq s_1 \leq \cdots \leq s_n \leq t,$$

where f is the density function

$$(5.3.2) \qquad f(x) = \begin{cases} \dfrac{\lambda e^{-\lambda(t-x)}}{1 - e^{-\lambda t}} & 0 \leq x \leq t \\ 0 & \text{otherwise.} \end{cases}$$

But since (5.3.1) is the joint density function of the order statistics of a sample of n random variables from the density f (see Section 2.3 of Chapter 2), we have thus proven the following.

PROPOSITION 5.3.1

Consider a Yule process with $X(0) = 1$. Then, given that $X(t) = n + 1$, the birth times $S_1, \ldots, S_n$ are distributed as the ordered values from a sample of size n from a population having density (5.3.2).

Proposition 5.3.1 can be used to establish results about the Yule process in the same way that the corresponding result for Poisson processes is used.

EXAMPLE 5.3(B) Consider a Yule process with $X(0) = 1$. Let us compute the expected sum of the ages of the members of the population at time t. The sum of the ages at time t, call it $A(t)$, can be expressed as

$$A(t) = a_0 + t + \sum_{i=1}^{X(t)-1} (t - S_i),$$

where a_0 is the age at $t = 0$ of the initial individual. To compute $E[A(t)]$ condition on $X(t)$,

$$\begin{aligned} E[A(t)|X(t) = n + 1] &= a_0 + t + E\left[\sum_{i=1}^{n} (t - S_i)|X(t) = n + 1\right] \\ &= a_0 + t + n \int_0^t (t - x) \frac{\lambda e^{-\lambda(t-x)}}{1 - e^{-\lambda t}} dx \end{aligned}$$

or

$$E[A(t)|X(t)] = a_0 + t + (X(t) - 1) \frac{1 - e^{-\lambda t} - \lambda t e^{-\lambda t}}{\lambda(1 - e^{-\lambda t})}.$$

Taking expectations and using the fact that $X(t)$ has mean $e^{\lambda t}$ yields

$$\begin{aligned} E[A(t)] &= a_0 + t + \frac{e^{\lambda t} - 1 - \lambda t}{\lambda} \\ &= a_0 + \frac{e^{\lambda t} - 1}{\lambda}. \end{aligned}$$

Other quantities related to $A(t)$, such as its generating function, can be computed by the same method.

The above formula for $E[A(t)]$ can be checked by making use of the following identity whose proof is left as an exercise:

$$A(t) = a_0 + \int_0^t X(s)\, ds. \tag{5.3.3}$$

Taking expectations gives

$$
\begin{aligned}
E[A(t)] &= a_0 + E\left[\int_0^t X(s)\,ds\right] \\
&= a_0 + \int_0^t E[X(s)]\,ds \quad \text{since } X(s) \ge 0 \\
&= a_0 + \int_0^t e^{\lambda s}\,ds \\
&= a_0 + \frac{e^{\lambda t} - 1}{\lambda}.
\end{aligned}
$$

The following example provides another illustration of a pure birth process.

Example 5.3(c) ***A Simple Epidemic Model.*** Consider a population of m individuals that at time 0 consists of one "infected" and $m - 1$ "susceptibles." Once infected an individual remains in that state forever and we suppose that in any time interval h any given infected person will cause, with probability $\alpha h + o(h)$, any given susceptible to become infected. If we let $X(t)$ denote the number of infected individuals in the population at time t, the $\{X(t), t \ge 0\}$ is a pure birth process with

$$
\lambda_n = \begin{cases} (m - n)n\alpha & n = 1, \ldots, m - 1 \\ 0 & \text{otherwise.} \end{cases}
$$

The above follows since when there are n infected individuals, then each of the $m - n$ susceptibles will become infected at rate $n\alpha$.

If we let T denote the time until the total population is infected, then T can be represented as

$$
T = \sum_{i=1}^{m-1} T_i,
$$

where T_i is the time to go from i infectives to $i + 1$ infectives. As the T_i are independent exponential random variables with respective rates $\lambda_i = (m - i)i\alpha$, $i = 1, \ldots, m - 1$, we see that

$$
E[T] = \frac{1}{\alpha} \sum_{i=1}^{m-1} \frac{1}{i(m - i)}
$$

and

$$
\operatorname{Var}(T) = \frac{1}{\alpha^2} \sum_{i=1}^{m-1} \left(\frac{1}{i(m - i)}\right)^2.
$$

For reasonably sized populations $E[T]$ can be approximated as follows:

$$E[T] = \frac{1}{m\alpha} \sum_{i=1}^{m-1} \left(\frac{1}{m-i} + \frac{1}{i} \right)$$

$$\approx \frac{1}{m\alpha} \int_{1}^{m-1} \left(\frac{1}{m-t} + \frac{1}{t} \right) dt = \frac{2 \log(m-1)}{m\alpha}.$$

5.4 The Kolmogorov Differential Equations

Recall that

$$P_{ij}(t) = P\{X(t+s) = j \mid X(s) = i\}$$

represents the probability that a process presently in state i will be in state j a time t later.

By exploiting the Markovian property, we will derive two sets of differential equations for $P_{ij}(t)$, which may sometimes be explicitly solved. However, before doing so we need the following lemmas.

Lemma 5.4.1

(i) $\lim_{t \to 0} \frac{1 - P_{ii}(t)}{t} = v_i.$

(ii) $\lim_{t \to 0} \frac{P_{ij}(t)}{t} = q_{ij}, \qquad i \neq j.$

Lemma 5.4.2

For all s, t,

$$P_{ij}(t+s) = \sum_{k=0}^{\infty} P_{ik}(t) P_{kj}(s).$$

Lemma 5.4.1 follows from the fact (which must be proven) that the probability of two or more transitions in time t is $o(t)$; whereas Lemma 5.4.2, which is the continuous-time version of the Chapman-Kolmogorov equations of discrete-time Markov chains, follows directly from the Markovian property. The details of the proof are left as exercises.

From Lemma 5.4.2 we obtain

$$P_{ij}(t+h) = \sum_k P_{ik}(h)P_{kj}(t),$$

or, equivalently,

$$P_{ij}(t+h) - P_{ij}(t) = \sum_{k \neq i} P_{ik}(h)P_{kj}(t) - [1 - P_{ii}(h)]P_{ij}(t).$$

Dividing by h and then taking the limit as $h \to 0$ yields, upon application of Lemma 5.4.1,

$$\lim_{h\to 0} \frac{P_{ij}(t+h) - P_{ij}(t)}{h} = \lim_{h\to 0} \sum_{k \neq i} \frac{P_{ik}(h)}{h} P_{kj}(t) - v_i P_{ij}(t). \tag{5.4.1}$$

Assuming that we can interchange the limit and summation on the right-hand side of (5.4.1), we thus obtain, again using Lemma 5.4.1, the following.

THEOREM 5.4.3 (Kolmogorov's Backward Equations).

For all i, j, and $t \geq 0$,

$$P'_{ij}(t) = \sum_{k \neq i} q_{ik} P_{kj}(t) - v_i P_{ij}(t).$$

Proof To complete the proof we must justify the interchange of limit and summation on the right-hand side of (5.4.1). Now, for any fixed N,

$$\begin{aligned}\liminf_{h\to 0} \sum_{k \neq i} \frac{P_{ik}(h)}{h} P_{kj}(t) &\geq \liminf_{h\to 0} \sum_{\substack{k \neq i \\ k<N}} \frac{P_{ik}(h)}{h} P_{kj}(t) \\ &= \sum_{\substack{k \neq i \\ k<N}} q_{ik} P_{kj}(t).\end{aligned}$$

Since the above holds for all N we see that

$$\liminf_{h\to 0} \sum_{k \neq i} \frac{P_{ik}(h)}{h} P_{kj}(t) \geq \sum_{k \neq i} q_{ik} P_{kj}(t). \tag{5.4.2}$$

To reverse the inequality note that for $N > i$, since $P_{kj}(t) \le 1$,

$$\limsup_{h\to 0} \sum_{k\neq i} \frac{P_{ik}(h)}{h} P_{kj}(t)$$

$$\le \limsup_{h\to 0} \left[\sum_{\substack{k\neq i\\ k<N}} \frac{P_{ik}(h)}{h} P_{kj}(t) + \sum_{k\ge N} \frac{P_{ik}(h)}{h} \right]$$

$$= \limsup_{h\to 0} \left[\sum_{\substack{k\neq i\\ k<N}} \frac{P_{ik}(h)}{h} P_{kj}(t) + \frac{1 - P_{ii}(h)}{h} - \sum_{\substack{k\neq i\\ k<N}} \frac{P_{ik}(h)}{h} \right]$$

$$= \sum_{\substack{k\neq i\\ k<N}} q_{ik} P_{kj}(t) + v_i - \sum_{\substack{k\neq i\\ k<N}} q_{ik},$$

where the last equality follows from Lemma 5.4.1. As the above inequality is true for all $N > i$, we obtain upon letting $N \to \infty$ and using the fact $\sum_{k\neq i} q_{ik} = v_i$,

$$\limsup_{h\to 0} \sum_{k\neq i} \frac{P_{ik}(h)}{h} P_{kj}(t) \le \sum_{k\neq i} q_{ik} P_{kj}(t).$$

The above combined with (5.4.2) shows that

$$\lim_{h\to 0} \sum_{k\neq i} \frac{P_{ik}(h)}{h} P_{kj}(t) = \sum_{k\neq i} q_{ik} P_{kj}(t),$$

which completes the proof of Theorem 5.4.3.

The set of differential equations for $P_{ij}(t)$ given in Theorem 5.4.3 are known as the Kolmogorov *backward equations.* They are called the backward equations because in computing the probability distribution of the state at time $t + h$ we conditioned on the state (all the way) back at time h. That is, we started our calculation with

$$P_{ij}(t+h) = \sum_k P\{X(t+h) = j \mid X(0) = i, X(h) = k\} P\{X(h) = k \mid X(0) = i\}$$

$$= \sum_k P_{kj}(t) P_{ik}(h).$$

We may derive another set of equations, known as the Kolmogorov's *forward equations,* by now conditioning on the state at time t. This yields

$$P_{ij}(t+h) = \sum_k P_{ik}(t) P_{kj}(h)$$

or

$$\begin{aligned} P_{ij}(t+h) - P_{ij}(t) &= \sum_k P_{ik}(t)P_{kj}(h) - P_{ij}(t) \\ &= \sum_{k \neq j} P_{ik}(t)P_{kj}(h) - [1 - P_{jj}(h)]P_{ij}(t). \end{aligned}$$

Therefore,

$$\lim_{h\to 0} \frac{P_{ij}(t+h) - P_{ij}(t)}{h} = \lim_{h\to 0} \left\{ \sum_{k \neq j} P_{ik}(t)\frac{P_{kj}(h)}{h} - \frac{1 - P_{jj}(h)}{h} P_{ij}(t) \right\}.$$

Assuming that we can interchange limit with summation, we obtain by Lemma 5.4.1 that

$$P'_{ij}(t) = \sum_{k \neq j} q_{kj} P_{ik}(t) - v_j P_{ij}(t).$$

Unfortunately, we cannot always justify the interchange of limit and summation, and thus the above is not always valid. However, they do hold in most models—including all birth and death processes and all finite-state models. We thus have

THEOREM 5.4.4 (Kolmogorov's Forward Equations).

Under suitable regularity conditions,

$$P'_{ij}(t) = \sum_{k \neq j} q_{kj} P_{ik}(t) - v_j P_{ij}(t).$$

Example 5.4(A) ***The Two-State Chain.*** Consider a two-state continuous-time Markov chain that spends an exponential time with rate λ in state 0 before going to state 1, where it spends an exponential time with rate μ before returning to state 0. The forward equations yield

$$\begin{aligned} P'_{00}(t) &= \mu P_{01}(t) - \lambda P_{00}(t) \\ &= -(\lambda + \mu)P_{00}(t) + \mu, \end{aligned}$$

where the last equation follows from $P_{01}(t) = 1 - P_{00}(t)$. Hence,

$$e^{(\lambda+\mu)t}[P'_{00}(t) + (\lambda+\mu)P_{00}(t)] = \mu e^{(\lambda+\mu)t}$$

or

$$\frac{d}{dt}[e^{(\lambda+\mu)t}P_{00}(t)] = \mu e^{(\lambda+\mu)t}.$$

Thus,

$$e^{(\lambda+\mu)t}P_{00}(t) = \frac{\mu}{\lambda+\mu}e^{(\lambda+\mu)t} + c.$$

Since $P_{00}(0) = 1$, we see that $c = \lambda/(\lambda+\mu)$, and thus

$$P_{00}(t) = \frac{\mu}{\lambda+\mu} + \frac{\lambda}{\lambda+\mu}e^{-(\lambda+\mu)t}.$$

Similarly (or by symmetry),

$$P_{11}(t) = \frac{\lambda}{\lambda+\mu} + \frac{\mu}{\lambda+\mu}e^{-(\lambda+\mu)t}.$$

EXAMPLE 5.4(B) The Kolmogorov forward equations for the birth and death process are

$$P'_{i0}(t) = \mu_1 P_{i1}(t) - \lambda_0 P_{i0}(t),$$

$$P'_{ij}(t) = \lambda_{j-1}P_{i,j-1}(t) + \mu_{j+1}P_{i,j+1}(t) - (\lambda_j + \mu_j)P_{ij}(t), \qquad j \neq 0.$$

EXAMPLE 5.4(C) For a pure birth process, the forward equations reduce to

$$\begin{aligned} P'_{ii}(t) &= -\lambda_i P_{ii}(t), \\ P'_{ij}(t) &= \lambda_{j-1}P_{i,j-1}(t) - \lambda_j P_{ij}(t), \qquad j > i. \end{aligned} \tag{5.4.3}$$

Integrating the top equation of (5.4.3) and then using $P_{ii}(0) = 1$ yields

$$P_{ii}(t) = e^{-\lambda_i t}.$$

The above, of course, is true as $P_{ii}(t)$ is the probability that the time until a transition from state i is greater than t. The other quantities $P_{ij}(t)$, $j > i$, can be obtained recursively from (5.4.3) as follows: From (5.4.3) we have, for $j > i$,

$$e^{\lambda_j t}\lambda_{j-1}P_{i,j-1}(t) = e^{\lambda_j t}[P'_{ij}(t) + \lambda_j P_{ij}(t)]$$
$$= \frac{d}{dt}[e^{\lambda_j t}P_{ij}(t)].$$

Integration, using $P_{ij}(0) = 0$, yields

$$P_{ij}(t) = \lambda_{j-1}e^{-\lambda_j t}\int_0^t e^{\lambda_j s}P_{i,j-1}(s)\,ds, \qquad j > i.$$

In the special case of a Yule process, where $\lambda_j = j\lambda$, we can use the above to verify the result of Section 5.3; namely,

$$P_{ij}(t) = \binom{j-1}{i-1}e^{-\lambda t i}(1 - e^{-\lambda t})^{j-i}, \qquad j \geq i \geq 1.$$

In many models of interest the Kolmogorov differential equations cannot be explicitly solved for the transition probabilities. However, we can often compute other quantities of interest, such as mean values, by first deriving and then solving a differential equation. Our next example illustrates the technique.

Example 5.4(d) ***A Two Sex Population Growth Model.*** Consider a population of males and females and suppose that each female in the population independently gives birth at an exponential rate λ; and that each birth is (independently of all else) a female with probability p and a male with probability $1 - p$. Individual females die at an exponential rate μ and males at an exponential rate ν. Thus, if $X(t)$ and $Y(t)$ denote, respectively, the numbers of females and males in the population at time t, then $\{[X(t), Y(t)], t \geq 0\}$ is a continuous-time Markov chain with infinitesimal rates:

$$q[(n,m),(n+1,m)] = n\lambda p, \qquad q[(n,m),(n,m+1)] = n\lambda(1-p)$$
$$q[(n,m),(n-1,m)] = n\mu \qquad q[(n,m),(n,m-1)] = m\nu.$$

If $X(0) = i$, and $Y(0) = j$, find (a) $E[X(t)]$, (b) $E[Y(t)]$, and (c) $\text{Cov}[X(t), Y(t)]$.

Solution. (a) To find $E[X(t)]$, note that $\{X(t), t \geq 0\}$ is itself a continuous-time Markov chain. Letting $M_X(t) = E[X(t)|X(0) = i]$ we will derive a differential equation satisfied by $M_X(t)$. To begin, note that, given $X(t)$:

$$X(t+h) = \begin{cases} X(t)+1 & \text{with probability } \lambda p X(t)h + o(h) \\ X(t)-1 & \text{with probability } \mu X(t)h + o(h) \\ X(t) & \text{with probability } 1-(\mu+\lambda p)X(t)h + o(h). \end{cases}$$

(5.4.4)

Hence, taking expectations yields

$$E[X(t+h)|X(t)] = X(t) + (\lambda p - \mu)X(t)h + o(h),$$

and taking expectations once again yields

$$M_X(t+h) = M_X(t) + (\lambda p - \mu)M_X(t)h + o(h)$$

or,

$$\frac{M_X(t+h) - M_X(t)}{h} = (\lambda p - \mu)M_X(t) + \frac{o(h)}{h}.$$

Letting $h \to 0$ gives

$$M_X'(t) = (\lambda p - \mu)M_X(t)$$

or,

$$M_X'(t)/M_X(t) = \lambda p - \mu$$

and, upon integration

$$\log M_X(t) = (\lambda p - \mu)t + C$$

or,

$$M_X(t) = Ke^{(\lambda p - \mu)t}.$$

Since $M(0) = K = i$, we obtain that

(5.4.5) $$E[X(t)] = M_X(t) = ie^{(\lambda p - \mu)t}.$$

(b) Since knowledge of $Y(s)$, $0 \le s < t$, is informative about the number of females in the population at time t, it follows that the conditional distribution of $Y(t+s)$ given $Y(u)$, $0 \le u \le t$, does not depend solely on $Y(t)$. Hence, $\{Y(t), t \geq 0\}$ is not a continuous-time Markov chain, and so we cannot derive a differential equation for $M_Y(t) = E[Y(t)|Y(0) = j]$ in the same manner as we did for $M_X(t)$. However, we can compute an expression for $M_Y(t + h)$ by conditioning on both $X(t)$ and $Y(t)$. Indeed, conditional on $X(t)$ and $Y(t)$,

$$Y(t+h) = \begin{cases} Y(t)+1 & \text{with probability } \lambda(1-p)X(t)h + o(h) \\ Y(t)-1 & \text{with probability } \nu Y(t)h + o(h) \\ Y(t) & \text{with probability } 1 - [\lambda(1-p)X(t) - \nu Y(t)]h + o(h). \end{cases}$$

Taking expectations gives

$$E[Y(t+h)|X(t), Y(t)] = Y(t) + [\lambda(1-p)X(t) - \nu Y(t)]h + o(h),$$

and taking expectations once again, yields

$$M_Y(t+h) = M_Y(t) + [\lambda(1-p)M_X(t) - \nu M_Y(t)]h + o(h).$$

Upon subtracting $M_Y(t)$, dividing by h, and letting $h \to 0$, we obtain

$$\begin{aligned} M_Y'(t) &= \lambda(1-p)M_X(t) - \nu M_Y(t) \\ &= i\lambda(1-p)e^{(\lambda p - \mu)t} - \nu M_Y(t). \end{aligned}$$

Adding $\nu M_Y(t)$ to both sides and then multiplying by $e^{\nu t}$ gives

$$e^{\nu t}[M_Y'(t) + \nu M_Y(t)] = i\lambda(1-p)e^{(\lambda p + \nu - \mu)t},$$

or

$$\frac{d}{dt}\{e^{\nu t}M_Y(t)\} = i\lambda(1-p)e^{(\lambda p + \nu - \mu)t}.$$

Integrating gives

$$e^{\nu t}M_Y(t) = \frac{i\lambda(1-p)}{\lambda p + \nu - \mu}e^{(\lambda p + \nu - \mu)t} + C.$$

Evaluating at $t = 0$ gives,

$$j = M_Y(0) = \frac{i\lambda(1-p)}{\lambda p + \nu - \mu} + C$$

and so,

$$(5.4.6)\quad M_Y(t) = \frac{i\lambda(1-p)}{\lambda p + \nu - \mu} e^{(\lambda p - \mu)t} + \left(j - \frac{i\lambda(1-p)}{\lambda p + \nu - \mu}\right) e^{-\nu t}.$$

(c) Before finding Cov$[X(t), Y(t)]$, let us first compute $E[X^2(t)]$. To begin, note that from (5.4.4) we have that:

$$\begin{aligned} E[X^2(t+h)|X(t)] &= [X(t)+1]^2\lambda p X(t)h + [X(t)-1]^2\mu X(t)h \\ &\quad + X^2(t)[1-(\mu+\lambda p)X(t)h] + o(h) \\ &= X^2(t) + 2(\lambda p - \mu)hX^2(t) \\ &\quad + (\lambda p + \mu)hX(t) + o(h). \end{aligned}$$

Now, letting $M_2(t) = E[X^2(t)]$, and taking expectations of the preceding yields that

$$M_2(t+h) - M_2(t) = 2(\lambda p - \mu)hM_2(t) + (\lambda p + \mu)hM_X(t) + o(h).$$

Dividing through by h and taking the limit as $h \to 0$ shows, upon using Equation (5.4.5), that

$$M_2'(t) = 2(\lambda p - \mu)M_2(t) + i(\lambda p + \mu)e^{(\lambda p - \mu)t}.$$

Hence,

$$e^{-2(\lambda p - \mu)t}\{M_2'(t) - 2(\lambda p - \mu)M_2(t)\} = i(\lambda p + \mu)e^{(\mu - \lambda p)t}$$

or,

$$\frac{d}{dt}\{e^{-2(\lambda p - \mu)t}M_2(t)\} = i(\lambda p + \mu)e^{(\mu - \lambda p)t}$$

or, equivalently,

$$e^{-2(\lambda p - \mu)t}M_2(t) = \frac{i(\mu + \lambda p)}{\mu - \lambda p}e^{(\mu - \lambda p)t} + C,$$

or

$$M_2(t) = \frac{i(\mu + \lambda p)}{\mu - \lambda p}e^{(\lambda p - \mu)t} + Ce^{2(\lambda p - \mu)t}.$$

As $M_2(0) = i^2$, we obtain the result

$$(5.4.7)\qquad M_2(t) = \frac{i(\mu + \lambda p)}{\mu - \lambda p}e^{(\lambda p - \mu)t} + \left[i^2 - \frac{i(\mu + \lambda p)}{\mu - \lambda p}\right]e^{2(\lambda p - \mu)t}.$$

Now, let $M_{XY}(t) = E[X(t)Y(t)]$. Since the probability of two or more transitions in a time h is $o(h)$, we have that:

$$X(t+h)Y(t+h)$$
$$= \begin{cases} (X(t)+1)Y(t) & \text{with prob. } \lambda p X(t)h + o(h) \\ X(t)(Y(t)+1) & \text{with prob. } \lambda(1-p)X(t)h + o(h) \\ (X(t)-1)Y(t) & \text{with prob. } \mu X(t)h + o(h) \\ X(t)(Y(t)-1) & \text{with prob. } \nu Y(t)h + o(h) \\ X(t)Y(t) & \text{with prob. } 1 - \text{the above.} \end{cases}$$

Hence,

$$\begin{aligned} E[X(t+h)Y(t+h) \mid X(t), Y(t)] = X(t)Y(t) & \\ + hX(t)Y(t)[\lambda p - \mu - \nu] & \\ + X^2(t)\lambda(1-p)h + o(h). & \end{aligned}$$

Taking expectations gives

$$\begin{aligned} M_{XY}(t+h) = M_{XY}(t) + h(\lambda p - \mu - \nu)M_{XY}(t) & \\ + \lambda(1-p)hE[X^2(t)] + o(h) & \end{aligned}$$

implying that,

$$M'_{XY}(t) = (\lambda p - \mu - \nu)M_{XY}(t) + \lambda(1-p)M_2(t)$$

or,

$$\begin{aligned} \frac{d}{dt}\{e^{-(\lambda p - \mu - \nu)t}M_{XY}(t)\} &= \lambda(1-p)e^{-(\lambda p - \mu - \nu)t}M_2(t) \\ &= \lambda(1-p)\frac{i(\mu + \lambda p)}{\mu - \lambda p}e^{\nu t} \\ &\quad + \lambda(1-p)\left[i^2 - \frac{i(\mu + \lambda p)}{\mu - \lambda p}\right]e^{(\lambda p - \mu + \nu)t}. \end{aligned}$$

Integration now yields

$$\begin{aligned} e^{-(\lambda p - \mu - \nu)t}M_{XY}(t) &= \frac{i\lambda(1-p)(\mu + \lambda p)}{\nu(\mu - \lambda p)}e^{\nu t} \\ &\quad + \frac{\lambda(1-p)}{\lambda p - \mu + \nu}\left[i^2 - \frac{i(\mu + \lambda p)}{\mu - \lambda p}\right]e^{(\lambda p - \mu + \nu)t} + C, \end{aligned}$$

or,

$$
(5.4.8) \quad M_{XY}(t) = \frac{i\lambda(1-p)(\mu+\lambda p)}{\nu(\mu-\lambda p)} e^{(\lambda p-\mu)t} + \frac{\lambda(1-p)}{\lambda p-\mu+\nu}\left[i^2 - \frac{i(\mu+\lambda p)}{\mu-\lambda p}\right] e^{2(\lambda p-\mu)t} + Ce^{(\lambda p-\mu-\nu)t}.
$$

Hence, from Equations (5.4.5), (5.4.6), and (5.4.8) we obtain after some algebraic simplifications that,

$$
\begin{aligned}
\operatorname{Cov}[X(t), Y(t)] = {} & \left[C - ij + \frac{i^2\lambda(1-p)}{\lambda p+\nu-\mu}\right] e^{(\lambda p-\mu-\nu)t} \\
& + \frac{i\lambda(1-p)(\mu+\lambda p)}{\nu(\mu-\lambda p)} e^{(\lambda p-\mu)t} \\
& - \frac{\lambda(1-p)}{\lambda p-\mu+\nu}\,\frac{i(\mu+\lambda p)}{\mu-\lambda p} e^{2(\lambda p-\mu)t}.
\end{aligned}
$$

Using that $\operatorname{Cov}[X(0), Y(0)] = 0$, gives

$$
C = ij - \frac{i^2\lambda(1-p)}{\lambda p+\nu-\mu} - \frac{i\lambda(1-p)(\mu+\lambda p)}{\nu(\mu-\lambda p)} + \frac{\lambda(1-p)}{\lambda p-\mu+\nu}\,\frac{i(\mu+\lambda p)}{\mu-\lambda p}.
$$

It also follows from Equations (5.4.5) and (5.4.7) that,

$$
\operatorname{Var}[X(t)] = \frac{i(\mu+\lambda p)}{\mu-\lambda p} e^{(\lambda p-\mu)t} - \frac{i(\mu+\lambda p)}{\mu-\lambda p} e^{2(\lambda p-\mu)t}.
$$

5.4.1 Computing the Transition Probabilities

If we define r_{ij} by

$$
r_{ij} = \begin{cases} q_{ij} & \text{if } i \neq j \\ -\nu_i & \text{if } i = j \end{cases}
$$

then the Kolmogorov backward equations can be written as

$$
P'_{ij}(t) = \sum_k r_{ik} P_{kj}(t)
$$

and the forward equations as

$$P'_{ij}(t) = \sum_k r_{kj} P_{ik}(t).$$

These equations have a particularly nice form in matrix notation. If we define the matrices $\boldsymbol{R}$, $\boldsymbol{P}(t)$, and $\boldsymbol{P}'(t)$ by letting the element in row i, column j of these matrices be, respectively, r_{ij}, $P_{ij}(t)$, and $P'_{ij}(t)$, then the backwards equations can be written as

$$\boldsymbol{P}'(t) = \boldsymbol{R}\boldsymbol{P}(t)$$

and the forward equations as

$$\boldsymbol{P}'(t) = \boldsymbol{P}(t)\boldsymbol{R}.$$

This suggests the solution

$$\boldsymbol{P}(t) = e^{\boldsymbol{R}t} \equiv \sum_{i=0}^{\infty} (\boldsymbol{R}t)^i/i! \tag{5.4.9}$$

where $\boldsymbol{R}^0$ is the identity matrix $\boldsymbol{I}$; and in fact it can be shown that (5.4.9) is valid provided that the ν_i are bounded.

The direct use of Equation (5.4.9) to compute $\boldsymbol{P}(t)$ turns out to be very inefficient for two reasons. First, since the matrix $\boldsymbol{R}$ contains both positive and negative values (recall that the off-diagonal elements are the q_{ij} while the ith diagonal element is $r_{ii} = -\nu_i$) there is a problem of computer round-off error when we compute the powers of $\boldsymbol{R}$. Secondly, we often have to compute many of the terms in the infinite sum (5.4.9) to arrive at a good approximation.

We can, however, use (5.4.9) to efficiently approximate the matrix $\boldsymbol{P}(t)$ by utilizing the matrix equivalent of the identity

$$e^x = \lim_{n \to \infty} (1 + x/n)^n$$

which states that

$$e^{\boldsymbol{R}t} = \lim_{n \to \infty} (\boldsymbol{I} + \boldsymbol{R}t/n)^n.$$

By letting n be a power of 2, say $n = 2^k$, we can thus approximate $\boldsymbol{P}(t)$ by raising the matrix $\boldsymbol{M} = \boldsymbol{I} + \boldsymbol{R}t/n$ to the nth power, which can be accomplished by k matrix multiplications. In addition, since only the diagonal elements of $\boldsymbol{R}$ are negative and the diagonal elements of $\boldsymbol{I}$ are 1 we can, by choosing n large enough, guarantee that all of the elements of $\boldsymbol{M}$ are nonnegative.

5.5 Limiting Probabilities

Since a continuous-time Markov chain is a semi-Markov process with

$$F_{ij}(t) = 1 - e^{-\nu_i t}$$

it follows, from the results of Section 4.8 of Chapter 4, that if the discrete-time Markov chain with transition probabilities P_{ij} is irreducible and positive recurrent, then the limiting probabilities $P_j = \lim_{t\to\infty} P_{ij}(t)$ are given by

$$P_j = \frac{\pi_j/\nu_j}{\sum_i \pi_i/\nu_i} \tag{5.5.1}$$

where the π_j are the unique nonnegative solution of

$$\begin{aligned} \pi_j &= \sum_i \pi_i P_{ij}, \\ \sum_i \pi_i &= 1. \end{aligned} \tag{5.5.2}$$

From (5.5.1) and (5.5.2) we see that the P_j are the unique nonnegative solution of

$$\begin{aligned} \nu_j P_j &= \sum_i \nu_i P_i P_{ij}, \\ \sum_j P_j &= 1, \end{aligned}$$

or, equivalently, using $q_{ij} = \nu_i P_{ij}$,

$$\begin{aligned} \nu_j P_j &= \sum_i P_i q_{ij}, \\ \sum_j P_j &= 1. \end{aligned} \tag{5.5.3}$$

Remarks

(1) It follows from the results given in Section 4.8 of Chapter 4 for semi-Markov processes that P_j also equals the long-run proportion of time the process is in state j.

(2) If the initial state is chosen according to the limiting probabilities $\{P_j\}$, then the resultant process will be stationary. That is,

$$\sum_i P_i P_{ij}(t) = P_j \qquad \text{for all } t.$$

The above is proven as follows:

$$\begin{aligned}\sum_i P_{ij}(t)P_i &= \sum_i P_{ij}(t) \lim_{s\to\infty} P_{ki}(s)\\ &= \lim_{s\to\infty} \sum_i P_{ij}(t)P_{ki}(s)\\ &= \lim_{s\to\infty} P_{kj}(t+s)\\ &= P_j.\end{aligned}$$

The above interchange of limit and summation is easily justified, and is left as an exercise.

(3) Another way of obtaining Equations (5.5.3) is by way of the forward equations

$$P'_{ij}(t) = \sum_{k\neq j} q_{kj}P_{ik}(t) - v_j P_{ij}(t).$$

If we assume that the limiting probabilities $P_j = \lim_{t\to\infty} P_{ij}(t)$ exist, then $P'_{ij}(t)$ would necessarily converge to 0 as $t \to \infty$. (Why?) Hence, assuming that we can interchange limit and summation in the above, we obtain upon letting $t \to \infty$,

$$0 = \sum_{k\neq j} P_k q_{kj} - v_j P_j.$$

It is worth noting that the above is a more formal version of the following heuristic argument—which yields an equation for P_j, the probability of being in state j at $t = \infty$—by conditioning on the state h units prior in time:

$$\begin{aligned}P_j &= \sum_i P_{ij}(h)P_i\\ &= \sum_{i\neq j} (q_{ij}h + o(h))P_i + (1 - v_j h + o(h))P_j\end{aligned}$$

or

$$0 = \sum_{i\neq j} P_i q_{ij} - v_j P_j + \frac{o(h)}{h},$$

and the result follows by letting $h \to 0$.

(4) Equation (5.5.3) has a nice interpretation, which is as follows: In any interval $(0, t)$, the number of transitions into state j must equal to within

1 the number of transitions out of state j. (Why?) Hence, in the long run the rate at which transitions into state j occur must equal the rate at which transitions out of state j occur. Now when the process is in state j it leaves at rate ν_j, and, since P_j is the proportion of time it is in state j, it thus follows that

$$\nu_j P_j = \text{rate at which the process leaves state } j.$$

Similarly, when the process is in state i it departs to j at rate q_{ij}, and, since P_i is the proportion of time in state i, we see that the rate at which transitions from i to j occur is equal to $q_{ij}P_i$. Hence,

$$\sum_i P_i q_{ij} = \text{rate at which the process enters state } j.$$

Therefore, (5.5.3) is just a statement of the equality of the rate at which the process enters and leaves state j. Because it balances (that is, equates) these rates, Equations (5.5.3) are sometimes referred to as *balance equations*.

(5) When the continuous-time Markov chain is irreducible and $P_j > 0$ for all j, we say that the chain is *ergodic*.

Let us now determine the limiting probabilities for a birth and death process. From Equations (5.5.3), or, equivalently, by equating the rate at which the process leaves a state with the rate at which it enters that state, we obtain

State	Rate Process Leaves		Rate Process Enters
0	$\lambda_0 P_0$	$=$	$\mu_1 P_1$
$n, n > 0$	$(\lambda_n + \mu_n)P_n$	$=$	$\mu_{n+1}P_{n+1} + \lambda_{n-1}P_{n-1}$

Rewriting these equations gives

$$\lambda_0 P_0 = \mu_1 P_1,$$
$$\lambda_n P_n = \mu_{n+1}P_{n+1} + (\lambda_{n-1}P_{n-1} - \mu_n P_n), \qquad n \geq 1,$$

or, equivalently,

$$\lambda_0 P_0 = \mu_1 P_1,$$
$$\lambda_1 P_1 = \mu_2 P_2 + (\lambda_0 P_0 - \mu_1 P_1) = \mu_2 P_2,$$
$$\lambda_2 P_2 = \mu_3 P_3 + (\lambda_1 P_1 - \mu_2 P_2) = \mu_3 P_3,$$
$$\lambda_n P_n = \mu_{n+1}P_{n+1} + (\lambda_{n-1}P_{n-1} - \mu_n P_n) = \mu_{n+1}P_{n+1}.$$

Solving in terms of P_0 yields

$$P_1 = \frac{\lambda_0}{\mu_1} P_0,$$

$$P_2 = \frac{\lambda_1}{\mu_2} P_1 = \frac{\lambda_1 \lambda_0}{\mu_2 \mu_1} P_0,$$

$$P_3 = \frac{\lambda_2}{\mu_3} P_2 = \frac{\lambda_2 \lambda_1 \lambda_0}{\mu_3 \mu_2 \mu_1} P_0.$$

$$P_n = \frac{\lambda_{n-1}}{\mu_n} P_{n-1} = \frac{\lambda_{n-1} \lambda_{n-2} \cdots \lambda_1 \lambda_0}{\mu_n \mu_{n-1} \cdots \mu_2 \mu_1} P_0.$$

Using $\sum_{n=0}^{\infty} P_n = 1$ we obtain

$$1 = P_0 + P_0 \sum_{n=1}^{\infty} \frac{\lambda_{n-1} \cdots \lambda_1 \lambda_0}{\mu_n \cdots \mu_2 \mu_1}$$

or

$$P_0 = \left[1 + \sum_{n=1}^{\infty} \frac{\lambda_0 \lambda_1 \cdots \lambda_{n-1}}{\mu_1 \mu_2 \cdots \mu_n}\right]^{-1},$$

and hence

$$P_n = \frac{\lambda_0 \lambda_1 \cdots \lambda_{n-1}}{\mu_1 \mu_2 \cdots \mu_n \left(1 + \sum_{n=1}^{\infty} \frac{\lambda_0 \lambda_1 \cdots \lambda_{n-1}}{\mu_1 \mu_2 \cdots \mu_n}\right)}, \qquad n \geq 1. \tag{5.5.4}$$

The above equations also show us what condition is needed for the limiting probabilities to exist. Namely,

$$\sum_{n=1}^{\infty} \frac{\lambda_0 \lambda_1 \cdots \lambda_{n-1}}{\mu_1 \mu_2 \cdots \mu_n} < \infty.$$

EXAMPLE 5.5(A) ***The M/M/1 Queue.*** In the $M/M/1$ queue $\lambda_n = \lambda$, $\mu_n = \mu$, and thus from (5.5.4)

$$P_n = \frac{(\lambda/\mu)^n}{1 + \sum_{n=1}^{\infty} \left(\frac{\lambda}{\mu}\right)^n} = \left(\frac{\lambda}{\mu}\right)^n \left(1 - \frac{\lambda}{\mu}\right), \qquad n \geq 0$$

provided that $\lambda/\mu < 1$. It is intuitive that λ must be less than μ for limiting probabilities to exist. Customers arrive at rate λ and are served at rate μ, and thus if $\lambda > \mu$, they will arrive at a faster rate than they can be served and the queue size will go to infinity. The case $\lambda = \mu$ behaves much like the symmetric random walk of Section 4.3 of Chapter 4, which is null recurrent and thus has no limiting probabilities.

EXAMPLE 5.5(B) Consider a job shop consisting of M machines and a single repairman, and suppose that the amount of time a machine runs before breaking down is exponentially distributed with rate λ and the amount of time it takes the repairman to fix any broken machine is exponential with rate μ. If we say that the state is n whenever there are n machines down, then this system can be modeled as a birth and death process with parameters

$$\mu_n = \mu, \quad n \geq 1$$

$$\lambda_n = \begin{cases} (M-n)\lambda & n \leq M \\ 0 & n > M. \end{cases}$$

From Equation (5.5.4) we have that P_n, the limiting probability that n machines will not be in use, is given by

$$P_0 = \frac{1}{1 + \sum_{n=1}^{M} \left(\frac{\lambda}{\mu}\right)^n \frac{M!}{(M-n)!}}$$

$$P_n = \frac{\frac{M!}{(M-n)!}\left(\frac{\lambda}{\mu}\right)^n}{1 + \sum_{n=1}^{M} \left(\frac{\lambda}{\mu}\right)^n \frac{M!}{(M-n)!}}, \qquad n = 0, \ldots, M.$$

Hence, the average number of machines not in use is given by

$$\sum_{n=0}^{M} nP_n = \frac{\sum_{n=0}^{M} n \frac{M!}{(M-n)!}\left(\frac{\lambda}{\mu}\right)^n}{1 + \sum_{n=1}^{M} \left(\frac{\lambda}{\mu}\right)^n \frac{M!}{(M-n)!}}.$$

Suppose we wanted to know the long-run proportion of time that a given machine is working. To determine this, we compute the equivalent limiting probability of its working:

$$P\{\text{machine is working}\}$$
$$= \sum_{n=0}^{M} P\{\text{machine is working} \mid n \text{ not working}\} P_n$$
$$= \sum_{n=0}^{M} \frac{M-n}{M} P_n.$$

Consider a positive recurrent, irreducible continuous-time Markov chain and suppose we are interested in the distribution of the number of visits to some state, say state 0, by time t. Since visits to state 0 constitute renewals, it follows from Theorem 3.3.5 that for t large the number of visits is approximately normally distributed with mean $t/E[T_{00}]$ and variance $t\text{Var}(T_{00})/E^3[T_{00}]$, where T_{00} denotes the time between successive entrances into state 0.

$E[T_{00}]$ can be obtained by solving for the stationary probabilities and then using the identity

$$P_0 = \frac{1/\nu_0}{E[T_{00}]}.$$

To calculate $E[T_{00}^2]$ suppose that, at any time t, a reward is being earned at a rate equal to the time from t until the next entrance into state 0. It then follows from the theory of renewal reward processes that, with a new cycle beginning each time state 0 is entered, the long-run average reward per unit time is given by

$$\text{Average Reward} = \frac{E[\text{reward earned during a cycle}]}{E[\text{cycle time}]}$$
$$= \frac{E\left[\int_0^{T_{00}} x\,dx\right]}{E[T_{00}]} = E[T_{00}^2]/(2E[T_{00}]).$$

But as P_i is the proportion of time the chain is in state i, it follows that the average reward per unit time can also be expressed as

$$\text{Average Reward} = \sum_i P_i E[T_{i0}]$$

where T_{i0} is the time to enter 0 given that the chain is presently in state i.

Therefore, equating the two expressions for the average reward gives that

$$E[T_{00}^2] = 2E[T_{00}] \sum_i P_i E[T_{i0}].$$

By expressing T_{i0} as the time it takes to leave state i plus any additional time after it leaves i before it enters state 0, we obtain that the quantities $E[T_{i0}]$ satisfy the following set of linear equations

$$E[T_{i0}] = 1/\nu_i + \sum_{j \neq 0} P_{ij} E[T_{j0}], \qquad i \geq 0.$$

5.6 Time Reversibility

Consider an ergodic continuous-time Markov chain and suppose that it has been in operation an infinitely long time; that is, suppose that it started at time $-\infty$. Such a process will be stationary, and we say that it is in *steady state*. (Another approach to generating a stationary continuous-time Markov chain is to let the initial state of this chain be chosen according to the stationary probabilities.) Let us consider this process going backwards in time. Now, since the forward process is a continuous-time Markov chain it follows that given the present state, call it $X(t)$, the past state $X(t - s)$ and the future states $X(y)$, $y > t$ are independent. Therefore,

$$P\{X(t-s) = j \mid X(t) = i, X(y), y > t\} = P\{X(t-s) = j \mid X(t) = i\}$$

and so we can conclude that the reverse process is also a continuous-time Markov chain. Also, since the amount of time spent in a state is the same whether one is going forward or backward in time it follows that the amount of time the reverse chain spends in state i on a visit is exponential with the same rate ν_i as in the forward process. To check this formally, suppose that the process is in i at time t. Then the probability that its (backwards) time in i exceeds s is as follows:

$$\begin{aligned}
&P\{\text{process is in state } i \text{ throughout } [t-s, t] \mid X(t) = i\} \\
&\quad = P\{\text{process is in state } i \text{ throughout } [t-s, t]\}/P\{X(t) = i\} \\
&\quad = \frac{P\{X(t-s) = i\}e^{-\nu_i s}}{P\{X(t) = i\}} \\
&\quad = e^{-\nu_i s}
\end{aligned}$$

since $P\{X(t-s) = i\} = P\{X(t) = i\} = P_i$.

In other words, going backwards in time, the amount of time that the process spends in state i during a visit is also exponentially distributed with rate ν_i. In addition, as was shown in Section 4.7, the sequence of states visited by the reverse process constitutes a discrete-time Markov chain with transition probabilities P_{ij}^* given by

$$P_{ij}^* = \frac{\pi_j P_{ji}}{\pi_i}$$

where $\{\pi_j, j \geq 0\}$ are the stationary probabilities of the embedded discrete-time Markov chain with transition probabilities P_{ij}. Hence, we see that the reverse process is also a continuous-time Markov chain with the same transition rates out of each state as the forward time process and with one-stage transition probabilities P^*_{ij}. Let

$$q^*_{ij} = \nu_i P^*_{ij}$$

denote the infinitesimal rates of the reverse chain. Using the preceding formula for P^*_{ij} we see that

$$q^*_{ij} = \frac{\nu_i \pi_j P_{ji}}{\pi_i}.$$

However, recalling that

$$P_k = \frac{\pi_k/\nu_k}{C} \quad \text{where} \quad C = \sum_i \pi_i/\nu_i,$$

we see that

$$\frac{\pi_j}{\pi_i} = \frac{\nu_j P_j}{\nu_i P_i}$$

and so,

$$q^*_{ij} = \frac{\nu_j P_j P_{ji}}{P_i}$$
$$= \frac{P_j q_{ji}}{P_i}.$$

That is,

$$P_i q^*_{ij} = P_j q_{ji}. \tag{5.6.1}$$

Equation (5.6.1) has a very nice and intuitive interpretation, but to understand it we must first argue that the P_j are not only the stationary probabilities for the original chain but also for the reverse chain. This follows because if P_j is the proportion of time that the Markov chain is in state j when looking in the forward direction of time then it will also be the proportion of time when looking backwards in time. More formally, we show that P_j, $j \geq 0$ are the stationary probabilities for the reverse chain by showing that they satisfy the balance equations for this chain. This is done by summing Equation (5.6.1) over all states i, to obtain

$$\sum_i P_i q_{ij}^* = P_j \sum_i q_{ji} = P_j v_j$$

and so $\{P_j\}$ do indeed satisfy the balance equations and so are the stationary probabilities of the reversed chain.

Now since q_{ij}^* is the rate, when in state i, that the reverse chain makes a transition into state j, and P_i is the proportion of time this chain is in state i, it follows that $P_i q_{ij}^*$ is the rate at which the reverse chain makes a transition from i to j. Similarly, $P_j q_{ji}$ is the rate at which the forward chain makes a transition from j to i. Thus, Equation (5.6.1) states that the rate at which the forward chain makes a transition from j to i is equal to the rate at which the reverse chain makes a transition from i to j. But this is rather obvious, because every time the chain makes a transition from j to i in the usual (forward) direction of time, the chain when looked at in the reverse direction is making a transition from i to j.

The stationary continuous-time Markov chain is said to be time reversible if the reverse process follows the same probabilistic law as the original process. That is, it is *time reversible* if for all i and j

$$q_{ij}^* = q_{ij}$$

which is equivalent to

$$P_i q_{ij} = P_j q_{ji} \qquad \text{for all } i, j.$$

Since P_i is the proportion of time in state i, and since when in state i the process goes to j with rate q_{ij}, the condition of time reversibility is that *the rate at which the process goes directly from state i to state j is equal to the rate at which it goes directly from j to i.* It should be noted that this is exactly the same condition needed for an ergodic discrete-time Markov chain to be time reversible (see Section 4.7 of Chapter 4).

An application of the above condition for time reversibility yields the following proposition concerning birth and death processes.

PROPOSITION 5.6.1

An ergodic birth and death process is in steady state time reversible.

Proof To prove the above we must show that the rate at which a birth and death process goes from state i to state $i + 1$ is equal to the rate at which it goes from $i + 1$ to i. Now in any length of time t the number of transitions from i to $i + 1$ must equal to within 1 the number from $i + 1$ to i (since between each transition from i to $i + 1$ the process must return to i, and this can only occur through $i + 1$, and vice versa). Hence as the number of such transitions goes to infinity as $t \to \infty$, it follows that the rate of transitions from i to $i + 1$ equals the rate from $i + 1$ to i.

Proposition 5.6.1 can be used to prove the important result that the output process of an $M/M/s$ queue is a Poisson process. We state this as a corollary.

Corollary 5.6.2

Consider an $M/M/s$ queue in which customers arrive in accordance with a Poisson process having rate λ and are served by any one of s servers—each having an exponentially distributed service time with rate μ. If $\lambda < s\mu$, then the output process of customers departing is, in steady state, a Poisson process with rate λ.

Proof Let $X(t)$ denote the number of customers in the system at time t. Since the $M/M/s$ process is a birth and death process, it follows from Proposition 5.6.1 that $\{X(t), t \geq 0\}$ is time reversible. Now going forward in time, the time points at which $X(t)$ increases by 1 constitute a Poisson process since these are just the arrival times of customers. Hence by time reversibility, the time points at which the $X(t)$ increases by 1 when we go backwards in time also constitute a Poisson process. But these latter points are exactly the points of time when customers depart. (See Figure 5.6.1.) Hence the departure times constitute a Poisson process with rate λ.

As in the discrete-time situation, if a set of probabilities $\{x_j\}$ satisfy the time reversibility equations then the stationary Markov chain is time reversible and the x_j's are the stationary probabilities. To verify this, suppose that the x_j's are nonnegative and satisfy

$$\sum_j x_j = 1$$

$$x_i q_{ij} = x_j q_{ji}, \qquad \text{for all } i, j.$$

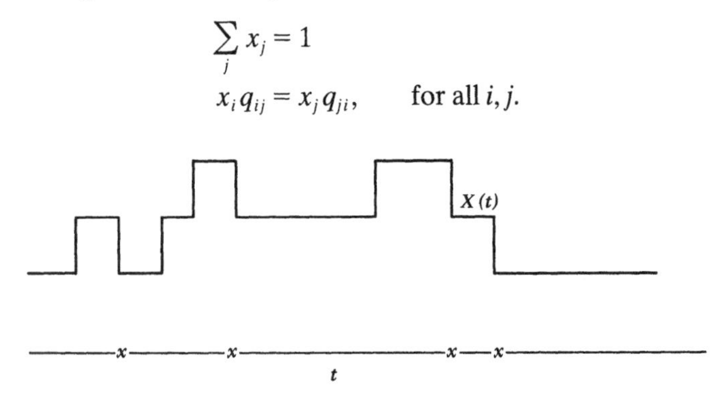

Figure 5.6.1. ***x equals the times at which, going backwards in time, X(t) increases; x also equals the times at which, going forward in time, X(t) decreases.***

Summing the bottom set of equations over all i gives

$$\sum_i x_i q_{ij} = x_j \sum_i q_{ji} = x_j v_j.$$

Thus, the $\{x_j\}$ satisfy the balance equations and so are the stationary probabilities, and since $x_i q_{ij} = x_j q_{ji}$ the chain is time reversible.

Consider a continuous-time Markov chain whose state space is S. We say that the Markov chain is *truncated* to the set $A \subset S$ if q_{ij} is changed to 0 for all $i \in A$, $j \notin A$. All other q_{ij} remain the same. Thus transitions out of the class of states A are not allowed. A useful result is that a truncated time-reversible chain remains time reversible.

PROPOSITION 5.6.3

A time-reversible chain with limiting probabilities P_j, $j \in S$, that is truncated to the set $A \subset S$ and remains irreducible is also time reversible and has limiting probabilities

$$P_j^A = P_j \Big/ \sum_{j \in A} P_j, \qquad j \in A. \tag{5.6.2}$$

Proof. We must show that

$$P_i^A q_{ij} = P_j^A q_{ji} \qquad \text{for } i \in A, j \in A,$$

or, equivalently,

$$P_i q_{ij} = P_j q_{ji} \qquad \text{for } i \in A, j \in A.$$

But the above follows since the original chain is time reversible by assumption.

EXAMPLE 5.6(A) Consider an $M/M/1$ queue in which arrivals finding N in the system do not enter but rather are lost. This finite capacity $M/M/1$ system can be regarded as a truncated version of the $M/M/1$ and so is time reversible with limiting probabilities given by

$$P_j = \frac{(\lambda/\mu)^j}{\sum_{i=0}^{N} \left(\frac{\lambda}{\mu}\right)^i}, \qquad 0 \le j \le N,$$

where we have used the results of Example 5.5(A) in the above.

Figure 5.6.2. A tandem queue.

5.6.1 Tandem Queues

The time reversibility of the $M/M/s$ queue has other important implications for queueing theory. For instance, consider a 2-server system in which customers arrive at a Poisson rate λ at server 1. After being served by server 1, they then join the queue in front of server 2. We suppose there is infinite waiting space at both servers. Each server serves one customer at a time with server i taking an exponential time with rate μ_i for a service, $i = 1, 2$. Such a system is called a tandem, or sequential, system (Figure 5.6.2).

Since the output from server 1 is a Poisson process, it follows that what server 2 faces is also an $M/M/1$ queue. However, we can use time reversibility to obtain much more. We need first the following Lemma.

Lemma 5.6.3

In an ergodic $M/M/1$ queue in steady state:

(i) the number of customers presently in the system is independent of the sequence of past departure times;

(ii) the waiting time spent in the system (waiting in queue plus service time) by a customer is independent of the departure process prior to his departure.

Proof (i) Since the arrival process is Poisson, it follows that the sequence of future arrivals is independent of the number presently in the system. Hence by time reversibility the number presently in the system must also be independent of the sequence of past departures (since, looking backwards in time, departures are seen as arrivals).

(ii) Consider a customer that arrives at time T_1 and departs at time T_2. Because the system is first come first served and has Poisson arrivals, it follows that the waiting time of the customer, $T_2 - T_1$, is independent of the arrival process after the time T_1. Now looking backward in time we will see that a customer arrives at time T_2 and the *same* customer departs at time T_1 (why the same customer?). Hence, by time reversibility, we see, by looking at the reversed process, that $T_2 - T_1$ will be independent of the (backward) arrival process after (in the backward direction) time T_2. But this is just the departure process before time T_2.

THEOREM 5.6.4

For the ergodic tandem queue in steady state:

(i) *the number of customers presently at server* 1 *and at server* 2 *are independent, and*

$$P\{n \text{ at server } 1, m \text{ at server } 2\} = \left(\frac{\lambda}{\mu_1}\right)^n \left(1 - \frac{\lambda}{\mu_1}\right) \left(\frac{\lambda}{\mu_2}\right)^m \left(1 - \frac{\lambda}{\mu_2}\right);$$

(ii) *the waiting time of a customer at server* 1 *is independent of its waiting time at server* 2.

Proof (i) By part (i) of Lemma 5.6.3 we have that the number of customers at server 1 is independent of the past departure times from server 1. Since these past departure times constitute the arrival process to server 2, the independence of the numbers of customers in the two systems follows. The formula for the joint probability follows from independence and the formula for the limiting probabilities of an $M/M/1$ queue given in Example 5.5(A).

(ii) By part (ii) of Lemma 5.6.3 we see that the time spent by a given customer at server 1 is independent of the departure process prior to his departing server 1. But this latter process, in conjunction with the service times at server 2, clearly determines the customer's wait at server 2. Hence the result follows.

Remarks

(1) For the formula in Theorem 5.6.4(i) to represent the joint probability, it is clearly necessary that $\lambda/\mu_i < 1$, $i = 1, 2$. This is the necessary and sufficient condition for the tandem queue to be ergodic.

(2) Though the waiting times of a given customer at the two servers are independent, it turns out somewhat surprisingly that the waiting times in queue of a customer are not independent. For a counterexample, suppose that λ is very small with respect to $\mu_1 = \mu_2$; and thus almost all customers have zero wait in queue at both servers. However, given that the wait in queue of a customer at server 1 is positive, his wait in queue at server 2 will also be positive with probability at least as large as $\frac{1}{2}$. (Why?) Hence, the waiting times in queue are not independent.

5.6.2 A Stochastic Population Model

Suppose that mutant individuals enter a population in accordance to a Poisson process with rate λ. Upon entrance each mutant becomes the initial ancestor of a family. All individuals in the population act independently and give birth at an exponential rate ν and die at an exponential rate μ, where we assume that $\nu < \mu$.

Let $N_j(t)$ denote the number of families at time t that consist of exactly j members, $j \geq 0$, and let

$$\underline{N}(t) = (N_1(t), N_2(t), \ldots).$$

Then $\{\underline{N}(t), t \geq 0\}$ is a continuous-time Markov chain.

For any state $\underline{n} = (n_1, n_2, \ldots, n_j, \ldots)$ with $n_j > 0$, define the states

$$B_j\underline{n} = (n_1, n_2, \ldots, n_{j-1}, n_j - 1, n_{j+1} + 1, \ldots), \qquad j \geq 1,$$

$$D_j\underline{n} = (n_1, n_2, \ldots, n_{j-1} + 1, n_j - 1, n_{j+1}, \ldots), \qquad j \geq 2,$$

and also define

$$B_0\underline{n} = (n_1 + 1, n_2, \ldots),$$

$$D_1\underline{n} = (n_1 - 1, n_2, \ldots).$$

Thus $B_j\underline{n}$ and $D_j\underline{n}$ represent the next state from n if there is respectively a birth or death in a family of size j, $j \geq 1$, whereas $B_0\underline{n}$ is the next state when a mutant appears.

If we let $q(\underline{n}, \underline{n}')$ denote the transition rates of the Markov chain, then the only nonzero rates are

$$q(\underline{n}, B_0\underline{n}) = \lambda,$$

$$q(\underline{n}, B_j\underline{n}) = jn_j\nu, \qquad j \geq 1,$$

$$q(\underline{n}, D_j\underline{n}) = jn_j\mu, \qquad j \geq 1.$$

To analyze this Markov chain it will be valuable to note that *families* enter the population in accordance to a Poisson process, and then change states in a random fashion that is independent of the activities of other families, where we say that a family is in state j if it consists of j individuals. Now let us suppose that the population is initially void—that is, $N(0) = \underline{0}$—and call an arriving mutant type j if its family will consist of j individuals at time t. Then it follows from the generalization of Proposition 2.3.2 of Chapter 2 to more than two types that $\{N_j(t), j \geq 1\}$ are independent Poisson random variables with respective means

$$E[N_j(t)] = \lambda \int_0^t P_j(s)\,ds, \tag{5.6.3}$$

where $P_j(s)$ is the probability that a family, originating at time s, will consist of j individuals at time t.

Letting $P(\underline{n})$ denote the limiting probabilities, it follows from the fact that $N_j(t)$, $j \geq 1$, are independent Poisson random variables when $N(0) = \underline{0}$ that the limiting probabilities will be of the form

$$P(\underline{n}) = \prod_{i=1}^{\infty} e^{-\alpha_i} \frac{\alpha_i^{n_i}}{n_i!} \tag{5.6.4}$$

for some set $\alpha_1, \alpha_2, \ldots$. We will now determine the α_i and show at the same time that the process is time reversible. For $P(\underline{n})$ of the form (5.6.4),

$$P(\underline{n})q(\underline{n}, B_0\underline{n}) = \lambda \prod_{i=1}^{\infty} e^{-\alpha_i} \frac{\alpha_i^{n_i}}{n_i!},$$

$$P(B_0\underline{n})q(B_0\underline{n}, \underline{n}) = (n_1 + 1)\mu \frac{e^{-\alpha_1}\alpha_1^{n_1+1}}{(n_1+1)!} \prod_{i=2}^{\infty} e^{-\alpha_i} \frac{\alpha_i^{n_i}}{n_i!}.$$

Equating $P(\underline{n})q(\underline{n}, B_0\underline{n})$ with $P(B_0\underline{n})q(B_0\underline{n}, \underline{n})$ yields

$$\alpha_1 = \lambda/\mu. \tag{5.6.5}$$

Similarly (5.6.4) gives that, for $j \geq 1$,

$$P(\underline{n})q(\underline{n}, B_j\underline{n}) \text{ will equal } P(B_j\underline{n})q(B_j\underline{n}, \underline{n}) \qquad \text{if} \quad j\nu\alpha_j = (j+1)\mu\alpha_{j+1}.$$

Using (5.6.5), this yields that

$$\alpha_j = \frac{\lambda}{j\nu}\left(\frac{\nu}{\mu}\right)^j.$$

Hence, for $P(\underline{n})$ given by (5.6.4) with $\alpha_j = \lambda(\nu/\mu)^j/j\nu$, we have shown

$$P(\underline{n})q(\underline{n}, B_j\underline{n}) = P(B_j(\underline{n}))q(B_j(\underline{n}), \underline{n}).$$

We can show a similar result for state $\underline{n}$ and $D_j\underline{n}$ by writing $\underline{n}$ as $B_{j-1}(D_j(\underline{n}))$ and then using the above. We thus have the following.

THEOREM 5.6.5

The continuous-time Markov chain $\{N(t), t \geq 0\}$ is, in steady state, time reversible with limiting probabilities,

$$P(\underline{n}) = \prod_{i=1}^{\infty} e^{-\alpha_i} \frac{\alpha_i^{n_i}}{n_i!},$$

where

$$\alpha_i = \frac{\lambda}{i\nu}\left(\frac{\nu}{\mu}\right)^i, \qquad i \geq 1.$$

In other words the limiting number of families that consist of i individuals are independent Poisson random variables with respective means

$$\frac{\lambda}{i\nu}\left(\frac{\nu}{\mu}\right)^i, \qquad i \geq 1.$$

There is an interesting interpretation of the α_i aside from it being the limiting mean number of families of size i. From (5.6.3) we see that

$$\begin{aligned} E[N_i(t)] &= \lambda \int_0^t q(t-s)\, ds \\ &= \lambda \int_0^t q(s)\, ds, \end{aligned}$$

where $q(s)$ is the probability that a family consists of i individuals a time s after its origination. Hence,

$$\lim_{t\to\infty} E[N_i(t)] = \lambda \int_0^{\infty} q(s)\, ds. \tag{5.6.6}$$

But consider an arbitrary family and let

$$I(s) = \begin{cases} 1 & \text{the family contains } i \text{ members a time } s \text{ after its origination} \\ 0 & \text{otherwise.} \end{cases}$$

Then

$$\begin{aligned} \int_0^{\infty} q(s)\, ds &= \int_0^{\infty} E[I(s)]\, ds \\ &= E\left[\int_0^{\infty} I(s)\, ds\right] \\ &= E[\text{amount of time the family has } i \text{ members}]. \end{aligned}$$

Hence, from (5.6.6),

$$\lim_{t\to\infty} E[N_i(t)] = \lambda E[\text{amount of time a family has } i \text{ members}],$$

and since

$$E[N_i(t)] \to \alpha_i = \frac{\lambda}{i\nu}\left(\frac{\nu}{\mu}\right)^i$$

we see that

$$E[\text{amount of time a family has } i \text{ members}] = \frac{(\nu/\mu)^i}{i\nu}.$$

Consider now the population model in steady state and suppose that the present state is $\underline{n}^*$. We would like to determine the probability that a given family of size i is the oldest (in the sense that it originated earliest) family in the population. It would seem that we would be able to use the time reversibility of the process to infer that this is the same as the probability that the given family will be the last surviving family of those presently in existence. Unfortunately, however, this conclusion does not immediately follow, for with our state space it is not possible, by observing the process throughout, to determine the exact time a given family becomes extinct. Thus we will need a more informative state space—one that enables us to follow the progress of a given family throughout time.

For technical reasons, it will be easiest to start by truncating the model and not allowing any more than M families to exist, where $M \geq \sum_i n_i^*$. That is, whenever there are M families in the population, no additional mutants are allowed. Note that, by Proposition 5.6.3, the truncated process with states $\underline{n}$ remains time reversible and has stationary probabilities

$$P(\underline{n}) = C\prod_{i=1}^{\infty} e^{-\alpha_i}\frac{\alpha_i^{n_i}}{n_i!}, \qquad \sum_{i=1} n_i \leq M,$$

where $\alpha_i = \lambda(\nu/\mu)^i/i\nu$.

To keep track of a given family as time progresses, we will have to label the different families. Let us use the labels $1, 2, \ldots, M$ and agree that whenever a new family originates (that is, a mutant appears) its label is uniformly chosen from the set of labels not being used at that time. If we let s_i denote the number of individuals in the family labeled i, $i = 1, \ldots, M$ (with $s_i = 0$ meaning that there is at present no family labeled i), then we can consider the process has states $\underline{s} = (s_1, \ldots, s_M)$, $s_i \geq 0$. For a given $\underline{s}$, let $\underline{n}(\underline{s}) = (n_1(\underline{s}), \ldots, n_k(\underline{s}) \ldots)$, where $n_i(\underline{s})$ is, as before, the number of families of size

i. That is,

$$n_i(\underline{s}) = \text{number of } j\text{: } s_j = i.$$

To obtain the stationary distribution of the Markov chain having states $\underline{s}$, note that

$$P(\underline{s}) = P(\underline{n})P(\underline{s}|\underline{n}) \qquad \text{for } \underline{n} = \underline{n}(\underline{s})$$

$$= P(\underline{s}|\underline{n})C\prod_{i=1}^{\infty} e^{-\alpha_i}\frac{\alpha_i^{n_i}}{n_i!}.$$

Since all labelings are chosen at random, it is intuitive that for a given vector $\underline{n}$ all of the

$$\frac{M!}{\left(M - \sum_1^{\infty} n_i\right)! \prod_{i=1}^{\infty} n_i!}$$

possible $\underline{s}$ vectors that are consistent with $\underline{n}$ are equally likely (that is, if $M = 3$, $n_1 = n_2 = 1$, $n_i = 0$, $i \geq 3$, then there are two families—one of size 1 and one of size 2—and it seems intuitive that the six possible states $\underline{s}$—all permutations of 0, 1, 2—are equally likely). Hence it seems intuitive that

$$P(\underline{s}) = \frac{\left(M - \sum_1^{\infty} n_i\right)! \prod_{i=1}^{\infty} n_i!}{M!} C\prod_{i=1}^{\infty} e^{-\alpha_i}\frac{\alpha_i^{n_i}}{n_i!}, \tag{5.6.7}$$

where $n_i = n_i(\underline{s})$ and $\alpha_i = \lambda(\nu/\mu)^i/i\nu$. We will now verify the above formula and show at the same time that the chain with states $\underline{s}$ is time reversible.

PROPOSITION 5.6.6

The chain with states $\underline{s} = (s_1, \ldots, s_M)$ is time reversible and has stationary probabilities given by (5.6.7).

Proof *For a vector* $\underline{s} = (s_1, \ldots, s_i, \ldots, s_M)$, let

$$B_i(\underline{s}) = (s_1, \ldots, s_i + 1, \ldots, s_M),$$

that is, $B_i(\underline{s})$ is the state following $\underline{s}$ if a member of the family labeled i gives birth. Now for $s_i > 0$

$$q(\underline{s}, B_i(\underline{s})) = s_i\nu, \qquad s_i > 0,$$

$$q(B_i(\underline{s}), \underline{s}) = (s_i + 1)\mu, \qquad s_i > 0.$$

Also, if

$$n(\underline{s}) = (n_1, \ldots, n_{s_i}, n_{s_i+1}, \ldots),$$

then

$$n(B_i(\underline{s})) = (n_1, \ldots, n_{s_i} - 1, n_{s_i+1} + 1, \ldots).$$

Hence for the $P(\underline{s})$ given by (5.6.7) and for $s_i > 0$,

(5.6.8) $$P(\underline{s})q(\underline{s}, B_i(\underline{s})) = P(B_i(\underline{s}))q(B_i(\underline{s}), \underline{s})$$

is equivalent to

$$\frac{n_{s_i}\alpha_{s_i}}{n_{s_i}!\, n_{s_i+1}!} s_i \nu = \frac{(n_{s_i+1} + 1)\alpha_{s_i+1}}{(n_{s_i} - 1)!(n_{s_i+1} + 1)!} (s_i + 1)\mu$$

or

$$\alpha_{s_i} s_i \nu = \alpha_{s_i+1}(s_i + 1)\mu$$

or, as $\alpha_i = \lambda(\nu/\mu)^i/i\nu$,

$$\lambda\left(\frac{\nu}{\mu}\right)^{s_i} = \lambda\left(\frac{\nu}{\mu}\right)^{s_i+1}\left(\frac{\mu}{\nu}\right),$$

which is clearly true.

Since there are $M - \sum n_i(\underline{s})$ available labelings to a mutant born when the state is $\underline{s}$, we see that

$$q(\underline{s}, B_i(\underline{s})) = \frac{\lambda}{M - \sum n_i(\underline{s})} \qquad \text{if } s_i = 0,$$

$$q(B_i(\underline{s}), \underline{s}) = \mu \qquad \text{if } s_i = 0.$$

The Equation (5.6.8) is easily shown to also hold in this case and the proof is thus complete.

Corollary 5.6.7

If in steady state there are n_i, $i > 0$, families of size i, then the probability that a given family of size i is the oldest family in the population is $i/\sum_j jn_j$.

Proof Consider the truncated process with states $\underline{s}$ and suppose that the state $\underline{s}$ is such that $n_i(\underline{s}) = n_i$. A given family of size i will be the oldest if, going backwards in time, it has been in existence the longest. But by time reversibility, the process going backward in time has the same probability laws as the one going forward, and hence this is the same as the probability that the specified family will survive the longest among all those presently in the population. But each individual has the same probability of having his or her descendants survive the longest, and since there are $\sum jn_j$ individuals in the population, of which i belong to the specified family, the result follows in this case. The proof now follows in general by letting $M \to \infty$.

Remark We have chosen to work with a truncated process since it makes it easy to guess at the limiting probabilities of the labeled process with states $\underline{s}$.

5.7 Applications of the Reversed Chain to Queueing Theory

The reversed chain can be quite a useful concept even when the process is not time reversible. To see this we start with the following continuous-time analogue of Theorem 4.7.2 of Chapter 4.

THEOREM 5.7.1

Let q_{ij} denote the transition rates of an irreducible continuous-time Markov chain. If we can find a collection of numbers q^*_{ij}, $i, j \geq 0$, $i \neq j$, and a collection of nonnegative numbers P_i, $i \geq 0$, summing to unity, such that

$$P_i q_{ij} = P_j q^*_{ji}, \qquad i \neq j,$$

and

$$\sum_{j \neq i} q_{ij} = \sum_{j \neq i} q^*_{ij}, \qquad i \geq 0,$$

then q^*_{ij} are the transition rates for the reversed chain and P_i are the limiting probabilities (for both chains).

The proof of Theorem 5.7.1 is left as an exercise.

Thus if we can guess at the reversed chain and the limiting probabilities, then we can use Theorem 5.7.1 to validate our guess. To illustrate this approach consider the following model in the network of queues that substantially generalizes the tandem queueing model of the previous section.

5.7.1 Network of Queues

We consider a system of k servers in which customers arrive, from outside the system, to each server i, $i = 1, \ldots, k$, in accordance with independent Poisson processes at rate r_i; they then join the queue at i until their turn at service comes. Once a customer is served by server i he then joins the queue in front of server j with probability P_{ij}, where $\sum_{j=1}^{k} P_{ij} \leq 1$, and $1 - \sum_{j=1}^{k} P_{ij}$ represents the probability that a customer departs the system after being served by server i.

If we let λ_j denote the total arrival rate of customers to server j, then the λ_j can be obtained as the solution of

$$\lambda_j = r_j + \sum_{i=1}^{k} \lambda_i P_{ij}, \qquad j = 1, \ldots, k. \tag{5.7.1}$$

Equation (5.7.1) follows since r_j is the arrival rate of customers to j coming from outside the system and, since λ_i is the rate at which customers depart server i (rate in must equal rate out), $\lambda_i P_{ij}$ is the arrival rate to j of those coming from server i.

This model can be analyzed as a continuous-time Markov chain with states $(n_1, n_2, \ldots, n_k)$, where n_i denotes the number of customers at server i. In accordance with the tandem queue results we might hope that the numbers of customers at each server are independent random variables. That is, letting the limiting probabilities be denoted by $P(n_1, n_2, \ldots, n_k)$, let us start by attempting to show that

$$P(n_1, n_2, \ldots, n_k) = P_1(n_1)P_2(n_2) \cdots P_k(n_k), \tag{5.7.2}$$

where $P_i(n_i)$ is thus the limiting probability that there are n_i customers at server i. To prove that the probabilities are indeed of the above form and to obtain the $P_i(n_i)$, $i = 1, \ldots, k$, we will need to first digress and speculate about the reversed process.

Now in the reversed process when a customer leaves server i, the customer will go to j with some probability that will hopefully not depend on the past. If that probability—let's call it $\overline{P}_{ij}$—indeed does not depend on the past, what would its value be? To answer this, note first that since the arrival rate to a server must equal the departure rate from the server, it follows that in both the forward and reverse processes the arrival rate at server j is λ_j. Since the rate at which customers go from j to i in the forward process must equal the rate at which they go from i to j in the reverse process, this means that

$$\lambda_j P_{ji} = \lambda_i \overline{P}_{ij}$$

or

$$\overline{P}_{ij} = \frac{\lambda_j P_{ji}}{\lambda_i}. \tag{5.7.3}$$

Thus we would hope that in the reversed process, when a customer leaves server i, he will go to server j with probability $\overline{P}_{ij} = \lambda_j P_{ji}/\lambda_i$.

Also, arrivals to i from outside the system in the reversed process correspond to departures from i that leave the system in the forward process, and hence occur at rate $\lambda_i(1 - \sum_j P_{ij})$. The nicest possibility would be if this was a Poisson process; then we make the following conjecture.

Conjecture The reversed stochastic process is a network process of the same type as the original. It has Poisson arrivals from outside the system to server i at rate $\lambda_i(1 - \sum_j P_{ij})$ and a departure from i goes to j with probability $\overline{P}_{ij}$ as given by (5.7.3). The service rate i is exponential with rate μ_i. In addition the limiting probabilities satisfy

$$P(n_1, n_2, \ldots n_k) = P_1(n_1)P_2(n_2) \cdots P_k(n_k).$$

To prove the conjecture and obtain the $P_i(n_i)$, consider first transitions resulting from an outside arrival. That is, consider the states $\underline{n} = (n_1, \ldots, n_i, \ldots, n_k)$ and $\underline{n}' = (n_1, \ldots, n_i + 1, \ldots, n_k)$. Now

$$q_{n,n'} = r_i,$$

and, if the conjecture is true,

$$\begin{aligned} q^*_{n',n} &= \mu_i\left(1 - \sum_j \overline{P}_{ij}\right) \\ &= \mu_i \frac{\left(\lambda_i - \sum_j \lambda_j P_{ji}\right)}{\lambda_i} \\ &= \frac{\mu_i r_i}{\lambda_i} \qquad \text{(from (5.7.1))} \end{aligned}$$

and

$$P(\underline{n}) = \prod_j P_j(n_j), \qquad P(\underline{n}') = P_i(n_i + 1) \prod_{j \neq i} P_j(n_j).$$

Hence from Theorem 5.7.1 we need that

$$r_i \prod_j P_j(n_j) = \frac{\mu_i r_i}{\lambda_i} P_i(n_i + 1) \prod_{j \neq i} P_j(n_j)$$

or

$$P_i(n_i + 1) = \frac{\lambda_i}{\mu_i} P_i(n_i).$$

That is,

$$\begin{aligned} P_i(n + 1) &= \frac{\lambda_i}{\mu_i} P_i(n) \\ &= \left(\frac{\lambda_i}{\mu_i}\right)^2 P_i(n - 1) \cdots = \left(\frac{\lambda_i}{\mu_i}\right)^{n+1} P_i(0), \end{aligned}$$

and using that $\sum_{n=0}^{\infty} P_i(n) = 1$ yields

$$P_i(n) = \left(\frac{\lambda_i}{\mu_i}\right)^n \left(1 - \frac{\lambda_i}{\mu_i}\right). \tag{5.7.4}$$

Thus λ_i/μ_i must be less than unity and the P_i must be as given above for the conjecture to be true.

To continue with our proof of the conjecture, consider those transitions that result from a departure from server j going to server i. That is, let $\underline{n} = (n_1, \ldots, n_i, \ldots, n_j, \ldots, n_k)$ and $\underline{n}' = (n_1, \ldots, n_i + 1, \ldots, n_j - 1, \ldots, n_k)$, where $n_j > 0$. Since

$$q_{\underline{n},\underline{n}'} = \mu_j P_{ji}$$

and the conjecture yields

$$q^*_{\underline{n}',\underline{n}} = \mu_i \overline{P}_{ij},$$

we need to show that

$$P(\underline{n})\mu_j P_{ji} = P(\underline{n}')\mu_i \overline{P}_{ij},$$

or, using (5.7.4), that

$$\lambda_j P_{ji} = \lambda_i \overline{P}_{ij},$$

which is the definition of $\overline{P}_{ij}$.

Since the addition verifications needed for Theorem 5.7.1 follow in the same manner, we have thus proven the following.

THEOREM 5.7.2

Assuming that $\lambda_i < \mu_i$ for all i, in steady state, the number of customers at service i are independent and the limiting probabilities are given by

$$P(n_1, \ldots, n_k) = \prod_{i=1}^{k} \left(\frac{\lambda_i}{\mu_i}\right)^{n_i} \left(1 - \frac{\lambda_i}{\mu_i}\right).$$

Also from the reversed chain we have the following.

Corollary 5.7.3

The processes of customers departing the system from server i, $i = 1, \ldots, k$, are independent Poisson processes having respective rates $\lambda_i(1 - \sum_j P_{ij})$.

Proof We've shown that in the reverse process, customers arrive to server i from outside the system according to independent Poisson processes having rates $\lambda_i(1 - \sum_j P_{ij})$, $i \geq 1$. Since an arrival from outside to server i in the reverse process corresponds to a departure out of the system from server i in the forward process, the result follows.

Remarks

(1) The result embodied in Theorem 5.7.2 is rather remarkable in that it says that the distribution of the number of customers at server i is the same as in an $M/M/1$ system with rates λ_i and μ_i. What is remarkable is that in the network model the arrival process at node i need *not* be a Poisson process. For if there is a possibility that a customer may visit a server more than once (a situation called feedback), then the arrival process will not be Poisson. An easy example illustrating this is to suppose that there is a single server whose service rate is very large with respect to the small arrival rate from outside. Suppose also that with probability $p = .9$ a customer upon completion of service is fed back into the system. Hence, at an arrival time epoch there is a large probability of another arrival in a short time (namely, the feedback arrival); whereas at an arbitrary time point there will be only a very slight chance of an arrival occurring shortly (since λ is small). Hence the arrival process does not possess independent increments and so cannot be Poisson.

(2) The model can be generalized to allow each service station to be a multi-server system (that is, server i operates as an $M/M/k_i$ rather than an $M/M/1$ system). The limiting numbers of customers at each station will again be independent and the number of customers at a server will have the same limiting distribution as if its arrival process was Poisson.

5.7.2 The Erlang Loss Formula

Consider a queueing loss model in which customers arrive at a k server system in accordance with a Poisson process having rate λ. It is a loss model in the sense that any arrival that finds all k servers busy does not enter but rather is lost to the system. The service times of the servers are assumed to have distribution G. We shall suppose that G is a continuous distribution having density g and hazard rate function $\lambda(t)$. That is, $\lambda(t) = g(t)/\overline{G}(t)$ is, loosely speaking, the instantaneous probability intensity that a t-unit-old service will end.

We can analyze the above system by letting the state at any time be the ordered ages of the customers in service at that time. That is, the state will be $\underline{x} = (x_1, x_2, \ldots, x_n)$, $x_1 \le x_2 \le \cdots \le x_n$, if there are n customers in service, the most recent one having arrived x_1 time units ago, the next most recent arrived being x_2 time units ago, and so on. The process of successive states will be a Markov process in the sense that the conditional distribution of any future state, given the present and all the past states, will depend only on the present state. Even though the process is not a continuous-time Markov *chain*, the theory we've developed for chains can be extended to cover this process, and we will analyze the model on this basis.

We will attempt to use the reverse process to obtain the limiting probability density $p(x_1, x_2, \ldots, x_n)$, $1 \le n \le k$, $x_1 \le x_2 \le \cdots \le x_n$, and $P(\phi)$ the limiting probability that the system is empty. Now since the age of a customer in service increases linearly from 0 upon her arrival to her service time upon her departure, it is clear that if we look backwards we will be following her excess or additional service time. As there will never be more than k in the system, we make the following conjecture.

Conjecture The reverse process is also a k-server loss system with service distribution G in which arrivals occur according to a Poisson process with rate λ. The state at any time represents the ordered residual service times of customers presently in service.

We shall now attempt to prove the above conjecture and at the same time obtain the limiting distribution. For any state $\underline{x} = (x_1, \ldots, x_i, \ldots, x_n)$ let $e_i(\underline{x}) = (x_1, \ldots, x_{i-1}, x_{i+1}, \ldots, x_n)$. Now in the original process when the state is $\underline{x}$ it will instantaneously go to $e_i(\underline{x})$ with a probability density equal to $\lambda(x_i)$ since the person whose time in service is x_i would have to instantaneously complete its service. Similarly, in the reversed process if the state is $e_i(\underline{x})$, then it will instantaneously go to $\underline{x}$ if a customer having service time x_i instantaneously arrives. So we see that

in forward: $\underline{x} \to e_i(\underline{x})$ with probability intensity $\lambda(x_i)$;

in reverse: $e_i(\underline{x}) \to \underline{x}$ with (joint) probability intensity $\lambda g(x_i)$.

Hence if $p(\underline{x})$ represents the limiting density, then in accordance with Theorem 5.7.1 we would need that

$$p(\underline{x})\lambda(x_i) = p(e_i(\underline{x}))\lambda g(x_i),$$

or, since $\lambda(x_i) = g(x_i)/\overline{G}(x_i)$,

$$p(\underline{x}) = p(e_i(\underline{x}))\lambda\overline{G}(x_i).$$

Letting $i = 1$ and iterating the above yields

$$\begin{aligned} p(\underline{x}) &= \lambda\overline{G}(x_1)p(e_1(\underline{x})) \\ &= \lambda\overline{G}(x_1)\lambda\overline{G}(x_2)p(e_1(e_1(\underline{x}))) \\ &\vdots \\ &= \prod_{i=1}^{n} \lambda\overline{G}(x_i)P(\phi). \end{aligned} \tag{5.7.5}$$

Integrating over all vectors $\underline{x}$ yields

$$\begin{aligned} P\{n \text{ in the system}\} &= P(\phi)\lambda^n \iint_{x_1 \le x_2 \le \cdots \le x_n} \cdots \int \prod_{i=1}^{n} \overline{G}(x_i)\, dx_1\, dx_2 \cdots dx_n \\ &= P(\phi)\frac{\lambda^n}{n!} \iint_{x_1, x_2, \ldots, x_n} \cdots \int \prod_{i=1}^{n} \overline{G}(x_i)\, dx_1\, dx_2 \cdots dx_n \\ &= P(\phi)\frac{(\lambda E[S])^n}{n!}, \qquad n = 1, 2, \ldots, k, \end{aligned} \tag{5.7.6}$$

where $E[S] = \int \overline{G}(x)\, dx$ is the mean service time. And, upon using

$$P(\phi) + \sum_{n=1}^{k} P\{n \text{ in the system}\} = 1,$$

we obtain

$$(5.7.7) \qquad P\{n \text{ in the system}\} = \frac{\dfrac{(\lambda E[S])^n}{n!}}{\displaystyle\sum_{i=0}^{k} \frac{(\lambda E[S])^i}{i!}}, \qquad n = 0, 1, \ldots, k.$$

From (5.7.5) we have that

$$(5.7.8) \qquad p(\underline{x}) = \frac{\lambda^n \displaystyle\prod_{i=1}^{n} \overline{G}(x_i)}{\displaystyle\sum_{i=0}^{k} \frac{(\lambda E[S])^i}{i!}}$$

and we see that the conditional distribution of the ordered ages given that there are n in the system is

$$\begin{aligned} p\{\underline{x} \mid n \text{ in the system}\} &= \frac{p(\underline{x})}{P\{n \text{ in the system}\}} \\ &= n! \prod_{i=1}^{n} \frac{\overline{G}(x_i)}{E[S]}. \end{aligned}$$

As $\overline{G}(x)/E[S]$ is just the density of G_e, the equilibrium distribution of G, we see that, if the conjecture is valid, then *the limiting distribution of the number in the system depends on G only through its mean and is given by* (5.7.7), *and, given that there are n in the system, their* (*unordered*) *ages are independent and are identically distributed according to the equilibrium distribution G_e of G.*

To complete our proof of the conjecture we must consider transitions of the forward process from $\underline{x}$ to $(0, \underline{x}) = (0, x_1, x_2, \ldots, x_n)$ when $n < k$. Now

in forward: $\underline{x} \rightarrow (0, \underline{x})$ with instantaneous intensity λ;

in reverse: $(0, \underline{x}) \rightarrow \underline{x}$ with probability 1

Hence in conjunction with Theorem 5.7.1 we must verify that

$$p(\underline{x})\lambda = p(0, \underline{x}),$$

which follows from (5.7.8) since $\overline{G}(0) = 1$.

Therefore, assuming the validity of the analogue of Theorem 5.7.1 we have thus proven the following.

THEOREM 5.7.4

The limiting distribution of the number of customers in the system is given by

$$P\{n \text{ in system}\} = \frac{\dfrac{(\lambda E[S])^n}{n!}}{\displaystyle\sum_{i=0}^{k} \frac{(\lambda E[S])^i}{i!}}, \qquad n = 0, 1, \ldots, k, \tag{5.7.9}$$

and given that there are n in the system the ages (or the residual times) of these n are independent and identically distributed according to the equilibrium distribution of G.

The model considered is often called the Erlang loss system, and Equation (5.7.9), the Erlang loss formula.

By using the reversed process we also have the following corollary.

Corollary 5.7.5

In the Erlang loss model the departure process (including both customers completing service and those that are lost) is a Poisson process at rate λ.

Proof The above follows since in the reversed process arrivals of all customers (including those that are lost) constitutes a Poisson process.

5.7.3 The $M/G/1$ Shared Processor System

Suppose that customers arrive in accordance with a Poisson process having rate λ. Each customer requires a random amount of work, distributed according to G. The server can process work at a rate of one unit of work per unit time, and divides his time equally among all of the customers presently in the system. That is, whenever there are n customers in the system, each will receive service work at a rate of $1/n$ per unit time.

Let $\lambda(t) = g(t)/\overline{G}(t)$ denote the failure rate function of the service distribution, and suppose that $\lambda E[S] < 1$, where $E[S]$ is the mean of G.

To analyze the above let the state at any time be the ordered vector of the amounts of work already performed on customers still in the system. That is, the state is $\underline{x} = (x_1, x_2, \ldots, x_n)$, $x_1 \le x_2 \le \cdots \le x_n$, if there are n customers in the system and $x_1, \ldots, x_n$ is the amount of work performed on these n

customers. Let $p(\underline{x})$ and $P(\phi)$ denote the limiting probability density and the limiting probability that the system is empty. We make the following conjecture regarding the reverse process.

Conjecture The reverse process is a system of the same type, with customers arriving at a Poisson rate λ, having workloads distributed according to G and with the state representing the ordered residual workloads of customers presently in the system.

To verify the above conjecture and at the same time obtain the limiting distribution let $e_i(\underline{x}) = (x_1, \ldots, x_{i-1}, x_{i+1}, \ldots, x_n)$ when $\underline{x} = (x_1, \ldots, x_n)$, $x_1 \le x_2 \le \cdots \le x_n$. Note that

in forward: $\underline{x} \to e_i(\underline{x})$ with probability intensity $\lambda(x_i)/n$;

in reverse: $e_i(x) \to \underline{x}$ with (joint) probability intensity $\lambda g(x_i)$.

The above follows as in the previous section with the exception that if there are n in the system then a customer who already had the amount of work x_i performed on it will instantaneously complete service with probability intensity $\lambda(x_i)/n$.

Hence, if $p(x)$ is the limiting density, then in accordance with Theorem 5.7.1 we need that

$$p(\underline{x})\frac{\lambda(x_i)}{n} = p(e_i(\underline{x}))\lambda g(x_i),$$

or, equivalently,

$$\begin{aligned} p(\underline{x}) &= n\overline{G}(x_i)p(e_i(\underline{x}))\lambda \qquad (5.7.10) \\ &= n\overline{G}(x_i)(n-1)\overline{G}(x_j)p(e_j(e_i(\underline{x})))\lambda^2, \qquad j \neq i \\ &\vdots \\ &= n!\lambda^n P(\phi)\prod_{i=1}^{n}\overline{G}(x_i). \end{aligned}$$

Integrating over all vectors $\underline{x}$ yields as in (5.7.6)

$$P\{n \text{ in system}\} = (\lambda E[S])^n P(\phi).$$

Using

$$P(\phi) + \sum_{n=1}^{\infty} P\{n \text{ in system}\} = 1$$

gives

$$P\{n \text{ in system}\} = (\lambda E[S])^n(1 - \lambda E[S]), \qquad n \geq 0.$$

Also, the conditional distribution of the ordered amounts of work already performed, given n in the system, is, from (5.7.10),

$$p(\underline{x}|n) = p(\underline{x})/P\{n \text{ in system}\}$$
$$= n! \prod_{i=1}^{n} \frac{\overline{G}(x_i)}{E[S]}.$$

That is, given n customers in the system the unordered amounts of work already performed are distributed independently according to G_e, the equilibrium distribution of G.

All of the above is based on the assumption that the conjecture is valid. To complete the proof of its validity we must verify that

$$p(\underline{x})\lambda = p(0, \underline{x})\frac{1}{n+1},$$

the above being the relevant equation since the reverse process when in state $(\varepsilon, \underline{x})$ will go to state $\underline{x}$ in time $(n + 1)\varepsilon$. Since the above is easily verified we have thus shown the following.

THEOREM 5.7.6

For the Processor Sharing Model the number of customers in the system has the distribution

$$P\{n \text{ in system}\} = (\lambda E[S])^n(1 - \lambda E[S]), \qquad n \geq 0.$$

Given n in the system, the completed (or residual) workloads are independent and have distribution G_e. The departure process is a Poisson process with rate λ.

If we let L denote the average number in the system, and W, the average time a customer spends in the system, then

$$L = \sum_{n=0}^{\infty} n(\lambda E[S])^n(1 - \lambda E[S])$$
$$= \frac{\lambda E[S]}{1 - \lambda E[S]}.$$

We can obtain W from the formula $L = \lambda W$ (see Section 3.6.1 of Chapter 3), and so

$$W = \frac{L}{\lambda} = \frac{E[S]}{1 - \lambda E[S]}.$$

Remark It is quite interesting that the *average* time a customer spends in the system depends on G only through its mean. For instance, consider two such systems: one in which the workloads are identically 1 and the other where they are exponentially distributed with mean 1. Then from Theorem 5.7.6 the distribution of the number in the system as seen by an arrival is the same for both. However, in the first system the remaining workloads of those found by an arrival are uniformly distributed over (0, 1) (this being the equilibrium distribution for the distribution of a deterministic random variable with mean 1), whereas in the second system the remaining workloads are exponentially distributed with mean 1. Hence it is quite surprising that the mean time in system for an arrival is the same in both cases. Of course the distribution of time spent by a customer in the system, as opposed to the mean of this distribution, depends on the entire distribution G and not just on its mean.

Another interesting computation in this model is the conditional mean time an arrival spends in the system given its workload is y. To compute this quantity fix y and say that a customer is "special" if its workload is between y and $y + \varepsilon$. By $L = \lambda W$ we thus have that

$$\begin{aligned}&\text{average number of special customers in the system}\\ &\quad = (\text{average arrival rate of special customer})\\ &\qquad \times (\text{average time a special customer spends in the system}).\end{aligned}$$

To determine the average number of special customers in the system let us first determine the density of the total workload of an arbitrary customer presently in the system. Suppose such a customer has already received the amount of work x. Then the conditional density of the customer's workload is

$$f(w|\text{has received } x) = g(w)/\overline{G}(x), \qquad x \le w.$$

But, from Theorem 5.7.6, the amount of work an arbitrary customer in the system has already received has the distribution G_e. Hence the density of the total workload of someone present in the system is

$$\begin{aligned}f(w) &= \int_0^w \frac{g(w)}{\overline{G}(x)}\, dG_e(x)\\ &= \int_0^w \frac{g(w)}{E[S]}\, dx \qquad \left(\text{since } dG_e(x) = \frac{\overline{G}(x)}{E[S]}\, dx\right)\\ &= wg(w)/E[S].\end{aligned}$$

Hence the average number of special customers in the system is

$$\begin{aligned}E&[\text{number in system having workload between } y \text{ and } y+\varepsilon]\\&= Lf(y)\varepsilon + o(\varepsilon)\\&= Lyg(y)\varepsilon/E[S] + o(\varepsilon).\end{aligned}$$

In addition, the average arrival rate of customers whose workload is between y and $y + \varepsilon$ is

$$\text{average arrival rate} = \lambda g(y)\varepsilon + o(\varepsilon).$$

Hence we see that

$$E[\text{time in system}|\text{workload in } (y, y+\varepsilon)] = \frac{g(y)\varepsilon}{\lambda g(y)\varepsilon}\frac{Ly}{E[S]} + \frac{o(\varepsilon)}{\varepsilon}.$$

Letting $\varepsilon \to 0$ we obtain

$$\text{(5.7.11)} \qquad \begin{aligned}E[\text{time in system}|\text{workload is } y] &= \frac{y}{\lambda E[S]}L\\&= \frac{y}{1-\lambda E[S]}.\end{aligned}$$

Thus the average time in the system of a customer needing y units of work also depends on the service distribution only through its mean. As a check of the above formula note that

$$\begin{aligned}W &= E[\text{time in system}]\\&= \int E[\text{time in system}|\text{workload is } y]\, dG(y)\\&= \frac{E[S]}{1-\lambda E[S]} \qquad \text{(from (5.7.11))},\end{aligned}$$

which checks.

5.8 Uniformization

Consider a continuous-time Markov chain in which the mean time spent in a state is the same for all states. That is, suppose that $\nu_i = \nu$ for all states i. In this case since the amount of time spent in each state during a visit is exponentially distributed with rate ν, it follows that if we let $N(t)$ denote the number of state transitions by time t, then $\{N(t), t \geq 0\}$ will be a Poisson process with rate ν.

To compute the transition probabilities $P_{ij}(t)$, we can condition on $N(t)$ as follows:

$$\begin{aligned} P_{ij}(t) &= P\{X(t) = j \mid X(0) = i\} \\ &= \sum_{n=0}^{\infty} P\{X(t) = j \mid X(0) = i, N(t) = n\} P\{N(t) = n \mid X(0) = i\} \\ &= \sum_{n=0}^{\infty} P\{X(t) = j \mid X(0) = i, N(t) = n\} e^{-\nu t} \frac{(\nu t)^n}{n!}. \end{aligned}$$

The fact that there have been n transitions by time t gives us some information about the amounts of time spent in each of the first n states visited, but as the distribution of time spent in each state is the same for all states, it gives us no information about which states were visited. Hence,

$$P\{X(t) = j \mid X(0) = i, N(t) = n\} = P_{ij}^n,$$

where P_{ij}^n is just the n-stage transition probability associated with the discrete-time Markov chain with transition probabilities P_{ij}; and so when $\nu_i \equiv \nu$

$$P_{ij}(t) = \sum_{n=0}^{\infty} P_{ij}^n e^{-\nu t} \frac{(\nu t)^n}{n!}. \tag{5.8.1}$$

The above equation is quite useful from a computational point of view since it enables us to approximate $P_{ij}(t)$ by taking a partial sum and then computing (by matrix multiplication of the transition probability matrix) the relevant n-stage probabilities P_{ij}^n.

Whereas the applicability of Equation (5.8.1) would appear to be quite limited since it supposes that $\nu_i \equiv \nu$, it turns out that most Markov chains can be put in that form by the trick of allowing fictitious transitions from a state to itself. To see how this works, consider any Markov chain for which the ν_i are bounded and let ν be any number such that

$$\nu_i \le \nu \qquad \text{for all } i. \tag{5.8.2}$$

Now when in state i the process actually leaves at rate ν_i; but this is equivalent to supposing that transitions occur at rate ν, but only the fraction ν_i/ν are real transitions out of i and the remainder are fictitious transitions that leave the process in state i. In other words, any Markov chain satisfying condition (5.8.2) can be thought of as being a process that spends an exponential amount of time with rate ν in state i and then makes a transition to j with probability P_{ij}^*, where

$$P_{ij}^* = \begin{cases} 1 - \dfrac{\nu_i}{\nu} & j = i \\[2ex] \dfrac{\nu_i}{\nu} P_{ij} & j \ne i. \end{cases} \tag{5.8.3}$$

Hence, from (5.8.1) we have that the transition probabilities can be computed by

$$P_{ij}(t) = \sum_{n=0}^{\infty} P_{ij}^{*n} e^{-\nu t} \frac{(\nu t)^n}{n!},$$

where P_{ij}^{*n} are the n-stage transition probabilities corresponding to (5.8.3). This technique of uniformizing the rate in which a transition occurs from each state by introducing transitions from a state to itself is known as *uniformization*.

EXAMPLE 5.8(A) Let us reconsider the two-state chain of Example 5.4(A), which has

$$P_{01} = P_{10} = 1,$$

$$\nu_0 = \lambda, \qquad \nu_1 = \mu.$$

Letting $\nu = \lambda + \mu$, the uniformized version of the above is to consider it a continuous-time Markov chain with

$$P_{00} = \frac{\mu}{\lambda + \mu} = 1 - P_{01},$$

$$P_{10} = \frac{\mu}{\lambda + \mu} = 1 - P_{11},$$

$$\nu_i = \lambda + \mu, \qquad i = 0, 1.$$

Since $P_{00} = P_{10}$, it follows that the probability of transition into state 0 is equal to $\mu/(\lambda + \mu)$ no matter what the present state. Since a similar result is true for state 1, it follows that the n-stage transition probabilities are given by

$$P_{i0}^{n} = \frac{\mu}{\lambda + \mu}, \qquad n \geq 1, \quad i = 0, 1.$$

Hence,

$$\begin{aligned} P_{00}(t) &= \sum_{n=0}^{\infty} P_{00}^{n} e^{-(\lambda+\mu)t} \frac{[(\lambda + \mu)t]^n}{n!} \\ &= e^{-(\lambda+\mu)t} + [1 - e^{-(\lambda+\mu)t}] \frac{\mu}{\lambda + \mu} \\ &= \frac{\mu}{\lambda + \mu} + \frac{\lambda}{\lambda + \mu} e^{-(\lambda+\mu)t}. \end{aligned}$$

Similarly,

$$\begin{aligned} P_{11}(t) &= \sum_{n=0}^{\infty} P_{11}^{n} e^{-(\lambda+\mu)t} \frac{[(\lambda + \mu)t]^n}{n!} \\ &= \frac{\lambda}{\lambda + \mu} + \frac{\mu}{\lambda + \mu} e^{-(\lambda+\mu)t}. \end{aligned}$$

Another quantity of interest is the total time up to t that the process has been in a given state. That is, if $X(t)$ denotes the state at t (either 0 or 1), then if we let $S_i(t)$, $i = 1, 0$, be given by

$$(5.8.4)\qquad \begin{aligned} S_1(t) &= \int_0^t X(s)\,ds \\ S_0(t) &= \int_0^t (1 - X(s))\,ds \end{aligned}$$

then $S_i(t)$ denotes the time spent in state i by time t. The stochastic process $\{S_i(t), t > 0\}$ is called the occupation time process for state i.

Let us now suppose that $X(0) = 0$ and determine the distribution of $S_0(t)$. Its mean is computed from (5.8.4):

$$\begin{aligned} E[S_0(t)|X(0) = 0] &= \int_0^t E[1 - X(s)]\,ds \\ &= \int_0^t P_{00}(s)\,ds \\ &= \frac{\mu}{\lambda + \mu}t + \frac{\lambda}{(\lambda + \mu)^2}[1 - e^{-(\lambda+\mu)t}]. \end{aligned}$$

We will not attempt to compute the variance of $S_0(t)$ (see Problem 5.36) but rather will directly consider its distribution.

To determine the conditional distribution of $S_0(t)$ given that $X(0) = 0$, we shall make use of the uniformized representation of $\{X(t), t \ge 0\}$ and start by conditioning on $N(t)$ to obtain, for $s < t$,

$$\begin{aligned} P\{S_0(t) \le s\} &= \sum_{n=0}^{\infty} e^{-(\lambda+\mu)t}\frac{((\lambda + \mu)t)^n}{n!} P\{S_0(t) \le s | N(t) = n\} \\ &= \sum_{n=1}^{\infty} e^{-(\lambda+\mu)t}\frac{((\lambda + \mu)t)^n}{n!} P\{S_0(t) \le s | N(t) = n\}. \end{aligned}$$

The last equality following since $N(t) = 0$ implies that $S_0(t) = t$ (since $X(0) = 0$). Now given that $N(t) = n$, the interval $(0, t)$ is broken into the $n + 1$ subintervals $(0, X_{(1)}), (X_{(1)}, X_{(2)}), \ldots, (X_{(n-1)}, X_{(n)}), (X_{(n)}, t)$ where $X_{(1)} \le X_{(2)} \le \cdots \le X_{(n)}$ are the n event times by t of $\{N(s)\}$. The process $\{X(t)\}$ will be in the same state throughout any given subinterval. It will be in state 0 for the first subinterval and afterwards it will, independently for each subinterval, be in state 0 with probability $\mu/(\lambda + \mu)$. Hence, given $N(t) = n$, the number of subintervals in which the process is in state 0 equals 1 plus a Binomial $(n, \mu/(\lambda + \mu))$ random variable. If this sum equals k (that is, if $1 + \text{Bin}(n, \mu/(\lambda + \mu) = k)$) then $S_0(t)$ will equal the sum of the lengths of k of the above subintervals. However, as $X_{(1)}, \ldots, X_{(n)}$ are distributed as the order statistics of a set of n independent uniform $(0, t)$ random variables it follows that the joint distribution of the $n + 1$ subinterval lengths $Y_1, Y_2, \ldots, Y_{n+1}$—where $Y_i = X_{(i)} - X_{(i-1)}, i = 1, \ldots, n + 1$ with $X_{(0)} = 0, X_{(n+1)} = t$—is exchangeable. That is, $P\{Y_i \leq y_i, i = 1, \ldots, n + 1\}$ is a symmetric function of the vector $(y_1, \ldots, y_{n+1})$. (See Problem 5.37.) Therefore, the distribution of the sum of any k of the Y_i's is the same no matter which ones are in the sum. Finally, as

$$X_{(k)} = Y_1 + \cdots + Y_k$$

we see that when $k \leq n$, this sum has the same distribution as the kth largest of a set of n independent uniform $(0, t)$ random variables. (When $k = n + 1$ the sum equals t.) Hence, for $s < t$,

$$P\{X_{(k)} \leq s\} = \sum_{i=k}^{n} \binom{n}{i} \left(\frac{s}{t}\right)^i \left(1 - \frac{s}{t}\right)^{n-i}.$$

Putting the above together gives, for $s < t$,

$$\begin{aligned} P\{S_0(t) \leq s | X(0) = 0\} \\ = \sum_{n=1}^{\infty} e^{-(\lambda+\mu)t} \frac{((\lambda + \mu)t)^n}{n!} \sum_{k=1}^{n} \binom{n}{k-1} \left(\frac{\mu}{\lambda + \mu}\right)^{k-1} \left(\frac{\lambda}{\lambda + \mu}\right)^{n-k+1} \\ \times \sum_{i=k}^{n} \binom{n}{i} \left(\frac{s}{t}\right)^i \left(1 - \frac{s}{t}\right)^{n-i}. \end{aligned}$$

Also,

$$P\{S_0(t) = t | X(0) = 0\} = e^{-\lambda t}.$$

Problems

5.1. A population of organisms consists of both male and female members. In a small colony any particular male is likely to mate with any particular female in any time interval of length h, with probability $\lambda h + o(h)$. Each mating immediately produces one offspring, equally likely to be male or female. Let $N_1(t)$ and $N_2(t)$ denote the number of males and females in the population at t. Derive the parameters of the continuous-time Markov chain $\{N_1(t), N_2(t)\}$.

5.2. Suppose that a one-celled organism can be in one of two states—either A or B. An individual in state A will change to state B at an exponential rate α; an individual in state B divides into two new individuals of type A at an exponential rate β. Define an appropriate continuous-time Markov chain for a population of such organisms and determine the appropriate parameters for this model.

5.3. Show that a continuous-time Markov chain is regular, given (a) that $\nu_i < M < \infty$ for all i or (b) that the discrete-time Markov chain with transition probabilities P_{ij} is irreducible and recurrent.

5.4. For a pure birth process with birth parameters λ_n, $n \geq 0$, compute the mean, variance, and moment generating function of the time it takes the population to go from size 0 to size N.

5.5. Consider a Yule process with $X(0) = i$. Given that $X(t) = i + k$, what can be said about the conditional distribution of the birth times of the k individuals born in $(0, t)$?

5.6. Verify the formula

$$A(t) = a_0 + \int_0^t X(s)\, ds,$$

given in Example 5.3(B).

5.7. Consider a Yule process starting with a single individual and suppose that with probability $P(s)$ an individual born at time s will be robust. Compute the distribution of the number of robust individuals born in $(0, t)$.

5.8. Prove Lemma 5.4.1.

5.9. Prove Lemma 5.4.2.

5.10. Let $P(t) = P_{00}(t)$.

(a) Find

$$\lim_{t \to 0} \frac{1 - P(t)}{t}.$$

(b) Show that

$$P(t)P(s) \leq P(t + s) \leq 1 - P(s) + P(s)P(t).$$

(c) Show

$$|P(t) - P(s)| \leq 1 - P(t - s), \qquad s < t$$

and conclude that P is continuous.

5.11. For the Yule process:

(a) verify that

$$P_{ij}(t) = \binom{j-1}{i-1} e^{-i\lambda t}(1 - e^{-\lambda t})^{j-i}$$

satisfies the forward and backward equations.

(b) Suppose that $X(0) = 1$ and that at time T the process stops and is replaced by an emigration process in which departures occur in a Poisson process of rate μ. Let τ denote the time taken after T for the population to vanish. Find the density function of τ and show that

$$E[\tau] = e^{\lambda T}/\mu.$$

5.12. Suppose that the "state" of the system can be modeled as a two-state continuous-time Markov chain with transition rates $\nu_0 = \lambda$, $\nu_1 = \mu$. When the state of the system is i, "events" occur in accordance with a Poisson process with rate α_i, $i = 0, 1$. Let $N(t)$ denote the number of events in $(0, t)$.

(a) Find $\lim_{t\to\infty} N(t)/t$.

(b) If the initial state is state 0, find $E[N(t)]$.

5.13. Consider a birth and death process with birth rates $\{\lambda_n\}$ and death rates $\{\mu_n\}$. Starting in state i, find the probability that the first k events are all births.

5.14. Consider a population of size n, some of whom are infected with a certain virus. Suppose that in an interval of length h any specified pair of individuals will independently interact with probability $\lambda h + o(h)$. If exactly one of the individuals involved in the interaction is infected then the other one becomes infected with probability α. If there is a single individual infected at time 0, find the expected time at which the entire population is infected.

5.15. Consider a population in which each individual independently gives birth at an exponential rate λ and dies at an exponential rate μ. In addition, new members enter the population in accordance with a Poisson process with rate θ. Let $X(t)$ denote the population size at time t.

(a) What type of process is $\{X(t), t \geq 0\}$?

(b) What are its parameters?

(c) Find $E[X(t)|X(0) = i]$.

5.16. In Example 5.4(D), find the variance of the number of males in the population at time t.

5.17. Let A be a specified set of states of a continuous-time Markov chain and let $T_i(t)$ denote the amount of time spent in A during the time interval $[0, t]$ given that the chain begins in state i. Let $Y_1, \ldots, Y_n$ be independent exponential random variables with mean λ. Suppose the Y_i are independent of the Markov chain, and set $t_i(n) = E[T_i(Y_1 + \cdots + Y_n)]$.

(a) Derive a set of linear equations for $t_i(1)$, $i \geq 0$.

(b) Derive a set of linear equations for $t_i(n)$ in terms of the other $t_j(n)$ and $t_i(n-1)$.

(c) When n is large, for what value of λ is $t_i(n)$ a good approximation of $E[T_i(t)]$?

5.18. Consider a continuous-time Markov chain with $X(0) = 0$. Let A denote a set of states that does not include 0 and set $T = \text{Min}\{t > 0: X(t) \in A\}$. Suppose that T is finite with probability 1. Set $q_i = \sum_{j \in A} q_{ij}$, and consider the random variable $H = \int_0^T q_{X(t)}\, dt$, called the random hazard.

(a) Find the hazard rate function of H. That is, find $\lim_{h \to 0} P\{s < H < s + h | H > s\}/h$. (*Hint:* Condition on the state of the chain at the time τ when $\int_0^T q_{X(t)}\, dt = s$.)

(b) What is the distribution of H?

5.19. Consider a continuous-time Markov chain with stationary probabilities $\{P_i, i \geq 0\}$, and let T denote the first time that the chain has been in state 0 for t consecutive time units. Find $E[T|X(0) = 0]$.

5.20. Each individual in a biological population is assumed to give birth at an exponential rate λ and to die at an exponential rate μ. In addition, there is an exponential rate of increase θ due to immigration. However, immigration is not allowed when the population size is N or larger.

(a) Set this up as a birth and death model.

(b) If $N = 3$, $1 = \theta = \lambda$, $\mu = 2$, determine the proportion of time that immigration is restricted.

5.21. A small barbershop, operated by a single barber, has room for at most two customers. Potential customers arrive at a Poisson rate of three per hour, and the successive service times are independent exponential random variables with mean $\frac{1}{4}$ hour.

(a) What is the average number of customers in the shop?

(b) What is the proportion of potential customers that enter the shop?

(c) If the barber could work twice as fast, how much more business would she do?

5.22. Find the limiting probabilities for the $M/M/s$ system and determine the condition needed for these to exist.

5.23. If $\{X(t), t \geq 0\}$ and $\{Y(t), t \geq 0\}$ are independent time-reversible Markov chains, show that the process $\{(X(t), Y(t), t \geq 0\}$ is also.

5.24. Consider two $M/M/1$ queues with respective parameters λ_i, μ_i, where $\lambda_i < \mu_i$, $i = 1, 2$. Suppose they both share the same waiting room, which has finite capacity N. (That is, whenever this room is full all potential arrivals to either queue are lost.) Compute the limiting probability that there will be n people at the first queue (1 being served and $n - 1$ in the waiting room when $n > 0$) and m at the second. (*Hint:* Use the result of Problem 5.23.)

5.25. What can you say about the departure process of the stationary $M/M/1$ queue having finite capacity?

5.26. In the stochastic population model of Section 5.6.2:

(a) Show that

$$P(\underline{n})q(\underline{n}, D_j\underline{n}) = P(D_j\underline{n})q(D_j\underline{n}, \underline{n})$$

when $P(\underline{n})$ is as given by (5.6.4) with $\alpha_j = (\lambda/j\nu)(\nu/\mu)^j$.

(b) Let $D(t)$ denote the number of families that die out in $(0, t)$. Assuming that the process is in steady state 0 at time $t = 0$, what type of stochastic process is $\{D(t), t \geq 0\}$? What if the population is initially empty at $t = 0$?

5.27. Complete the proof of the conjecture in the queueing network model of Section 5.7.1.

5.28. N customers move about among r servers. The service times at server i are exponential at rate μ_i and when a customer leaves server i it joins the queue (if there are any waiting—or else it enters service) at server j, $j \neq i$, with probability $1/(r - 1)$. Let the state be $(n_1, \ldots, n_r)$ when there are n_i customers at server i, $i = 1, \ldots, r$. Show the corresponding continuous-time Markov chain is time reversible and find the limiting probabilities.

5.29. Consider a time-reversible continuous-time Markov chain having parameters ν_i, P_{ij} and having limiting probabilities P_j, $j \geq 0$. Choose some state—say state 0—and consider the new Markov chain, which makes state 0 an absorbing state. That is, reset ν_0 to equal 0. Suppose now at time points chosen according to a Poisson process with rate λ, Markov chains—all of the above type (having 0 as an absorbing state)—are started with the initial states chosen to be j with probability P_{0j}. All the existing chains are assumed to be independent. Let $N_j(t)$ denote the number of chains in state j, $j > 0$, at time t.

(a) Argue that if there are no initial chains, then $N_j(t)$, $j > 0$, are independent Poisson random variables.

(b) In steady state argue that the vector process $\{(N_1(t), N_2(t), \ldots)\}$ is time reversible with stationary probabilities

$$P(\underline{n}) = \prod_{j=1}^{\infty} e^{-\alpha_j} \frac{\alpha_j^{n_j}}{n_j!} \qquad \text{for } \underline{n} = (n_1, n_2, \ldots),$$

where $\alpha_j = \lambda P_j / P_0 \nu_0$.

5.30. Consider an $M/M/\infty$ queue with channels (servers) numbered $1, 2, \ldots$. On arrival, a customer will choose the lowest numbered channel that is free. Thus, we can think of all arrivals as occurring at channel 1. Those who find channel 1 busy overflow and become arrivals at channel 2. Those finding both channels 1 and 2 busy overflow channel 2 and become arrivals at channel 3, and so on.

(a) What fraction of time is channel 1 busy?

(b) By considering the corresponding $M/M/2$ loss system, determine the fraction of time that channel 2 is busy.

(c) Write down an expression for the fraction of time channel c is busy for arbitrary c.

(d) What is the overflow rate from channel c to channel $c + 1$? Is the corresponding overflow process a Poisson process? A renewal process? Explain briefly.

(e) If the service distribution were general rather than exponential, which (if any) of your answers to (a)–(d) would change? Briefly explain.

5.31. Prove Theorem 5.7.1.

5.32. **(a)** Prove that a stationary Markov process is reversible if, and only if, its transition rates satisfy

$$q(j_1, j_2)q(j_2, j_3) \cdots q(j_{n-1}, j_n)q(j_n, j_1)$$
$$= q(j_1, j_n)q(j_n, j_{n-1}) \cdots q(j_3, j_2)q(j_2, j_1)$$

for any finite sequence of states $j_1, j_2, \ldots, j_n$.

(b) Argue that it suffices to verify that the equality in (a) holds for sequences of distinct states.

(c) Suppose that the stream of customers arriving at a queue forms a Poisson process of rate ν and that there are two servers who possibly differ in efficiency. Specifically, suppose that a customer's service time at server i is exponentially distributed with rate μ_i, for $i = 1, 2$, where $\mu_1 + \mu_2 > \nu$. If a customer arrives to find both servers free, he is equally likely to be allocated to either server. Define an appropriate continuous-time Markov chain for this model, show that it is time reversible, and determine the limiting probabilities.

5.33. The work in a queueing system at any time is defined as the sum of the remaining service times of all customers in the system at that time. For the $M/G/1$ in steady state compute the mean and variance of the work in the system.

5.34. Consider an ergodic continuous-time Markov chain, with transition rates q_{ij}, in steady state. Let P_j, $j \geq 0$, denote the stationary probabilities. Suppose the state space is partitioned into two subsets B and $B^c = G$.

(a) Compute the probability that the process is in state i, $i \in B$, given that it is in B. That is, compute

$$P\{X(t) = i | X(t) \in B\}.$$

(b) Compute the probability that the process is in state i, $i \in B$, given that it has just entered B. That is, compute

$$P\{X(t) = i | X(t) \in B, X(t^-) \in G\}.$$

(c) For $i \in B$, let T_i denote the time until the process enters G given that it is in state i, and let $\tilde{F}_i(s) = E[e^{-sT_i}]$. Argue that

$$\tilde{F}_i(s) = \frac{\nu_i}{\nu_i + s}\left[\sum_{j \in B} \tilde{F}_j(s) P_{ij} + \sum_{j \in G} P_{ij}\right],$$

where $P_{ij} = q_{ij}/\sum_j q_{ij}$.

(d) Argue that

$$\sum_{i \in G}\sum_{j \in B} P_i q_{ij} = \sum_{i \in B}\sum_{j \in G} P_i q_{ij}.$$

(e) Show by using (c) and (d) that

$$s \sum_{i \in B} P_i \tilde{F}_i(s) = \sum_{i \in G}\sum_{j \in B} P_i q_{ij}(1 - \tilde{F}_j(s)).$$

(f) Given that the process has just entered B from G, let T_ν denote the time until it leaves B. Use part (b) to conclude that

$$E[e^{-sT_\nu}] = \frac{\sum_{i \in B} \sum_{j \in G} \tilde{F}_i(s) P_j q_{ji}}{\sum_{j \in G} \sum_{k \in B} P_j q_{jk}}.$$

(g) Using (e) and (f) argue that

$$\sum_{j \in B} P_j = \left[\sum_{i \in G} \sum_{j \in B} P_i q_{ij} \right] E[T_\nu].$$

(h) Given that the process is in a state in B, let T_x denote the time until it leaves B. Use (a), (e), (f), and (g) to show that

$$E[e^{-sT_x}] = \frac{1 - E[e^{-sT_\nu}]}{sE[T_\nu]}.$$

(i) Use (h) and the uniqueness of Laplace transforms to conclude that

$$P\{T_x \le t\} = \int_0^t \frac{P\{T_\nu > s\}\, ds}{E[T_\nu]}.$$

(j) Use (i) to show that

$$E[T_x] = \frac{E[T_\nu^2]}{2E[T_\nu]} \ge \frac{E[T_\nu]}{2}.$$

The random variable T_ν is called the visit or sojourn time in the set of states B. It represents the time spent in B during a visit. The random variable T_x, called the exit time from B, represents the remaining time in B given that the process is presently in B. The results of the above problem show that the distributions of T_x and T_ν possess the same structural relation as the distributions of the excess or residual life of a renewal process at steady state and the distribution of the time between successive renewals.

5.35. Consider a renewal process whose interarrival distribution F is a mixture of two exponentials. That is, $\overline{F}(x) = pe^{-\lambda_1 x} + qe^{-\lambda_2 x}$, $q = 1 - p$. Compute the renewal function $E[N(t)]$.

Hint: Imagine that at each renewal a coin, having probability p of landing heads, is flipped. If heads appears, the next interarrival is exponential with rate λ_1, and if tails, it is exponential with rate λ_2. Let $R(t) = i$ if the exponential rate at time t is λ_i. Then:

(a) determine $P\{R(t) = i\}$, $i = 1, 2$;

(b) argue that

$$E[N(t)] = \sum_{i=1}^{2} \lambda_i \int_0^t P\{R(s) = i\}\, ds = E\left[\int_0^t \Lambda(s)\, ds\right]$$

where $\Lambda(s) = \lambda_{R(s)}$.

5.36. Consider the two-state Markov chain of Example 5.8(A), with $X(0) = 0$.

(a) Compute $\text{Cov}(X(s), X(y))$.

(b) Let $S_0(t)$ denote the occupation time of state 0 by t. Use (a) and (5.8.4) to compute $\text{Var}(S(t))$.

5.37. Let $Y_i = X_{(i)} - X_{(i-1)}$, $i = 1, \ldots, n + 1$ where $X(0) = 0$, $X_{(n+1)} = t$, and $X_{(1)} \leq X_{(2)} \leq \cdots \leq X_{(n)}$ are the ordered values of a set of n independent uniform $(0, t)$ random variables. Argue that $P\{Y_i \leq y_i, i = 1, \ldots, n + 1\}$ is a symmetric function of $y_1, \ldots, y_n$.

References

References 1, 2, 3, 7, and 8 provide many examples of the uses of continuous-time Markov chains. For additional material on time reversibility the reader should consult References 5, 6, and 10. Additional material on queueing networks can be found in References 6, 9, 10, and 11.

1. N. Bailey, *The Elements of Stochastic Processes with Application to the Natural Sciences*, Wiley, New York, 1964.
2. D. J. Bartholomew, *Stochastic Models for Social Processes*, 2nd ed., Wiley, London, 1973.
3. M. S. Bartlett, *An Introduction to Stochastic Processes*, 3rd ed., Cambridge University Press, Cambridge, England, 1978.
4. D. R. Cox and H. D. Miller, *The Theory of Stochastic Processes*, Chapman and Hall, London, 1965.
5. J. Keilson, *Markov Chain Models—Rarity and Exponentiality*, Springer-Verlag, Berlin, 1979.
6. F. Kelly, *Reversibility and Stochastic Networks*, Wiley, New York, 1979.
7. Renshaw, *Modelling Biological Populations in Space and Time*, Cambridge University Press, Cambridge, England, 1991.
8. H. C. Tijms, *Stochastic Models, An Algorithmic Approach*, Wiley, Chichester, England, 1994.
9. N. Van Dijk, *Queueing Networks and Product Forms*, Wiley, New York, 1993.
10. J. Walrand, *Introduction to Queueing Networks*, Prentice-Hall, Englewood Cliffs, NJ, 1988.
11. P. Whittle, *Systems in Stochastic Equilibrium*, Wiley, Chichester, England, 1986.

CHAPTER 6

Martingales

Introduction

In this chapter we consider a type of stochastic process, known as a martingale, whose definition formalizes the concept of a fair game. As we shall see, not only are these processes inherently interesting, but they also are powerful tools for analyzing a variety of stochastic processes.

Martingales are defined and examples are presented in Section 6.1. In Section 6.2 we introduce the concept of a stopping time and prove the useful martingale stopping theorem. In Section 6.3 we derive and apply Azuma's inequality, which yields bounds on the tail probabilities of a martingale whose incremental changes can be bounded. In Section 6.4 we present the important martingale convergence theorem and, among other applications, show how it can be utilized to prove the strong law of large numbers. Finally, in Section 6.5 we derive an extension of Azuma's inequality.

6.1 Martingales

A stochastic process $\{Z_n, n \geq 1\}$ is said to be a *martingale* process if

$$E[|Z_n|] < \infty \qquad \text{for all } n$$

and

$$E[Z_{n+1} | Z_1, Z_2, \ldots, Z_n] = Z_n. \tag{6.1.1}$$

A martingale is a generalized version of a fair game. For if we interpret Z_n as a gambler's fortune after the nth gamble, then (6.1.1) states that his expected fortune after the $(n + 1)$st gamble is equal to his fortune after the nth gamble no matter what may have previously occurred.

Taking expectations of (6.1.1) gives

$$E[Z_{n+1}] = E[Z_n],$$

and so

$$E[Z_n] = E[Z_1] \qquad \text{for all } n.$$

Some Examples of Martingales (1) Let $X_1, X_2, \ldots$ be independent random variables with 0 mean; and let $Z_n = \sum_{i=1}^n X_i$. Then $\{Z_n, n \geq 1\}$ is a martingale since

$$\begin{aligned} E[Z_{n+1}|Z_1, \ldots, Z_n] &\\ &= E[Z_n + X_{n+1}|Z_1, \ldots, Z_n] \\ &= E[Z_n|Z_1, \ldots, Z_n] + E[X_{n+1}|Z_1, \ldots, Z_n] \\ &= Z_n + E[X_{n+1}] \\ &= Z_n. \end{aligned}$$

(2) If $X_1, X_2, \ldots$ are independent random variables with $E[X_i] = 1$, then $\{Z_n, n \geq 1\}$ is a martingale when $Z_n = \prod_{i=1}^n X_i$. This follows since

$$\begin{aligned} E[Z_{n+1}|Z_1, \ldots, Z_n] &= E[Z_n X_{n+1}|Z_1, \ldots, Z_n] \\ &= Z_n E[X_{n+1}|Z_1, \ldots, Z_n] \\ &= Z_n E[X_{n+1}] \\ &= Z_n. \end{aligned}$$

(3) Consider a branching process (see Section 4.5 of Chapter 4) and let X_n denote the size of the nth generation. If m is the mean number of offspring per individual, then $\{Z_n, n \geq 1\}$ is a martingale when

$$Z_n = X_n/m^n.$$

We leave the verification as an exercise.

Since the conditional expectation $E[X|\boldsymbol{U}]$ satisfies all the properties of ordinary expectations, except that all probabilities are now computed conditional on the value of $\boldsymbol{U}$, it follows from the identity $E[X] = E[E[X|\boldsymbol{Y}]]$ that

$$E[X|\boldsymbol{U}] = E[E[X|\boldsymbol{Y}, \boldsymbol{U}]|\boldsymbol{U}]. \tag{6.1.2}$$

It is sometimes convenient to show that $\{Z_n, n \geq 1\}$ is a martingale by considering the conditional expectation of Z_{n+1} given not only $Z_1, \ldots, Z_n$ but also some other random vector $\boldsymbol{Y}$. If we can then show that

$$E[Z_{n+1}|Z_1, \ldots, Z_n, \boldsymbol{Y}] = Z_n,$$

then it follows that Equation (6.1.1) is satisfied. This is so because if the preceding equation holds, then

$$\begin{aligned} E[Z_{n+1}|Z_1, \ldots, Z_n] &= E[E[Z_{n+1}|Z_1, \ldots, Z_n, \mathbf{Y}]|Z_1, \ldots, Z_n] \\ &= E[Z_n|Z_1, \ldots, Z_n] \\ &= Z_n. \end{aligned}$$

Additional Examples of Martingales **(4)** Let $X, Y_1, Y_2, \ldots$ be arbitrary random variables such that $E[|X|] < \infty$, and let

$$Z_n = E[X|Y_1, \ldots, Y_n].$$

It then follows that $\{Z_n, n \geq 1\}$ is a martingale. To show this we will compute the conditional expectation of Z_{n+1} given not $Z_1, \ldots, Z_n$ but the more informative random variables $Y_1, \ldots, Y_n$ (which is equivalent to conditioning on $Z_1, \ldots, Z_n, Y_1, \ldots, Y_n$). This yields

$$\begin{aligned} E[Z_{n+1}|Y_1, \ldots, Y_n] &= E[E[X|Y_1, \ldots, Y_n, Y_{n+1}]|Y_1, \ldots, Y_n] \\ &= E[X|Y_1, \ldots, Y_n] \qquad \text{(from (6.1.2))} \\ &= Z_n \end{aligned}$$

and the result follows. This martingale, often called a *Doob type martingale*, has important applications. For instance, suppose X is a random variable whose value we want to predict and suppose that data $Y_1, Y_2 \ldots$ are accumulated sequentially. Then, as shown in Section 1.9, the predictor of X that minimizes the expected squared error given the data $Y_1, \ldots, Y_n$ is just $E[X|Y_1, \ldots, Y_n]$ and so the sequence of optimal predictors constitutes a Doob type martingale.

(5) Our next example generalizes the fact that the partial sums of independent random variables having mean 0 is a martingale. For any random variables $X_1, X_2, \ldots$, the random variables $X_i - E[X_i|X_1, \ldots, X_{i-1}]$, $i \geq 1$, have mean 0. Even though they need not be independent, their partial sums constitute a martingale. That is, if

$$Z_n = \sum_{i=1}^{n} \{X_i - E[X_i|X_1, \ldots, X_{i-1}]\}$$

then, provided $E[|Z_n|] < \infty$ for all n, $\{Z_n, n \geq 1\}$ is a martingale with mean 0. To verify this, note that

$$Z_{n+1} = Z_n + X_{n+1} - E[X_{n+1}|X_1, \ldots, X_n].$$

Conditioning on $X_1, \ldots, X_n$, which is more informative than $Z_1, \ldots, Z_n$ (since the latter are all functions of $X_1, \ldots, X_n$), yields that

$$\begin{aligned} E[Z_{n+1}|X_1, \ldots, X_n] &\\ &= Z_n + E[X_{n+1}|X_1, \ldots, X_n] - E[X_{n+1}|X_1, \ldots, X_n] \\ &= Z_n \end{aligned}$$

thus showing that $\{Z_n, n \geq 1\}$ is a martingale.

6.2 Stopping Times

Definition

The positive integer-valued, possibly infinite, random variable N is said to be a *random time* for the process $\{Z_n, n \geq 1\}$ if the event $\{N = n\}$ is determined by the random variables $Z_1, \ldots, Z_n$. That is, knowing $Z_1, \ldots, Z_n$ tells us whether or not $N = n$. If $P\{N < \infty\} = 1$, then the random time N is said to be a *stopping time*.

Let N be a random time for the process $\{Z_n, n \geq 1\}$ and let

$$\overline{Z}_n = \begin{cases} Z_n & \text{if } n \leq N \\ Z_N & \text{if } n > N. \end{cases}$$

$\{\overline{Z}_n, n \geq 1\}$ is called the *stopped* process.

PROPOSITION 6.2.1

If N is a random time for the martingale $\{Z_n\}$, then the stopped process $\{\overline{Z}_n\}$ is also a martingale.

Proof Let

$$I_n = \begin{cases} 1 & \text{if } N \geq n \\ 0 & \text{if } N < n. \end{cases}$$

That is, I_n equals 1 if we haven't yet stopped after observing $Z_1, \ldots, Z_{n-1}$. We claim that

$$\overline{Z}_n = \overline{Z}_{n-1} + I_n(Z_n - Z_{n-1}). \tag{6.2.1}$$

To verify (6.2.1) consider two cases:

(i) $N \geq n$: In this case, $\bar{Z}_n = Z_n, \bar{Z}_{n-1} = Z_{n-1}$, and $I_n = 1$, and (6.2.1) follows.
(ii) $N < n$: In this case, $\bar{Z}_{n-1} = \bar{Z}_n = Z_N$, $I_n = 0$, and (6.2.1) follows.

Now

$$(6.2.2) \qquad \begin{aligned} E[\bar{Z}_n | Z_1, \ldots, Z_{n-1}] &= E[\bar{Z}_{n-1} + I_n(Z_n - Z_{n-1}) | Z_1, \ldots, Z_{n-1}] \\ &= \bar{Z}_{n-1} + I_n E[Z_n - Z_{n-1} | Z_1, \ldots, Z_{n-1}] \\ &= \bar{Z}_{n-1}, \end{aligned}$$

where the next to last equality follows since both $\bar{Z}_{n-1}$ and I_n are determined by $Z_1, \ldots, Z_{n-1}$, and the last since $\{Z_n\}$ is a martingale.

We must prove that $E[\bar{Z}_n | \bar{Z}_1, \ldots, \bar{Z}_{n-1}] = \bar{Z}_{n-1}$. However, (6.2.2) implies this result since, if we know the values of $Z_1, \ldots, Z_{n-1}$, then we also know the values of $\bar{Z}_1, \ldots, \bar{Z}_{n-1}$.

Since the stopped process is also a martingale, and since $\bar{Z}_1 = Z_1$, we have

$$(6.2.3) \qquad E[\bar{Z}_n] = E[Z_1] \qquad \text{for all } n.$$

Now let us suppose that the random time N is a stopping time, that is, $P\{N < \infty\} = 1$. Since

$$\bar{Z}_n = \begin{cases} Z_n & \text{if } n \leq N \\ Z_N & \text{if } n > N, \end{cases}$$

it follows that $\bar{Z}_n$ equals Z_N when n is sufficiently large. Hence,

$$\bar{Z}_n \to Z_N \qquad \text{as } n \to \infty, \text{ with probability 1.}$$

Is it also true that

$$(6.2.4) \qquad E[\bar{Z}_n] \to E[Z_N] \qquad \text{as } n \to \infty.$$

Since $E[\bar{Z}_n] = E[Z_1]$ for all n, (6.2.4) states that

$$E[Z_N] = E[Z_1].$$

It turns out that, subject to some regularity conditions, (6.2.4) is indeed valid. We state the following theorem without proof.

THEOREM 6.2.2 The Martingale Stopping Theorem

If either:

(i) $\overline{Z}_n$ are uniformly bounded, or;
(ii) N is bounded, or;
(iii) $E[N] < \infty$, and there is an $M < \infty$ such that

$$E[|Z_{n+1} - Z_n||Z_1, \ldots, Z_n] < M,$$

then (6.2.4) is valid. Thus

$$E[Z_N] = E[Z_1].$$

Theorem 6.2.2 states that in a fair game if a gambler uses a stopping time to decide when to quit, then his expected final fortune is equal to his expected initial fortune. Thus in the expected value sense, no successful gambling system is possible provided one of the sufficient conditions of Theorem 6.2.2 is satisfied. The martingale stopping theorem provides another proof of Wald's equation (Theorem 3.3.2).

Corollary 6.2.3 (Wald's Equation)

If X_i, $i \geq 1$, are independent and identically distributed (iid) with $E[|X|] < \infty$ and if N is a stopping time for $X_1, X_2, \ldots$ with $E[N] < \infty$, then

$$E\left[\sum_{i=1}^{N} X_i\right] = E[N]E[X].$$

Proof Let $\mu = E[X]$. Since

$$Z_n = \sum_{i=1}^{n} (X_i - \mu)$$

is a martingale, it follows, if Theorem 6.2.2 is applicable, that

$$E[Z_N] = E[Z_1] = 0.$$

But

$$E[Z_N] = E\left[\sum_{i=1}^{N} (X_i - \mu)\right]$$

$$= E\left[\sum_{i=1}^{N} X_i - N\mu\right]$$

$$= E\left[\sum_{i=1}^{N} X_i\right] - E[N]\mu.$$

To show that Theorem 6.2.2 is applicable, we verify condition (iii). Now $Z_{n+1} - Z_n = X_{n+1} - \mu$ and thus

$$E[|Z_{n+1} - Z_n||Z_1, \ldots, Z_n] = E[|X_{n+1} - \mu||Z_1, \ldots, Z_n]$$
$$= E[|X_{n+1} - \mu|]$$
$$\leq E[|X|] + |\mu|.$$

In Example 3.5(A) we showed how to use Blackwell's theorem to compute the expected time until a specified pattern appears. The next example presents a martingale approach to this problem.

Example 6.2(A) ***Computing the Mean Time Until a Given Pattern Occurs.*** Suppose that a sequence of independent and identically distributed discrete random variables is observed sequentially, one at each day. What is the expected number that must be observed until some given sequence appears? More specifically, suppose that each outcome is either 0, 1, or 2 with respective probabilities $\frac{1}{2}$, $\frac{1}{3}$, and $\frac{1}{6}$, and we desire the expected time until the run 0 2 0 occurs. For instance, if the sequence of outcomes is 2, 1, 2, 0, 2, 1, 0, 1, 0, 0, 2, 0, then the required number N would equal 12.

To compute $E[N]$ imagine a sequence of gamblers, each initially having 1 unit, playing at a fair gambling casino. Gambler i begins betting at the beginning of day i and bets her 1 unit that the value on that day will equal 0. If she wins (and thus has 2 units), she then bets the 2 units on the next outcome being 2, and if she wins this bet (and thus has 12 units), then all 12 units are bet on the next outcome being 0. Hence each gambler will lose 1 unit if any of her bets fails and will win 23 if all three of her bets succeed. At the beginning of each day another gambler starts playing. If we let X_n denote the total winnings of the casino after the nth day, then since all bets are fair it follows that $\{X_n, n \geq 1\}$ is a martingale with mean 0. Let N denote the time until the sequence 0 2 0 appears. Now at the end of day N each of the gamblers $1, \ldots, N - 3$ would have lost 1 unit; gambler $N - 2$ would have won 23, gambler $N - 1$ would have lost 1, and gambler N would have won 1 (since the outcome on day N is 0). Hence,

$$X_N = N - 3 - 23 + 1 - 1 = N - 26$$

and, since $E[X_N] = 0$ (it is easy to verify condition (iii) of Theorem 6.2.2) we see that

$$E[N] = 26.$$

In the same manner as in the above, we can compute the expected time until any given pattern of outcomes appears. For instance in coin tossing the mean time until HHTTHH occurs is $p^{-4}q^{-2} + p^{-2} + p^{-1}$, where $p = P\{H\} = 1 - q$.

Example 6.2(b) Consider an individual who starts at position 0 and at each step either moves 1 position to the right with probability p or one to the left with probability $1 - p$. Assume that the successive movements are independent. If $p > 1/2$ find the expected number of steps it takes until the individual reaches position i, $i > 0$.

Solution. Let X_j equal 1 or -1 depending on whether step j is to the right or the left. If N is the number of steps it takes to reach i, then

$$\sum_{j=1}^{N} X_j = i.$$

Hence, since $E[X_j] = 2p - 1$, we obtain from Wald's equation that

$$E[N](2p - 1) = i$$

or,

$$E[N] = \frac{i}{2p - 1}.$$

Example 6.2(c) Players X, Y, and Z contest the following game. At each stage two of them are randomly chosen in sequence, with the first one chosen being required to give 1 coin to the other. All of the possible choices are equally likely and successive choices are independent of the past. This continues until one of the players has no remaining coins. At this point that player departs and the other two continue playing until one of them has all the coins. If the players initially have x, y, and z coins, find the expected number of plays until one of them has all the $s \equiv x + y + z$ coins.

Solution. Suppose that the game does not end when one of the players has all s coins but rather that the final two contestants continue to play, keeping track of their winnings by allowing for negative fortunes. Let X_n, Y_n, and Z_n denote, respectively, the amount of money that X, Y, and Z have after the nth stage of play. Thus, for instance, $X_n = 0$, $Y_n = -4$, $Z_n = s + 4$ indicates that X was the first player to go broke and that after the nth play Y had lost 4 more than he began with. If we let T denote the first time that two of the values X_n, Y_n, and Z_n are equal to 0, then the problem is to find $E[T]$.

To find $E[T]$ we will show that

$$M_n = X_nY_n + X_nZ_n + Y_nZ_n + n$$

is a martingale. It will then follow from the martingale stopping theorem (condition (iii) is easily shown to be satisfied) that

$$E[M_T] = E[M_0] = xy + xz + yz.$$

But, since two of X_T, Y_T, and Z_T are 0, it follows that

$$M_T = T$$

and so,

$$E[T] = xy + xz + yz.$$

To show that $\{M_n, n \geq 0\}$ is a martingale, consider

$$E[M_{n+1}|X_i, Y_i, Z_i, i = 0, 1, \ldots, n]$$

and consider two cases.

Case 1: $X_nY_nZ_n > 0$
In this case X, Y, and Z are all in competition after stage n. Hence,

$$\begin{aligned} E[X_{n+1}Y_{n+1}|X_n = x, Y_n = y] \\ &= [(x+1)y + (x+1)(y-1) + x(y+1) + x(y-1) \\ &\quad + (x-1)y + (x-1)(y+1)]/6 \\ &= xy - 1/3. \end{aligned}$$

As the conditional expectations of $X_{n+1}Z_{n+1}$ and $Y_{n+1}Z_{n+1}$ are similar, we see that in this case

$$E[M_{n+1}|X_i, Y_i, Z_i, i = 0, 1, \ldots, n] = M_n.$$

Case 2: One of the players has been eliminated by stage n, say player X. In this case $X_{n+1} = X_n = 0$ and

$$E[Y_{n+1}Z_{n+1}|Y_n = y, Z_n = z] = [(y+1)(z-1) + (y-1)(z+1)]/2$$
$$= yz - 1.$$

Hence, in this case we again obtain that

$$E[M_{n+1}|X_i, Y_i, Z_i, i = 0, 1, \ldots, n] = M_n.$$

Therefore, $\{M_n, n \geq 0\}$ is a martingale and

$$E[T] = xy + xz + yz.$$

Example 6.2(d) In Example 1.5(C) we showed, by mathematical induction, that the expected number of rounds that it takes for all n people in the matching problem to obtain their own hats is equal to n. We will now present a martingale argument for this result. Let R denote the number of rounds until all people have a match. Let X_i denote the number of matches on the ith round, for $i = 1, \ldots, R$, and define X_i to equal 1 for $i > R$.

We will use the zero-mean martingale $\{Z_k, k \geq 1\}$, where

$$Z_k = \sum_{i=1}^{k} (X_i - E[X_i|X_1, \ldots, X_{i-1}])$$
$$= \sum_{i=1}^{k} (X_i - 1)$$

where the last equality follows since, when any number of individuals are to randomly choose from among their own hats, the expected number of matches is 1. As R is a stopping time for this martingale (it is the smallest k for which $\sum_{i=1}^{k} X_i = n$), we obtain by the martingale stopping theorem that

$$0 = E[Z_R] = E\left[\sum_{i=1}^{R} (X_i - 1)\right]$$
$$= E\left[\sum_{i=1}^{R} X_i\right] - E[R]$$
$$= n - E[R]$$

where the final equality used the identity $\sum_{i=1}^{R} X_i = n$.

6.3 Azuma's Inequality for Martingales

Let Z_i, $i \geq 0$ be a martingale sequence. In situations where these random variables do not change too rapidly, Azuma's inequality enables us to obtain useful bounds on their probabilities. Before stating it, we need some lemmas.

Lemma 6.3.1

Let X be such that $E[X] = 0$ and $P\{-\alpha \leq X \leq \beta\} = 1$. Then for any convex function f

$$E[f(X)] \leq \frac{\beta}{\alpha+\beta} f(-\alpha) + \frac{\alpha}{\alpha+\beta} f(\beta).$$

Proof Since f is convex it follows that, in the region $-\alpha \leq x \leq \beta$, it is never above the line segment connecting the points $(-\alpha, f(-\alpha))$ and $(\beta, f(\beta))$. (See Figure 6.3.1.) As the formula for this line segment is

$$y = \frac{\beta}{\alpha+\beta} f(-\alpha) + \frac{\alpha}{\alpha+\beta} f(\beta) + \frac{1}{\alpha+\beta}[f(\beta) - f(-\alpha)]x$$

it follows, since $-\alpha \leq X \leq \beta$, that

$$f(X) \leq \frac{\beta}{\alpha+\beta} f(-\alpha) + \frac{\alpha}{\alpha+\beta} f(\beta) + \frac{1}{\alpha+\beta}[f(\beta) - f(-\alpha)]X.$$

Taking expectations gives the result.

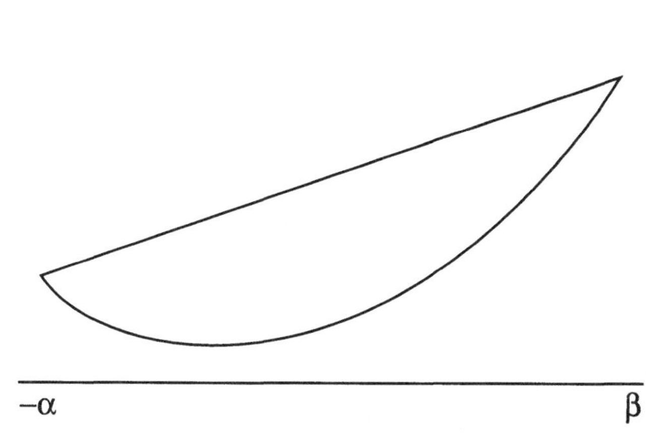

Figure 6.3.1. A convex function.

Lemma 6.3.2

For $0 \leq \theta \leq 1$

$$\theta e^{(1-\theta)x} + (1-\theta)e^{-\theta x} \leq e^{x^2/8}.$$

Proof Letting $\theta = (1+\alpha)/2$ and $x = 2\beta$, we must show that for $-1 \leq \alpha \leq 1$,

$$(1+\alpha)e^{\beta(1-\alpha)} + (1-\alpha)e^{-\beta(1+\alpha)} \leq 2e^{\beta^2/2}$$

or, equivalently,

$$e^{\beta} + e^{-\beta} + \alpha(e^{\beta} - e^{-\beta}) \leq 2\exp\{\alpha\beta + \beta^2/2\}.$$

Now the preceding inequality is true when $\alpha = -1$ or $+1$ and when β is large (say when $|\beta| \geq 100$). Thus, if Lemma 6.3.2 were false then the function

$$f(\alpha, \beta) = e^{\beta} + e^{-\beta} + \alpha(e^{\beta} - e^{-\beta}) - 2\exp\{\alpha\beta + \beta^2/2\},$$

would assume a strictly positive maximum in the interior of the region $R = \{(\alpha, \beta): |\alpha| \leq 1, |\beta| \leq 100\}$. Setting the partial derivatives of f equal to 0, gives that

$$e^{\beta} - e^{-\beta} + \alpha(e^{\beta} + e^{-\beta}) = 2(\alpha + \beta)\exp\{\alpha\beta + \beta^2/2\}$$
$$e^{\beta} - e^{-\beta} = 2\beta\exp\{\alpha\beta + \beta^2/2\}.$$

Assuming a solution in which $\beta \neq 0$ implies, upon division, that

$$1 + \alpha\frac{e^{\beta} + e^{-\beta}}{e^{\beta} - e^{-\beta}} = 1 + \frac{\alpha}{\beta}.$$

As it can be shown that there is no solution for which $\alpha = 0$ and $\beta \neq 0$ (see Problem 6.14) we see that

$$\beta(e^{\beta} + e^{-\beta}) = e^{\beta} - e^{-\beta}$$

or, expanding in a Taylor series,

$$\sum_{i=0}^{\infty} \beta^{2i+1}/(2i)! = \sum_{i=0}^{\infty} \beta^{2i+1}/(2i+1)!$$

which is clearly not possible when $\beta \neq 0$. Hence, if the lemma is not true, we can conclude that the strictly positive maximal value of $f(\alpha, \beta)$ occurs when $\beta = 0$. However, $f(\alpha, 0) = 0$ and thus the lemma is proven.

THEOREM 6.3.3 Azuma's Inequality

Let Z_n, $n \geq 1$ be a martingale with mean $\mu = E[Z_n]$. Let $Z_0 = \mu$ and suppose that for nonnegative constants α_i, β_i, $i \geq 1$,

$$-\alpha_i \leq Z_i - Z_{i-1} \leq \beta_i.$$

Then for any $n \geq 0$, $a > 0$:

(i) $P\{Z_n - \mu \geq a\} \leq \exp\{-2a^2 \Big/ \sum_{i=1}^{n} (\alpha_i + \beta_i)^2\}$.

(ii) $P\{Z_n - \mu \leq -a\} \leq \exp\{-2a^2 \Big/ \sum_{i=1}^{n} (\alpha_i + \beta_i)^2\}$.

Proof Suppose first that $\mu = 0$. Now, for any $c > 0$

$$\begin{aligned} P\{Z_n \geq a\} &= P\{\exp\{cZ_n\} \geq e^{ca}\} \\ &\leq E[\exp\{cZ_n\}]e^{-ca} \qquad \text{(by Markov's inequality)} \end{aligned} \tag{6.3.1}$$

To obtain a bound on $E[\exp\{cZ_n\}]$, let $W_n = \exp\{cZ_n\}$. Note that $W_0 = 1$, and that for $n > 0$

$$W_n = \exp\{cZ_{n-1}\} \exp\{c(Z_n - Z_{n-1})\}.$$

Therefore,

$$\begin{aligned} E[W_n|Z_{n-1}] &= \exp\{cZ_{n-1}\}E[\exp\{c(Z_n - Z_{n-1})\}|Z_{n-1}] \\ &\leq W_{n-1}[\beta_n \exp\{-c\alpha_n\} + \alpha_n \exp\{c\beta_n\}] / (\alpha_n + \beta_n) \end{aligned}$$

where the inequality follows from Lemma 6.3.1 since

(i) $f(x) = e^{cx}$ is convex,
(ii) $-\alpha_n \leq Z_n - Z_{n-1} \leq \beta_n$, and
(iii) $E[Z_n - Z_{n-1}|Z_{n-1}] = E[Z_n|Z_{n-1}] - E[Z_{n-1}|Z_{n-1}] = 0$.

Taking expectations gives

$$E[W_n] \leq E[W_{n-1}](\beta_n \exp\{-c\alpha_n\} + \alpha_n \exp\{c\beta_n\}) / (\alpha_n + \beta_n).$$

Iterating this inequality gives, since $E[W_0] = 1$, that

$$E[W_n] \leq \prod_{i=1}^{n} \{(\beta_i \exp\{-c\alpha_i\} + \alpha_i \exp\{c\beta_i\}) / (\alpha_i + \beta_i)\}.$$

Thus, from Equation (6.3.1), we obtain that for any $c > 0$

$$(6.3.2) \qquad P\{Z_n \geq a\} \leq e^{-ca} \prod_{i=1}^{n} \{(\beta_i \exp\{-c\alpha_i\} + \alpha_i \exp\{c\beta_i\}) / (\alpha_i + \beta_i)\}$$

$$\leq e^{-ca} \prod_{i=1}^{n} \exp\{c^2(\alpha_i + \beta_i)^2 / 8\},$$

where the preceding inequality follows from Lemma 6.3.2 upon setting $\theta = \alpha_i/(\alpha_i + \beta_i)$ and $x = c(\alpha_i + \beta_i)$. Hence for any $c > 0$,

$$P\{Z_n \geq a\} \leq \exp\left\{-ca + c^2 \sum_{i=1}^{n} (\alpha_i + \beta_i)^2 / 8\right\}.$$

Letting $c = 4a \Big/ \sum_{i=1}^{n} (\alpha_i + \beta_i)^2$ (which is the value that minimizes $-ca + c^2 \sum_{i=1}^{n} (\alpha_i + \beta_i)^2 / 8$) gives that

$$P\{Z_n \geq a\} \leq \exp\left\{-2a^2 \Big/ \sum_{i=1}^{n} (\alpha_i + \beta_i)^2\right\}.$$

Parts (i) and (ii) of Azuma's inequality now follow from applying the preceding, first to the zero-mean martingale $\{Z_n - \mu\}$ and secondly to the zero-mean martingale $\{\mu - Z_n\}$.

Remark From a computational point of view, one should make use of the inequality (6.3.2) rather than the final Azuma inequality. For instance, evaluating the right side of (6.3.2) with $c = 4a \Big/ \sum_{i=1}^{n} (\alpha_i + \beta_i)^2$ will give a sharper bound than the Azuma inequality.

Example 6.3(a) Let $X_1, \ldots, X_n$ be random variables such that $E[X_1] = 0$ and $E[X_i | X_1, \ldots, X_{i-1}] = 0$, $i \geq 1$. Then, from Example 5 of Section 6.1, we see that $\sum_{i=1}^{j} X_i$, $j = 1, \ldots, n$ is a zero-mean martingale. Hence, if $-\alpha_i \leq X_i \leq \beta_i$ for all i, then we obtain from Azuma's inequality that for $a > 0$,

$$P\left\{\sum_{i=1}^{n} X_i \geq a\right\} \leq \exp\left\{-2a^2 \Big/ \sum_{i=1}^{n} (\alpha_i + \beta_i)^2\right\}$$

and

$$P\left\{\sum_{i=1}^{n} X_i \le -a\right\} \le \exp\left\{-2a^2 \Big/ \sum_{i=1}^{n} (\alpha_i + \beta_i)^2\right\}.$$

Azuma's inequality is often applied in conjunction with a Doob type martingale.

Example 6.3(b) Suppose that n balls are put in m urns in such a manner that each ball, independently, is equally likely to go into any of the urns. We will use Azuma's inequality to obtain bounds on the tail probabilities of X, the number of empty urns. To begin, letting $I\{A\}$ be the indicator variable for the event A, we have that,

$$X = \sum_{i=1}^{m} I\{\text{urn } i \text{ is empty}\}$$

and thus,

$$E[X] = mP\{\text{urn } i \text{ is empty}\} = m(1 - 1/m)^n \equiv \mu.$$

Now, let X_j denote the urn in which the jth ball is placed, $j = 1, \ldots, n$, and define a Doob type martingale by letting $Z_0 = E[X]$, and for $i > 0$, $Z_i = E[X|X_1, \ldots, X_i]$. We will now obtain our result by using Azuma's inequality along with the fact that $X = Z_n$. To determine a bound on $|Z_i - Z_{i-1}|$ first note that $|Z_1 - Z_0| = 0$. Now, for $i \ge 2$, let D denote the number of distinct values in the set $X_1, \ldots, X_{i-1}$. That is, D is the number of urns having at least one ball after the first $i - 1$ balls are distributed. Then, since each of the $m - D$ urns that are presently empty will end up empty with probability $(1 - 1/m)^{n-i+1}$, we have that

$$E[X|X_1, \ldots, X_{i-1}] = (m - D)(1 - 1/m)^{n-i+1}.$$

On the other hand,

$$E[X|X_1, \ldots, X_i] = \begin{cases} (m - D)(1 - 1/m)^{n-i} & \text{if } X_i \in (X_1, \ldots, X_{i-1}) \\ (m - D - 1)(1 - 1/m)^{n-i} & \text{if } X_i \notin (X_1, \ldots, X_{i-1}). \end{cases}$$

Hence, the two possible values of $Z_i - Z_{i-1}$, $i \ge 2$, are

$$\frac{m - D}{m}(1 - 1/m)^{n-i} \quad \text{and} \quad \frac{-D}{m}(1 - 1/m)^{n-i}.$$

Since $1 \leq D \leq \min(i - 1, m)$, we thus obtain that

$$-\alpha_i \leq Z_i - Z_{i-1} \leq \beta_i,$$

where

$$\alpha_i = \min\left(\frac{i-1}{m}, 1\right)(1 - 1/m)^{n-i}, \ \beta_i = (1 - 1/m)^{n-i+1}.$$

From Azuma's inequality we thus obtain that for $a > 0$,

$$P\{X - \mu \geq a\} \leq \exp\left\{-2a^2 \Big/ \sum_{i=2}^{n} (\alpha_i + \beta_i)^2\right\}$$

and,

$$P\{X - \mu \leq -a\} \leq \exp\left\{-2a^2 \Big/ \sum_{i=2}^{n} (\alpha_i + \beta_i)^2\right\}$$

where,

$$\sum_{i=2}^{n} (\alpha_i + \beta_i)^2 = \sum_{i=2}^{m} (m + i - 2)^2(1 - 1/m)^{2(n-i)}/m^2 + \sum_{i=m+1}^{n} (1 - 1/m)^2(2 - 1/m)^2.$$

Azuma's inequality is often applied to a Doob type martingale for which $|Z_i - Z_{i-1}| \leq 1$. The following corollary gives a sufficient condition for a Doob type martingale to satisfy this condition.

Corollary 6.3.4

Let h be a function such that if the vectors $\boldsymbol{x} = (x_1, \ldots, x_n)$ and $\boldsymbol{y} = (y_1, \ldots, y_n)$ differ in at most one coordinate (that is, for some k, $x_i = y_i$ for all $i \neq k$) then $|h(\boldsymbol{x}) - h(\boldsymbol{y})| \leq 1$. Let $X_1, \ldots, X_n$ be independent random variables. Then, with $\boldsymbol{X} = (X_1, \ldots, X_n)$, we have for $a > 0$ that

(i) $P\{h(\boldsymbol{X}) - E[h(\boldsymbol{X})] \geq a\} \leq e^{-a^2/2n}$

(ii) $P\{h(\boldsymbol{X}) - E[h(\boldsymbol{X})] \leq -a\} \leq e^{-a^2/2n}$.

Proof Consider the martingale $Z_i = E[h(\boldsymbol{X})|X_1, \ldots, X_i]$, $i = 1, \ldots, n$. Now,

$$\begin{aligned}
&|E[h(\boldsymbol{X})|X_1 = x_1, \ldots, X_i = x_i] - E[h(\boldsymbol{X})|X_1 = x_1, \ldots, X_{i-1} = x_{i-1}]| \\
&\quad = |E[h(x_1, \ldots, x_i, X_{i+1}, \ldots, X_n)] - E[h(x_1, \ldots, x_{i-1}, X_i, \ldots, X_n)]| \\
&\quad = |E[h(x_1, \ldots, x_i, X_{i+1}, \ldots, X_n) - h(x_1, \ldots, x_{i-1}, X_i, \ldots, X_n)]| \le 1.
\end{aligned}$$

Hence, $|Z_i - Z_{i-1}| \le 1$ and so the result follows from Azuma's inequality with $\alpha_i = \beta_i = 1$.

Example 6.3(c) Suppose that n balls are to be placed in m urns, with each ball independently going into urn j with probability p_j, $j = 1, \ldots, m$. Let Y_k denote the number of urns with exactly k balls, $0 \le k < n$, and use the preceding corollary to obtain a bound on its tail probabilities.

Solution. To begin, note that

$$\begin{aligned}
E[Y_k] &= E\left[\sum_{i=1}^{m} I\{\text{urn } i \text{ has exactly } k \text{ balls}\}\right] \\
&= \sum_{i=1}^{m} \binom{n}{k} p_i^k (1 - p_i)^{n-k}.
\end{aligned}$$

Now, let X_i denote the urn in which ball i is put, $i = 1, \ldots, n$. Also, let $h_k(x_1, x_2, \cdots, x_n)$ denote the number of urns with exactly k balls when $X_i = x_i, i = 1, \ldots, n$, and note that $Y_k = h_k(X_1, \ldots, X_n)$.

Suppose first that $k = 0$. In this case it is easy to see that h_0 satisfies the condition that if $\boldsymbol{x}$ and $\boldsymbol{y}$ differ in at most one coordinate then $|h_0(\boldsymbol{x}) - h_0(\boldsymbol{y})| \le 1$. (That is, suppose n x-balls and n y-balls are put in m urns so that the ith x-ball and the ith y-ball are put in the same urn for all but one i. Then the number of urns empty of x-balls and the number empty of y-balls can clearly differ by at most 1.) Therefore, we see from Corollary 6.3.4 that

$$P\left\{Y_0 - \sum_{i=1}^{m} (1 - p_i)^n \ge a\right\} \le \exp\{-a^2/2n\}$$

$$P\left\{Y_0 - \sum_{i=1}^{m} (1 - p_i)^n \le -a\right\} \le \exp\{-a^2/2n\}.$$

Now, suppose that $0 < k < n$. In this case if $\boldsymbol{x}$ and $\boldsymbol{y}$ differ in at most one coordinate then it is not necessarily true that $|h_k(\boldsymbol{x}) - h_k(\boldsymbol{y})| \le 1$, for the one different value could result in one of the vectors having 1 more and the other 1 less urn with k balls than they would have had if that coordinate was not included. But from this we can see that if $\boldsymbol{x}$ and $\boldsymbol{y}$ differ in at most one coordinate then

$$|h_k(\boldsymbol{x}) - h_k(\boldsymbol{y})| \leq 2.$$

Hence, $h_k^*(\boldsymbol{x}) = h_k(\boldsymbol{x})/2$ satisfies the condition of Corollary 6.3.4 and so we can conclude that for $0 < k < n$, $a > 0$,

$$P\left\{Y_k - \sum_{i=1}^{m} \binom{n}{k} p_i^k (1 - p_i)^{n-k} \geq 2a\right\} \leq \exp\{-a^2/2n\}$$

and,

$$P\left\{Y_k - \sum_{i=1}^{m} \binom{n}{k} p_i^k (1 - p_i)^{n-k} \leq -2a\right\} \leq \exp\{-a^2/2n\}.$$

Example 6.3(d) Consider a set of n components that are to be used in performing certain experiments. Let X_i equal 1 if component i is in functioning condition and let it equal 0 otherwise, and suppose that the X_i are independent with $E[X_i] = p_i$. Suppose that in order to perform experiment j, $j = 1, \ldots, m$, all of the components in the set A_j must be functioning. If any particular component is needed in at most three experiments, show that for $a > 0$

$$P\left\{X - \sum_{j=1}^{m} \prod_{i \in A_j} p_i \geq 3a\right\} \leq \exp\{-a^2/2n\}$$

$$P\left\{X - \sum_{j=1}^{m} \prod_{i \in A_j} p_i \leq -3a\right\} \leq \exp\{-a^2/2n\}$$

where X is the number of experiments that can be performed.

Solution. Since

$$X = \sum_{j=1}^{m} I\{\text{experiment } j \text{ can be performed}\},$$

we see that

$$E[X] = \sum_{j=1}^{m} \prod_{i \in A_j} p_i.$$

If we let $h(\boldsymbol{X})$ equal the number of experiments that can be performed, then h itself does not satisfy the condition of Corollary

6.3.4 because changing the value of one of the X_i can change the value of h by as much as 3. However, $h(X)/3$ does satisfy the conditions of the corollary and so we obtain that

$$P\{X/3 - E[X]/3 \geq a\} \leq \exp\{-a^2/2n\}$$
$$P\{X/3 - E[X]/3 \leq -a\} \leq \exp\{-a^2/2n\}.$$

6.4 Submartingales, Supermartingales, and the Martingale Convergence Theorem

A stochastic process $\{Z_n, n \geq 1\}$ having $E[|Z_n|] < \infty$ for all n is said to be a *submartingale* if

$$E[Z_{n+1}|Z_1, \ldots, Z_n] \geq Z_n \tag{6.4.1}$$

and is said to be a *supermartingale* if

$$E[Z_{n+1}|Z_1, \ldots, Z_n] \leq Z_n. \tag{6.4.2}$$

Hence, a submartingale embodies the concept of a superfair and a supermartingale a subfair game.

From (6.4.1) we see that for a submartingale

$$E[Z_{n+1}] \geq E[Z_n]$$

with the inequality reversed for a supermartingale. The analogues of Theorem 6.2.2, the martingale stopping theorem, remain valid for submartingales and supermartingales. That is, the following result, whose proof is left as an exercise, can be established.

THEOREM 6.4.1

If N is a stopping time for $\{Z_n, n \geq 1\}$ such that any one of the sufficient conditions of Theorem 6.2.2 is satisfied, then

$$E[Z_N] \geq E[Z_1] \qquad \text{for a submartingale}$$

and

$$E[Z_N] \leq E[Z_1] \qquad \text{for a supermartingale.}$$

The most important martingale result is the martingale convergence theorem. Before presenting it we need some preliminaries.

Lemma 6.4.2

If $\{Z_i, i \geq 1\}$ is a submartingale and N a stopping time such that $P\{N \leq n\} = 1$ then

$$E[Z_1] \leq E[Z_N] \leq E[Z_n].$$

Proof It follows from Theorem 6.4.1 that, since N is bounded, $E[Z_N] \geq E[Z_1]$. Now,

$$\begin{aligned} E[Z_n | Z_1, \ldots, Z_N, N = k] &= E[Z_n | Z_1, \ldots, Z_k, N = k] \\ &= E[Z_n | Z_1, \ldots, Z_k] \quad \text{(Why?)} \\ &\geq Z_k \\ &= Z_N. \end{aligned}$$

Hence the result follows by taking expectations of the above.

Lemma 6.4.3

If $\{Z_n, n \geq 1\}$ is a martingale and f a convex function, then $\{f(Z_n), n \geq 1\}$ is a submartingale.

Proof

$$\begin{aligned} E[f(Z_{n+1}) | Z_1, \ldots, Z_n] &\geq f(E[Z_{n+1} | Z_1, \ldots, Z_n]) \qquad \text{(by Jensen's inequality)} \\ &= f(Z_n). \end{aligned}$$

THEOREM 6.4.4 (Kolmogorov's Inequality for Submartingales)

If $\{Z_n, n \geq 1\}$ is a nonnegative submartingale, then

$$P\{\max(Z_1, \ldots, Z_n) > a\} \leq \frac{E[Z_n]}{a} \qquad \textit{for } a > 0.$$

Proof Let N be the smallest value of i, $i \leq n$, such that $Z_i > a$, and define it to equal n if $Z_i \leq a$ for all $i = 1, \ldots, n$. Note that $\max(Z_1, \ldots, Z_n) > a$ is equivalent to $Z_N > a$. Therefore,

$$\begin{aligned} P\{\max(Z_1, \ldots, Z_n) > a\} &= P\{Z_N > a\} \\ &\leq \frac{E[Z_N]}{a} \qquad \text{(by Markov's inequality)} \\ &\leq \frac{E[Z_n]}{a}, \end{aligned}$$

where the last inequality follows from Lemma 6.4.2 since $N \leq n$.

Corollary 6.4.5

Let $\{Z_n, n \geq 1\}$ be a martingale. Then, for $a > 0$:

(i) $P\{\max(|Z_1|, \ldots, |Z_n|) > a\} \leq E[|Z_n|]/a$;
(ii) $P\{\max(|Z_1|, \ldots, |Z_n|) > a\} \leq E[Z_n^2]/a^2$.

Proof Parts (i) and (ii) follow from Lemma 6.4.3 and Kolmogorov's inequality since the functions $f(x) = |x|$ and $f(x) = x^2$ are both convex.

We are now ready for the martingale convergence theorem.

THEOREM 6.4.6 (The Martingale Convergence Theorem)

If $\{Z_n, n \geq 1\}$ is a martingale such that for some $M < \infty$

$$E[|Z_n|] \leq M, \text{ for all } n$$

then, with probability 1, $\lim_{n\to\infty} Z_n$ exists and is finite.

Proof We shall prove the theorem under the stronger assumption that $E[Z_n^2]$ is bounded (stronger since $E[|Z_n|] \leq (E[Z_n^2])^{1/2}$). Since $f(x) = x^2$ is convex, it follows from Lemma 6.4.3 that $\{Z_n^2, n \geq 1\}$ is a submartingale; hence $E[Z_n^2]$ is nondecreasing. Since $E[Z_n^2]$ is bounded, it follows that it converges as $n \to \infty$. Let $\mu < \infty$ be given by

$$\mu = \lim_{n\to\infty} E[Z_n^2].$$

We shall show that $\lim_n Z_n$ exists and is finite by showing that $\{Z_n, n \geq 1\}$ is, with probability 1, a Cauchy sequence. That is, we will show that with probability 1

$$|Z_{m+k} - Z_m| \to 0 \qquad \text{as } k, m \to \infty.$$

Now

$$(6.4.3) \qquad P\{|Z_{m+k} - Z_m| > \varepsilon \text{ for some } k \leq n\}$$
$$\leq E[(Z_{m+n} - Z_m)^2]/\varepsilon^2 \qquad \text{(by Kolmogorov's inequality)}$$
$$= E[Z_{m+n}^2 - 2Z_m Z_{m+n} + Z_m^2]/\varepsilon^2.$$

But

$$\begin{aligned} E[Z_m Z_{m+n}] &= E[E[Z_m Z_{m+n}|Z_m]] \\ &= E[Z_m E[Z_{m+n}|Z_m]] \\ &= E[Z_m^2]. \end{aligned}$$

Hence, from (6.4.3),

$$P\{|Z_{m+k} - Z_m| > \varepsilon \text{ for some } k \le n\} \le \frac{E[Z_{m+n}^2] - E[Z_m^2]}{\varepsilon^2}.$$

Letting $n \to \infty$ and recalling the definition of μ yields

$$P\{|Z_{m+k} - Z_m| > \varepsilon \text{ for some } k\} \le \frac{\mu - E[Z_m^2]}{\varepsilon^2}.$$

And, therefore,

$$P\{|Z_{m+k} - Z_m| > \varepsilon \text{ for some } k\} \to 0 \qquad \text{as } m \to \infty.$$

Thus, with probability 1, Z_n will be a Cauchy sequence, and thus $\lim_{n\to\infty} Z_n$ will exist and be finite.

Corollary 6.4.7

If $\{Z_n, n \ge 0\}$ is a nonnegative martingale, then, with probability 1, $\lim_{n\to\infty} Z_n$ exists and is finite.

Proof Since Z_n is nonnegative,

$$E[|Z_n|] = E[Z_n] = E[Z_1].$$

Example 6.4(a) ***Branching Processes.*** If X_n is the population size of the nth generation in a branching process whose mean number of offspring per individual is m, then $Z_n = X_n/m^n$ is a nonnegative martingale. Hence from Corollary 6.4.7 it will converge as $n \to \infty$. From this we can conclude that either $X_n \to 0$ or else it goes to ∞ at an exponential rate.

Example 6.4(b) ***A Gambling Result.*** Consider a gambler playing a fair game; that is, if Z_n is the gambler's fortune after the nth play, then $\{Z_n, n \ge 1\}$ is a martingale. Now suppose that no credit is given, and so the gambler's fortune is not allowed to become negative, and on each gamble at least 1 unit is either won or lost. Let

$$N = \min\{n: \quad Z_n = Z_{n+1}\}$$

denote the number of plays until the gambler is forced to quit. (Since $Z_n - Z_{n+1} = 0$, she did not gamble on the $n + 1$ play.)

Since $\{Z_n\}$ is a nonnegative martingale we see by the convergence theorem that

(6.4.4) $\lim_{n\to\infty} Z_n$ exists and is finite, with probability 1.

But $|Z_{n+1} - Z_n| \geq 1$ for $n < N$, and so (6.4.4) implies that

$$N < \infty \quad \text{with probability 1.}$$

That is, with probability 1, the gambler will eventually go broke.

We will now use the martingale convergence theorem to prove the strong law of large numbers.

THEOREM 6.4.8 The Strong Law of Large Numbers

Let $X_1, X_2, \ldots$ be a sequence of independent and identically distributed random variables having a finite mean μ, and let $S_n = \sum_{i=1}^{n} X_i$. Then

$$P\left\{\lim_{n\to\infty} S_n/n = \mu\right\} = 1.$$

Proof We will prove the theorem under the assumption that the moment generating function $\Psi(t) = E[e^{tX}]$ exists.

For a given $\varepsilon > 0$, let $g(t)$ be defined by

$$g(t) = e^{t(\mu+\varepsilon)}/\Psi(t).$$

Since

$$g(0) = 1,$$

$$g'(0) = \left.\frac{\Psi(t)(\mu+\varepsilon)e^{t(\mu+\varepsilon)} - \Psi'(t)e^{t(\mu+\varepsilon)}}{\Psi^2(t)}\right|_{t=0} = \varepsilon > 0,$$

there exists a value $t_0 > 0$ such that $g(t_0) > 1$. We now show that S_n/n can be as large as $\mu + \varepsilon$ only finitely often. For, note that

$$\frac{S_n}{n} \geq \mu + \varepsilon \Rightarrow \frac{e^{t_0 S_n}}{\Psi^n(t_0)} \geq \left(\frac{e^{t_0(\mu+\varepsilon)}}{\Psi(t_0)}\right)^n = (g(t_0))^n \tag{6.4.5}$$

However, $e^{t_0 S_n}/\Psi^n(t_0)$, being the product of independent random variables with unit means (the ith being $e^{t_0 X_i}/\Psi(t)$), is a martingale. Since it is also nonnegative, the convergence theorem shows that, with probability 1,

$$\lim_{n\to\infty} e^{t_0 S_n}/\Psi^n(t_0) \qquad \text{exists and is finite.}$$

Hence, since $g(t_0) > 1$, it follows from (6.4.5) that

$$P\{S_n/n > \mu + \varepsilon \text{ for an infinite number of } n\} = 0.$$

Similarly, by defining the function $f(t) = e^{t(\mu-\varepsilon)}/\Psi(t)$ and noting that since $f(0) = 1$, $f'(0) = -\varepsilon$, there exists a value $t_0 < 0$ such that $f(t_0) > 1$, we can prove in the same manner that

$$P\{S_n/n \le \mu - \varepsilon \text{ for an infinite number of } n\} = 0.$$

Hence,

$$P\{\mu - \varepsilon \le S_n/n \le \mu + \varepsilon \text{ for all but a finite number of } n\} = 1,$$

or, since the above is true for all $\varepsilon > 0$,

$$P\left\{\lim_{n\to\infty} S_n/n = \mu\right\} = 1.$$

We will end this section by characterizing Doob martingales. To begin we need a definition.

Definition

The sequence of random variables X_n, $n \ge 1$, is said to be *uniformly integrable* if for every $\varepsilon > 0$ there is a y_ε such that

$$\int_{|x|>y_\varepsilon} |x|\, dF_n(x) < \varepsilon, \qquad \text{for all } n$$

where F_n is the distribution function of X_n.

Lemma 6.4.9

If X_n, $n \ge 1$, is uniformly integrable then there exists $M < \infty$ such that $E[|X_n|] < M$ for all n.

Proof Let y_1 be as in the definition of uniformly integrable. Then

$$E[|X_n|] = \int_{|x| \le y_1} |x|\, dF_n(x) + \int_{|x| > y_1} |x|\, dF_n(x)$$
$$\le y_1 + 1.$$

Thus, using the preceding, it follows from the martingale convergence theorem that any uniformly integrable martingale has a finite limit. Now, let $Z_n = E[X|Y_1, \ldots, Y_n]$, $n \ge 1$, be a Doob martingale. As it can be shown that a Doob martingale is always uniformly integrable, it thus follows that $\lim_{n \to \infty} E[X|Y_1, \ldots, Y_n]$ exists. As might be expected, this limiting value is equal to the conditional expectation of X given the entire sequence of the Y_i. That is,

$$\lim_{n \to \infty} E[X|Y_1, \ldots, Y_n] = E[X|Y_1, Y_2, \ldots].$$

Not only is a Doob martingale uniformly integrable but it turns out that every uniformly integrable martingale can be represented as a Doob martingale. For suppose that $\{Z_n, n \ge 1\}$ is a uniformly integrable martingale. Then, by the martingale convergence theorem, it has a limit, say $\lim_n Z_n = Z$. Now, consider the Doob martingale $\{E[Z|Z_1, \ldots, Z_k], k \ge 1\}$ and note that

$$\begin{aligned} E[Z|Z_1, \ldots, Z_k] &= E[\lim_{n \to \infty} Z_n | Z_1, \ldots, Z_k] \\ &= \lim_{n \to \infty} E[Z_n | Z_1, \ldots, Z_k] \\ &= Z_k, \end{aligned}$$

where the interchange of expectation and limit can be shown to be justified by the uniformly integrable assumption. Thus, we see that any uniformly integrable martingale can be represented as a Doob martingale.

Remark Under the conditions of the martingale convergence theorem, if we let $Z = \lim_{n \to \infty} Z_n$ then it can be shown that $E[|Z|] < \infty$.

6.5 A Generalized Azuma Inequality

The supermartingale stopping time theorem can be used to generalize Azuma's inequality when we have the same bound on all the $Z_i - Z_{i-1}$. We start with the following proposition which is of independent interest.

PROPOSITION 6.5.1

Let $\{Z_n, n \geq 1\}$ be a martingale with mean $Z_0 = 0$, for which

$$-\alpha \leq Z_n - Z_{n-1} \leq \beta \qquad \text{for all } n \geq 1.$$

Then, for any positive values a and b

$$P\{Z_n \geq a + bn \text{ for some } n\} \leq \exp\{-8ab/(\alpha + \beta)^2\}.$$

Proof Let, for $n \geq 0$,

$$W_n = \exp\{c(Z_n - a - bn)\}$$

and note that, for $n \geq 1$,

$$W_n = W_{n-1}e^{-cb} \exp\{c(Z_n - Z_{n-1})\}.$$

Using the preceding, plus the fact that knowledge of $W_1, \ldots, W_{n-1}$ is equivalent to that of $Z_1, \ldots, Z_{n-1}$, we obtain that

$$\begin{aligned} E[W_n|W_1, \ldots, W_{n-1}] &= W_{n-1}e^{-cb}E[\exp\{c(Z_n - Z_{n-1})\}|Z_1, \ldots, Z_{n-1}] \\ &\leq W_{n-1}e^{-cb}[\beta e^{-c\alpha} + \alpha e^{c\beta}]/(\alpha + \beta) \\ &\leq W_{n-1}e^{-cb}e^{c^2(\alpha+\beta)^2/8} \end{aligned}$$

where the first inequality follows from Lemma 6.3.1, and the second from applying Lemma 6.3.2 with $\theta = \alpha/(\alpha + \beta)$, $x = c(\alpha + \beta)$. Hence, fixing the value of c as $c = 8b/(\alpha + \beta)^2$ yields

$$E[W_n|W_1, \ldots, W_{n-1}] \leq W_{n-1},$$

and so $\{W_n, n \geq 0\}$ is a supermartingale. For a fixed positive integer k, define the bounded stopping time N by

$$N = \text{Minimum}\{n\text{: either } Z_n \geq a + bn \qquad \text{or} \qquad n = k\}.$$

Now,

$$\begin{aligned} P\{Z_N \geq a + bN\} &= P\{W_N \geq 1\} \\ &\leq E[W_N] \qquad \text{(by Markov's inequality)} \\ &\leq E[W_0] \end{aligned}$$

where the final equality follows from the supermartingale stopping theorem. But the preceding is equivalent to

$$P\{Z_n \geq a + bn \text{ for some } n \leq k]\} \leq e^{-8ab/(\alpha+\beta)^2}.$$

Letting $k \to \infty$ gives the result.

THEOREM 6.5.2 The Generalized Azuma Inequality

Let $\{Z_n, n \geq 1\}$ be a martingale with mean $Z_0 = 0$. If $-\alpha \leq Z_n - Z_{n-1} \leq \beta$ for all $n \geq 1$ then, for any positive constant c and positive integer m:

(i) $P\{Z_n \geq nc \text{ for some } n \geq m\} \leq \exp\{-2mc^2/(\alpha + \beta)^2\}$
(ii) $P\{Z_n \leq -nc \text{ for some } n \geq m\} \leq \exp\{-2mc^2/(\alpha + \beta)^2\}$.

Proof To begin, note that if there is an n such that $n \geq m$ and $Z_n \geq nc$ then, for that n, $Z_n \geq nc \geq mc/2 + nc/2$. Hence,

$$P\{Z_n \geq nc \text{ for some } n \geq m\} \leq P\{Z_n \geq mc/2 + (c/2)n \text{ for some } n\}$$

$$\leq \exp\{-8(mc/2)(c/2)/(\alpha+\beta)^2\}$$

where the final inequality follows from Proposition 6.5.1.
Part (ii) follows from part (i) upon consideration of the martingale $-Z_n, n \geq 0$.

Remark Note that Azuma's inequality states that the probability that Z_m/m is at least c is such that

$$P\{Z_m/m \geq c\} = P\{Z_m \geq mc\} \leq \exp\{-2mc^2/(\alpha + \beta)^2\}$$

whereas the generalized Azuma gives the *same* bound for the larger probability that Z_n/n is at least c for *any* $n \geq m$.

EXAMPLE 6.5(A) Let S_n equal the number of heads in the first n independent flips of a coin that lands heads with probability p, and let us consider the probability that after some specified number of flips the proportion of heads will ever differ from p by more than ε. That is, consider

$$P\{|S_n/n - p| > \varepsilon \text{ for some } n \geq m\}.$$

Now, if we let X_i equal 1 if the ith flip lands heads and 0 otherwise, then

$$Z_n \equiv S_n - np = \sum_{i=1}^{n} (X_i - p)$$

is a martingale with 0 mean. As,

$$-p \leq Z_n - Z_{n-1} \leq 1 - p$$

it follows that $\{Z_n, n \geq 0\}$ is a zero-mean martingale that satisfies the conditions of Theorem 6.5.2 with $\alpha = p$, $\beta = 1 - p$. Hence,

$$P\{Z_n \geq n\varepsilon \text{ for some } n \geq m\} \leq \exp\{-2m\varepsilon^2\}$$

or, equivalently,

$$P\{S_n/n - p \geq \varepsilon \text{ for some } n \geq m\} \leq \exp\{-2m\varepsilon^2\}.$$

Similarly,

$$P\{S_n/n - p \leq -\varepsilon \text{ for some } n \geq m\} \leq \exp\{-2m\varepsilon^2\}$$

and thus,

$$P\{|S_n/n - p| \geq \varepsilon \text{ for some } n \geq m\} \leq 2\exp\{-2m\varepsilon^2\}.$$

For instance, the probability that after the first 99 flips the proportion of heads ever differs from p by as much as .1 satisfies

$$P\{|S_n/n - p| \geq .1 \text{ for some } n \geq 100\} \leq 2e^{-2} \approx .2707.$$

Problems

6.1. If $\{Z_n, n \geq 1\}$ is a martingale show that, for $1 \leq k < n$,

$$E[Z_n | Z_1, \ldots, Z_k] = Z_k.$$

6.2. For a martingale $\{Z_n, n \geq 1\}$, let $X_i = Z_i - Z_{i-1}$, $i \geq 1$, where $Z_0 \equiv 0$. Show that

$$\text{Var}(Z_n) = \sum_{i=1}^{n} \text{Var}(X_i).$$

6.3. Verify that X_n/m^n, $n \geq 1$, is a martingale when X_n is the size of the nth generation of a branching process whose mean number of offspring per individual is m.

6.4. Consider the Markov chain which at each transition either goes up 1 with probability p or down 1 with probability $q = 1 - p$. Argue that $(q/p)^{S_n}$, $n \geq 1$, is a martingale.

6.5. Consider a Markov chain $\{X_n, n \geq 0\}$ with $P_{NN} = 1$. Let $P(i)$ denote the probability that this chain eventually enters state N given that it starts in state i. Show that $\{P(X_n), n \geq 0\}$ is a martingale.

6.6. Let $X(n)$ denote the size of the nth generation of a branching process, and let π_0 denote the probability that such a process, starting with a single individual, eventually goes extinct. Show that $\{\pi_o^{X(n)}, n \geq 0\}$ is a martingale.

6.7. Let $X_1, \ldots$ be a sequence of independent and identically distributed random variables with mean 0 and variance σ^2. Let $S_n = \sum_{i=1}^{n} X_i$ and show that $\{Z_n, n \geq 1\}$ is a martingale when

$$Z_n = S_n^2 - n\sigma^2.$$

6.8. If $\{X_n, n \geq 0\}$ and $\{Y_n, n \geq 0\}$ are independent martingales, is $\{Z_n, n \geq 0\}$ a martingale when
(a) $Z_n = X_n + Y_n$?
(b) $Z_n = X_n Y_n$?
Are these results true without the independence assumption? In each case either present a proof or give a counterexample.

6.9. A process $\{Z_n, n \geq 1\}$ is said to be a *reverse*, or *backwards*, martingale if $E|Z_n| < \infty$ for all n and

$$E[Z_n | Z_{n+1}, Z_{n+2}, \ldots] = Z_{n+1}.$$

Show that if $X_i, i \geq 1$, are independent and identically distributed random variables with finite expectation, then $Z_n = (X_1 + \cdots + X_n)/n, n \geq 1$, is a reverse martingale.

6.10. Consider successive flips of a coin having probability p of landing heads. Use a martingale argument to compute the expected number of flips until the following sequences appear:
(a) HHTTHHT
(b) HTHTHTH

6.11. Consider a gambler who at each gamble is equally likely to either win or lose 1 unit. Suppose the gambler will quit playing when his winnings are either A or $-B$, $A > 0$, $B > 0$. Use an appropriate martingale to show that the expected number of bets is AB.

6.12. In Example 6.2(C), find the expected number of stages until one of the players is eliminated.

6.13. Let $Z_n = \prod_{i=1}^{n} X_i$, where $X_i, i \geq 1$ are independent random variables with

$$P\{X_i = 2\} = P\{X_i = 0\} = 1/2.$$

Let $N = \text{Min}\{n: Z_n = 0\}$. Is the martingale stopping theorem applicable? If so, what would you conclude? If not, why not?

6.14. Show that the equation

$$e^{\beta} - e^{-\beta} = 2\beta e^{\beta^2/2}$$

has no solution when $\beta \neq 0$.
(*Hint*: Expand in a power series.)

6.15. Let X denote the number of heads in n independent flips of a fair coin. Show that:

(a) $P\{X - n/2 \geq a\} \leq \exp\{-2a^2/n\}$.

(b) $P\{X - n/2 \leq -a\} \leq \exp\{-2a^2/n\}$.

6.16. Let X denote the number of successes in n independent Bernoulli trials, with trial i resulting in a success with probability p_i. Show that

$$P\left\{\left|X - \sum_{i=1}^{n} p_i\right| \geq a\right\} \leq 2\exp\{-2a^2/n\}.$$

6.17. Suppose that 100 balls are to be randomly distributed among 20 urns. Let X denote the number of urns that contain at least five balls. Derive an upper bound for $P\{X \geq 15\}$.

6.18. Let p denote the probability that a random selection of 88 people will contain at least three with the same birthday. Use Azuma's inequality to obtain an upper bound on p. (It can be shown that $p \approx .50$.)

6.19. For binary n-vectors $\boldsymbol{x}$ and $\boldsymbol{y}$ (meaning that each coordinate of these vectors is either 0 or 1) define the distance between them by

$$\rho(\boldsymbol{x}, \boldsymbol{y}) = \sum_{i=1}^{n} |x_i - y_i|.$$

(This is called the Hamming distance.) Let A be a finite set of such vectors, and let $X_1, \ldots, X_n$ be independent random variables that are each equally likely to be either 0 or 1. Set

$$D = \min_{\boldsymbol{y} \in A} \rho(\boldsymbol{X}, \boldsymbol{y})$$

and let $\mu = E[D]$. In terms of μ, find an upper bound for $P\{D \geq b\}$ when $b > \mu$.

6.20. Let $\boldsymbol{X}_1, \ldots, \boldsymbol{X}_n$ be independent random vectors that are all uniformly distributed in the circle of radius 1 centered at the origin. Let $T =$

$T(X_1, \ldots, X_n)$ denote the length of the shortest path connecting these n points. Argue that

$$P\{|T - E[T]| \geq a\} \leq 2 \exp\{-a^2/(32n)\}.$$

6.21. A group of $2n$ people, consisting of n men and n women, are to be independently distributed among m rooms. Each woman chooses room j with probability p_j while each man chooses it with probability q_j, $j = 1, \ldots, m$. Let X denote the number of rooms that will contain exactly one man and one woman.

(a) Find $\mu = E[X]$.

(b) Bound $P\{|X - \mu| > b\}$ for $b > 0$.

6.22. Let $\{X_n, n \geq 0\}$ be a Markov process for which X_0 is uniform on $(0, 1)$ and, conditional on X_n,

$$X_{n+1} = \begin{cases} \alpha X_n + 1 - \alpha & \text{with probability } X_n \\ \alpha X_n & \text{with probability } 1 - X_n \end{cases}$$

where $0 < \alpha < 1$. Discuss the limiting properties of the sequence X_n, $n \geq 1$.

6.23. An urn initially contains one white and one black ball. At each stage a ball is drawn and is then replaced in the urn along with another ball of the same color. Let Z_n denote the fraction of balls in the urn that are white after the nth replication.

(a) Show that $\{Z_n, n \geq 1\}$ is a martingale.

(b) Show that the probability that the fraction of white balls in the urn is ever as large as 3/4 is at most 2/3.

6.24. Consider a sequence of independent tosses of a coin and let $P\{\text{head}\}$ be the probability of a head on any toss. Let A be the hypothesis that $P\{\text{head}\} = a$ and let B be the hypothesis that $P\{\text{head}\} = b$, $0 < a, b < 1$. Let X_i denote the outcome of the ith toss and let

$$Z_n = \frac{P\{X_1, \ldots, X_n | A\}}{P\{X_1, \ldots, X_n | B\}}.$$

Show that if B is true, then:

(a) Z_n is a martingale, and

(b) $\lim_{n\to\infty} Z_n$ exists with probability 1.

(c) If $b \neq a$, what is $\lim_{n\to\infty} Z_n$?

6.25. Let Z_n, $n \geq 1$, be a sequence of random variables such that $Z_1 \equiv 1$ and given $Z_1, \ldots, Z_{n-1}$, Z_n is a Poisson random variable with mean Z_{n-1}, $n > 1$. What can we say about Z_n for n large?

6.26. Let $X_1, X_2, \ldots$ be independent and such that

$$P\{X_i = -1\} = 1 - 1/2^i,$$
$$P\{X_i = 2^i - 1\} = 1/2^i, \qquad i \geq 1.$$

Use this sequence to construct a zero mean martingale Z_n such that $\lim_{n\to\infty} Z_n = -\infty$ with probability 1. (*Hint:* Make use of the Borel–Cantelli lemma.)

A continuous-time process $\{X(t), t \geq 0\}$ is said to be a martingale if $E[|X(t)|] < \infty$ for all t and, for all $s < t$,

$$E[X(t) | X(u), 0 \leq u \leq s] = X(s).$$

6.27. Let $\{X(t), t \geq 0\}$ be a continuous-time Markov chain with infinitesimal transition rates q_{ij}, $i \neq j$. Give the conditions on the q_{ij} that result in $\{X(t), t \geq 0\}$ being a continuous-time martingale.

Do Problems 6.28–6.30 under the assumptions that (a) the continuous-time analogue of the martingale stopping theorem is valid, and (b) any needed regularity conditions are satisfied.

6.28. Let $\{N(t), t \geq 0\}$ be a nonhomogeneous Poisson process with intensity function $\lambda(t)$, $t \geq 0$. Let T denote the time at which the nth event occurs. Show that

$$n = E\left[\int_0^T \lambda(t)\, dt\right].$$

6.29. Let $\{X(t), t \geq 0\}$ be a continuous-time Markov chain that will, in finite expected time, enter an absorbing state N. Suppose that $X(0) = 0$ and let m_i denote the expected time the chain is in state i. Show that for $j \neq 0$, $j \neq N$:

(a) $E[\text{number of times the chain leaves state } j] = v_j m_j$, where $1/v_j$ is the mean time the chain spends in j during a visit.

(b) $E[\text{number of times it enters state } j] = \sum_{i \neq j} m_i q_{ij}$.

(c) Argue that

$$v_j m_j = \sum_{i \neq j} m_i q_{ij}, \qquad j \neq 0$$
$$v_0 m_0 = 1 + \sum_{i \neq 0} m_i q_{i0}.$$

6.30. Let $\{X(t), t \geq 0\}$ be a compound Poisson process with Poisson rate λ and component distribution F. Define a continuous-time martingale related to this process.

REFERENCES

Martingales were developed by Doob and his text (Reference 2) remains a standard reference. References 6 and 7 give nice surveys of martingales at a slightly more advanced level than that of the present text. Example 6.2(A) is taken from Reference 3 and Example 6.2(C) from Reference 5. The version of the Azuma inequality we have presented is, we believe, more general than has previously appeared in the literature. Whereas the usual assumption on the increments is the symmetric condition $|X_i - X_{i-1}| \leq \alpha_i$, we have allowed for a nonsymmetric bound. The treatment of the generalized Azuma inequality presented in Section 6.5 is taken from Reference 4. Additional material on Azuma's inequality can be found in Reference 1.

1. N. Alon, J. Spencer, and P. Erdos, *The Probabilistic Method*, John Wiley, New York, 1992.
2. J. Doob, *Stochastic Processes*, John Wiley, New York, 1953.
3. S. Y. R. Li, "A Martingale Approach to the Study of the Occurrence of Pattern in Repeated Experiments," *Annals of Probability*, 8, No. 6 (1980), pp. 1171–1175.
4. S. M. Ross, "Generalizing Blackwell's Extension of the Azuma Inequality," *Probability in the Engineering and Informational Science*, 9, No. 3 (1995), pp. 493–496.
5. D. Stirzaker, "Tower Problems and Martingales," *The Mathematical Scientist*, 19, No. 1 (1994), pp. 52–59.
6. P. Whittle, *Probability via Expectation*, 3rd ed., Springer-Verlag, Berlin, 1992.
7. D. Williams, *Probability with Martingales*, Cambridge University Press, Cambridge, England, 1991.

CHAPTER 7

Random Walks

INTRODUCTION

Let $X_1, X_2, \ldots$ be independent and identically distributed (iid) with $E[|X_i|] < \infty$. Let $S_0 = 0$, $S_n = \sum_1^n X_i$, $n \geq 1$. The process $\{S_n, n \geq 0\}$ is called a *random walk* process.

Random walks are quite useful for modeling various phenomena. For instance, we have previously encountered the *simple random walk*—$P\{X_i = 1\} = p = 1 - P\{X_i = -1\}$—in which S_n can be interpreted as the winnings after the nth bet of a gambler who either wins or loses 1 unit on each bet. We may also use random walks to model more general gambling situations; for instance, many people believe that the successive prices of a given company listed on the stock market can be modeled as a random walk. As we will see, random walks are also useful in the analysis of queueing and ruin systems.

In Section 7.1 we present a duality principle that is quite useful in obtaining various probabilities concerning random walks. One of the examples in this section deals with the $G/G/1$ queueing system, and in analyzing it we are led to the consideration of the probability that a random walk whose mean step is negative will ever exceed a given constant.

Before dealing with this probability we, however, digress in Section 7.2 to a discussion of exchangeability, which is the condition that justifies the duality principle. We present De Finetti's theorem which provides a characterization of an infinite sequence of exchangeable Bernoulli random variables. In Section 7.3 we return to our discussion of random walks and show how martingales can be effectively utilized. For instance, using martingales we show how to approximate the probability that a random walk with a negative drift ever exceeds a fixed positive value. In Section 7.4 we apply the results of the preceding section to $G/G/1$ queues and to certain ruin problems.

Random walks can also be thought of as generalizations of renewal processes. For if X_i is constrained to be a nonnegative random variable, then S_n could be interpreted as the time of the nth event of a renewal process. In Section 7.5 we present a generalization of Blackwell's theorem when the X_i

are not required to be nonnegative and indicate a proof based on results in renewal theory.

7.1 Duality in Random Walks

Let

$$S_n = \sum_1^n X_i, \qquad n \geq 1$$

denote a random walk. In computing probabilities concerning $\{S_n, n \geq 1\}$, there is a duality principle that, though obvious, is quite useful.

Duality Principle

$(X_1, X_2, \ldots, X_n)$ has the same joint distributions as $(X_n, X_{n-1}, \ldots, X_1)$. The validity of the duality principle is immediate since the $X_i, i > 1$, are independent and identically distributed. We shall now illustrate its use in a series of propositions.

Proposition 7.1.1 states that if $E(X) > 0$, then the random walk will become positive in a finite expected number of steps.

PROPOSITION 7.1.1

Suppose $X_1, X_2, \ldots$ are independent and identically distributed random variables with $E[X] > 0$. If

$$N = \min\{n: X_1 + \cdots + X_n > 0\},$$

then

$$E[N] < \infty.$$

Proof

$$\begin{aligned} E[N] &= \sum_{n=0}^{\infty} P\{N > n\} \\ &= \sum_{n=0}^{\infty} P\{X_1 \leq 0, X_1 + X_2 \leq 0, \ldots, X_1 + \cdots + X_n \leq 0\} \\ &= \sum_{n=0}^{\infty} P\{X_n \leq 0, X_n + X_{n-1} \leq 0, \ldots, X_n + \cdots + X_1 \leq 0\}, \end{aligned}$$

where the last equality follows from duality. Therefore,

$$E[N] = \sum_{n=0}^{\infty} P\{S_n \le S_{n-1}, S_n \le S_{n-2}, \ldots, S_n \le 0\}. \tag{7.1.1}$$

Now let us say that a renewal takes place at time n if $S_n \le S_{n-1}$, $S_n \le S_{n-2}, \ldots, S_n \le 0$; that is, a renewal takes place each time the random walk hits a low. (A little thought should convince us that the times between successive renewals are indeed independent and identically distributed.) Hence, from Equation (7.1.1),

$$\begin{aligned} E[N] &= \sum_{n=0}^{\infty} P\{\text{renewal occurs at time } n\} \\ &= 1 + E[\text{number of renewals that occur}]. \end{aligned}$$

Now by the strong law of large numbers it follows, since $E[X] > 0$, that $S_n \to \infty$, and so the number of renewals that occurs will be finite (with probability 1). But the number of renewals that occurs is either infinite with probability 1 if $F(\infty)$—the probability that the time between successive renewals is finite—equals 1, or has a geometric distribution with finite mean if $F(\infty) < 1$. Hence it follows that

$$E[\text{number of renewals that occurs}] < \infty$$

and so,

$$E[N] < \infty.$$

Our next proposition deals with the expected rate at which a random walk assumes new values. Let us define R_n, called the *range* of $(S_0, S_1, \ldots, S_n)$, by the following.

Definition

R_n is the number of distinct values of $(S_0, \ldots, S_n)$.

PROPOSITION 7.1.2

$$\lim_{n\to\infty} \frac{E[R_n]}{n} = P\{\text{random walk never returns to } 0\}.$$

Proof Letting

$$I_k = \begin{cases} 1 & \text{if } S_k \neq S_{k-1}, S_k \neq S_{k-2}, \ldots, S_k \neq S_0 \\ 0 & \text{otherwise,} \end{cases}$$

then

$$R_n = 1 + \sum_{k=1}^{n} I_k,$$

and so

$$\begin{aligned} E[R_n] &= 1 + \sum_{k=1}^{n} P\{I_k = 1\} \\ &= 1 + \sum_{k=1}^{n} P\{S_k \neq S_{k-1}, S_k \neq S_{k-2}, \ldots, S_k \neq 0\} \\ &= 1 + \sum_{k=1}^{n} P\{X_k \neq 0, X_k + X_{k-1} \neq 0, \ldots, X_k + X_{k-1} + \cdots + X_1 \neq 0\} \\ &= 1 + \sum_{k=1}^{n} P\{X_1 \neq 0, X_1 + X_2 \neq 0, \ldots, X_1 + \cdots + X_k \neq 0\}, \end{aligned}$$

where the last equality follows from duality. Hence,

$$\begin{aligned} E[R_n] &= 1 + \sum_{k=1}^{n} P\{S_1 \neq 0, S_2 \neq 0, \ldots, S_k \neq 0\} \\ &= \sum_{k=0}^{n} P\{T > k\}, \end{aligned} \tag{7.1.2}$$

where T is the time of the first return to 0. Now, as $k \to \infty$,

$$P\{T > k\} \to P\{T = \infty\} = P\{\text{no return to } 0\},$$

and so, from (7.1.2) we see that

$$E[R_n]/n \to P\{\text{no return to } 0\}.$$

Example 7.1(a) ***The Simple Random Walk.*** In the simple random walk $P\{X_i = 1\} = p = 1 - P\{X_i = -1\}$. Now when $p = \frac{1}{2}$ (the symmetric simple random walk), the random walk is recurrent and thus

$$P\{\text{no return to } 0\} = 0 \qquad \text{when } p = \tfrac{1}{2}.$$

Hence,

$$E[R_n/n] \to 0 \qquad \text{when } p = \tfrac{1}{2}.$$

When $p > \frac{1}{2}$, let $\alpha = P\{\text{return to } 0 | X_1 = 1\}$. Since $P\{\text{return to } 0 | X_1 = -1\} = 1$ (why?), we have

$$P\{\text{return to } 0\} = \alpha p + 1 - p.$$

Also, by conditioning on X_2,

$$\alpha = \alpha^2 p + 1 - p,$$

or, equivalently,

$$(\alpha - 1)(\alpha p - 1 + p) = 0.$$

Since $\alpha < 1$ from transience, we see that

$$\alpha = (1 - p)/p,$$

and so

$$E[R_n/n] \to 2p - 1 \qquad \text{when } p > \tfrac{1}{2}.$$

Similarly,

$$E[R_n/n] \to 2(1 - p) - 1 \qquad \text{when } p \le \tfrac{1}{2}.$$

Our next application of the utility of duality deals with the symmetric random walk.

PROPOSITION 7.1.3

In the symmetric simple random walk the expected number of visits to state k before returning to the origin is equal to 1 for all $k \neq 0$.

Proof For $k > 0$ let Y denote the number of visits to state k prior to the first return to the origin. Then Y can be expressed as

$$Y = \sum_{n=1}^{\infty} I_n,$$

where

$$I_n = \begin{cases} 1 & \text{if a visit to state } k \text{ occurs at time } n \text{ and there is no} \\ & \text{return to the origin before } n \\ 0 & \text{otherwise,} \end{cases}$$

or, equivalently,

$$I_n = \begin{cases} 1 & \text{if } S_n > 0, S_{n-1} > 0, S_{n-2} > 0, \ldots, S_1 > 0, S_n = k \\ 0 & \text{otherwise.} \end{cases}$$

Thus,

$$\begin{aligned} E[Y] &= \sum_{n=1}^{\infty} P\{S_n > 0, S_{n-1} > 0, \ldots, S_1 > 0, S_n = k\} \\ &= \sum_{n=1}^{\infty} P\{X_n + \cdots + X_1 > 0, X_{n-1} + \cdots + X_1 > 0, \ldots, X_1 > 0, X_n + \cdots + X_1 = k\} \\ &= \sum_{n=1}^{\infty} P\{X_1 + \cdots + X_n > 0, X_2 + \cdots + X_n > 0, \ldots, X_n > 0, X_1 + \cdots + X_n = k\}, \end{aligned}$$

where the last equality follows from duality. Hence,

$$\begin{aligned} E[Y] &= \sum_{n=1}^{\infty} P\{S_n > 0, S_n > S_1, \ldots, S_n > S_{n-1}, S_n = k\} \\ &= \sum_{n=1}^{\infty} P\{\text{symmetric random walk hits } k \text{ for the first time at time } n\} \\ &= P\{\text{symmetric random walk ever hits } k\} \\ &= 1 \qquad \text{(by recurrence)}, \end{aligned}$$

and the proof is complete.*

Our final application of duality is to the $G/G/1$ queueing model. This is a single-server model that assumes customers arrive in accordance with a renewal process having an arbitrary interarrival distribution F, and the service distribution is G. Let the interarrival times be $X_1, X_2, \ldots,$ and let the service times be $Y_1, Y_2, \ldots,$ and let D_n denote the delay (or wait) in queue of the nth customer. Since customer n spends a time $D_n + Y_n$ in the system and customer $n + 1$ arrives a time X_{n+1} after customer n, it follows (see Figure 7.1.1) that

$$D_{n+1} = \begin{cases} D_n + Y_n - X_{n+1} & \text{if } D_n + Y_n \geq X_{n+1} \\ 0 & \text{if } D_n + Y_n < X_{n+1}, \end{cases}$$

or, equivalently, letting $U_n = Y_n - X_{n+1}$, $n \geq 1$,

(7.1.3) $$D_{n+1} = \max\{0, D_n + U_n\}, \qquad n \geq 0.$$

* The reader should compare this proof with the one outlined in Problem 4.46 of Chapter 4.

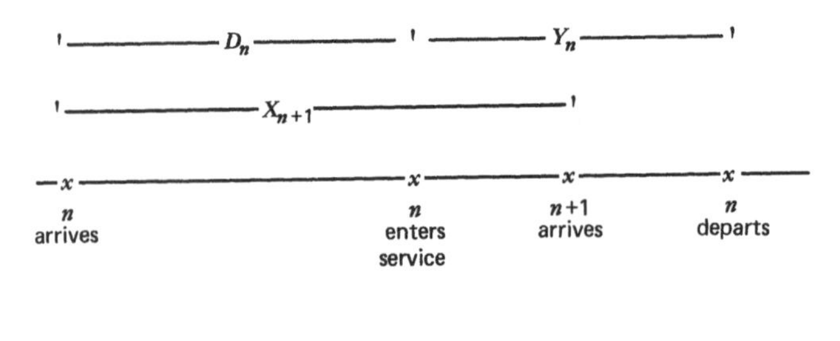

or

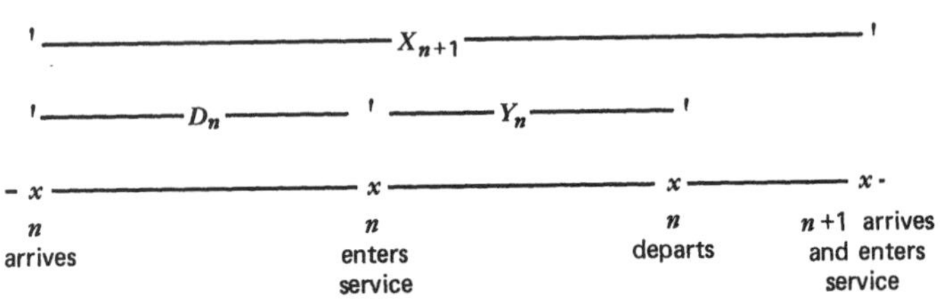

Figure 7.1.1. $\boldsymbol{D_{n+1} = \max\{D_n + Y_n - X_{n+1}, 0\}}$.

Iterating the relation (7.1.3) yields

$$\begin{aligned} D_{n+1} &= \max\{0, D_n + U_n\} \\ &= \max\{0, U_n + \max\{0, D_{n-1} + U_{n-1}\}\} \\ &= \max\{0, U_n, U_n + U_{n-1} + D_{n-1}\} \\ &= \max\{0, U_n, U_n + U_{n-1} + \max\{0, U_{n-2} + D_{n-2}\}\} \\ &= \max\{0, U_n, U_n + U_{n-1}, U_n + U_{n-1} + U_{n-2} + D_{n-2}\} \\ &\;\;\vdots \\ &= \max\{0, U_n, U_n + U_{n-1}, \ldots, U_n + U_{n-1} + \cdots + U_1\}, \end{aligned}$$

where the last step uses the fact that $D_1 = 0$. Hence, for $c > 0$,

$$\begin{aligned} P\{D_{n+1} \geq c\} &= P\{\max(0, U_n, U_n + U_{n-1}, \ldots, U_n + \cdots + U_1) \geq c\} \\ &= P\{\max(0, U_1, U_1 + U_2, \ldots, U_1 + \cdots + U_n) \geq c\}, \end{aligned}$$

where the last equality follows from duality. Hence we have the following.

PROPOSITION 7.1.4

If D_n is the delay in queue of the nth customer in a $G/G/1$ queue with interarrival times X_i, $i \geq 1$, and service times Y_i, $i \geq 1$, then

(7.1.4) $\qquad P\{D_{n+1} \geq c\} = P\{\text{the random walk } S_j, j \geq 1, \text{crosses } c \text{ by time } n\},$

where

$$S_j = \sum_{i=1}^{j} (Y_i - X_{i+1}).$$

We also note from Proposition 7.1.4 that $P\{D_{n+1} \geq c\}$ is nondecreasing in n. Letting

$$P\{D_\infty \geq c\} = \lim_{n\to\infty} P\{D_n \geq c\},$$

we have from (7.1.4)

$$P\{D_\infty \geq c\} = P\{\text{the random walk } S_j,\ j \geq 1, \text{ ever crosses } c\}. \tag{7.1.5}$$

If $E[U] = E[Y] - E[X]$ is positive, then the random walk will, by the strong law of large numbers, converge to infinity, and so

$$P\{D_\infty \geq c\} = 1 \qquad \text{for all } c \text{ if } E[Y] > E[X].$$

The above will also be true when $E[Y] = E[X]$, and thus it is only when $E[Y] < E[X]$ that a limiting delay distribution exists. To compute $P\{D_\infty > c\}$ in this case we need to compute the probability that a random walk whose mean change is negative will ever exceed a constant. However, before attacking this problem we present a result, known as Spitzer's identity, that will enable us to explicitly compute $E[D_n]$ in certain special cases.

Spitzer's identity is concerned with the expected value of the maximum of the random walk up to a specified time. Let $M_n = \max(0, S_1, \ldots, S_n)$, $n \geq 1$.

PROPOSITION 7.1.5 Spitzer's Identity

$$E[M_n] = \sum_{k=1}^{n} \frac{1}{k} E[S_k^+]$$

Proof For any event A, let $I(A)$ equal 1 if A occurs and 0 otherwise. We will use the representation

$$M_n = I(S_n > 0)M_n + I(S_n \leq 0)M_n$$

Now,

$$I(S_n > 0)M_n = I(S_n > 0) \max_{1\leq i\leq n} S_i = I(S_n > 0)(X_1 + \max(0, X_2, \ldots, X_2 + \cdots + X_n))$$

Hence,

$$(7.1.6)\quad E[I(S_n > 0)M_n] = E[I(S_n > 0)X_1] + E[I(S_n > 0)\max(0, X_2, \ldots, X_2 + \cdots + X_n)]$$

But, since $X_1, X_2, \ldots, X_n$ and $X_n, X_1, X_2, \ldots, X_{n-1}$ have the same joint distribution, we see that

$$(7.1.7)\quad E[I(S_n > 0)\max(0, X_2, \ldots, X_2 + \cdots + X_n) = E[I(S_n > 0)M_{n-1}]$$

Also, since X_i, S_n has the same joint distribution for all i,

$$E[S_n I(S_n > 0)] = E\left[\sum_{i=1}^{n} X_i I(S_n > 0)\right] = nE[X_1 I(S_n > 0)]$$

implying that

$$(7.1.8)\qquad E[X_1 I(S_n > 0)] = \frac{1}{n}E[S_n I(S_n > 0)] = \frac{1}{n}E[S_n^+]$$

Thus, from (7.1.6), (7.1.7), and (7.1.8) we have that

$$E[I(S_n > 0)M_n] = E[I(S_n > 0)M_{n-1}] + \frac{1}{n}E[S_n^+]$$

In addition, since $S_n \le 0$ implies that $M_n = M_{n-1}$ it follows that

$$I(S_n \le 0)M_n = I(S_n \le 0)M_{n-1}$$

which, combined with the preceding, yields that

$$E[M_n] = E[I(S_n > 0)M_{n-1}] + \frac{1}{n}E[S_n^+] + E[I(S_n \le 0)M_{n-1}]$$

$$= E[M_{n-1}] + \frac{1}{n}E[S_n^+]$$

Reusing the preceding equation, this time with $n - 1$ substituting for n, gives that

$$E[M_n] = \frac{1}{n}E[S_n^+] + \frac{1}{n-1}E[S_{n-1}^+] + E[M_{n-2}]$$

and, upon continual repetition of this argument, we obtain

$$E[M_n] = \sum_{k=2}^{n} \frac{1}{k}E[S_k^+] + E[M_1]$$

which proves the result since $M_1 = S_1^+$.

It follows from Proposition 7.1.4 that, with $M_n = \max(0, S_1, \ldots, S_n)$,

$$P\{D_{n+1} \geq c\} = P\{M_n \geq c\}$$

which implies, upon integrating from 0 to ∞, that

$$E[D_{n+1}] = E[M_n]$$

Hence, from Spitzer's identity we see that

$$E[D_{n+1}] = \sum_{k=1}^{n} \frac{1}{k} E[S_k^+]$$

Using the preceding, an explicit formula for $E[D_{n+1}]$ can be obtained in certain special cases.

Example 7.1(b) Consider a single server queueing model where arrivals occur according to a renewal process with interarrival distribution $G(s, \lambda)$ and the service times have distribution $G(r, \mu)$, where $G(a, b)$ is the gamma distribution with parameters a and b (and thus has mean a/b). We will use Spitzer's identity to calculate a formula for $E[D_{n+1}]$ in the case where at least one of s or r is integral.

To begin, suppose that r is an integer. To compute

$$E[S_k^+] = E\left[\left(\sum_{i=1}^{k} Y_i - \sum_{i=1}^{k} X_{i+1}\right)^+\right]$$

first condition on $\sum_{i=1}^{k} X_{i+1}$. Then, use the fact that $\sum_{i=1}^{k} Y_i$ is distributed as the sum of kr independent exponentials with rate μ to obtain, upon conditioning on the number of events of a Poisson process with rate μ that occur by time $\sum_{i=1}^{k} X_{i+1}$, that

$$E\left[\left(\sum_{i=1}^{k} Y_i - \sum_{i=1}^{k} X_{i+1}\right)^+ \,\middle|\, \sum_{i=1}^{k} X_{i+1} = t\right] = \sum_{i=0}^{kr-1} e^{-\mu t} \frac{(\mu t)^i}{i!} \frac{kr - i}{\mu}$$

Hence, letting

$$W_k = \sum_{i=1}^{k} X_{i+1}$$

we have that

$$E[S_k^+] = \sum_{i=0}^{kr-1} \frac{kr - i}{i!\,\mu} E[e^{-\mu W_k}(\mu W_k)^i]$$

Since W_k is gamma with parameters ks and λ a simple computation yields that

$$E[e^{-\mu W_k}(\mu W_k)^i] = \frac{(ks+i-1)!\lambda^{ks}\mu^i}{(ks-1)!(\lambda+\mu)^{ks+i}}$$

where $a! = \int_0^\infty e^{-x}x^a dx$ for nonintegral a.

Hence, we have that when r is integral

$$E[D_{n+1}] = \sum_{k=1}^{n} \frac{1}{k} \sum_{i=0}^{kr-1} \frac{kr-i}{\mu} \binom{ks+i-1}{i} \left(\frac{\lambda}{\lambda+\mu}\right)^{ks} \left(\frac{\mu}{\lambda+\mu}\right)^{i}$$

where $\binom{a+b}{a} = (a+b)!/(a!b!)$.

When s is an integer, we use the identity $S_k^+ = S_k + (-S_k)^+$ to obtain that

$$E[S_k^+] = k(E[Y] - E[X]) + E\left[\left(\sum_{i=1}^{k} X_{i+1} - \sum_{i=1}^{k} Y_i\right)^+\right]$$

We can now use the preceding analysis to obtain the following.

$$E[D_{n+1}] = n(r/\mu - s/\lambda) + \sum_{k=1}^{n} \frac{1}{k} \sum_{i=0}^{ks-1} \frac{ks-i}{\lambda} \binom{kr+i-1}{i} \left(\frac{\mu}{\lambda+\mu}\right)^{kr} \left(\frac{\lambda}{\lambda+\mu}\right)^{i}$$

7.2 Some Remarks Concerning Exchangeable Random Variables

It is not necessary to assume that the random variables $X_1, \ldots, X_n$ are independent and identically distributed to obtain the duality relationship. A weaker general condition is that the random variables are *exchangeable*, where we say that $X_1, \ldots, X_n$ is exchangeable if $X_{i_1}, \ldots, X_{i_n}$ has the same joint distribution for all permutations $(i_1, \ldots, i_n)$ of $(1, 2, \ldots, n)$.

Example 7.2(a) Suppose balls are randomly selected, without replacement, from an urn originally consisting of n balls of which k are white. If we let

$$X_i = \begin{cases} 1 & \text{if the } i\text{th selection is white} \\ 0 & \text{otherwise,} \end{cases}$$

then $X_1, \ldots, X_n$ will be exchangeable but not independent.

As an illustration of the use of exchangeability, suppose X_1 and X_2 are exchangeable and let $f(x)$ and $g(x)$ be increasing functions. Then for all x_1, x_2

$$(f(x_1) - f(x_2))(g(x_1) - g(x_2)) \geq 0,$$

implying that

$$E[(f(X_1) - f(X_2))(g(X_1) - g(X_2)] \geq 0.$$

But as exchangeability implies that

$$E[f(X_1)g(X_1)] = E[f(X_2)g(X_2)],$$
$$E[f(X_1)g(X_2)] = E[f(X_2)g(X_1)],$$

we see upon expanding the above inequality that

$$E[f(X_1)g(X_1)] \geq E[f(X_1)g(X_2)].$$

Specializing to the case where X_1 and X_2 are independent, we have the following.

PROPOSITION 7.2.1

If f and g are both increasing functions, then

$$E[f(X)g(X)] \geq E[f(X)]E[g(X)].$$

The infinite sequence of random variables $X_1, X_2, \ldots$ is said to be exchangeable if every finite subsequence $X_1, \ldots, X_n$ is exchangeable.

EXAMPLE 7.2(B) Let Λ denote a random variable having distribution G and suppose that conditional on the event that $\Lambda = \lambda$, $X_1, X_2, \ldots$ are independent and identically distributed with distribution F_λ—that is,

$$P\{X_1 \leq x_1, \ldots, X_n \leq x_n | \Lambda = \lambda\} = \prod_{i=1}^{n} F_\lambda(x_i).$$

The random variables $X_1, X_2, \ldots$ are exchangeable since

$$P\{X_1 \leq x_1, \ldots, X_n \leq x_n\} = \int \prod_{i=1}^{n} F_\lambda(x_i)\, dG(\lambda),$$

which is symmetric in $(x_1, \ldots, x_n)$. They are, however, not independent.

There is a famous result known as De Finetti's theorem, which states that every infinite sequence of exchangeable random variables is of the form specified by Example 7.2(B). We will present a proof when the X_i are $0 - 1$ (that is, Bernoulli) random variables.

THEOREM 7.2.2

(De Finetti's Theorem). *To every infinite sequence of exchangeable random variables* $X_1, X_2, \ldots$ *taking values either* 0 *or* 1, *there corresponds a probability distribution* G *on* $[0, 1]$ *such that, for all* $0 \le k \le n$,

$$(7.2.2) \qquad P\{X_1 = X_2 = \cdots = X_k = 1, X_{k+1} = \cdots = X_n = 0\} = \int_0^1 \lambda^k (1-\lambda)^{n-k}\, dG(\lambda).$$

Proof Let $m \ge n$. We start by computing the above probability by first conditioning on

$$S_m = \sum_{i=1}^{m} X_i.$$

This yields

$$(7.2.2) \quad P\{X_1 = \cdots = X_k = 1, X_{k+1} = \cdots = X_n = 0\}$$

$$= \sum_{j=0}^{m} P\{X_1 = \cdots = X_k = 1, X_{k+1} = \cdots = X_n = 0 \mid S_m = j\} P\{S_m = j\}$$

$$= \sum_j \frac{j(j-1)\cdots(j-k+1)(m-j)(m-j-1)\cdots(m-j-(n-k)+1)}{m(m-1)\cdots(m-n+1)} P\{S_m = j\}.$$

This last equation follows since, given $S_m = j$, by exchangeability each subset of size j of $X_1, \ldots, X_m$ is equally likely to be the one consisting of all 1's.

If we let $Y_m = S_m/m$, then (7.2.2) may be written as

$$(7.2.3) \quad P\{X_1 = \cdots = X_k = 1, X_{k+1} = \cdots = X_n = 0\}$$

$$= E\left[\frac{(mY_m)(mY_m - 1)\cdots(mY_m - k + 1)[m(1-Y_m)][m(1-Y_m)-1]\cdots[m(1-Y_m)-n+k+1]}{m(m-1)\cdots(m-n+1)}\right].$$

The above is, for large m, roughly equal to $E[Y_m^k(1 - Y_m)^{n-k}]$, and the theorem should follow upon letting $m \to \infty$. Indeed it can be shown by a result called Helly's theorem that for some subsequence m' converging to ∞, the distribution of $Y_{m'}$ will converge to a distribution G and (7.2.3) will converge to

$$E[Y_\infty^k (1 - Y_\infty)^{n-k}] = \int_0^1 y^k (1-y)^{n-k}\, dG(y).$$

Remark De Finetti's theorem is not valid for a *finite* sequence of exchangeable random variables. For instance, if $n = 2$, $k = 1$ in Example 7.2(A), then $P\{X_1 = 1, X_2 = 0\} = P\{X_1 = 0, X_2 = 1\} = \frac{1}{2}$, which cannot be put in the form (7.2.1).

7.3 Using Martingales to Analyze Random Walks

Let

$$S_n = \sum_{i=1}^{n} X_i, \qquad n \geq 1$$

denote a random walk. Our first result is to show that if the X_i are finite integer-valued random variables, then S_n is recurrent if $E[X] = 0$.

THEOREM 7.3.1

Suppose X_i can only take on one of the values $0, \pm 1, \ldots, \pm M$ for some $M < \infty$. Then $\{S_n, n \geq 0\}$ is a recurrent Markov chain if, and only if, $E[X] = 0$.

Proof It is clear that the random walk is transient when $E[X] \neq 0$, since it will either converge to $+\infty$ (if $E[X] > 0$), or $-\infty$ (if $E[X] < 0$). So suppose $E[X] = 0$ and note that this implies that $\{S_n, n \geq 1\}$ is a martingale.

Let A denote the set of states from $-M$ up to -1—that is, $A = \{-M, -(M - 1), \ldots, -1\}$. Suppose the process starts in state i, where $i \geq 0$. For $j > i$, let A_j denote the set of states $A_j = \{j, j + 1, \ldots, j + M\}$ and let N denote the first time that the process is in either A or A_j. By Theorem 6.2.2,

$$E[S_N] = E[S_0] = i$$

and so

$$\begin{aligned} i &= E[S_N | S_N \in A]P\{S_N \in A\} + E[S_N | S_N \in A_j]P\{S_N \in A_j\} \\ &\geq -MP\{S_N \in A\} + j(1 - P\{S_N \in A\}) \end{aligned}$$

or

$$P\{S_N \in A\} \geq \frac{j - i}{j + M}.$$

Hence,

$$P\{\text{process ever enters } A\} \geq P\{S_N \in A\} \geq \frac{j - i}{j + M},$$

and letting $j \to \infty$, we see that

$$P\{\text{process ever enters } A \mid \text{start at } i\} = 1, \qquad i \geq 0.$$

Now let $B = \{1, 2, \ldots, M\}$. By the same argument we can show that for $i \leq 0$

$$P\{\text{process ever enters } B \mid \text{start at } i\} = 1, \qquad i \leq 0.$$

Therefore,

$$P\{\text{process ever enters } A \cup B \mid \text{start at } i\} = 1, \text{ for all } i.$$

It is easy to see that the above implies that the finite set of states $A \cup B$ will be visited infinitely often. However, if the process is transient, then any finite set of states is only visited finitely often. Hence, the process is recurrent.

Once again, let

$$S_n = \sum_{i=1}^{n} X_i, \qquad n \geq 1$$

denote a random walk and suppose $\mu = E[X] \neq 0$. For given $A, B > 0$, we shall attempt to compute P_A, the probability that S_n reaches a value at least A before it reaches a value less than or equal to $-B$. To start, let $\theta \neq 0$ be such that

$$E[e^{\theta X}] = 1.$$

We shall suppose that such a θ exists (and is usually unique). Since

$$Z_n \equiv e^{\theta S_n}$$

is the product of independent unit mean random variables, it follows that $\{Z_n\}$ is a martingale with mean 1. Define the stopping time N by

$$N = \min\{n: S_n \geq A \text{ or } S_n \leq -B\}.$$

Since condition (iii) of Theorem 6.2.2 can be shown to be satisfied, we have

$$E[e^{\theta S_N}] = 1.$$

Therefore,

$$1 = E[e^{\theta S_N} \mid S_N \geq A]P_A + E[e^{\theta S_N} \mid S_N \leq -B](1 - P_A). \tag{7.3.1}$$

We can use Equation (7.3.1) to obtain an approximation for P_A as follows. If we neglect the excess (or overshoot past A or $-B$), we have the following approximations:

$$E[e^{\theta S_N} | S_N \geq A] \approx e^{\theta A},$$
$$E[e^{\theta S_N} | S_N \leq -B] \approx e^{-\theta B}.$$

Hence, from (7.3.1),

$$1 \approx e^{\theta A} P_A + e^{-\theta B}(1 - P_A)$$

or

$$P_A \approx \frac{1 - e^{-\theta B}}{e^{\theta A} - e^{-\theta B}}. \tag{7.3.2}$$

We can also approximate $E[N]$ by using Wald's equation and then neglecting the excess. That is,

$$E[S_N] = E[S_N | S_N \geq A] P_A + E[S_N | S_N \leq -B](1 - P_A).$$

Using the approximation

$$E[S_N | S_N \geq A] \approx A,$$
$$E[S_N | S_N \leq -B] \approx -B,$$

we have

$$E[S_N] \approx A P_A - B(1 - P_A),$$

and since

$$E[S_N] = E[N] E[X],$$

we see that

$$E[N] \approx \frac{A P_A - B(1 - P_A)}{E[X]}.$$

Using the approximation (7.3.2) for P_A, we obtain

$$E[N] \approx \frac{A(1 - e^{-\theta B}) - B(e^{\theta A} - 1)}{(e^{\theta A} - e^{-\theta B}) E[X]}. \tag{7.3.3}$$

Example 7.3(a) ***The Gambler's Ruin Problem.*** Suppose

$$X_i = \begin{cases} 1 & \text{with probability } p \\ -1 & \text{with probability } q = 1 - p. \end{cases}$$

We leave it as an exercise to show that $E[(q/p)^X] = 1$ and so $e^\theta = q/p$. If we assume that A and B are integers, there is no overshoot, and so the approximations (7.3.2) and (7.3.3) are exact. Therefore,

$$P_A = \frac{1 - (q/p)^{-B}}{(q/p)^A - (q/p)^{-B}} = \frac{(q/p)^B - 1}{(q/p)^{A+B} - 1}$$

and

$$E[N] = \frac{A(1 - (q/p)^{-B}) - B((q/p)^A - 1)}{((q/p)^A - (q/p)^{-B})(2p - 1)}.$$

Suppose now that $E[X] < 0$ and we are interested in the probability that the random walk ever crosses A.* We will attempt to compute this by using the results so far obtained and then letting $B \to \infty$. Equation (7.3.1) states

$$1 = E[e^{\theta S_N} | S_N \geq A] P\{\text{process crosses } A \text{ before } -B\} + E[e^{\theta S_N} | S_N \leq -B] P\{\text{process crosses } -B \text{ before } A\}. \tag{7.3.4}$$

$\theta \neq 0$ was defined to be such that $E[e^{\theta X}] = 1$. Since $E[X] < 0$, it can be shown (see Problem 7.9) that $\theta > 0$. Hence, from (7.3.4), we have

$$1 \geq e^{\theta A} P\{\text{process crosses } A \text{ before } -B\},$$

and, upon letting $B \to \infty$, we obtain

$$P\{\text{random walk ever crosses } A\} \leq e^{-\theta A}. \tag{7.3.5}$$

7.4 Applications to G/G/1 Queues and Ruin Problems

7.4.1 The G/G/1 Queue

For the $G/G/1$ queue, the limiting distribution of the delay in queue of a customer is by (7.1.5) given by

$$P\{D_\infty \geq A\} = P\{S_n \geq A \text{ for some } n\}, \tag{7.4.1}$$

where

$$S_n = \sum_{i=1}^{n} U_i, \qquad U_i = Y_i - X_{i+1},$$

* By crosses A we mean "either hits or exceeds A."

and where Y_i is the service time of the ith customer and X_{i+1} the interarrival time between the ith and $(i + 1)$st arrival.

Hence, when $E[U] = E[Y] - E[X] < 0$, letting $\theta > 0$ be such that

$$E[e^{\theta U}] = E[e^{\theta(Y-X)}] = 1,$$

we have from (7.3.5)

$$P\{D_\infty \geq A\} \leq e^{-\theta A}. \tag{7.4.2}$$

There is one situation in which we can obtain the exact distribution of D_∞, and that is when the service distribution is exponential.

So suppose Y_i is exponential with rate μ. Recall that for N defined as the time it takes for S_n to either cross A or $-B$, we showed in Equation (7.3.4) that

$$\begin{aligned} 1 = {} & E[e^{\theta S_N} | S_N \geq A] P\{S_n \text{ crosses } A \text{ before } -B\} \\ & + E[e^{\theta S_N} | S_N \leq -B] P\{S_n \text{ crosses } -B \text{ before } A\}. \end{aligned} \tag{7.4.3}$$

Now, $S_n = \sum_{i=1}^n (Y_i - X_{i+1})$ and let us consider the conditional distribution of S_N given that $S_N \geq A$. This is the conditional distribution of

$$\sum_{i=1}^N (Y_i - X_{i+1}) \quad \text{given that} \quad \sum_{i=1}^N (Y_i - X_{i+1}) > A. \tag{7.4.4}$$

Conditioning on the value of N (say $N = n$) and on the value of

$$X_{n+1} - \sum_{i=1}^{n-1} (Y_i - X_{i+1}) \qquad \text{(say it equals } c\text{)},$$

note that the conditional distribution given by (7.4.4) is just the conditional distribution of

$$Y_n - c \quad \text{given that} \quad Y_n - c > A.$$

But by the lack of memory of the exponential, it follows that the conditional distribution of Y given that $Y > c + A$ is just $c + A$ plus an exponential with rate μ. Hence, the conditional distribution of $Y_n - c$ given that $Y_n - c > A$ is just the distribution of A plus an exponential with rate μ. Since this is true for all n and c, we see

$$\begin{aligned} E[e^{\theta S_N} | S_N \geq A] &= E[e^{\theta(A+Y)}] \\ &= e^{\theta A} \int e^{\theta y} \mu e^{-\mu y}\, dy \\ &= \frac{\mu}{\mu - \theta} e^{\theta A}. \end{aligned}$$

Hence, from (7.4.3),

$$1 = \frac{\mu}{\mu - \theta} e^{\theta A} P\{S_n \text{ crosses } A \text{ before } -B\}$$
$$+ E[e^{\theta S_N} | S_N \leq -B] P\{S_n \text{ crosses } -B \text{ before } A\}.$$

Since $\theta > 0$, we obtain, by letting $B \to \infty$,

$$1 = \frac{\mu}{\mu - \theta} e^{\theta A} P\{S_n \text{ ever crosses } A\},$$

and thus from (7.4.1)

$$P\{D_\infty \geq A\} = \frac{\mu - \theta}{\mu} e^{-\theta A}, \qquad A > 0.$$

Summing up, we have shown the following.

THEOREM 7.4.1

For the G/G/1 queue with iid service times Y_i, $i \geq 1$, and iid interarrival times X_1, $X_2, \ldots$, when $E[Y] < E[X]$,

$$P\{D_\infty \geq A\} \leq e^{-\theta A},$$

where $\theta > 0$ is such that

$$E[e^{\theta Y}] E[e^{-\theta X}] = 1.$$

In addition, if Y_i is exponential with rate μ, then

$$P\{D_\infty \geq A\} = \frac{\mu - \theta}{\mu} e^{-\theta A}, \qquad A > 0,$$

$$P\{D_\infty = 0\} = \frac{\theta}{\mu},$$

where in this case θ is such that

$$E[e^{-\theta X}] = \frac{\mu - \theta}{\mu}.$$

7.4.2 A Ruin Problem

Suppose that claims are made to an insurance company in accordance with a renewal process with interarrival times $X_1, X_2, \ldots$. Suppose that the values of the successive claims are iid and are independent of the renewal process of when they occurred. Let Y_i denote the value of the ith claim. Thus if $N(t)$ denotes the number of claims by time t, then the total value of claims made to the insurance company by time t is $\sum_1^{N(t)} Y_i$. On the other hand, suppose that the company receives money at a constant rate of c per unit time, $c > 0$. We are interested in determining the probability that the insurance company, starting with an initial capital of A, will eventually be wiped out. That is, we want

$$p = P\left\{\sum_{i=1}^{N(t)} Y_i > ct + A \text{ for some } t \geq 0\right\}.$$

Now it is clear that the company will eventually be wiped out with probability 1 if $E[Y] \geq cE[X]$. (Why is that?) So we'll assume

$$E[Y] < cE[X].$$

It is also fairly obvious that if the company is to be wiped out, that event will occur when a claim occurs (since it is only when claims occur that the insurance company's assets decrease). Now at the moment after the nth claim occurs the company's fortune is

$$A + c\sum_{i=1}^{n} X_i - \sum_{i=1}^{n} Y_i.$$

Thus the probability we want, call it $p(A)$, is

$$p(A) = P\left\{A + c\sum_{i=1}^{n} X_i - \sum_{i=1}^{n} Y_i < 0 \text{ for some } n\right\},$$

or, equivalently,

$$p(A) = P\{S_n > A \text{ for some } n\},$$

where

$$S_n = \sum_{i=1}^{n} (Y_i - cX_i)$$

is a random walk. From (7.4.1), we see that

$$p(A) = P\{D_\infty > A\} \tag{7.4.5}$$

where D_∞ is the limiting delay in queue of a $G/G/1$ queue with interarrival times cX_i and service times Y_i. Thus from Theorem 7.4.1 we have the following.

THEOREM 7.4.2

(i) *The probability of the insurance company ever being wiped out, call it $p(A)$, is such that*

$$p(A) \le e^{-\theta A},$$

where θ is such that

$$E[\exp\{\theta(Y_i - cX_i)\}] = 1.$$

(ii) *If the claim values are exponentially distributed with rate μ, then*

$$p(A) = \frac{\mu - \theta}{\mu} e^{-\theta A},$$

where θ is such that

$$E[e^{-\theta c X}] = \frac{\mu - \theta}{\mu}.$$

(iii) *If the arrival process of claims is a Poisson process with rate λ, then*

$$p(0) = \lambda E[Y]/c.$$

Proof Parts (i) and (ii) follow immediately from Theorem 7.4.1. In part (iii), cX will be exponential with rate λ/c; thus from (7.4.5) $p(0)$ will equal the probability that the limiting customer delay in an $M/G/1$ queue is positive. But this is just the probability that an arriving customer in an $M/G/1$ finds the system nonempty. Since it is a system with Poisson arrivals, the limiting distribution of what an arrival sees is identical to the limiting distribution of the system state at time t. (This is so since the distribution of the system state at time t, given that a customer has arrived at t is, due to the Poisson arrival assumption, identical to the unconditional distribution of the state at t.) Hence the (limiting) probability that an arrival will find the system nonempty is equal to the limiting probability that the system is nonempty and that, as we have shown by many different ways, is equal to the arrival rate times the mean service time (see Example 4.3(A) of Chapter 4).

7.5 Blackwell's Theorem on the Line

Let $\{S_n, n \geq 1\}$ denote a random walk for which $0 < \mu = E[X] < \infty$. Let $U(t)$ denote the number of n for which $S_n \leq t$. That is,

$$U(t) = \sum_{n=1}^{\infty} I_n, \quad \text{where } I_n = \begin{cases} 1 & \text{if } S_n \leq t \\ 0 & \text{otherwise.} \end{cases}$$

If the X_i were nonnegative, then $U(t)$ would just be $N(t)$, the number of renewals by time t.

Let $u(t) = E[U(t)]$. In this section we will prove the analog of Blackwell's theorem.

BLACKWELL'S THEOREM

If $\mu > 0$ and the X_i are not lattice, then

$$u(t + a) - u(t) \to a/\mu \qquad \text{as } t \to \infty \quad \text{for } a > 0.$$

Before presenting a proof of the above, we will find it useful to introduce the concept of ascending and descending ladder variables. We say that an *ascending ladder variable* of *ladder height* S_n occurs at time n if

$$S_n > \max(S_0, S_1, \ldots, S_{n-1}), \qquad \text{where } S_0 \equiv 0.$$

That is, an ascending ladder variable occurs whenever the random walk reaches a new high. For instance, the initial one occurs the first time the random walk becomes positive. If a ladder variable of height S_n occurs at time n, then the next ladder variable will occur at the first value of $n + j$ for which

$$S_{n+j} > S_n,$$

or, equivalently, at the first $n + j$ for which

$$X_{n+1} + \cdots + X_{n+j} > 0.$$

Since the X_i are independent and identically distributed, it thus follows that the changes in the random walk between ladder variables are all probabilistic replicas of each other. That is, if N_i denotes the time between the $(i - 1)$st and ith ladder variable, then the random vectors $(N_i, S_{N_i} - S_{N_{i-1}})$, $i \geq 1$, are independent and identically distributed (where $S_{N_0} \equiv 0$).

Similarly we can define the concept of descending ladder variables by saying that they occur when the random walk hits a new low. Let $p(p_*)$ denote the probability of ever achieving an ascending (descending) ladder variable. That is,

$$p = P\{S_n > 0 \text{ for some } n\},$$

$$p_* = P\{S_n < 0 \text{ for some } n\}.$$

Now at each ascending (descending) ladder variable there will again be the same probability $p(p_*)$ of ever achieving another one. Hence there will be exactly n such ascending [descending] ladder variables, $n \geq 0$, with probability $p^n(1 - p)$ $[p_*^n(1 - p_*)]$. Therefore, the number of ascending [descending] ladder variables will be finite and have a finite mean if, and only if, $p[p_*]$ is less than 1. Since $E[X] > 0$, it follows by the strong law of large numbers that, with probability 1, $S_n \to \infty$ as $n \to \infty$; and so, with probability 1 there will be an infinite number of ascending ladder variables but only a finite number of descending ones. Thus, $p = 1$ and $p_* < 1$.

We are now ready to prove Blackwell's theorem. The proof will be in parts. First we will argue that $u(t + a) - u(t)$ approaches a limit as $t \to \infty$. Then we will show that this limiting value is equal to a constant times a; and finally we will prove the generalization of the elementary renewal theorem, which will enable us to identify this constant as $1/\mu$.

PROOF OF BLACKWELL'S THEOREM

The successive ascending ladder heights constitute a renewal process. Let $Y(t)$ denote the excess at t of this renewal process. That is, $t + Y(t)$ is the first value of the random walk that exceeds t. Now it is easy to see that given the value of $Y(t)$, say $Y(t) = y$, the distribution of $U(t + a) - U(t)$ does not depend on t. That is, if we know that the first point of the random walk that exceeds t occurs at a distance y past t, then the number of points in $(t, t + a)$ has the same distribution as the number of points in $(0, a)$ given that the first positive value is y. Hence, for some function g,

$$E[U(t + a) - U(t) | Y(t)] = g(Y(t)),$$

and so, taking expectations,

$$u(t + a) - u(t) = E[g(Y(t))].$$

Now $Y(t)$, being the excess at t of a nonlattice renewal process, converges to a limiting distribution (namely, the equilibrium interarrival distribution). Hence, $E[g(Y(t))]$ will converge to $E[g(Y(\infty))]$ where $Y(\infty)$ has the limiting distribution of $Y(t)$. Thus we have shown the existence of

$$\lim_{t \to \infty} [u(t + a) - u(t)].$$

Now let

$$h(a) = \lim_{t\to\infty} [u(t + a) - u(t)].$$

Then

$$\begin{aligned} h(a + b) &= \lim_{t\to\infty} [u(t + a + b) - u(t + b) + u(t + b) - u(t)] \\ &= \lim_{t\to\infty} [u(t + b + a) - u(t + b)] \\ &\quad + \lim_{t\to\infty} [u(t + b) - u(t)] \\ &= h(a) + h(b), \end{aligned}$$

which implies that for some constant c

$$h(a) = \lim_{t\to\infty} [u(t + a) - u(t)] = ca. \tag{7.5.1}$$

To identify the value of c let N_t denote the first n for which $S_n > t$. If the X_i are bounded, say by M, then

$$t < \sum_{i=1}^{N_t} X_i \le t + M.$$

Taking expectations and using Wald's equation ($E[N_t] < \infty$ by an argument similar to that used in Proposition 7.1.1) yields

$$t < E[N_t]\mu \le t + M,$$

and so

$$\frac{E[N_t]}{t} \to \frac{1}{\mu} \qquad \text{as } t \to \infty. \tag{7.5.2}$$

If the X_i are not bounded, then we can use a truncation argument (exactly as in the proof of the elementary renewal theorem) to establish (7.5.2). Now $U(t)$ can be expressed as

$$U(t) = N_t - 1 + N_t^* \tag{7.5.3}$$

where N_t^* is the number of times S_n lands in $(-\infty, t]$ after having gone past t. Since the random variable N_t^* will be no greater than the number of points occurring after time N_t for which the random walk is less than S_{N_t}, it follows that

$$E[N_t^*] \le E[\text{number of } n \text{ for which } S_n < 0]. \tag{7.5.4}$$

We will now argue that the right-hand side of the above equation is finite, and so from (7.5.2) and (7.5.3)

$$\frac{u(t)}{t} \to \frac{1}{\mu} \qquad \text{as } t \to \infty. \tag{7.5.5}$$

The argument that the right-hand side of (7.5.4) is finite runs as follows: we note from Proposition 7.1.1 that $E[N] < \infty$ when N is the first value of n for which $S_n > 0$. At time N there is a positive probability $1 - p^*$ that no future value of the random walk will ever fall below S_N. If a future value does fall below S_N, then again by an argument similar to that used in Proposition 7.1.1, the expected additional time until the random walk again becomes positive is finite. At that point there will again be a positive probability $1 - p^*$ that no future value will fall below the present positive one, and so on. We can use this as the basis of a proof showing that

$$E[\text{number of } n \text{ for which } S_n < 0] \le \frac{E[N|X_1 < 0]}{1 - p^*} < \infty.$$

Thus we have shown (7.5.5).

We will now complete our proof by appealing to (7.5.1) and (7.5.5). From (7.5.1) we have

$$u(i + 1) - u(i) \to c \qquad \text{as } i \to \infty,$$

implying that

$$\sum_{i=1}^{n} \frac{u(i+1) - u(i)}{n} \to c \text{ as } n \to \infty,$$

or, equivalently,

$$\frac{u(n+1) - u(1)}{n} \to c,$$

which, from (7.5.5) implies that $c = 1/\mu$, and the proof is complete.

Remark The proof given lacks rigor in one place. Namely, even though the distribution of $Y(t)$ converges to that of $Y(\infty)$, it does not necessarily follow that $E[g(Y(t))]$ will converge to $E[g(Y(\infty))]$. We should have proven this convergence directly.

Problems

7.1. Consider the following model for the flow of water in and out of a dam. Suppose that, during day n, Y_n units of water flow into the dam from outside sources such as rainfall and river flow. At the end of each day water is released from the dam according to the following rule: If the water content of the dam is greater than a, then the amount a is released. If it is less than or equal to a, then the total contents of the dam are released. The capacity of the dam is C, and once at capacity any additional water that attempts to enter the dam is assumed lost. Thus, for instance, if the water level at the beginning of day n is x, then the level at the end of the day (before any water is released) is $\min(x + Y_n, C)$. Let S_n denote the amount of water in the dam immediately after the water has been released at the end of day n. Assuming that the Y_n, $n \geq 1$, are independent and identically distributed, show that $\{S_n, n \geq 1\}$ is a random walk with reflecting barriers at 0 and $C - a$.

7.2. Let $X_1, \ldots, X_n$ be equally likely to be any of the $n!$ permutations of $(1, 2, \ldots, n)$. Argue that

$$P\left\{\sum_{j=1}^{n} jX_j \leq a\right\} = P\left\{\sum_{j=1}^{n} jX_j \geq n(n+1)^2/2 - a\right\}$$

7.3. For the simple random walk compute the expected number of visits to state k.

7.4. Let $X_1, X_2, \ldots, X_n$ be exchangeable. Compute $E[X_1 | X_{(1)}, X_{(2)}, \ldots, X_{(n)}]$, where $X_{(1)} \leq X_{(2)} \leq \cdots \leq X_{(n)}$ are the X_i in ordered arrangement.

7.5. If $X_1, X_2, \ldots$ is an infinite sequence of exchangeable random variables, with $E[X_1^2] < \infty$, show that $\text{Cov}(X_1, X_2) \geq 0$. (*Hint:* Look at $\text{Var}(\sum_1^n X_i)$.) Give a counterexample when the set of exchangeable random variables is finite.

7.6. An ordinary deck of cards is randomly shuffled and then the cards are exposed one at a time. At some time before all the cards have been exposed you must say "next," and if the next card exposed is a spade then you win and if not then you lose. For any strategy, show that at the moment you call "next" the conditional probability that you win is equal to the conditional probability that the last card is a spade. Conclude from this that the probability of winning is 1/4 for all strategies.

7.7. Argue that the random walk for which X_i only assumes the values $0, \pm 1, \ldots, \pm M$ and $E[X_i] = 0$ is null recurrent.

7.8. Let $S_n, n \geq 0$ denote a random walk for which

$$\mu = E[S_{n+1} - S_n] \neq 0.$$

Let, for $A > 0$, $B > 0$,

$$N = \min\{n: \quad S_n \geq A \text{ or } S_n \leq -B\}.$$

Show that $E[N] < \infty$. (*Hint:* Argue that there exists a value k such that $P\{S_k > A + B\} > 0$. Then show that $E[N] \leq kE[G]$, where G is an appropriately defined geometric random variable.)

7.9. Use Jensen's inequality, which states that

$$E[f(X)] \geq f(E[X])$$

whenever f is convex to prove that if $\theta \neq 0$, $E[X] < 0$, and $E[e^{\theta X}] = 1$, then $\theta > 0$.

7.10. In the insurance ruin problem of Section 7.4 explain why the company will eventually be ruined with probability 1 if $E[Y] \geq cE[X]$.

7.11. In the ruin problem of Section 7.4 let F denote the interarrival distribution of claims and let G be the distribution of the size of a claim. Show that $p(A)$, the probability that a company starting with A units of assets is ever ruined, satisfies

$$p(A) = \int_0^\infty \int_0^{A+ct} p(A + ct - x)\, dG(x)\, dF(t) + \int_0^\infty \overline{G}(A + ct)\, dF(t).$$

7.12. For a random walk with $\mu = E[X] > 0$ argue that, with probability 1,

$$\frac{u(t)}{t} \to \frac{1}{\mu} \qquad \text{as } t \to \infty,$$

where $u(t)$ equals the number of n for which $0 \leq S_n \leq t$.

7.13. Let $S_n = \sum_1^n X_i$ be a random walk and let λ_i, $i > 0$, denote the probability that a ladder height equals i—that is, $\lambda_i = P\{\text{first positive value of } S_n \text{ equals } i\}$.

(a) Show that if

$$P\{X_i = j\} = \begin{cases} q, & j = -1 \\ \alpha_j, & j \geq 1, \end{cases} \qquad q + \sum_{j=1}^\infty \alpha_j = 1,$$

then λ_i satisfies

$$\lambda_i = \alpha_i + q(\lambda_{i+1} + \lambda_1 \lambda_i), \qquad i > 0.$$

(b) If $P\{X_i = j\} = 1/5$, $j = -2, -1, 0, 1, 2$, show that

$$\lambda_1 = \frac{1 + \sqrt{5}}{3 + \sqrt{5}}, \qquad \lambda_2 = \frac{2}{3 + \sqrt{5}}.$$

7.14. Let S_n, $n \geq 0$, denote a random walk in which X_i has distribution F. Let $G(t, s)$ denote the probability that the first value of S_n that exceeds t is less than or equal to $t + s$. That is,

$$G(t, s) = P\{\text{first sum exceeding } t \text{ is} \leq t + s\}.$$

Show that

$$G(t, s) = F(t + s) - F(t) + \int_{-\infty}^{t} G(t - y, s)\, dF(y).$$

References

Reference 5 is the standard text on random walks. Useful chapters on this subject can be found in References 1, 2, and 4. Reference 3 should be consulted for additional results both on exchangeable random variables and on some special random walks. Our proof of Spitzer's identity is from Reference 1.

1. S. Asmussen, *Applied Probability and Queues*, Wiley, New York, 1985.
2. D. R. Cox and H. D. Miller, *Theory of Stochastic Processes*, Methuen, London, 1965.
3. B. DeFinetti, *Theory of Probability*, Vols. I and II, Wiley, New York, 1970.
4. W. Feller, *An Introduction to Probability Theory and its Applications*, Wiley, New York, 1966.
5. F. Spitzer, *Principles of Random Walks*, Van Nostrand, Princeton NJ, 1964.

CHAPTER 8

Brownian Motion and Other Markov Processes

8.1 Introduction and Preliminaries

Let us start by considering the symmetric random walk that in each time unit is equally likely to take a unit step either to the left or to the right. Now suppose that we speed up this process by taking smaller and smaller steps in smaller and smaller time intervals. If we now go to the limit in the correct manner, what we obtain is Brownian motion.

More precisely suppose that each Δt time units we take a step of size Δx either to the left or to the right with equal probabilities. If we let $X(t)$ denote the position at time t, then

$$X(t) = \Delta x(X_1 + \cdots + X_{[t/\Delta t]}), \tag{8.1.1}$$

where

$$X_i = \begin{cases} +1 & \text{if the } i\text{th step of length } \Delta x \text{ is to the right} \\ -1 & \text{if it is to the left,} \end{cases}$$

and where the X_i are assumed independent with

$$P\{X_i = 1\} = P\{X_i = -1\} = \tfrac{1}{2}.$$

Since $E[X_i] = 0$, $\text{Var}(X_i) = E[X_i^2] = 1$, we see from (8.1.1) that

$$\begin{aligned} &E[X(t)] = 0, \\ &\text{Var}(X(t)) = (\Delta x)^2 \left[\frac{t}{\Delta t}\right]. \end{aligned} \tag{8.1.2}$$

We shall now let Δx and Δt go to 0. However, we must do it in a way to keep the resulting limiting process nontrivial (for instance, if we let $\Delta x = \Delta t$ and then let $\Delta t \to 0$, then from the above we see that $E[X(t)]$ and $\text{Var}(X(t))$ would both converge to 0 and thus $X(t)$ would equal 0 with probability 1). If we let $\Delta x = c\sqrt{\Delta t}$ for some positive constant c, then from (8.1.2) we see that as $\Delta t \to 0$

$$E[X(t)] = 0,$$
$$\text{Var}(X(t)) \to c^2 t.$$

We now list some intuitive properties of this limiting process obtained by taking $\Delta x = c\sqrt{\Delta t}$ and then letting $\Delta t \to 0$. From (8.1.1) and the central limit theorem we see that:

(i) $X(t)$ is normal with mean 0 and variance $c^2 t$.

In addition, as the changes of value of the random walk in nonoverlapping time intervals are independent, we have:

(ii) $\{X(t), t \geq 0\}$ has independent increments.

Finally, as the distribution of the change in position of the random walk over any time interval depends only on the length of that interval, it would appear that:

(iii) $\{X(t), t \geq 0\}$ has stationary increments.

We are now ready for the following definition.

Definition

A stochastic process $[X(t), t \geq 0]$ is said to be a *Brownian motion process* if:

(i) $X(0) = 0$;
(ii) $\{X(t), t \geq 0\}$ has stationary independent increments;
(iii) for every $t > 0$, $X(t)$ is normally distributed with mean 0 and variance $c^2 t$.

The Brownian motion process, sometimes called the Wiener process, is one of the most useful stochastic processes in applied probability theory. It originated in physics as a description of Brownian motion. This phenomenon, named after the English botanist Robert Brown, who discovered it, is the motion exhibited by a small particle that is totally immersed in a liquid or gas. Since its discovery, the process has been used beneficially in such areas as statistical testing of goodness of fit, analyzing the price levels on the stock market, and quantum mechanics.

The first explanation of the phenomenon of Brownian motion was given by Einstein in 1905. He showed that Brownian motion could be explained by assuming that the immersed particle was continually being subject to bombardment by the molecules of the surrounding medium. However, the above concise definition of this stochastic process underlying Brownian motion was given by Wiener in a series of papers originating in 1918.

When $c = 1$, the process is often called standard Brownian motion. As any Brownian motion can always be converted to the standard process by looking at $X(t)/c$, we shall suppose throughout that $c = 1$.

The interpretation of Brownian motion as the limit of the random walks (8.1.1) suggests that $X(t)$ should be a continuous function of t. This turns out to be the case, and it may be proven that, with probability 1, $X(t)$ is indeed a continuous function of t. This fact is quite deep, and no proof shall be attempted. Also, we should note in passing that while the sample path $X(t)$ is always continuous, it is in no way an ordinary function. For, as we might expect from its limiting random walk interpretation, $X(t)$ is always pointy and thus never smooth, and, in fact, it can be proven (though it's deep) that, with probability 1, $X(t)$ is nowhere differentiable.

The independent increment assumption implies that the change in position between time points s and $t + s$—that is, $X(t + s) - X(s)$—is independent of all process values before time s. Hence

$$\begin{aligned} &P\{X(t+s) \le a \mid X(s) = x, X(u), 0 \le u < s\} \\ &\quad = P\{X(t+s) - X(s) \le a - x \mid X(s) = x, X(u), 0 \le u < s\} \\ &\quad = P\{X(t+s) - X(s) \le a - x\} \\ &\quad = P\{X(t+s) \le a \mid X(s) = x\}, \end{aligned}$$

which states that the conditional distribution of a future state $X(t + s)$ given the present $X(s)$ and the past $X(u)$, $0 < u < s$, depends only on the present. A process satisfying this condition is called a *Markov* process.

Since $X(t)$ is normal with mean 0 and variance t, its density function is given by

$$f_t(x) = \frac{1}{\sqrt{2\pi t}} e^{-x^2/2t}.$$

From the stationary independent increment assumption, it easily follows that the joint density of $X(t_1), \ldots, X(t_n)$ is given by

$$\text{(8.1.3)} \quad f(x_1, x_2, \ldots, x_n) = f_{t_1}(x_1) f_{t_2 - t_1}(x_2 - x_1) \cdots f_{t_n - t_{n-1}}(x_n - x_{n-1}).$$

By using (8.1.3), we may compute in principle any desired probabilities. For instance suppose we require the conditional distribution of $X(s)$ given that $X(t) = B$, where $s < t$. The conditional density is

$$\begin{aligned} f_{s|t}(x \mid B) &= \frac{f_s(x) f_{t-s}(B - x)}{f_t(B)} \\ &= K_1 \exp\left\{\frac{-x^2}{2s} - \frac{(B - x)^2}{2(t - s)}\right\} \\ &= K_2 \exp\left\{-\frac{t(x - Bs/t)^2}{2s(t - s)}\right\}. \end{aligned}$$

Hence the conditional distribution of $X(s)$ given that $X(t) = B$ is, for $s < t$, normal with mean and variance given by

$$E[X(s) \mid X(t) = B] = Bs/t, \tag{8.1.4a}$$

$$\mathrm{Var}(X(s) \mid X(t) = B) = s(t - s)/t. \tag{8.1.4b}$$

It is interesting to note that the conditional variance of $X(s)$, given that $X(t) = B$, $s < t$, does not depend on B! That is, if we let $s/t = \alpha$, $0 < \alpha < 1$, then the conditional distribution of $X(s)$ given $X(t)$ is normal with mean $\alpha X(t)$ and variance $\alpha(1 - \alpha)t$.

It also follows from (8.1.3) that $X(t_1), \ldots, X(t_n)$ has a joint distribution that is multivariate normal, and thus a Brownian motion process is a *Gaussian* process where we have made use of the following definition.

Definition

A stochastic process $\{X(t), t \geq 0\}$ is called a *Gaussian* process if $X(t_1), \ldots, X(t_n)$ has a multivariate normal distribution for all $t_1, \ldots, t_n$.

Since a multivariate normal distribution is completely determined by the marginal mean values and the covariance values, it follows that Brownian motion could also be defined as a Gaussian process having $E[X(t)] = 0$ and, for $s \leq t$,

$$\begin{aligned} \mathrm{Cov}(X(s), X(t)) &= \mathrm{Cov}(X(s), X(s) + X(t) - X(s)) \\ &= \mathrm{Cov}(X(s), X(s)) + \mathrm{Cov}(X(s), X(t) - X(s)) \\ &= s, \end{aligned}$$

where the last equality follows from independent increments and $\mathrm{Var}(X(s)) = s$.

Let $\{X(t), t \geq 0\}$ be a Brownian motion process and consider the process values between 0 and 1 conditional on $X(1) = 0$. That is, consider the conditional stochastic process $\{X(t), 0 \leq t \leq 1 | X(1) = 0\}$. By the same argument as we used in establishing (8.1.4), we can show that this process, known as the *Brownian Bridge* (as it is tied down both at 0 and 1), is a Gaussian process. Let us compute its covariance function. Since, from (8.1.4),

$$E[X(s)|X(1) = 0] = 0 \qquad \text{for } s < 1,$$

we have that, for $s \leq t \leq 1$,

$$\begin{aligned}
\text{Cov}[(X(s), X(t))|X(1) = 0] &= E[X(s)X(t)|X(1) = 0] \\
&= E[E[X(s)X(t)|X(t), X(1) = 0]|X(1) = 0] \\
&= E[X(t)E[X(s)|X(t)]|X(1) = 0] \\
&= E\left[X(t)\frac{s}{t}X(t)|X(1) = 0\right] \qquad \text{(by (8.1.4a))} \\
&= \frac{s}{t}E[X^2(t)|X(1) = 0] \\
&= \frac{s}{t}t(1 - t) \qquad \text{(by (8.1.4b))} \\
&= s(1 - t).
\end{aligned}$$

Thus the Brownian Bridge can be defined as a Gaussian process with mean value 0 and covariance function $s(1 - t)$, $s \leq t$. This leads to an alternative approach to obtaining such a process.

PROPOSITION 8.1.1

If $\{X(t), t \geq 0\}$ is Brownian motion, then $\{Z(t), 0 \leq t \leq 1\}$ is a Brownian Bridge process when $Z(t) = X(t) - tX(1)$.

Proof Since it is immediate that $\{Z(t), t \geq 0\}$ is a Gaussian process, all we need verify is that $E[Z(t)] = 0$ and $\text{Cov}(Z(s), Z(t)) = s(1 - t)$ when $s \leq t$. The former is immediate and the latter follows from

$$\begin{aligned}
\text{Cov}(Z(s), Z(t)) &= \text{Cov}(X(s) - sX(1), X(t) - tX(1)) \\
&= \text{Cov}(X(s), X(t)) - t\,\text{Cov}(X(s), X(1)) \\
&\quad - s\,\text{Cov}(X(1), X(t)) + st\,\text{Cov}(X(1), X(1)) \\
&= s - st - st + st \\
&= s(1 - t),
\end{aligned}$$

and the proof is complete.

The Brownian Bridge plays a pivotal role in the study of empirical distribution functions. To see this let $X_1, X_2, \ldots$ be independent uniform (0, 1) random variables and define $N_n(s)$, $0 < s < 1$, as the number of the first n that are less than or equal to s. That is,

$$N_n(s) = \sum_{i=1}^{n} I_i(s),$$

where

$$I_i(s) = \begin{cases} 1 & \text{if } X_i \leq s \\ 0 & \text{otherwise.} \end{cases}$$

The random function $F_n(s) = N_n(s)/n$, $0 \leq s \leq 1$, is called the *Empirical Distribution Function.* Let us study its limiting properties as $n \to \infty$.

Since $N_n(s)$ is a binomial random variable with parameters n and s, it follows from the strong law of large numbers that, for fixed s,

$$F_n(s) \to s \qquad \text{as } n \to \infty \text{ with probability 1.}$$

In fact, it can be proven (the so-called Glivenko–Cantelli theorem) that this convergence is uniform in s. That is, with probability 1,

$$\sup_{0<s<1} |F_n(s) - s| \to 0 \qquad \text{as } n \to \infty.$$

It also follows, by the central limit theorem, that for fixed s, $\sqrt{n}(F_n(s) - s)$ has an asymptotic normal distribution with mean 0 and variance $s(1 - s)$. That is,

$$P\{\alpha_n(s) < x\} \to \frac{1}{\sqrt{2\pi s(1-s)}} \int_{-\infty}^{x} \exp\left\{\frac{-y^2}{2s(1-s)}\right\} ds,$$

where

$$\alpha_n(s) = \sqrt{n}(F_n(s) - s).$$

Let us study the limiting properties, as $n \to \infty$, of the stochastic process $\{\alpha_n(s), 0 \leq s \leq 1\}$. To start with, note that, for $s < t$, the conditional distribution of $N_n(t) - N_n(s)$, given $N_n(s)$, is just the binomial distribution with parameters $n - N_n(s)$ and $(t - s)/(1 - s)$. Hence it would seem, using the central limit theorem, that the asymptotic joint distribution of $\alpha_n(s)$ and $\alpha_n(t)$ should be a bivariate normal distribution. In fact, similar reasoning makes it plausible to expect the limiting process (if one exists) to be a Gaussian process. To see which one, we need to compute $E[\alpha_n(s)]$ and $\text{Cov}(\alpha_n(s), \alpha_n(t))$. Now,

$$E[\alpha_n(s)] = 0,$$

and, for $0 \le s \le t \le 1$,

$$\begin{aligned}
\text{Cov}(\alpha_n(s), \alpha_n(t)) &= n\,\text{Cov}(F_n(s), F_n(t)) \\
&= \frac{1}{n}\text{Cov}(N_n(s), N_n(t)) \\
&= \frac{E[E[N_n(s)N_n(t) \mid N_n(s)]] - n^2 st}{n} \\
&= \frac{E\left[N_n(s)\left(N_n(s) + (n - N_n(s))\dfrac{t-s}{1-s}\right)\right] - n^2 st}{n} \\
&= s(1-t),
\end{aligned}$$

where the last equality follows, upon simplification, from using that $N_n(s)$ is binomial with parameters n, s.

Hence it seems plausible (and indeed can be rigorously shown) that the limiting stochastic process is a Gaussian process having a mean value function equal to 0 and a covariance function given by $s(1 - t)$, $0 \le s \le t \le 1$. But this is just the Brownian Bridge process.

Whereas the above analysis has been done under the assumption that the X_i have a uniform (0, 1) distribution, its scope can be widened by noting that if the distribution function is F, then, when F is continuous, the random variables $F(X_i)$ are uniformly distributed over (0, 1). For instance, suppose we want to study the limiting distribution of

$$\sqrt{n} \sup_x |F_n(x) - F(x)|$$

for an arbitrary continuous distribution F, where $F_n(x)$ is the proportion of the first n of the X_i, independent random variables each having distribution F, that are less than or equal to x. From the preceding it follows that if we let

$$\begin{aligned}
\alpha_n(s) &= \sqrt{n}[(\text{number of } X_i, i = 1, \ldots, n\text{: } F(X_i) \le s) - s] \\
&= \sqrt{n}[(\text{number of } X_i, i = 1, \ldots, n\text{: } X_i \le F^{-1}(s)) - s] \\
&= \sqrt{n}[F_n(F^{-1}(s)) - s] \\
&= \sqrt{n}[F_n(y_s) - F(y_s)],
\end{aligned}$$

where $y_s = F^{-1}(s)$, then $\{\alpha_n(s), 0 \le s \le 1\}$ converges to the Brownian Bridge process. Hence the limiting distribution of $\sqrt{n}\,\sup_x(F_n(x) - F(x))$ is that of the supremum (or maximum, by continuity) of the Brownian Bridge. Thus,

$$\lim_{n\to\infty} P\left\{\sqrt{n}\sup_x |F_n(x) - F(x)| < a\right\} = P\left\{\max_{0\le t\le 1} |Z(t)| < a\right\},$$

where $\{Z(t), t \ge 0\}$ is the Brownian Bridge process.

8.2 Hitting Times, Maximum Variable, and Arc Sine Laws

Let us denote by T_a the first time the Brownian motion process hits a. When $a > 0$ we will compute $P\{T_a \le t\}$ by considering $P\{X(t) \ge a\}$ and conditioning on whether or not $T_a \le t$. This gives

$$\begin{aligned}(8.2.1) \qquad P\{X(t) \ge a\} &= P\{X(t) \ge a \mid T_a \le t\}P\{T_a \le t\} \\ &\quad + P\{X(t) \ge a \mid T_a > t\}P\{T_a > t\}.\end{aligned}$$

Now if $T_a \le t$, then the process hits a at some point in $[0, t]$ and, by symmetry, it is just as likely to be above a or below a at time t. That is,

$$P\{X(t) \ge a \mid T_a \le t\} = \tfrac{1}{2}.$$

Since the second right-hand term of (8.2.1) is clearly equal to 0 (since by continuity, the process value cannot be greater than a without having yet hit a), we see that

$$\begin{aligned}(8.2.2) \qquad P\{T_a \le t\} &= 2P\{X(t) \ge a\} \\ &= \frac{2}{\sqrt{2\pi t}}\int_a^\infty e^{-x^2/2t}\,dx \\ &= \frac{2}{\sqrt{2\pi}}\int_{a/\sqrt{t}}^\infty e^{-y^2/2}\,dy, \qquad a > 0.\end{aligned}$$

Hence, we see that

$$P\{T_a < \infty\} = \lim_{t\to\infty} P\{T_a \le t\} = \frac{2}{\sqrt{2\pi}}\int_0^\infty e^{-y^2/2}\,dy = 1.$$

In addition, using (8.2.2) we obtain

$$
\begin{aligned}
E[T_a] &= \int_0^\infty P\{T_a > t\}\,dt \\
&= \int_0^\infty \left(1 - \frac{2}{\sqrt{2\pi}} \int_{a/\sqrt{t}}^\infty e^{-y^2/2}\,dy\right) dt \\
&= \frac{2}{\sqrt{2\pi}} \int_0^\infty \int_0^{a/\sqrt{t}} e^{-y^2/2}\,dy\,dt \\
&= \frac{2}{\sqrt{2\pi}} \int_0^\infty \int_0^{a^2/y^2} dt\, e^{-y^2/2}\,dy \\
&= \frac{2a^2}{\sqrt{2\pi}} \int_0^\infty \frac{1}{y^2} e^{-y^2/2}\,dy \\
&\geq \frac{2a^2 e^{-1/2}}{\sqrt{2\pi}} \int_0^1 \frac{1}{y^2}\,dy \\
&= \infty.
\end{aligned}
$$

Thus it follows that T_a, though finite with probability 1, has an infinite expectation. That is, with probability 1, the Brownian motion process will eventually hit a, but its mean time is infinite. (Is this intuitive? Think of the symmetric random walk.)

For $a < 0$, the distribution of T_a is, by symmetry, the same as that of T_{-a}. Hence, from (8.2.2) we obtain

$$
P\{T_a \leq t\} = \frac{2}{\sqrt{2\pi}} \int_{|a|/\sqrt{t}}^\infty e^{-y^2/2}\,dy.
$$

Another random variable of interest is the maximum value the process attains in $[0, t]$. Its distribution is obtained as follows. For $a > 0$

$$
\begin{aligned}
P\left\{\max_{0 \leq s \leq t} X(s) \geq a\right\} &= P\{T_a \leq t\} \qquad \text{(by continuity)} \\
&= 2P\{X(t) \geq a\} \\
&= \frac{2}{\sqrt{2\pi}} \int_{a/\sqrt{t}}^\infty e^{-y^2/2}\,dy.
\end{aligned}
$$

Let $0(t_1, t_2)$ denote the event that the Brownian motion process takes on the value 0 at least once in the interval (t_1, t_2). To compute $P\{0(t_1, t_2)\}$, we condition on $X(t_1)$ as follows:

$$P\{0(t_1, t_2)\} = \frac{1}{\sqrt{2\pi t_1}} \int_{-\infty}^{\infty} P\{0(t_1, t_2) | X(t_1) = x\} e^{-x^2/2t_1} \, dx.$$

Using the symmetry of Brownian motion about the origin and its path continuity gives

$$P\{0(t_1, t_2) | X(t_1) = x\} = P\{T_{|x|} \leq t_2 - t_1\}.$$

Hence, using (8.2.2) we obtain

$$P\{0(t_1, t_2)\} = \frac{1}{\pi\sqrt{t_1(t_2 - t_1)}} \int_0^{\infty} \int_x^{\infty} e^{-y^2/2(t_2 - t_1)} \, dy \, e^{-x^2/2t_1} \, dx.$$

The above integral can be explicitly evaluated to yield

$$P\{0(t_1, t_2)\} = 1 - \frac{2}{\pi} \text{arc sine} \sqrt{t_1/t_2}.$$

Hence we have shown the following.

PROPOSITION 8.2.1

For $0 < x < 1$,

$$P\{\text{no zeros in } (xt, t)\} = \frac{2}{\pi} \text{arc sine} \sqrt{x}.$$

Remarks Proposition 8.2.1 does not surprise us. For we know by the remark at the end of Section 3.7 of Chapter 3 that for the symmetric random walk

$$P\{\text{no zeros in } (nx, n)\} \approx \frac{2}{\pi} \text{arc sine} \sqrt{x}$$

with the approximation becoming exact as $n \to \infty$. Since Brownian motion is the limiting case of symmetric random walk when the jumps come quicker and quicker (and have smaller and smaller sizes), it seems intuitive that for

Brownian motion the above approximation should hold with equality. Proposition 8.2.1 verifies this.

The other arc sine law of Section 3.7 of Chapter 3—namely, that the proportion of time the symmetric random walk is positive, obeys, in the limit, the arc sine distribution—can also be shown to be exactly true for Brownian motion. That is, the following proposition can be proven.

PROPOSITION 8.2.2

For Brownian motion, let $A(t)$ denote the amount of time in $[0, t]$ the process is positive. Then, for $0 < x < 1$,

$$P\{A(t)/t \le x\} = \frac{2}{\pi} \text{arc sine} \sqrt{x}.$$

8.3 Variations on Brownian Motion

In this section we consider four variations on Brownian motion. The first supposes that Brownian motion is absorbed upon reaching some given value, and the second assumes that it is not allowed to become negative. The third variation deals with a geometric, and the fourth, an integrated version.

8.3.1 Brownian Motion Absorbed at a Value

Let $\{X(t)\}$ be Brownian motion and recall that T_x is the first time it hits x, $x > 0$. Define $Z(t)$ by

$$Z(t) = \begin{cases} X(t) & \text{if } t < T_x \\ x & \text{if } t \ge T_x, \end{cases}$$

then $\{Z(t), t \ge 0\}$ is Brownian motion that when it hits x remains there forever.

The random variable $Z(t)$ has a distribution that has both discrete and continuous parts. The discrete part is

$$\begin{aligned} P\{Z(t) = x\} &= P\{T_x \le t\} \\ &= \frac{2}{\sqrt{2\pi t}} \int_x^\infty e^{-y^2/2t}\, dy \qquad \text{(from (8.2.2))}. \end{aligned}$$

For the continuous part we have for $y < x$

$$P\{Z(t) \le y\} = P\left\{X(t) \le y, \max_{0\le s\le t} X(s) < x\right\} \tag{8.3.1}$$

$$= P\{X(t) \le y\} - P\left\{X(t) \le y, \max_{0\le s\le t} X(s) > x\right\}.$$

We compute the second term on the right-hand side as follows:

$$P\left\{X(t) \le y, \max_{0\le s\le t} X(s) > x\right\} \tag{8.3.2}$$

$$= P\left\{X(t) \le y \,\middle|\, \max_{0\le s\le t} X(s) > x\right\} P\left\{\max_{0\le s\le t} X(s) > x\right\}.$$

Now the event that $\max_{0\le s\le t} X(s) > x$ is equivalent to the event that $T_x < t$; and if the Brownian motion process hits x at time T_x, where $T_x < t$, then in order for it to be below y at time t it must decrease by at least $x - y$ in the additional time $t - T_x$. By symmetry it is just as likely to increase by that amount. Hence,

(8.3.3)

$$P\left\{X(t) \le y \,\middle|\, \max_{0\le s\le t} X(s) > x\right\} = P\left\{X(t) \ge 2x - y \,\middle|\, \max_{0\le s\le t} X(s) > x\right\}.$$

From (8.3.2) and (8.3.3) we have

$$P\left\{X(t) \le y, \max_{0\le s\le t} X(s) > x\right\} = P\left\{X(t) \ge 2x - y, \max_{0\le s\le t} X(s) > x\right\}$$

$$= P\{X(t) \ge 2x - y\} \qquad (\text{since } y < x),$$

and from (8.3.1)

$$P\{Z(t) \le y\} = P\{X(t) \le y\} - P\{X(t) \ge 2x - y\}$$

$$= P\{X(t) \le y\} - P\{X(t) \le y - 2x\} \qquad \text{(by symmetry of the normal distribution)}$$

$$= \frac{1}{\sqrt{2\pi t}} \int_{y-2x}^{y} e^{-u^2/2t}\, du.$$

8.3.2 Brownian Motion Reflected at the Origin

If $\{X(t), t \geq 0\}$ is Brownian motion, then the process $\{Z(t), t \geq 0\}$, where

$$Z(t) = |X(t)|, \qquad t \geq 0,$$

is called Brownian motion reflected at the origin.

The distribution of $Z(t)$ is easily obtained. For $y > 0$,

$$\begin{aligned} P\{Z(t) \leq y\} &= P\{X(t) \leq y\} - P\{X(t) \leq -y\} \\ &= 2P\{X(t) \leq y\} - 1 \\ &= \frac{2}{\sqrt{2\pi t}} \int_{-\infty}^{y} e^{-x^2/2t}\, dx - 1, \end{aligned}$$

where the next to last equality follows since $X(t)$ is normal with mean 0.

The mean and variance of $Z(t)$ are easily computed and

$$E[Z(t)] = \sqrt{2t/\pi}, \qquad \operatorname{Var}(Z(t)) = \left(1 - \frac{2}{\pi}\right) t. \tag{8.3.4}$$

8.3.3 Geometric Brownian Motion

If $\{X(t), t \geq 0\}$ is Brownian motion, then the process $\{Y(t), t \geq 0\}$, defined by

$$Y(t) = e^{X(t)},$$

is called *geometric Brownian motion.*

Since $X(t)$ is normal with mean 0 and variance t, its moment generating function is given by

$$E[e^{sX(t)}] = e^{ts^2/2},$$

and so

$$\begin{aligned} E[Y(t)] &= E[e^{X(t)}] = e^{t/2} \\ \operatorname{Var}(Y(t)) &= E[Y^2(t)] - (E[Y(t)])^2 \\ &= E[e^{2X(t)}] - e^{t} \\ &= e^{2t} - e^{t}. \end{aligned}$$

Geometric Brownian motion is useful in modeling when one thinks that the percentage changes (and not the absolute changes) are independent and identically distributed. For instance, suppose that $Y(n)$ is the price of some commodity at time n. Then it might be reasonable to suppose that $Y(n)/Y(n-1)$ (as opposed to $Y_n - Y_{n-1}$) are independent and identically distributed. Letting

$$X_n = Y(n)/Y(n-1),$$

then, taking $Y(0) = 1$,

$$Y(n) = X_1 X_2 \cdots X_n,$$

and so

$$\log Y(n) = \sum_{i=1}^{n} \log X_i,$$

Since the X_i are independent and identically distributed, log $Y(n)$, when suitably normalized, would be approximately Brownian motion, and so $\{Y(n)\}$ would be approximately geometric Brownian motion.

Example 8.3(a) ***The Value of a Stock Option.*** Suppose one has the option of purchasing, at a time T in the future, one unit of a stock at a fixed price K. Supposing that the present value of the stock is y and that its price varies according to geometric Brownian motion, let us compute the expected worth of owning the option. As the option will be exercised if the stocks price at time T is K or higher, its expected worth is

$$\begin{aligned} E[\max(Y(T) - K, 0)] &= \int_0^\infty P\{Y(T) - K > a\}\, da \\ &= \int_0^\infty P\{ye^{X(T)} - K > a\}\, da \\ &= \int_0^\infty P\left\{X(T) > \log \frac{K+a}{y}\right\} da \\ &= \frac{1}{\sqrt{2\pi T}} \int_0^\infty \int_{\log[(K+a)/y]}^\infty e^{-x^2/2T}\, dx\, da. \end{aligned}$$

8.3.4 Integrated Brownian Motion

If $\{X(t), t \geq 0\}$ is Brownian motion, then the process $\{Z(t), t \geq 0\}$ defined by

$$Z(t) = \int_0^t X(s)\, ds \tag{8.3.5}$$

is called *integrated Brownian motion.* As an illustration of how such a process may arise in practice, suppose we are interested in modeling the price of a commodity throughout time. Letting $Z(t)$ denote the price at t, then, rather than assuming that $\{Z(t)\}$ is Brownian motion (or geometric Brownian motion), we might want to assume that the rate of change of $Z(t)$ follows a Brownian motion. For instance, we might suppose that the rate of change of the commodity's price is the current inflation rate, which is imagined to vary as Brownian motion. Hence,

$$\frac{d}{dt} Z(t) = X(t)$$

or

$$Z(t) = Z(0) + \int_0^t X(s)\, ds.$$

It follows from the fact that Brownian motion is a Gaussian process that $\{Z(t), t \geq 0\}$ is also Gaussian. To prove this first recall that $W_1, \ldots, W_n$ are said to have a joint normal distribution if they can be represented as

$$W_i = \sum_{j=1}^{m} a_{ij} U_j, \qquad i = 1, \ldots, n,$$

where $U_j, j = 1, \ldots, m$ are independent normal random variables. From this it follows that any set of partial sums of $W_1, \ldots, W_n$ are also jointly normal. The fact that $Z(t_1), \ldots, Z(t_n)$ is jointly normal can then be shown by writing the integral in (8.3.5) as a limit of approximating sums.

Since $\{Z(t), t \geq 0\}$ is Gaussian, it follows that its distribution is characterized by its mean value and covariance function. We now compute these:

$$\begin{aligned} E[Z(t)] &= E\left[\int_0^t X(s)\, ds\right] \\ &= \int_0^t E[X(s)]\, ds \\ &= 0. \end{aligned}$$

For $s \leq t$,

$$\begin{aligned} \text{Cov}[Z(s), Z(t)] &= E[Z(s)Z(t)] \\ &= E\left[\int_0^s X(y)\, dy \int_0^t X(u)\, du\right] \\ &= E\left[\int_0^s \int_0^t X(y)X(u)\, dy\, du\right] \\ &= \int_0^s \int_0^t E[X(y)X(u)]\, dy\, du \\ &= \int_0^s \int_0^t \min(y, u)\, dy\, du \\ &= \int_0^s \left(\int_0^u y\, dy + \int_u^t u\, dy\right) du \\ &= s^2 \left(\frac{t}{2} - \frac{s}{6}\right). \end{aligned} \tag{8.3.6}$$

The process $\{Z(t), t \geq 0\}$ defined by (8.3.5) is not a Markov process. (Why not?) However, the vector process $\{(Z(t), X(t)), t \geq 0\}$ is a Markov process. (Why?) We can compute the joint distribution of $Z(t)$, $X(t)$ by first noting, by the same reasoning as before, that they are jointly normal. To compute their covariance we use (8.3.6) as follows:

$$\begin{aligned}\operatorname{Cov}(Z(t), Z(t) - Z(t-h)) &= \operatorname{Cov}(Z(t), Z(t)) - \operatorname{Cov}(Z(t), Z(t-h)) \\ &= \frac{t^3}{3} - (t-h)^2\left[\frac{t}{2} - \frac{t-h}{6}\right] \\ &= t^2h/2 + o(h).\end{aligned}$$

However,

$$\begin{aligned}\operatorname{Cov}(Z(t), Z(t) - Z(t-h)) &= \operatorname{Cov}\left(Z(t), \int_{t-h}^{t} X(s)\, ds\right) \\ &= \operatorname{Cov}(Z(t), hX(t) + o(h)) \\ &= h\operatorname{Cov}(Z(t), X(t)) + o(h),\end{aligned}$$

and so

$$\operatorname{Cov}(Z(t), X(t)) = t^2/2.$$

Hence, $Z(t)$, $X(t)$ has a bivariate normal distribution with

$$\begin{aligned}E[Z(t)] = E[X(t)] &= 0, \\ \operatorname{Cov}(X(t), Z(t)) &= t^2/2.\end{aligned}$$

Another form of integrated Brownian motion is obtained if we suppose that the percent rate of change of price follows a Brownian motion process. That is, if $W(t)$ denotes the price at t, then

$$\frac{d}{dt} W(t) = X(t) W(t)$$

or

$$W(t) = W(0) \exp\left\{\int_0^t X(s)\, ds\right\},$$

where $\{X(t)\}$ is Brownian motion. Taking $W(0) = 1$, we see that

$$W(t) = e^{Z(t)}.$$

Since $Z(t)$ is normal with mean 0 and variance $t^2(t/2 - t/6) = t^3/3$, we see that

$$E[W(t)] = \exp\{t^6/6\}.$$

8.4 Brownian Motion With Drift

We say that $\{X(t), t \geq 0\}$ is a Brownian motion process with drift coefficient μ if:

(i) $X(0) = 0$;
(ii) $\{X(t), t \geq 0\}$ has stationary and independent increments;
(iii) $X(t)$ is normally distributed with mean μt and variance t.

We could also define it by saying that $X(t) = B(t) + \mu t$, where $\{B(t)\}$ is standard Brownian motion.

Thus a Brownian motion with drift is a process that tends to drift off at a rate μ. It can, as Brownian motion, also be defined as a limit of random walks. To see this, suppose that for every Δt time unit the process either goes one step of length Δx in the positive direction or in the negative direction, with respective probabilities p and $1 - p$. If we let

$$X_i = \begin{cases} 1 & \text{if the } i\text{th step is in the positive direction} \\ -1 & \text{otherwise,} \end{cases}$$

then $X(t)$, the position at time t, is

$$X(t) = \Delta x(X_1 + \cdots + X_{[t/\Delta t]}).$$

Now

$$E[X(t)] = \Delta x\,[t/\Delta t](2p - 1),$$
$$\text{Var}(X(t)) = (\Delta x)^2[t/\Delta t][1 - (2p - 1)^2].$$

Thus if we let $\Delta x = \sqrt{\Delta t}$, $p = \frac{1}{2}(1 + \mu\sqrt{\Delta t})$, and let $\Delta t \to 0$, then

$$E[X(t)] \to \mu t,$$
$$\text{Var}(X(t)) \to t,$$

and indeed $\{X(t)\}$ converges to Brownian motion with drift coefficient μ.

We now compute some quantities of interest for this process. We start with the probability that the process will hit A before $-B$, $A, B > 0$. Let $P(x)$ denote this probability conditionally on the event that we are now at x, $-B < x < A$. That is,

$$P(x) = P\{X(t) \text{ hits } A \text{ before } -B \mid X(0) = x\}.$$

We shall obtain a differential equation by conditioning on $Y = X(h) - X(0)$, the change in the process between time 0 and time h. This yields

$$P(x) = E[P(x + Y)] + o(h),$$

where $o(h)$ in the above refers to the probability that the process would have already hit one of the barriers, A or $-B$, by time h. Proceeding formally and assuming that $P(y)$ has a Taylor series expansion about x yields

$$P(x) = E[P(x) + P'(x)Y + P''(x)Y^2/2 + \cdots] + o(h).$$

Since Y is normal with mean μh and variance h, we obtain

$$P(x) = P(x) + P'(x)\mu h + P''(x)\frac{\mu^2 h^2 + h}{2} + o(h) \tag{8.4.1}$$

since the sum of the means of all the terms of differential order greater than 2 is $o(h)$. From (8.4.1) we have

$$P'(x)\mu + \frac{P''(x)}{2} = \frac{o(h)}{h},$$

and letting $h \to 0$,

$$P'(x)\mu + \frac{P''(x)}{2} = 0.$$

Integrating the above we obtain

$$2\mu P(x) + P'(x) = c_1,$$

or, equivalently,

$$e^{2\mu x}(2\mu P(x) + P'(x)) = c_1 e^{2\mu x}$$

or

$$\frac{d}{dx}(e^{2\mu x}P(x)) = c_1 e^{2\mu x},$$

or, upon integration,

$$e^{2\mu x}P(x) = C_1 e^{2\mu x} + C_2.$$

Thus

$$P(x) = C_1 + C_2 e^{-2\mu x}.$$

Using the boundary conditions that $P(A) = 1$, $P(-B) = 0$, we can solve for C_1 and C_2 to obtain

$$C_1 = \frac{e^{2\mu B}}{e^{2\mu B} - e^{-2\mu A}}, \qquad C_2 = \frac{-1}{e^{2\mu B} - e^{-2\mu A}},$$

and thus

$$P(x) = \frac{e^{2\mu B} - e^{-2\mu x}}{e^{2\mu B} - e^{-2\mu A}}. \tag{8.4.2}$$

Starting at $x = 0$, $P(0)$, the probability of reaching A before $-B$ is thus

$$P\{\text{process goes up } A \text{ before down } B\} = \frac{e^{2\mu B} - 1}{e^{2\mu B} - e^{-2\mu A}}. \tag{8.4.3}$$

Remarks

(1) Equation (8.4.3) could also have been obtained by using the limiting random walk argument. For by the gamblers ruin problem (see Example 4.4(A) of Chapter 4) it follows that the probability of going up A before going down B, when each gamble goes up or down Δx units with respective probabilities p and $1 - p$, is

$$P\{\text{up } A \text{ before down } B\} = \frac{1 - \left(\dfrac{1-p}{p}\right)^{B/\Delta x}}{1 - \left(\dfrac{1-p}{p}\right)^{(A+B)/\Delta x}}. \tag{8.4.4}$$

When $p = (\frac{1}{2})(1 + \mu\Delta x)$, we have

$$\lim_{\Delta x \to 0} \left(\frac{1-p}{p}\right)^{1/\Delta x} = \lim_{\Delta x \to 0} \left(\frac{1 - \mu\Delta x}{1 + \mu\Delta x}\right)^{1/\Delta x} = \frac{e^{-\mu}}{e^{\mu}}.$$

Hence, by letting $\Delta x \to 0$ we see from (8.4.4) that

$$P\{\text{up } A \text{ before down } B\} = \frac{1 - e^{-2\mu B}}{1 - e^{-2\mu(A+B)}},$$

which agrees with (8.4.3).

(2) If $\mu < 0$ we see from (8.4.3), by letting B approach infinity that

$$P\{\text{process ever goes up } A\} = e^{2\mu A}. \tag{8.4.5}$$

Thus, in this case, the process drifts off to negative infinity and its maximum is an exponential random variable with rate -2μ.

Example 8.4(a) ***Exercising a Stock Option.*** Suppose we have the option of buying, at some time in the future, one unit of a stock at a fixed price A, independent of its current market price. The current market price of the stock is taken to be 0, and we suppose that it changes in accordance with a Brownian motion process having a negative drift coefficient $-d$, where $d > 0$. The question is, when, if ever, should we exercise our option?

Let us consider the policy that exercises the option when the market price is x. Our expected gain under such a policy is

$$P(x)(x - A),$$

where $P(x)$ is the probability that the process will ever reach x. From (8.4.5) we see that

$$P(x) = e^{-2dx}, \qquad x > 0,$$

the optimal value of x is the one maximizing $(x - A)e^{-2dx}$, and this is easily seen to be

$$x = A + 1/2d.$$

For Brownian motion we obtain by letting $\mu \to 0$ in Equation (8.4.3)

$$P\{\text{Brownian motion goes up } A \text{ before down } B\} = \frac{B}{A + B}. \tag{8.4.6}$$

Example 8.4(b) ***Optimal Doubling in Backgammon.*** Consider two individuals that for a stake are playing some game of chance that eventually ends with one of the players being declared the winner. Initially one of the players is designated as the "doubling player," which means that at any time he has the option of doubling the stakes. If at any time he exercises his option, then the other player can either quit and pay the present stake to the doubling player or agree to continue playing for twice the old stakes. If the other player decides to continue playing, then that player becomes the "doubling player." In other words, each time the doubling player exercises the option, the option then switches to the other player. A popular game often played with a doubling option is backgammon.

We will suppose that the game consists of watching Brownian motion starting at the value $\frac{1}{2}$. If it hits 1 before 0, then player I wins, and if the reverse occurs, then player II wins. From (8.4.6) it follows that if the present state is x, then, if the game is to be continued until conclusion, player I will win with probability x. Also each player's objective is to maximize his expected return, and we will suppose that each player plays optimally in the game theory sense (this means, for instance, that a player can announce his strategy and the other player could not do any better even knowing this information).

It is intuitively clear that the optimal strategies should be of the following type.

OPTIMAL STRATEGIES

Suppose player I(II) has the option of doubling, then player I(II) should double at time t iff (if and only if) $X(t) \geq p^*$ ($X(t) \leq 1 - p^*$). Player II's (I's) optimal strategy is to accept a double at t iff $X(t) \leq p^{**}$ ($X(t) \geq 1 - p^{**}$). It remains to compute p^* and p^{**}.

Lemma 1

$$p^* \leq p^{**}$$

Proof For any $p > p^{**}$, it follows that player II would quit if $X(t) = p$ and player I doubles. Hence at $X(t) = p$ player I can guarantee himself an expected gain of the present stake by doubling, and since player II can always guarantee that I never receives more (by quitting if player I ever doubles), it follows that it must be optimal for player I to double. Hence $p^* \leq p^{**}$.

Lemma 2

$$p^* = p^{**}$$

Proof Suppose $p^* < p^{**}$. We will obtain a contradiction by showing that player I has a better strategy than p^*. Specifically, rather than doubling at p^*, player I can do better by waiting for $X(t)$ to either hit 0 or p^{**}. If it hits p^{**}, then he can double, and since player II will accept, player I will be in the same position as if he doubled at p^*. On the other hand, if 0 is hit before p^{**}, then under the new policy he will only lose the original stake whereas under the p^* policy he would have lost double the stake.

Thus from Lemmas 1 and 2 there is a single critical value p^* such that if player I has the option, then his optimal strategy is to double at t iff $X(t) \geq p^*$. Similarly, player II's optimal strategy is to accept at t iff $X(t) \leq p^*$. By continuity it follows that both players are indifferent as to their choices when the state is p^*. To compute p^*, we will take advantage of their indifference.

Let the stake be 1 unit. Now if player I doubles at p^*, then player II is indifferent as to quitting or accepting the double. Hence, since player I wins 1 under the former alternative, we have

$$1 = E[\text{gain to player I if player II accepts at } p^*].$$

Now if player II accepts at p^*, then II has the option of the next double, which he will exercise if $X(t)$ ever hits $1 - p^*$. If it never hits $1 - p^*$ (that is, if it hits 1 first), then II will lose 2 units. Hence, since the probability of hitting $1 - p^*$ before 1, when starting at p^* is, by (8.4.6), $(1 - p^*)/p^*$, we have

$$1 = E[\text{gain to I}|\text{hits } 1 - p^*]\frac{1 - p^*}{p^*} + 2\frac{2p^* - 1}{p^*}.$$

Now if it hits $1 - p^*$, then II will double the stakes to 4 units and I will be indifferent about accepting or not. Hence, as I will lose 2 if he quits, we have

$$E[\text{gain to I}|\text{hits } 1 - p^*] = -2,$$

and so

$$1 = -2\frac{1 - p^*}{p^*} + 2\frac{2p^* - 1}{p^*}$$

or

$$p^* = \tfrac{4}{5}.$$

Example 8.4(c) ***Controlling a Production Process.*** In this example, we consider a production process that tends to deteriorate with time. Specifically, we suppose that the production process changes its state in accordance with a Wiener process with drift coefficient μ, $\mu > 0$. When the state of the process is B, the process is assumed to break down and a cost R must be paid to return the process back to state 0. On the other hand, we may attempt to repair the process before it reaches the breakdown point B. If the state is x and an attempt to repair the process is made, then this attempt will succeed with probability α_x and fail with probability $1 - \alpha_x$. If the attempt is successful, then the process returns to state 0, and if it is unsuccessful, then we assume that the process goes to B (that is, it breaks down). The cost of attempting a repair is C.

We shall attempt to determine the policy that minimizes the long-run average cost per time, and in doing so we will restrict attention to policies that attempt a repair when the state of the process is x, $0 < x < B$. For these policies, it is clear that returns to state 0 constitute renewals, and thus by Theorem 3.6.1 of Chapter

3 the average cost is just

$$\text{(8.4.7)} \qquad \frac{E[\text{cost of a cycle}]}{E[\text{length of a cycle}]} = \frac{C + R(1 - \alpha_x)}{E[\text{time to reach } x]}.$$

Let $f(x)$ denote the expected time that it takes the process to reach x. We derive a differential equation for $f(x)$ by conditioning on $Y = X(h) - X(0)$ the change in time h. This yields

$$f(x) = h + E[f(x - Y)] + o(h),$$

where the $o(h)$ term represents the probability that the process would have already hit x by time h. Expanding in a Taylor series gives

$$\begin{aligned} f(x) &= h + E\left[f(x) - Yf'(x) + \frac{Y^2}{2}f''(x) + \cdots\right] + o(h) \\ &= h + f(x) - \mu h f'(x) + \frac{h}{2}f''(x) + o(h), \end{aligned}$$

or, equivalently,

$$1 = \mu f'(x) - \frac{f''(x)}{2} + \frac{o(h)}{h},$$

and letting $h \to 0$

$$\text{(8.4.8)} \qquad 1 = \mu f'(x) - f''(x)/2.$$

Rather than attempting to solve the above directly note that

$$\begin{aligned} f(x + y) &= E[\text{time to reach } x + y \text{ from } 0] \\ &= E[\text{time to reach } x] + E[\text{time to reach } x + y \text{ from } x] \\ &= f(x) + f(y). \end{aligned}$$

Hence, $f(x)$ is of the form $f(x) = cx$, and from (8.4.8) we see that $c = 1/\mu$. Therefore,

$$f(x) = x/\mu.$$

Hence, from (8.4.7), the policy that attempts to repair when the state is x, $0 < x < B$, has a long-run average cost of

$$\frac{\mu[C + R(1 - \alpha_x)]}{x}$$

while the policy that never attempts to repair has a long-run average cost of

$$R\mu/B.$$

For a given function $\alpha(x)$, we can then use calculus to determine the policy that minimizes the long-run average cost.

Let T_x denote the time it takes the Brownian motion process with drift coefficient μ to hit x when $\mu > 0$. We shall compute $E[e^{-\theta T_x}]$, $\theta > 0$, its moment generating function, for $x > 0$, in much the same manner as $E[T_x]$ was computed in Example 8.4(C). We start by noting

$$\begin{aligned}
(8.4.9) \qquad E[\exp\{-\theta T_{x+y}\}] &= E[\exp\{-\theta(T_x + T_{x+y} - T_x)\}] \\
&= E[\exp\{-\theta T_x\}]E[\exp\{-\theta(T_{x+y} - T_x)\}] \\
&= E[\exp\{-\theta T_x\}]E[\exp\{-\theta T_y\}],
\end{aligned}$$

where the last equality follows from stationary and the next to last from independent increments. But (8.4.9) implies

$$E[e^{-\theta T_x}] = e^{-cx}$$

for some $c > 0$. To determine c let

$$f(x) = E[e^{-\theta T_x}].$$

We will obtain a differential equation satisfied by f by conditioning on $Y = X(h) - X(0)$. This yields

$$\begin{aligned}
f(x) &= E[\exp\{-\theta(h + T_{x-Y})\}] + o(h) \\
&= e^{-\theta h}E[f(x - Y)] + o(h),
\end{aligned}$$

where the term $o(h)$ results from the possibility that the process hits x by time h. Expanding the above in a Taylor series about x yields

$$\begin{aligned}
f(x) &= e^{-\theta h}E\left[f(x) - Yf'(x) + \frac{Y^2}{2!}f''(x) + \cdots\right] + o(h) \\
&= e^{-\theta h}\left[f(x) - \mu h f'(x) + \frac{h}{2}f''(x)\right] + o(h).
\end{aligned}$$

Using $e^{-\theta h} = 1 - \theta h + o(h)$ now gives

$$f(x) = f(x)(1 - \theta h) - \mu h f'(x) + \frac{h}{2}f''(x) + o(h).$$

Dividing by h and letting $h \to 0$ yields

$$\theta f(x) = -\mu f'(x) + \tfrac{1}{2}f''(x),$$

and, using $f(x) = e^{-cx}$,

$$\theta e^{-cx} = \mu c e^{-cx} + \frac{c^2}{2} e^{-cx}$$

or

$$c^2 + 2\mu c - 2\theta = 0.$$

Hence, we see that either

$$c = -\mu + \sqrt{\mu^2 + 2\theta} \quad \text{or} \quad c = -\mu - \sqrt{\mu^2 + 2\theta}. \tag{8.4.10}$$

And, since $c > 0$, we see that when $\mu \geq 0$

$$c = \sqrt{\mu^2 + 2\theta} - \mu.$$

Thus we have the following.

PROPOSITION 8.4.1

Let T_x denote the time that Brownian motion with drift coefficient μ hits x. Then for $\theta > 0$, $x > 0$,

$$E[\exp\{-\theta T_x\}] = \exp\{-x(\sqrt{\mu^2 + 2\theta} - \mu)\} \text{ if } \mu \geq 0. \tag{8.4.11}$$

We will end this section by studying the limiting average value of the maximum variable. Specifically we have the following.

PROPOSITION 8.4.2

If $\{X(t), t \geq 0\}$ is a Brownian motion process with drift coefficient μ, $\mu \geq 0$, then, with probability 1,

$$\lim_{t \to \infty} \frac{\max_{0 \leq s \leq t} X(s)}{t} = \mu.$$

Proof Let $T_0 = 0$, and for $n > 0$ let T_n denote the time at which the process hits n. It follows, from the assumption of stationary, independent increments, that $T_n - T_{n-1}$, $n \geq 1$, are independent and identically distributed. Hence, we may think of the T_n as being the times at which events occur in a renewal process. Letting $N(t)$ be the number of such renewals by t, we have

$$N(t) \leq \max_{0 \leq s \leq t} X(s) \leq N(t) + 1. \tag{8.4.12}$$

Now, from the results of Example 8.4(C), we have $ET_1 = 1/\mu$, and hence the result follows from (8.4.12) and the well-known renewal result $N(t)/t \to 1/ET_1$.

8.4.1 Using Martingales to Analyze Brownian Motion

Brownian motion with drift can also be analyzed by using martingales. There are three important martingales associated with standard Brownian motion.

PROPOSITION 8.4.3

Let $\{B(t), t \geq 0\}$ be standard Brownian motion. Then $\{Y(t), t \geq 0\}$ is a martingale when

(a) $Y(t) = B(t)$,
(b) $Y(t) = B^2(t) - t$, and
(c) $Y(t) = \exp\{cB(t) - c^2t/2\}$,

where c is any constant. The martingales in parts (a) and (b) have mean 0, and the one in part (c) has mean 1.

Proof In all cases we write $B(t)$ as $B(s) + [B(t) - B(s)]$ and utilize independent increments.

(a)
$$\begin{aligned} E[B(t)|B(u), 0 \leq u \leq s] &= E[B(s)|B(u), 0 \leq u \leq s] \\ &\quad + E[B(t) - B(s)|B(u), 0 \leq u \leq s] \\ &= B(s) + E[B(t) - B(s)] \\ &= B(s) \end{aligned}$$

where the next to last equality made use of the independent increments property of Brownian motion.

(b) $E[B^2(t)|B(u), 0 \le u \le s] = E[\{B(s) + B(t) - B(s)\}^2|B(u), 0 \le u \le s]$

$$= B^2(s) + 2B(s)E[B(t) - B(s)|B(u), 0 \le u \le s] + E[\{B(t) - B(s)\}^2|B(u), 0 \le u \le s]$$

$$= B^2(s) + 2B(s)E[B(t) - B(s)] + E[\{B(t) - B(s)\}^2]$$

$$= B^2(s) + E[B^2(t - s)]$$

$$= B^2(s) + t - s$$

which verifies that $B^2(t) - t$ is a martingale.

We leave the verification that $\exp\{cB(t) - c^2t/2\}, t \ge 0$ is a martingale as an exercise.

Now, let $X(t) = B(t) + \mu t$ and so $\{X(t), t \ge 0\}$ is Brownian motion with drift μ. For positive A and B, define the stopping time T by

$$T = \min\{t\colon X(t) = A \quad \text{or} \quad X(t) = -B\}.$$

We will find $P_A \equiv P\{X(T) = A\}$ by making use of the martingale in Proposition 8.4.3(C), namely, $Y(t) = \exp\{cB(t) - c^2t/2\}$. Since this martingale has mean 1, it follows from the martingale stopping theorem that

$$E[\exp\{cB(T) - c^2T/2] = 1$$

or, since $B(T) = X(T) - \mu T$,

$$E[\exp\{cX(T) - c\mu T - c^2T/2] = 1.$$

Letting $c = -2\mu$ gives

$$E[\exp\{-2\mu X(T)\}] = 1.$$

But, $X(T)$ is either A or $-B$, and so we obtain that

$$e^{-2\mu A}P_A + e^{2\mu B}(1 - P_A) = 1$$

and so,

$$P_A = \frac{e^{2\mu B} - 1}{e^{2\mu B} - e^{-2\mu A}}$$

thus verifying the result of Equation (8.4.3).

If we now use the fact that $\{B(t), t \geq 0\}$ is a zero-mean martingale then, by the stopping theorem,

$$\begin{aligned} 0 = E[B(T)] &= E[X(T) - \mu T] \\ &= E[X(T)] - \mu E[T] \\ &= AP_A - B(1 - P_A) - \mu E[T]. \end{aligned}$$

Using the preceding formula for P_A gives that

$$E[T] = \frac{Ae^{2\mu B} + Be^{-2\mu A} - A - B}{\mu[e^{2\mu B} - e^{-2\mu A}]}.$$

8.5 Backward and Forward Diffusion Equations

The derivation of differential equations is a powerful technique for analyzing Markov processes. There are two general techniques for obtaining differential equations: the backwards and the forwards technique. For instance suppose we want the density of the random variable $X(t)$. The backward approach conditions on the value of $X(h)$—that is, it looks all the way back to the process at time h. The forward approach conditions on $X(t - h)$.

As an illustration, consider a Brownian motion process with drift coefficient μ and let $p(x, t; y)$ denote the probability density of $X(t)$, given $X(0) = y$. That is,

$$p(x, t; y) = \lim_{\Delta x \to 0} P\{x < X(t) < x + \Delta x | X(0) = y\}/\Delta x.$$

The backward approach is to condition on $X(h)$. Acting formally as if $p(x, t; y)$ were actually a probability, we have

$$p(x, t; y) = E[P\{X(t) = x | X(0) = y, X(h)\}].$$

Now

$$P\{X(t) = x | X(0) = y, X(h) = x_h\} = P\{X(t - h) = x | X(0) = x_h\},$$

and so

$$p(x, t; y) = E[p(x, t - h; X(h))],$$

where the expectation is with respect to $X(h)$, which is normal with mean $\mu h + y$ and variance h. Assuming that we can expand the right-hand side of the above in the Taylor series about $(x, t; y)$, we obtain

$$\begin{aligned}
p(x,t;y) &= E\Bigg[p(x,t;y) - h\frac{\partial}{\partial t}p(x,t;y) \\
&\quad + (X(h)-y)\frac{\partial}{\partial y}p(x,t;y) + \frac{h^2}{2}\frac{\partial^2}{\partial t^2}p(x,t;y) \\
&\quad + \frac{(X(h)-y)^2}{2}\frac{\partial^2}{\partial y^2}p(x,t;y) + \cdots\Bigg] \\
&= p(x,t;y) - h\frac{\partial}{\partial t}p(x,t;y) + \mu h\frac{\partial}{\partial y}p(x,t;y) \\
&\quad + \frac{h}{2}\frac{\partial^2}{\partial y^2}p(x,t;y) + o(h).
\end{aligned}$$

Dividing by h and then letting it approach 0 gives

$$\frac{1}{2}\frac{\partial^2}{\partial y^2}p(x,t;y) + \mu\frac{\partial}{\partial y}p(x,t;y) = \frac{\partial}{\partial t}p(x,t;y). \tag{8.5.1}$$

Equation (8.5.1) is called the backward diffusion equation.

The forward equation is obtained by conditioning on $X(t-h)$. Now

$$\begin{aligned}
P\{X(t) = x | X(0) = y, X(t-h) = a\} &= P\{X(h) = x | X(0) = a\} \\
&= P\{W = x - a\},
\end{aligned}$$

where W is a normal random variable with mean μh and variance h. Letting its density be f_W, we thus have

$$\begin{aligned}
p(x,t;y) &= \int f_W(x-a)p(a,t-h;y)\,da \\
&= \int\Bigg[p(x,t;y) + (a-x)\frac{\partial}{\partial x}p(x,t;y) - h\frac{\partial}{\partial t}p(x,t;y) \\
&\quad + \frac{(a-x)^2}{2}\frac{\partial^2}{\partial x^2}p(x,t;y) + \cdots\Bigg] f_W(x-a)\,da \\
&= p(x,t;y) - \mu h\frac{\partial}{\partial x}p(x,t;y) - h\frac{\partial}{\partial t}p(x,t;y) \\
&\quad + \frac{h}{2}\frac{\partial^2}{\partial x^2}p(x,t;y) + o(h).
\end{aligned}$$

Dividing by h and then letting it go to zero yields

$$\frac{1}{2}\frac{\partial^2}{\partial x^2}p(x, t; y) = \mu\frac{\partial}{\partial x}p(x, t; y) + \frac{\partial}{\partial t}p(x, t; y). \tag{8.5.2}$$

Equation (8.5.2) is called the forward diffusion equation.

8.6 Applications of the Kolmogorov Equations to Obtaining Limiting Distributions

The forward differential equation approach, which was first employed in obtaining the distribution of $N(t)$ for a Poisson process, is useful for obtaining limiting distributions in a wide variety of models. This approach derives a differential equation by computing the probability distribution of the system state at time $t + h$ in terms of its distribution at time t, and then lets $t \to \infty$. We will now illustrate its use in some models, the first of which has previously been studied by other methods.

8.6.1 Semi-Markov Processes

A semi-Markov process is one that, when it enters state i, spends a random time having distribution H_i and mean μ_i in that state before making a transition. If the time spent in state i is x, then the transition will be into state j with probability $P_{ij}(x)$, $i, j \geq 0$. We shall suppose that all the distributions H_i are continuous and we define the hazard rate function $\lambda_i(t)$ by

$$\lambda_i(t) = h_i(t)/\overline{H}_i(t),$$

where h_i is the density of H_i. Thus the conditional probability that the process will make a transition within the next dt time units, given that it has spent t units in state i, is $\lambda_i(t)\,dt + o(dt)$.

We can analyze the semi-Markov process as a Markov process by letting the "state" at any time be the pair (i, x) with i being the present state and x the amount of time the process has spent in state i since entering. Let

$$p_t(i, x) = \lim_{h \to 0} \frac{P\left\{\begin{array}{l}\text{at time } t \text{ state is } i \text{ and time since} \\ \text{entering is between } x - h \text{ and } x\end{array}\right\}}{h}.$$

That is, $p_t(i, x)$ is the probability density that the state at time t is (i, x).

For $x > 0$ we have that

$$p_{t+h}(i, x + h) = p_t(i, x)(1 - \lambda_i(x)h) + o(h) \tag{8.6.1}$$

since in order to be in state $(i, x + h)$ at time $t + h$ the process must have been in state (i, x) at time t and no transitions must have occurred in the next h time units. Assuming that the limiting density $p(i, x) = \lim_{t\to\infty} p_t(i, x)$ exists, we have from (8.6.1), by letting $t \to \infty$,

$$\frac{p(i, x + h) - p(i, x)}{h} = -\lambda_i(x)p(i, x) + \frac{o(h)}{h}.$$

And letting $h \to 0$, we have

$$\frac{d}{dx}p(i, x) = -\lambda_i(x)p(i, x).$$

Dividing by $p(i, x)$ and integrating yields

$$\log\left(\frac{p(i, x)}{p(i, 0)}\right) = -\int_0^x \lambda_i(y)\, dy$$

or

$$p(i, x) = p(i, 0)\exp\left(-\int_0^x \lambda_i(y)\, dy\right).$$

The identity (see Section 1.6 of Chapter 1)

$$\overline{H}_i(x) = \exp\left(-\int_0^x \lambda_i(y)\, dy\right)$$

thus yields

$$p(i, x) = p(i, 0)\overline{H}_i(x). \tag{8.6.2}$$

In addition, since the process will instantaneously go from state (j, x) to state $(i, 0)$ with probability intensity $\lambda_j(x)P_{ji}(x)$, we also have

$$\begin{aligned}
p(i, 0) &= \sum_j \int_x p(j, x)\lambda_j(x)P_{ji}(x)\, dx \\
&= \sum_j p(j, 0)\int_x \overline{H}_j(x)\lambda_j(x)P_{ji}(x)\, dx \qquad \text{(from (8.6.2))} \\
&= \sum_j p(j, 0)\int_x h_j(x)P_{ji}(x)\, dx.
\end{aligned}$$

Now $\int_x h_j(x)P_{ji}(x)\,dx$ is just the probability that when the process enters state j it will next enter i. Hence, calling that probability P_{ji}, we have

$$P(i,0) = \sum_j p(j,0)P_{ji}.$$

If we now suppose that the Markov chain of successive states, which has transition probabilities P_{ji}, is ergodic and has limiting probabilities π_i, $i \geq 0$, then since $p(i, 0)$, $i \geq 0$, satisfy the stationarity equations, it follows that, for some constant c,

$$p(i,0) = c\pi_i, \qquad \text{all } i. \tag{8.6.3}$$

From (8.6.2) we obtain, by integrating over x,

$$\begin{aligned} P\{\text{state is } i\} &= \int p(i,x)\,dx \\ &= p(i,0)\mu_i \qquad \text{(from (8.6.2))} \\ &= c\pi_i\mu_i \qquad \text{(from (8.6.3))}. \end{aligned} \tag{8.6.4}$$

Since $\sum_i P\{\text{state is } i\} = 1$, we see that

$$c = \frac{1}{\sum_i \pi_i\mu_i},$$

and so, from (8.6.2) and (8.6.3),

$$p(i,x) = \frac{\pi_i\mu_i}{\sum_i \pi_i\mu_i}\frac{\overline{H}_i(x)}{\mu_i}. \tag{8.6.5}$$

From (8.6.4) we note that

$$P\{\text{state is } i\} = \frac{\pi_i\mu_i}{\sum_i \pi_i\mu_i}, \tag{8.6.6}$$

and, from (8.6.5),

$$P\{\text{time in state} \leq y|\text{state is } i\} = \int_0^y \frac{\overline{H}_i(y)}{\mu_i}\,dy.$$

Thus the limiting probability of being in state i is as given by (8.6.6) and agrees with the result of Chapter 4; and, given that the state is i, the time already in that state has the equilibrium distribution of H_i.

8.6.2 The $M/G/1$ Queue

Consider the $M/G/1$ queue where arrivals are at a Poisson rate λ, and there is a single server whose service distribution is G, and suppose that G is continuous and has hazard rate function $\lambda(t)$. This model can be analyzed as a Markov process by letting the state of any time be the pair (n, x) with n denoting the number in the system at that time and x the amount of time the person being served has already been in service.

Letting $p_t(n, x)$ denote the density of the state at time t, we have, when $n \geq 1$, that

(8.6.7)
$$p_{t+h}(n, x + h) = p_t(n, x)(1 - \lambda(x)h)(1 - \lambda h) + p_t(n - 1, x)\lambda h + o(h).$$

The above follows since the state at time $t + h$ will be $(n, x + h)$ if either (a) the state at time t is n, x and in the next h time units there are no arrivals and no service completions, or if (b) the state at time t is $(n - 1, x)$ and there is a single arrival and no departures in the next h time units.

Assuming that the limiting density $p(n, x) = \lim_{t\to\infty} p_t(n, x)$ exists, we obtain from (8.6.7)

$$\frac{p(n, x + h) - p(n, x)}{h} = -(\lambda + \lambda(x))p(n, x) + \lambda p(n - 1, x) + \frac{o(h)}{h},$$

and upon letting $h \to 0$

$$\text{(8.6.8)} \qquad \frac{d}{dx}p(n, x) = -(\lambda + \lambda(x))p(n, x) + \lambda p(n - 1, x), \qquad n \geq 1.$$

Let us now define the generating function $G(s, x)$ by

$$G(s, x) = \sum_{n=1}^{\infty} s^n p(n, x).$$

Differentiation yields

$$\begin{aligned}
\frac{\partial}{\partial x} G(s, x) &= \sum_{n=1}^{\infty} s^n \frac{d}{dx} p(n, x) \\
&= \sum_{n=1}^{\infty} s^n [(-\lambda - \lambda(x))p(n, x) + \lambda p(n - 1, x)] \qquad \text{(from (8.6.8))} \\
&= (\lambda s - \lambda - \lambda(x))G(s, x).
\end{aligned}$$

Dividing both sides by $G(s, x)$ and integrating yields

$$\log\left(\frac{G(s, x)}{G(s, 0)}\right) = (\lambda s - \lambda)x - \int_0^x \lambda(y)\, dy$$

or

$$G(s, x) = G(s, 0)e^{-\lambda(1-s)x}\overline{G}(x), \tag{8.6.9}$$

where the above has made use of the identity

$$\overline{G}(x) = \exp\left\{-\int_0^x \lambda(y)\,dy\right\}.$$

To obtain $G(s, 0)$, note that the equation for $p(n, 0)$, $n > 0$, is

$$p(n, 0) = \begin{cases} \int p(n+1, x)\lambda(x)\,dx & n > 1 \\ \int p(n+1, x)\lambda(x)\,dx + P(0)\lambda & n = 1, \end{cases}$$

where

$$P(0) = P\{\text{system is empty}\}.$$

Thus

$$\sum_{n=1}^{\infty} s^{n+1} p(n, 0) = \int \sum_{n=1}^{\infty} s^{n+1} p(n+1, x)\lambda(x)\,dx + s^2\lambda P(0)$$

or

$$\begin{aligned} sG(s, 0) &= \int (G(s, x) - sp(1, x))\lambda(x)\,dx + s^2\lambda P(0) \\ &= G(s, 0)\int e^{-\lambda(1-s)x} g(x)\,dx - s\int p(1, x)\lambda(x)\,dx + s^2\lambda P(0), \end{aligned} \tag{8.6.10}$$

where the last equality is a result of Equation (8.6.9). To evaluate the second term on the right-hand side of (8.6.10), we derive an equation for $P(0)$ as follows:

$$P\{\text{empty at } t + h\} = P\{\text{empty at } t\}(1 - \lambda h) + \int \lambda(x) h p_t(1, x)\,dx + o(h).$$

Letting $t \to \infty$ and then letting $h \to 0$ yields

$$\lambda P(0) = \int \lambda(x) p(1, x)\,dx. \tag{8.6.11}$$

Substituting this back in (8.6.10), we obtain

$$sG(s, 0) = G(s, 0)\tilde{G}(\lambda(1 - s)) - s\lambda(1 - s)P(0)$$

or

$$G(s, 0) = \frac{s\lambda(1-s)P(0)}{\tilde{G}(\lambda(1-s)) - s} \tag{8.6.12}$$

where $\tilde{G}(s) = \int e^{-xs}\, dG(x)$ is the Laplace transform of the service distribution.

To obtain the marginal probability generating function of the number in the system, integrate (8.6.9) as follows:

$$\begin{aligned}
\sum_{n=1}^{\infty} s^n P\{n \text{ in system}\} &= \int_0^\infty G(s, x)\, dx \\
&= G(s, 0) \int_0^\infty e^{-\lambda(1-s)x} \overline{G}(x)\, dx \\
&= G(s, 0) \int_0^\infty e^{-\lambda(1-s)x} \int_x^\infty dG(y)\, dx \\
&= G(s, 0) \int_0^\infty \int_0^y e^{-\lambda(1-s)x}\, dx\, dG(y) \\
&= \frac{G(s, 0)}{\lambda(1-s)} \int_0^\infty (1 - e^{-\lambda(1-s)y})\, dG(y) \\
&= \frac{G(s, 0)(1 - \tilde{G}(\lambda(1-s)))}{\lambda(1-s)}.
\end{aligned}$$

Hence, from (8.6.12)

$$\begin{aligned}
\sum_{n=0}^{\infty} s^n P\{n \text{ in system}\} &= P(0) + \frac{sP(0)(1 - \tilde{G}(\lambda(1-s)))}{\tilde{G}(\lambda(1-s)) - s} \\
&= \frac{P(0)(1-s)\tilde{G}(\lambda(1-s))}{\tilde{G}(\lambda(1-s)) - s}.
\end{aligned}$$

To obtain the value of $P(0)$, let s approach 1 in the above. This yields

$$\begin{aligned}
1 &= \sum_{n=0}^{\infty} P\{n \text{ in system}\} \\
&= P(0) \lim_{s \to 1} \frac{(1-s)\tilde{G}(\lambda(1-s))}{\tilde{G}(\lambda(1-s)) - s} \\
&= P(0) \frac{\lim_{s \to 1} \frac{d}{ds}(1-s)}{\lim_{s \to 1} \frac{d}{ds}[\tilde{G}(\lambda(1-s)) - s]} \qquad \text{(by L'hopital's rule since } \tilde{G}(0) = 1\text{)} \\
&= \frac{P(0)}{1 - \lambda E[S]}
\end{aligned}$$

or

$$P_0 = 1 - \lambda E[S],$$

where $E[S] = \int x\, dG(x)$ is the mean service time.

Remarks

(1) We can also attempt to obtain the functions $p(n, x)$ recursively starting with $n = 1$ and then using (8.6.8) for the recursion. For instance, when $n = 1$, the second right-hand term in (8.6.8) disappears and so we end up with

$$\frac{d}{dx}p(1,x) = -(\lambda + \lambda(x))p(1,x).$$

Solving this equation yields

$$p(1, x) = p(1, 0)e^{-\lambda x}\overline{G}(x). \tag{8.6.13}$$

Thus

$$\int \lambda(x)p(1,x)\, dx = p(1,0) \int e^{-\lambda x}g(x)\, dx,$$

and, using (8.6.11), we obtain

$$\lambda P(0) = p(1, 0)\tilde{G}(\lambda).$$

Since $P(0) = 1 - \lambda E[S]$, we see that

$$p(1,0) = \frac{\lambda(1 - \lambda E[S])}{\tilde{G}(\lambda)}.$$

Finally, using (8.6.13), we have

$$p(1,x) = \frac{\lambda e^{-\lambda x}(1 - \lambda E[S])\overline{G}(x)}{\tilde{G}(\lambda)}. \tag{8.6.14}$$

This formula may now be substituted into (8.6.8) and the differential equation for $p(2, x)$ can be solved, at least in theory. We could then attempt to use $p(2, x)$ to solve for $p(3, x)$, and so on.

(2) It follows from (8.6.14) that $p(y|1)$, the conditional density of time the person being served has already spent in service, given a single customer in the system, is given by

$$p(y|1) = \frac{e^{-\lambda y}\overline{G}(y)}{\int e^{-\lambda y}\overline{G}(y)\,dy}.$$

In the special case where $G(y) = 1 - e^{-\mu y}$, we have

$$p(y|1) = (\lambda + \mu)e^{-(\lambda+\mu)y}.$$

Hence, the conditional distribution is exponential with rate $\lambda + \mu$ and thus is not the equilibrium distribution (which, of course, is exponential with rate μ).

(3) Of course, the above analysis requires that $\lambda E[S] < 1$ for the limiting distributions to exist.

Generally, if we are interested in computing a limiting probability distribution of a Markov process $\{X(t)\}$, then the appropriate approach is by way of the forward equations. On the other hand, if we are interested in a first passage time distribution, then it is usually the backward equation that is most valuable. That is in such problems we condition on what occurs the first h time units.

8.6.3 A Ruin Problem in Risk Theory

Assume that $N(t)$, the number of claims received by an insurance company by time t, is a Poisson process with rate λ. Suppose also that the dollar amount of successive claims are independent and have distribution G. If we assume that cash is received by the insurance company at a constant rate of 1 per unit time, then its cash balance at time t can be expressed as

$$\text{cash balance at } t = x + t - \sum_{i=1}^{N(t)} Y_i,$$

where x is the initial capital of the company and Y_i, $i \geq 1$, are the successive claims. We are interested in the probability, as a function of the initial capital x, that the company always remains solvent. That is, we wish to determine

$$R(x) = P\left\{x + t - \sum_{i=1}^{N(t)} Y_i > 0 \text{ for all } t\right\}.$$

To obtain a differential equation for $R(x)$ we shall use the backward approach and condition on what occurs the first h time units. If no claims are made, then the company's assets are $x + h$; whereas if a single claim is made, they are $x + h - Y$. Hence,

$$R(x) = R(x + h)(1 - \lambda h) + E[R(x + h - Y)]\lambda h + o(h),$$

and so

$$\frac{R(x + h) - R(x)}{h} = \lambda R(x + h) - \lambda E[R(x + h - Y)] + \frac{o(h)}{h}.$$

Letting $h \to 0$ yields

$$R'(x) = \lambda R(x) - \lambda E[R(x - Y)]$$

or

$$R'(x) = \lambda R(x) - \lambda \int_0^x R(x - y)\, dG(y).$$

This differential equation can sometimes be solved for R.

8.7 A Markov Shot Noise Process

Suppose that shocks occur in accordance to a Poisson process with rate λ. Associated with the ith shock is a random variable X_i, $i \geq 1$, which represents the "value" of that shock. The values are assumed to be additive and we also suppose that they decrease over time at a deterministic exponential rate. That is, let us denote by

$N(t)$, the number of shocks by t,

X_i, the value of the ith shock,

S_i, the time of the ith shock.

Then the total shock value at time t, call it $X(t)$, can be expressed as

$$X(t) = \sum_{i=1}^{N(t)} X_i e^{-\alpha(t - S_i)},$$

where α is a constant that determines the exponential rate of decrease.

When the X_i, $i \geq 1$, are assumed to be independent and identically distributed and $\{X_i, i \geq 1\}$ is independent of the Poisson process $\{N(t), t \geq 0\}$, we call $\{X(t), t \geq 0\}$ a *shot noise* process.

A shot noise process possesses the Markovian property that given the present state the future is conditionally independent of the past.

We can compute the moment generating function of $X(t)$ by first conditioning on $N(t)$ and then using Theorem 2.3.1 of Chapter 2, which states that given $N(t) = n$, the unordered set of arrival times are independent uniform $(0, t)$ random variables. This gives

$$E[\exp\{sX(t)\}|N(t) = n] = E\left[\exp\left\{s\sum_{i=0}^{n} X_i e^{-\alpha(t-U_i)}\right\}\right],$$

where $U_1, \ldots, U_n$ are independent uniform $(0, t)$ random variables. Continuing the equality and using independence gives

$$\begin{aligned} E[\exp\{sX(t)\}|N(t) = n] &= (E[\exp\{sX_i e^{-\alpha(t-U_i)}\}])^n \\ &= \left[\int_0^t \phi(se^{-\alpha y})\,dy/t\right]^n \equiv \beta^n, \end{aligned}$$

where $\phi(u) = E[e^{uX}]$ is the moment generating function of X. Hence,

$$\begin{aligned} E[\exp\{sX(t)\}] &= \sum_{n=0}^{\infty} \beta^n e^{-\lambda t}\frac{(\lambda t)^n}{n!} \\ &= e^{-\lambda t}e^{\lambda t\beta} \\ &= \exp\left\{\lambda\int_0^t [\phi(se^{-\alpha y}) - 1]\,dy\right\}. \end{aligned} \tag{8.7.1}$$

The moments of $X(t)$ can be obtained by differentiation of the above, and we leave it for the reader to verify that

$$\begin{aligned} E[X(t)] &= \lambda E[X](1 - e^{-\alpha t})/\alpha, \\ \operatorname{Var}[X(t)] &= \lambda E[X^2](1 - e^{-2\alpha t})/2\alpha. \end{aligned} \tag{8.7.2}$$

To obtain $\operatorname{Cov}(X(t), X(t + s))$, we use the representation

$$X(t + s) = e^{-\alpha s}X(t) + \overline{X}(s),$$

where $\overline{X}(s)$ has the same distribution as $X(s)$ and is independent of $X(t)$. That is, $\overline{X}(s)$ is the contribution at time $t + s$ of events arising in $(t, t + s)$. Hence,

$$\begin{aligned} \operatorname{Cov}(X(t), X(t + s)) &= e^{-\alpha s}\operatorname{Var}(X(t)) \\ &= e^{-\alpha s}\lambda E[X^2](1 - e^{-2\alpha t})/2\alpha. \end{aligned}$$

The limiting distribution of $X(t)$ can be obtained by letting $t \to \infty$ in (8.7.1). This gives

$$\lim_{t\to\infty} E[\exp\{sX(t)\}] = \exp\left\{\lambda\int_0^\infty [\phi(se^{-\alpha y}) - 1]\,dy\right\}.$$

Let us consider now the special case where the X_i are exponential random variables with rate θ. Hence,

$$\phi(u) = \frac{\theta}{\theta - u},$$

and so, in this case,

$$\begin{aligned}\lim_{t\to\infty} E[\exp\{sX(t)\}] &= \exp\left\{\lambda \int_0^\infty \left(\frac{\theta}{\theta - se^{-\alpha y}} - 1\right) dy\right\} \\ &= \exp\left\{\frac{\lambda}{\alpha}\int_0^s \frac{dx}{\theta - x}\right\} \\ &= \left(\frac{\theta}{\theta - s}\right)^{\lambda/\alpha}.\end{aligned}$$

But $(\theta/(\theta - s))^{\lambda/\alpha}$ is the moment generating function of a gamma random variable with parameters λ/α and θ. Hence the limiting density of $X(t)$, when the X_i are exponential with rate θ, is

$$f(y) = \frac{\theta e^{-\theta y}(\theta y)^{\lambda/\alpha - 1}}{\Gamma(\lambda/\alpha)}, \qquad 0 < y < \infty. \tag{8.7.3}$$

Let us suppose for the remainder of this section that the X_i are indeed exponential with rate θ and that the process is in steady state. For this latter requirement, we can either imagine that $X(0)$ is chosen according to the distribution (8.7.3) or (better yet) that the process originated at $t = -\infty$.

Suppose that $X(t) = y$. An interesting computation is to determine the distribution of the time since the last increase—that is, the time since the last Poisson event prior to t. Calling this random variable $A(t)$, we have

$$\begin{aligned}&P\{A(t) > s | X(t) = y\} \\ &= \lim_{h\to 0} P\{A(t) > s | y < X(t) < y + h\} \\ &= \lim_{h\to 0} \frac{P\{ye^{\alpha s} < X(t-s) < (y+h)e^{\alpha s}, 0 \text{ events in } (t-s,t)\}}{P\{y < X(t) < y+h\}} \\ &= \lim_{h\to 0} \frac{f(ye^{\alpha s})e^{\alpha s}he^{-\lambda s} + o(h)}{f(y)h + o(h)} \\ &= \exp\{-\theta y(e^{\alpha s} - 1)\}.\end{aligned} \tag{8.7.4}$$

It should be noted that we have made use of the assumption that the process is in steady state to conclude that the distributions of $X(t)$ and $X(t - s)$ are both given by (8.7.3). From (8.7.4), we see that the conditional hazard rate function of $A(t)$—given $X(t) = y$; call it $\lambda(s|y)$—is given by

$$\lambda(s|y) = \frac{\frac{d}{ds} P\{A(t) \le s | X(t) = y\}}{P\{A(t) > s | X(t) = y\}}$$
$$= \theta\alpha y e^{\alpha s}$$

From this we see that given $X(t) = y$, the hazard rate of the backwards time to the last event starts at t with rate $\theta\alpha y$ (that is, $\lambda(0|y) = \theta\alpha y$) and increases exponentially as we go backwards in time until an event happens. It should be noted that this differs markedly from the time at t until the next event (which, of course, is independent of $X(t)$ and is exponential with rate λ).

8.8 Stationary Processes

A stochastic process $\{X(t), t \ge 0\}$ is said to be a *stationary process* if for all $n, s, t_1, \ldots, t_n$ the random vectors $X(t_1), \ldots, X(t_n)$ and $X(t_1 + s), \ldots, X(t_n + s)$ have the same joint distribution. In other words, a process is stationary if choosing any fixed point as the origin, the ensuing process has the same probability law. Some examples of stationary processes are:

(i) An ergodic continuous-time Markov chain $\{X(t), t \ge 0\}$ when

$$P\{X(0) = j\} = P_j, \qquad j \ge 0,$$

where $\{P_j, j \ge 0\}$ are the stationary probabilities.

(ii) $\{X(t), t \ge 0\}$ when $X(t)$ is the age at time t of an equilibrium renewal process.

(iii) $\{X(t), t \ge 0\}$ when $X(t) = N(t + L) - N(t)$, $t \ge 0$, where $L > 0$ is a fixed constant and $\{N(t), t \ge 0\}$ is a Poisson process having rate λ.

The first two of the above processes are stationary for the same reason: they are Markov processes whose initial state is chosen according to the limiting state distribution (and thus they can be thought of as ergodic Markov processes that have already been in operation an infinite time). That the third example—where $X(t)$ represents the number of events of a Poisson process that occur between t and $t + L$—is stationary follows from the stationary and independent increment assumption of the Poisson process.

The condition for a process to be stationary is rather stringent, and so we define the process $\{X(t), t \ge 0\}$ to be a *second-order stationary*, or a *covariance stationary*, process if $E[X(t)] = c$ and $\text{Cov}(X(t), X(t + s))$ does not depend on t. That is, a process is second-order stationary (a further name sometimes seen in the literature is *weakly stationary*) if the first two moments of $X(t)$ are the same for all t and the covariance between $X(s)$ and $X(t)$ depends only on $|t - s|$. For a second-order stationary process, let

$$R(s) = \text{Cov}(X(t), X(s + t)).$$

As the finite-dimensional distributions of a Gaussian process (being multivariate normal) are determined by their means and covariances, it follows that a second-order stationary Gaussian process is stationary. However, there are many examples of second-order stationary processes that are not stationary.

EXAMPLE 8.8(A) ***An Auto Regressive Process.*** Let $Z_0, Z_1, Z_2, \ldots$ be uncorrelated random variables with $E[Z_n] = 0$, $n \geq 0$, and

$$\text{Var}(Z_n) = \begin{cases} \sigma^2/(1 - \lambda^2) & n = 0 \\ \sigma^2 & n \geq 1, \end{cases}$$

where $\lambda^2 < 1$. Define

$$X_0 = Z_0, \tag{8.8.1}$$

$$X_n = \lambda X_{n-1} + Z_n, \qquad n \geq 1. \tag{8.8.2}$$

The process $\{X_n, n \geq 0\}$ is called a *first-order auto-regressive process.* It says that the state at time $n(X_n)$ is a constant multiple of the state at time $n - 1$ plus a random error term (Z_n).

Iterating (8.8.2) yields

$$\begin{aligned} X_n &= \lambda(\lambda X_{n-2} + Z_{n-1}) + Z_n \\ &= \lambda^2 X_{n-2} + \lambda Z_{n-1} + Z_n \\ &\;\;\vdots \\ &= \sum_{i=0}^{n} \lambda^{n-i} Z_i, \end{aligned}$$

and so

$$\begin{aligned} \text{Cov}(X_n, X_{n+m}) &= \text{Cov}\left(\sum_{i=0}^{n} \lambda^{n-i} Z_i, \sum_{i=0}^{n+m} \lambda^{n+m-i} Z_i\right) \\ &= \sum_{i=0}^{n} \lambda^{n-i} \lambda^{n+m-i} \text{Cov}(Z_i, Z_i) \\ &= \sigma^2 \lambda^{2n+m} \left(\frac{1}{1 - \lambda^2} + \sum_{i=1}^{n} \lambda^{-2i}\right) \\ &= \frac{\sigma^2 \lambda^m}{1 - \lambda^2}, \end{aligned}$$

where the above uses the fact that Z_i and Z_j are uncorrelated when $i \neq j$. Since $E[X_n] = 0$, we see that $\{X_n, n \geq 0\}$ is weakly stationary (the definition for a discrete-time process is the obvious analogue of that given for continuous-time processes).

EXAMPLE 8.8(B) ***A Moving Average Process.*** Let $W_0, W_1, W_2, \ldots$ be uncorrelated with $E[W_n] = \mu$ and $\text{Var}(W_n) = \sigma^2$, $n \geq 0$, and for some positive integer k define

$$X_n = \frac{W_n + W_{n-1} + \cdots + W_{n-k}}{k+1}, \qquad n \geq k.$$

The process $\{X_n, n \geq k\}$, which at each time keeps track of the arithmetic average of the most recent $k + 1$ values of the W's, is called a moving average process. Using the fact that the W_n, $n \geq 0$, are uncorrelated, we see that

$$\text{Cov}(X_n, X_{n+m}) = \begin{cases} \dfrac{(k+1-m)\sigma^2}{(k+1)^2} & \text{if } 0 \leq m \leq k \\ 0 & \text{if } m > k. \end{cases}$$

Hence, $\{X_n, n \geq k\}$ is a second-order stationary process.

Let $\{X_n, n \geq 1\}$ be a second-order stationary process with $E[X_n] = \mu$. An important question is when, if ever, does $\overline{X}_n \equiv \sum_{i=1}^n X_i/n$ converge to μ. The following proposition shows that $E[(\overline{X}_n - \mu)^2] \to 0$ if, and only if, $\sum_{i=1}^n R(i)/n \to 0$. That is, the expected square of the difference between $\overline{X}_n$ and μ will converge to 0 if, and only if, the limiting average value of $R(i)$ converges to 0.

PROPOSITION 8.8.1

Let $\{X_n, n \geq 1\}$ be a second-order stationary process having mean μ and covariance function $R(i) = \text{Cov}(X_n, X_{n+i})$, and let $\overline{X}_n \equiv \sum_{i=1}^n X_i/n$. Then

$$\lim_{n\to\infty} E[(\overline{X}_n - \mu)^2] = 0$$

if, and only if,

$$\lim_{n\to\infty} \sum_{i=1}^n \frac{R(i)}{n} = 0.$$

Proof Let $Y_i = X_i - \mu$ and $\overline{Y}_n = \sum_{i=1}^n Y_i/n$ and suppose that $\sum_{i=1}^n R(i)/n \to 0$. We want to show that this implies that $E[\overline{Y}_n^2] \to 0$. Now

$$E[\overline{Y}_n^2] = \frac{1}{n^2} E\left[\sum_{i=1}^n Y_i^2 + 2 \sum\sum_{i<j\leq n} Y_i Y_j\right]$$

$$= \frac{R(0)}{n} + \frac{2 \sum\sum_{i<j\leq n} R(j-i)}{n^2}.$$

We leave it for the reader to verify that the right-hand side of the above goes to 0 when $\Sigma_1^n R(i)/n \to 0$.

To go the other way, suppose that $E[\bar{Y}_n^2] \to 0$, then

$$\begin{aligned}\left(\sum_{i=0}^{n-1} \frac{R(i)}{n}\right)^2 &= \left[\frac{1}{n}\sum_{i=1}^{n} \text{Cov}(Y_1, Y_i)\right]^2 \\ &= [\text{Cov}(Y_1, \bar{Y}_n)]^2 \\ &= [E(Y_1\bar{Y}_n)]^2 \\ &\le E[Y_1^2]E[\bar{Y}_n^2],\end{aligned}$$

which shows that $\Sigma_{i=0}^{n-1} R(i)/n \to 0$ as $n \to \infty$. The reader should note that the above makes use of the Cauchy-Schwarz inequality, which states that for random variables X and Y $(E[XY])^2 \le E[X^2]E[Y^2]$ (see Exercise 8.28).

PROBLEMS

In Problems 8.1, 8.2, and 8.3, let $\{X(t), t \ge 0\}$ denote a Brownian motion process.

8.1. Let $Y(t) = tX(1/t)$.

(a) What is the distribution of $Y(t)$?

(b) Compute $\text{Cov}(Y(s), Y(t))$.

(c) Argue that $\{Y(t), t \ge 0\}$ is also Brownian motion.

(d) Let

$$T = \inf\{t > 0: \quad X(t) = 0\}.$$

Using (c) present an argument that

$$P\{T = 0\} = 1.$$

8.2. Let $W(t) = X(a^2 t)/a$ for $a > 0$. Verify that $W(t)$ is also Brownian motion.

8.3. Compute the conditional distribution of $X(s)$ given that $X(t_1) = A$, $X(t_2) = B$, where $t_1 < s < t_2$.

8.4. Let $\{Z(t), t \ge 0\}$ denote a Brownian Bridge process. Show that if

$$X(t) = (t + 1)Z(t/(t + 1)),$$

then $\{X(t), t \ge 0\}$ is a Brownian motion process.

8.5. A stochastic process $\{X(t), t \geq 0\}$ is said to be *stationary* if $X(t_1), \ldots, X(t_n)$ has the same joint distribution as $X(t_1 + a), \ldots, X(t_n + a)$ for all $n, a, t_1, \ldots, t_n$.

(a) Prove that a necessary and sufficient condition for a Gaussian process to be stationary is that $\text{Cov}(X(s), X(t))$ depends only on $t - s$, $s \leq t$, and $E[X(t)] = c$.

(b) Let $\{X(t), t \geq 0\}$ be Brownian motion and define

$$V(t) = e^{-\alpha t/2} X(\alpha e^{\alpha t}).$$

Show that $\{V(t), t \geq 0\}$ is a stationary Gaussian process. It is called the Ornstein-Uhlenbeck process.

8.6. Let $\{X(t), t \geq 0\}$ denote a birth and death process that is allowed to go negative and that has constant birth and death rates $\lambda_n \equiv \lambda$, $\mu_n \equiv \mu$, $n = 0, \pm 1, \pm 2, \ldots$. Define μ and c as functions of λ in such a way that $\{cX(t), t \geq u\}$ converges to Brownian motion as $\lambda \to \infty$.

In Problems 8.7 through 8.12, let $\{X(t), t \geq 0\}$ denote Brownian motion.

8.7. Find the distribution of:

(a) $|X(t)|$.

(b) $\left| \min_{0 \leq s \leq t} X(s) \right|$.

(c) $\max_{0 \leq s \leq t} X(s) - X(t)$.

8.8. Suppose $X(1) = B$. Characterize, in the manner of Proposition 8.1.1, $\{X(t), 0 \leq t \leq 1\}$ given that $X(1) = B$.

8.9. Let $M(t) = \max_{0 \leq s \leq t} X(s)$ and show that

$$P\{M(t) > a | M(t) = X(t)\} = e^{-a^2/2t}, \qquad a > 0.$$

8.10. Compute the density function of T_x, the time until Brownian motion hits x.

8.11. Let T_1 denote the largest zero of $X(t)$ that is less than t and let T_2 be the smallest zero greater than t. Show that:

(a) $P\{T_2 < s\} = (2/\pi) \text{ arc cos } \sqrt{t/s}$, $s > t$.

(b) $P\{T_1 < s, T_2 > y\} = (2/\pi) \text{ arc sine } \sqrt{s/y}$, $s < t < y$.

8.12. Verify the formulas given in (8.3.4) for the mean and variance of $|X(t)|$.

8.13. For Brownian motion with drift coefficient μ, show that for $x > 0$

$$P\left\{\max_{0 \le s \le h} |X(s)| > x\right\} = o(h).$$

8.14. Let T_x denote the time until Brownian motion hits x. Compute

$$P\{T_1 < T_{-1} < T_2\}.$$

8.15. For a Brownian motion process with drift coefficient μ let

$$f(x) = E[\text{time to hit either } A \text{ or } -B | X_0 = x],$$

where $A > 0$, $B > 0$, $-B < x < A$.

(a) Derive a differential equation for $f(x)$.

(b) Solve this equation.

(c) Use a limiting random walk argument (see Problem 4.22 of Chapter 4) to verify the solution in part (b).

8.16. Let T_a denote the time Brownian motion process with drift coefficient μ hits a.

(a) Derive a differential equation which $f(a, t) \equiv P\{T_a \le t\}$ satisfies.

(b) For $\mu > 0$, let $g(x) = \text{Var}(T_x)$ and derive a differential equation for $g(x)$, $x > 0$.

(c) What is the relationship between $g(x)$, $g(y)$, and $g(x + y)$ in (b)?

(d) Solve for $g(x)$.

(e) Verify your solution by differentiating (8.4.11).

8.17. In Example 8.4(B), suppose $X_0 = x$ and player I has the doubling option. Compute the expected winnings of I for this situation.

8.18. Let $\{X(t), t \ge 0\}$ be a Brownian motion with drift coefficient μ, $\mu < 0$, which is not allowed to become negative. Find the limiting distribution of $X(t)$.

8.19. Consider Brownian motion with reflecting barriers of $-B$ and A, $A > 0$, $B > 0$. Let $p_t(x)$ denote the density function of X_t.

(a) Compute a differential equation satisfied by $p_t(x)$.

(b) Obtain $p(x) = \lim_{t\to\infty} p_t(x)$.

8.20. Prove that, with probability 1, for Brownian motion with drift μ

$$\frac{X(t)}{t} \to \mu \qquad \text{as } t \to \infty.$$

8.21. Verify that if $\{B(t), t \geq 0\}$ is standard Brownian motion then $\{Y(t), t \geq 0\}$ is a martingale with mean 1, when $Y(t) = \exp\{cB(t) - c^2t/2\}$.

8.22. In Problem 8.16, find $Var(T_a)$ by using a martingale argument.

8.23. Show that

$$p(x, t; y) \equiv \frac{1}{\sqrt{2\pi t}} e^{-(x-y-\mu t)^2/2t}$$

satisfies the backward and forward diffusion Equations (8.5.1) and (8.5.2).

8.24. Verify Equation (8.7.2).

8.25. Verify that $\{X(t) = N(t + L) - N(t), t \geq 0\}$ is stationary when $\{N(t)\}$ is a Poisson process.

8.26. Let U be uniformly distributed over $(-\pi, \pi)$, and let $X_n = \cos(nU)$. By using the trigonometric identity

$$\cos x \cos y = \tfrac{1}{2}[\cos(x + y) + \cos(x - y)],$$

verify that $\{X_n, n \geq 1\}$ is a second-order stationary process.

8.27. Show that

$$\sum_{i=1}^{n} \frac{R(i)}{n} \to 0 \quad \text{implies} \quad \sum\sum_{i<j<n} \frac{R(j-i)}{n^2} \to 0$$

thus completing the proof of Proposition 8.8.1.

8.28. Prove the Cauchy–Schwarz inequality:

$$(E[XY])^2 \leq E[X^2]E[Y^2].$$

(*Hint:* Start with the inequality $2|xy| \leq x^2 + y^2$ and then substitute $X/\sqrt{E[X^2]}$ for x and $Y/\sqrt{E[Y^2]}$ for y.)

8.29. For a second-order stationary process with mean μ for which $\sum_{i=0}^{n-1} R(i)/n \to 0$, show that for any $\varepsilon > 0$

$$\sum_{i=0}^{n-1} P\{|\bar{X}_n - \mu| > \varepsilon\} \to 0 \qquad \text{as } n \to \infty.$$

References

For a rigorous proof of the convergence of the empirical distribution function to the Brownian Bridge process, the interested (and mathematically advanced) reader should see Reference 1. Both References 3 and 5 provide nice treatments of Brownian motion. The former emphasizes differential equations and the latter martingales in many of their computations. For additional material on stationary processes the interested reader should see Reference 2 or Reference 4.

1. P. Billingsley, *Convergence of Probability Measures*, Wiley, New York, 1968.
2. G. Box and G. Jenkins, *Time Series Analysis—Forecasting and Control*, Holden-Day, San Francisco, 1970.
3. D. R. Cox and H. D. Miller, *Theory of Stochastic Processes*, Methuen, London, 1965.
4. H. Cramer and M. Leadbetter, *Stationary and Related Stochastic Processes*, Wiley, New York, 1966.
5. S. Karlin and H. Taylor, *A First Course in Stochastic Processes*, 2nd ed., Academic Press, New York, 1975.

CHAPTER 9

Stochastic Order Relations

Introduction

In this chapter we introduce some stochastic order relations between random variables. In Section 9.1 we consider the concept of one random variable being stochastically larger than another. Applications to random variables having a monotone hazard rate function are presented. We continue our study of the stochastically larger concept in Section 9.2, where we introduce the coupling approach and illustrate its usefulness. In particular, we use coupling to establish, in Section 9.2.1, some stochastic monotonicity properties of birth and death processes, and to prove, in Section 9.2.2, that the n step transition probabilities in a finite-state ergodic Markov chain converge exponentially fast to their limiting probabilities.

In Section 9.3 we consider hazard rate orderings, which are stronger than stochastically larger orderings, between random variables. We show how to use this idea to compare certain counting processes and, in fact, we use it to prove Blackwell's theorem of renewal theory when the interarrival distribution is continuous. Some monotonicity properties of renewal processes, having interarrival distributions that are decreasing failure rate, are also presented. In Section 9.4 we consider likelihood ratio orderings.

In Section 9.5 we consider the concept of one random variable having more variability than another; and, in Section 9.6, we present applications of this to comparison of (i) queueing systems (Section 9.6.1), (ii) a renewal process and a Poisson process (Section 9.6.2), and (iii) branching processes. In Section 9.7 we consider associated random variables.

9.1 Stochastically Larger

We say that the random variable X is stochastically larger than the random variable Y, written $X \geq_{\text{st}} Y$, if

$$P\{X > a\} \geq P\{Y > a\} \qquad \text{for all } a. \tag{9.1.1}$$

If X and Y have distributions F and G, respectively, then (9.1.1) is equivalent to

$$\overline{F}(a) \geq \overline{G}(a) \qquad \text{for all } a.$$

Lemma 9.1.1

If $X \geq_{st} Y$, then $E[X] \geq E[Y]$.

Proof Assume first that X and Y are nonnegative random variables. Then

$$E[X] = \int_0^\infty P\{X > a\}\, da \geq \int_0^\infty P\{Y > a\}\, da = E[Y].$$

In general we can write any random variable Z as the difference of two nonnegative random variables as follows:

$$Z = Z^+ - Z^-,$$

where

$$Z^+ = \begin{cases} Z & \text{if } Z \geq 0 \\ 0 & \text{if } Z < 0, \end{cases} \qquad Z^- = \begin{cases} 0 & \text{if } Z \geq 0 \\ -Z & \text{if } Z < 0. \end{cases}$$

Now we leave it as an exercise to show that

$$X \underset{st}{\geq} Y \Rightarrow X^+ \underset{st}{\geq} Y^+, \qquad X^- \underset{st}{\leq} Y^-.$$

Hence,

$$E[X] = E[X^+] - E[X^-] \geq E[Y^+] - E[Y^-] = E[Y].$$

The next proposition gives an alternative definition of stochastically larger.

PROPOSITION 9.1.2

$$X \underset{st}{\geq} Y \Leftrightarrow E[f(X)] \geq E[f(Y)] \qquad \text{for all increasing functions } f.$$

Proof Suppose first that $X \geq_{st} Y$ and let f be an increasing function. We show that $f(X) \geq_{st} f(Y)$ as follows. Letting $f^{-1}(a) = \inf\{x: f(x) \geq a\}$, then

$$P\{f(X) > a\} = P\{X > f^{-1}(a)\} \geq P\{Y > f^{-1}(a)\} = P\{f(Y) > a\}.$$

Hence, $f(X) \geq_{st} f(Y)$ and so, from Lemma 9.1.1, $E[f(X)] \geq E[f(Y)]$.

Now suppose that $E[f(X)] \geq E[f(Y)]$ for all increasing functions f. For any a let f_a denote the increasing function

$$f_a(x) = \begin{cases} 1 & \text{if } x > a \\ 0 & \text{if } x \leq a, \end{cases}$$

then

$$E[f_a(X)] = P\{X > a\}, \qquad E[f_a(Y)] = P\{Y > a\},$$

and we see that $X \geq_{st} Y$.

EXAMPLE 9.1(A) ***Increasing and Decreasing Failure Rate.*** Let X be a nonnegative random variable with distribution F and density f. Recall that the failure (or hazard) rate function of X is defined by

$$\lambda(t) = \frac{f(t)}{\bar{F}(t)}.$$

We say that X is an increasing failure rate (IFR) random variable if

$$\lambda(t) \uparrow t,$$

and we say that it is a decreasing failure rate (DFR) random variable if

$$\lambda(t) \downarrow t.$$

If we think of X as the life of some item, then since $\lambda(t)\,dt$ is the probability that a t-unit-old item fails in the interval $(t, t + dt)$, we see that X is IFR (DFR) means that the older the item is the more (less) likely it is to fail in a small time dt.

Suppose now that the item has survived to time t and let X_t denote its additional life from t onward. X_t will have distribution $\bar{F}_t$ given by

$$\begin{aligned} \bar{F}_t(a) &= P\{X_t > a\} \\ &= P\{X - t > a \mid X > t\} \\ &= \bar{F}(t + a)/\bar{F}(t). \end{aligned} \tag{9.1.2}$$

PROPOSITION 9.1.3

X is IFR $\Leftrightarrow$ X_t is stochastically decreasing in t,

X is DFR $\Leftrightarrow$ X_t is stochastically increasing in t.

Proof It can be shown that the hazard rate function of X_t—call it λ_t—is given by

$$\lambda_t(a) = \lambda(t + a), \tag{9.1.3}$$

where λ is the hazard rate function of X. Equation (9.1.3) can be formally proven by using (9.1.2), or can, more intuitively, be argued as follows:

$$\begin{aligned}
\lambda_t(a) &= \lim_{h\to 0} P\{a < X_t < a + h | X_t \geq a\}/h \\
&= \lim_{h\to 0} P\{a < X - t < a + h | X \geq t, X - t \geq a\}/h \\
&= \lim_{h\to 0} P\{t + a < X < t + a + h | X \geq t + a\}/h \\
&= \lambda(t + a).
\end{aligned}$$

Since

$$\begin{aligned}
\overline{F}_t(s) &= \exp\left\{-\int_0^s \lambda_t(a)\, da\right\} \\
&= \exp\left\{-\int_t^{t+s} \lambda(y)\, dy\right\},
\end{aligned} \tag{9.1.4}$$

it follows that if $\lambda(y)$ is increasing (decreasing), then $\overline{F}_t(s)$ is decreasing (increasing) in t. Similarly, if $\overline{F}_t(s)$ is decreasing (increasing) in t, then (9.1.4) implies that $\lambda(y)$ is increasing (decreasing) in y.

Thus the lifetime of an item is IFR (DFR) if the older the item is then the stochastically smaller (larger) is its additional life.

A common class of DFR distributions is the one consisting of mixtures of exponentials where we say that the distribution F is a *mixture* of the distributions F_α, $0 < \alpha < \infty$, if, for some distribution G,

$$F(x) = \int_0^\infty F_\alpha(x)\, dG(\alpha).$$

Mixtures occur when we sample from a population made up of distinct types. The value of an item from the type characterized by α has the distribution F_α. G is the distribution of the characterization quantities.

Consider now a mixture of two exponential distributions having rates λ_1 and λ_2, where $\lambda_1 < \lambda_2$. To show that this mixture distribution is DFR, note that if the item selected has survived up to time t, then its distribution of remaining life is still a mixture of the two exponential distributions. This is so since its remaining life will still be exponential with rate λ_1 if it is a type-1 item or with rate λ_2 if it is a type-2 item. However, the probability that it is a type-1 item is no longer the (prior) probability p but is now a conditional probability given that it has survived to time t. In fact, its probability of being a type-1 item is

$$P\{\text{type } 1 | \text{life} > t\} = \frac{P\{\text{type } 1, \text{life} > t\}}{P\{\text{life} > t\}}$$

$$= \frac{pe^{-\lambda_1 t}}{pe^{-\lambda_1 t} + (1-p)e^{-\lambda_2 t}}.$$

Since the above is increasing in t, it follows that the larger t is the more likely it is that the item in use is a type 1 (the better one, since $\lambda_1 < \lambda_2$). Hence, the older the item is, the less likely it is to fail, and thus the mixture of exponentials is DFR.

It turns out that the class of DFR distributions are closed under mixtures (which implies the above since the exponential distribution, as it has a constant hazard rate function, is both IFR and DFR). To prove this we need the following well-known lemma whose proof is left as an exercise.

Lemma 9.1.4 The Cauchy–Schwarz Inequality

For any distribution G and functions $h(t)$, $k(t)$, $t \geq 0$,

$$\left(\int h(t)k(t)\, dG(t)\right)^2 \leq \left(\int h^2(t)\, dG(t)\right)\left(\int k^2(t)\, dG(t)\right)$$

provided the integrals exist.

We may now state the following.

PROPOSITION 9.1.5

If F_α is a DFR distribution for all $0 < \alpha < \infty$ and G a distribution function on $(0, \infty)$, then F is DFR, where

$$F(t) = \int_0^\infty F_\alpha(t)\, dG(\alpha).$$

Proof

$$\lambda_F(t) = \frac{\frac{d}{dt}F(t)}{\overline{F}(t)} = \frac{\int_0^\infty f_\alpha(t)\, dG(\alpha)}{\overline{F}(t)}$$

We will argue that $\lambda_F(t)$ decreases in t by first assuming that all derivatives exist and then showing that $(d/dt)\lambda_F(t) \le 0$. Now

$$\frac{d}{dt}[\lambda_F(t)] = \frac{\overline{F}(t)\int f'_\alpha(t)\,dG(\alpha) + \left(\int f_\alpha(t)\,dG(\alpha)\right)^2}{\overline{F}^2(t)}.$$

Since $\overline{F}(t) = \int \overline{F}_\alpha(t)\,dG(\alpha)$, it follows from the above that to prove that $(d/dt)\lambda_F(t) \le 0$, we need to show

$$\left(\int f_\alpha(t)\,dG(\alpha)\right)^2 \le \left(\int \overline{F}_\alpha(t)\,dG(\alpha)\right)\left(\int -f'_\alpha(t)\,dG(\alpha)\right). \tag{9.1.5}$$

By letting $h(\alpha) = (\overline{F}_\alpha(t))^{1/2}$, $k(\alpha) = (-f'_\alpha(t))^{1/2}$, and applying the Cauchy–Schwarz inequality, we see

$$\left(\int (-\overline{F}_\alpha(t)f'_\alpha(t))^{1/2}\,dG(\alpha)\right)^2 \le \int \overline{F}_\alpha(t)\,dG(\alpha)\int -f'_\alpha(t)\,dG(\alpha).$$

Hence to prove (9.1.5) it suffices to show

$$\left(\int f_\alpha(t)\,dG(\alpha)\right)^2 \le \left(\int (-\overline{F}_\alpha(t)f'_\alpha(t))^{1/2}\,dG(\alpha)\right)^2. \tag{9.1.6}$$

Now F_α is, by assumption, DFR and thus

$$0 \ge \frac{d}{dt}\frac{f_\alpha(t)}{\overline{F}_\alpha(t)} = \frac{\overline{F}_\alpha(t)f'_\alpha(t) + f_\alpha^2(t)}{\overline{F}_\alpha^2(t)},$$

implying

$$-\overline{F}_\alpha(t)f'_\alpha(t) \ge f_\alpha^2(t),$$

which proves (9.1.6) and established the result. (The above also shows $f'_\alpha(t) \le 0$, and so $k(\alpha) \equiv (-f'_\alpha(t))^{1/2}$ was well defined.) ∎

9.2 Coupling

If $X \ge_{st} Y$, then there exists random variables X^* and Y^* having the same distributions of X and Y and such that X^* is, with probability 1, at least as large as Y^*. Before proving this we need the following lemma.

Lemma 9.2.1

Let F and G be continuous distribution functions. If X has distribution F then the random variable $G^{-1}(F(X))$ has distribution G.

Proof

$$\begin{aligned} P\{G^{-1}(F(X)) \le a\} &= P\{F(X) \le G(a)\} \\ &= P\{X \le F^{-1}(G(a))\} \\ &= F(F^{-1}(G(a)) \\ &= G(a). \end{aligned}$$

PROPOSITION 9.2.2

If F and G are distributions such that $\overline{F}(a) \ge \overline{G}(a)$, then there exists random variables X and Y having distributions F and G respectively such that

$$P\{X \ge Y\} = 1.$$

Proof We'll present a proof when F and G are continuous distribution functions. Let X have distribution F and define Y by $Y = G^{-1}(F(X))$. Then by Lemma 9.2.1 Y has distribution G. But as $F \le G$, it follows that $F^{-1} \ge G^{-1}$, and so

$$Y = G^{-1}(F(X)) \le F^{-1}(F(X)) = X,$$

which proves the result.

Oftentimes the easiest way to prove $\overline{F} \ge \overline{G}$ is to let X be a random variable having distribution F and then define a random variable Y in terms of X such that (i) Y has distribution G, and (ii) $Y \le X$. We illustrate this method, known as *coupling*, by some examples.

Example 9.2(a) ***Stochastic Ordering of Vectors.*** Let $X_1, \ldots, X_n$ be independent and $Y_1, \ldots, Y_n$ be independent. If $X_i \ge_{st} Y_i$, then for any increasing f

$$f(X_1, \ldots, X_n) \underset{st}{\ge} f(Y_1, \ldots, Y_n).$$

Proof Let $X_1, \ldots, X_n$ be independent and use Proposition 9.2.2 to generate independent $Y_1^*, \ldots, Y_n^*$, where Y_i^* has the distribution of Y_i and $Y_i^* \le X_i$. Then $f(X_1, \ldots, X_n) \ge f(Y_1^*, \ldots, Y_n^*)$ since f is increasing. Hence for any a

$$f(Y_1^*, \ldots, Y_n^*) > a \Rightarrow f(X_1, \ldots, X_n) > a,$$

and so

$$P\{f(Y_1^*, \ldots, Y_n^*) > a\} \le P\{f(X_1, \ldots, X_n) > a\}.$$

Since the left-hand side of the above is equal to $P\{f(Y_1, \ldots, Y_n) > a\}$, the result follows.

Example 9.2(b) ***Stochastic Ordering of Poisson Random Variables.*** We show that a Poisson random variable is stochastically increasing in its mean. Let N denote a Poisson random variable with mean λ. For any p, $0 < p < 1$, let $I_1, I_2, \ldots$ be independent of each other and of N and such that

$$I_j = \begin{cases} 1 & \text{with probability } p \\ 0 & \text{with probability } 1 - p. \end{cases}$$

Then

$$\sum_{j=1}^{N} I_j \qquad \text{is Poisson with mean } \lambda p. \quad \text{(Why?)}$$

Since

$$\sum_{j=1}^{N} I_j \le N,$$

the result follows.

We will make use of Proposition 9.1.2 to present a stochastically greater definition first for vectors and then for stochastic processes.

Definition

We say that the random vector $\underline{X} = (X_1, \ldots, X_n)$ is stochastically greater than the random vector $\underline{Y} = (Y_1, \ldots, Y_n)$, written $\underline{X} \ge_{st} \underline{Y}$ if for all increasing functions f

$$E[f(\underline{X})] \ge E[f(\underline{Y})].$$

We say that the stochastic process $\{X(t), t \geq 0\}$ is stochastically greater than $\{Y(t), t \geq 0\}$ if

$$(X(t_1), \ldots, X(t_n)) \underset{\text{st}}{\geq} (Y(t_1), \ldots, Y(t_n))$$

for all $n, t_1, \ldots, t_n$.

It follows from Example 9.2(A) that if $\underline{X}$ and $\underline{Y}$ are vectors of independent components such that $X_i \geq_{\text{st}} Y_i$, then $\underline{X} \geq_{\text{st}} \underline{Y}$. We leave it as an exercise for the reader to present a counterexample when the independence assumption is dropped.

In proving that one stochastic process is stochastically larger than another, coupling is once again often the key.

EXAMPLE 9.2(C) ***Comparing Renewal Processes.*** Let $N_i = \{N_i(t), t \geq 0\}$, $i = 1, 2$, denote renewal processes having interarrival distributions F and G, respectively. If $\overline{F} \geq \overline{G}$, then

$$\{N_1(t), t \geq 0\} \underset{\text{st}}{\leq} \{N_2(t), t \geq 0\}.$$

To prove the above we use coupling as follows. Let $X_1, X_2, \ldots$ be independent and distributed according to F. Then the renewal process generated by the X_i—call it N_1^*—has the same probability distributions as N_1. Now generate independent random variables $Y_1, Y_2, \ldots$ having distribution G and such that $Y_i \leq X_i$. Then the renewal process generated by the interarrival times Y_i—call it N_2^*—has the same probability distributions as N_2. However, as $Y_i \leq X_i$ for all i, it follows that

$$N_1^*(t) \leq N_2^*(t) \qquad \text{for all } t,$$

which proves the result.

Our next example uses coupling in conjunction with the strong law of large numbers.

EXAMPLE 9.2(D) Let $X_1, X_2, \ldots$ be a sequence of independent Bernoulli random variables, and let $p_i = P\{X_i = 1\}$, $i \geq 1$. If $p_i \geq p$ for all i, show that with probability 1

$$\liminf_n \sum_{i=1}^{n} X_i/n \geq p.$$

(The preceding means that for any $\varepsilon > 0$, $\sum_{i=1}^{n} X_i/n \le p - \varepsilon$ for only a finite number of n.)

Solution. We start by coupling the sequence X_i, $i \ge 1$ with a sequence of independent and identically distributed Bernoulli random variables Y_i, $i \ge 1$, such that $P\{Y_i = 1\} = p$ and $X_i \ge Y_i$ for all i. To accomplish this, let U_i, $i \ge 1$, be an independent sequence of uniform (0, 1) random variables. Now, for $i = 1, \ldots, n$, set

$$X_i = \begin{cases} 1 & \text{if } U_i \le p_i \\ 0 & \text{otherwise,} \end{cases} \quad \text{and} \quad Y_i = \begin{cases} 1 & \text{if } U_i \le p \\ 0 & \text{otherwise.} \end{cases}$$

Since $p \le p_i$ it follows that $Y_i \le X_i$. Hence,

$$\liminf_n \sum_{i=1}^{n} X_i/n \ge \liminf_n \sum_{i=1}^{n} Y_i/n.$$

But it follows from the strong law of large numbers that, with probability 1, $\liminf_n \sum_{i=1}^{n} Y_i/n = p$.

Example 9.2(e) ***Bounds on the Coupon Collector's Problem.*** Suppose that there are m distinct types of coupons and that each coupon collected is type j with probability P_j, $j = 1, \ldots, m$. Letting N denote the number of coupons one needs to collect in order to have at least one of each type, we are interested in obtaining bounds for $E[N]$.

To begin, let $i_1, \ldots, i_m$ be a permutation of $1, \ldots, m$. Let T_1 denote the number of coupons it takes to obtain a type i_1, and for $j > 1$, let T_j denote the number of additional coupons after having at least one of each type $i_1, \ldots, i_{j-1}$ until one also has at least one of type i_j. (Thus, if a type i_j coupon is obtained before at least one of the types $i_1, \ldots, i_{j-1}$, then $T_j = 0$, and if not then T_j is a geometric random variable with parameter P_{i_j}). Now,

$$N = \sum_{j=1}^{m} T_j$$

and so,

$$E[N] = \sum_{j=1}^{m} P\{i_j \text{ is the last of } i_1, \ldots, i_j\}/P_{i_j}. \tag{9.2.1}$$

Now, rather than supposing that a coupon is collected at fixed time points, it clearly would make no difference if we supposed that they are collected at random times distributed according to a Poisson process with rate 1. Given this assumption, we can assert that the times until the first occurrences of the different types of coupons are independent exponential random variables with respective (for a type j coupon) rates P_j, $j = 1, \ldots, m$. Hence, if we let X_j, $j = 1, \ldots, m$ be independent exponential random variables with rates 1, then X_j/P_j will be independent exponential random variables with rates P_j, $j = 1, \ldots, m$, and so

$$
\begin{aligned}
&P\{i_j \text{ is the last of } i_1, \ldots, i_j \text{ to be collected}\} \\
(9.2.2) \qquad &= P\{X_{i_j}/P_{i_j} = \max(X_{i_1}/P_{i_1}, \ldots, X_{i_j}/P_{i_j})\}.
\end{aligned}
$$

Now, renumber the coupon types so that $P_1 \leq P_2 \leq \cdots \leq P_m$. We will use a coupling argument to show that

$$(9.2.3) \qquad P\{j \text{ is the last of } 1, \ldots, j\} \leq 1/j,$$

$$(9.2.4) \qquad P\{j \text{ is the last of } m, m-1, \ldots, j\} \geq 1/(m-j+1).$$

To verify the inequality (9.2.3), note the following:

$$
\begin{aligned}
&P\{j \text{ is the last of } 1, \ldots, j \text{ to be obtained}\} \\
&\quad = P\{X_j/P_j = \max_{1 \leq i \leq j} X_i/P_i\} \\
&\quad \leq P\{X_j/P_j = \max_{1 \leq i \leq j} X_i/P_j\} \qquad \text{since } P_i \leq P_j \\
&\quad = 1/j.
\end{aligned}
$$

Thus, inequality (9.2.3) is proved. By a similar argument (left as an exercise) inequality (9.2.4) is also established.

Hence, upon utilizing Equation (9.2.1), first with the permutation 1, 2, ..., m (to obtain an upper bound) and then with m, $m - 1, \ldots, 1$ (to obtain a lower bound) we see that

$$\sum_{j=1}^{m} \frac{1}{(m-j+1)P_j} \leq E[N] \leq \sum_{j=1}^{m} \frac{1}{jP_j}.$$

Another lower bound for $E[N]$ is given in Problem 9.17.

Example 9.2(F) ***A Bin Packing Problem.*** Suppose that n items, whose weights are independent and uniformly distributed on (0, 1), are to be put into a sequence of bins that can each hold at most one unit of weight. Items are successively put in bin 1 until an item is reached whose additional weight when added to those presently in the bin would exceed the bin capacity of one unit. At that point, bin 1 is packed away, the item is put in bin 2, and the process continues. Thus, for instance, if the weights of the first four items are .45, .32, .92, and .11 then items one and two would be in bin 1, item three would be the only item in bin 2, and item four would be the initial item in bin 3. We are interested in $E[B]$, the expected number of bins needed.

To begin, suppose that there are an infinite number of items, and let N_i denote the number of items that go in bin i. Now, if W_i denotes the weight of the initial item in bin i (that is, the item that would not fit in bin $i - 1$) then

$$N_i \overset{d}{=} \max\{j: W_i + U_1 + \cdots + U_{j-1} \le 1\}, \tag{9.2.5}$$

where $X \overset{d}{=} Y$ means that X and Y have the same distribution, and where $U_1, U_2, \ldots$ is a sequence of independent uniform (0, 1) random variables that are independent of W_i. Let A_{i-1} be the amount of unused capacity in bin $i - 1$; that is, the weight of all items in that bin is $1 - A_{i-1}$. Now, the conditional distribution of W_i, given A_{i-1}, is the same as the conditional distribution of a uniform (0, 1) random variable given that it exceeds A_{i-1}. That is,

$$P\{W_i > x | A_{i-1}\} = P\{U > x | U > A_{i-1}\}$$

where U is a uniform (0, 1) random variable. As

$$P\{U > x | U > A_{i-1}\} > P\{U > x\},$$

we see that W_i is stochastically larger than a uniform (0, 1) random variable. Hence, from (9.2.5) independently of $N_1, \ldots, N_{i-1}$

$$N_i \underset{\text{st}}{\le} \max\{j: U_1 + \cdots + U_j \le 1\} \tag{9.2.6}$$

Note that the right-hand side of the preceding has the same distribution as the number of renewals by time 1 of the renewal process whose interarrival distribution is the uniform (0, 1) distribution.

The number of bins needed to store the n items can be expressed as

$$B = \min\left\{m: \sum_{i=1}^{m} N_i \ge n\right\}.$$

However, if we let $X_i, i \ge 1$, be a sequence of independent random variables having the same distribution as $N(1)$, the number of renewals by time 1 of the uniform (0, 1) renewal process, then from (9.2.6) we obtain that

$$B \underset{\text{st}}{\ge} N,$$

where

$$N \equiv \min\left\{m: \sum_{i=1}^{m} X_i \geq n\right\}.$$

But, by Wald's equation,

$$E\left[\sum_{i=1}^{N} X_i\right] = E[N]E[X_i].$$

Also, Problem 3.7 (whose solution can be found in the Answers and Solutions to Selected Problems Appendix) asked one to show that

$$E[X_i] = E[N(1)] = e - 1$$

and thus, since $\sum_{i=1}^{N} X_i \geq n$, we can conclude that

$$E[N] \geq \frac{n}{e-1}.$$

Finally, using that $B \underset{\text{st}}{\geq} N$, yields that

$$E[B] \geq \frac{n}{e-1}.$$

If the successive weights have an arbitrary distribution F concentrated on [0, 1] then the same argument shows that

$$E[B] \geq \frac{n}{m(1)}$$

where $m(1)$ is the expected number of renewals by time 1 of the renewal process with interarrival distribution F.

9.2.1 Stochastic Monotonicity Properties of Birth and Death Processes

Let $\{X(t), t \geq 0\}$ be a birth and death process. We will show two stochastic monotonicity properties of $\{X(t), t \geq 0\}$. The first is that the birth and death process is stochastically increasing in the initial state $X(0)$.

PROPOSITION 9.2.3

$\{X(t), t \geq 0\}$ is stochastically increasing in $X(0)$. That is, $E[f(X(t_1), \ldots, X(t_n))| X(0) = i]$ is increasing in i for all $t_1, \ldots, t_n$ and increasing functions f.

Proof Let $\{X_1(t), t \geq 0\}$ and $\{X_2(t), t \geq 0\}$ be independent birth and death processes having identical birth and death rates, and suppose $X_1(0) = i + 1$ and $X_2(0) = i$. Now since $X_1(0) > X_2(0)$, and the two processes always go up or down only by 1, and never at the same time (since this possibility has probability 0), it follows that either the $X_1(t)$ process is always larger than the $X_2(t)$ process or else they are equal at some time. Let us denote by T the first time at which they become equal. That is,

$$T = \begin{cases} \infty & \text{if } X_1(t) > X_2(t) \text{ for all } t \\ \text{1st } t: & X_1(t) = X_2(t) \text{ otherwise.} \end{cases}$$

Now if $T < \infty$, then the two processes are equal at time T and so, by the Markovian property, their continuation after time T has the same probabilistic structure. Thus if we define a third stochastic process—call it $\{X_3(t)\}$—by

$$X_3(t) = \begin{cases} X_1(t) & \text{if } t < T \\ X_2(t) & \text{if } t \geq T, \end{cases}$$

then $\{X_3(t)\}$ will also be a birth and death process, having the same parameters as the other two processes, and with $X_3(0) = X_1(0) = i + 1$. However, since by the definition of T

$$X_1(t) > X_2(t) \qquad \text{for } t < T,$$

we see

$$X_3(t) \geq X_2(t) \qquad \text{for all } t,$$

which proves the result.

Our second stochastic monotonicity property says that if the initial state is 0, then the state at time t increases stochastically in t.

PROPOSITION 9.2.4

$P\{X(t) \geq j | X(0) = 0\}$ increases in t for all j.

Proof For $s < t$:

$$
\begin{aligned}
&P\{X(t) \geq j | X(0) = 0\} \\
&\quad = \sum_i P\{X(t) \geq j | X(0) = 0, X(t-s) = i\} P\{X(t-s) = i | X(0) = 0\} \\
&\quad = \sum_i P\{X(t) \geq j | X(t-s) = i\} P_{0i}(t-s) \qquad \text{(by the Markovian property)} \\
&\quad = \sum_i P\{X(s) \geq j | X(0) = i\} P_{0i}(t-s) \\
&\quad \geq \sum_i P\{X(s) \geq j | X(0) = 0\} P_{0i}(t-s) \qquad \text{(by Proposition 9.2.3)} \\
&\quad = P\{X(s) \geq j | X(0) = 0\} \sum_i P_{0i}(t-s) \\
&\quad = P\{X(s) \geq j | X(0) = 0\}.
\end{aligned}
$$

Remark Besides providing a nice qualitative property about the transition probabilities of a birth and death process, Proposition 9.2.4 is also useful in applications. For whereas it is often quite difficult to determine explicitly the values $P_{0j}(t)$ for fixed t, it is a simple matter to obtain the limiting probabilities P_j. Now from Proposition 9.2.4 we have

$$P\{X(t) \geq j | X(0) = 0\} \leq \lim_{t \to \infty} P\{X(t) \geq j | X(0) = 0\} = \sum_{i=j}^{\infty} P_i,$$

which says that $X(t)$ is stochastically smaller than the random variable—call it $X(\infty)$—having the limiting distribution, thus supplying a bound on the distribution of $X(t)$.

9.2.2 Exponential Convergence in Markov Chains

Consider a finite-state irreducible Markov chain having transition probabilities P_{ij}^n. We will use a coupling argument to show that P_{ij}^n converges exponentially fast, as $n \to \infty$, to a limit that is independent of i. To prove this we will make use of the result that if an ergodic Markov chain has a finite number—say M—of states, then there must exist N, $\varepsilon > 0$, such that

$$P_{ij}^N > \varepsilon \qquad \text{for all } i, j. \tag{9.2.7}$$

Consider now two independent versions of the Markov chain, say $\{X_n, n \geq 0\}$ and $\{X_n', n \geq 0\}$, where one starts in i, say $P\{X_0 = i\} = 1$, and the other such that $P\{X'_0 = j\} = \pi_j$, $j = 1, \ldots, M$, where the π_j are a set of stationary probabilities. That is, they are a nonnegative solution of

$$\pi_j = \sum_{i=1}^{M} \pi_i P_{ij}, \qquad j = 1, \ldots, M,$$

$$\sum_{j=1}^{M} \pi_j = 1.$$

Let T denote the first time both processes are in the same state. That is,

$$T = \min\{n: \quad X_n = X'_n\}.$$

Now

$$T > mN \Rightarrow X_N \neq X'_N, X_{2N} \neq X'_{2N}, \ldots, X_{mN} \neq X'_{mN},$$

and so

$$P\{T > mN\} \leq P(A_1)P(A_2|A_1) \cdots P(A_m|A_1, \ldots, A_{m-1}), \tag{9.2.8}$$

where A_i is the event that $X_{Ni} \neq X'_{Ni}$. From (9.2.7) it follows that no matter what the present state is, there is a probability of at least ε that the state a time N in the future will be j. Hence no matter what the past, the probability that the two chains will both be in state j a time N in the future is at least ε^2, and thus the probability that they will be in the same state is at least $M\varepsilon^2$. Hence all the conditional probabilities on the right-hand side of (9.2.8) are no greater than $1 - M\varepsilon^2$. Thus

$$P\{T > mN\} \leq (1 - M\varepsilon^2)^m = (1 - \alpha)^m, \tag{9.2.9}$$

where $\alpha \equiv M\varepsilon^2$.

Let us now define a third Markov chain—call it $\{\overline{X}_n, n \geq 0\}$—that is equal to X' up to time T and is equal to X thereafter. That is,

$$\overline{X}_n = \begin{cases} X'_n & \text{for } n \leq T \\ X_n & \text{for } n \geq T. \end{cases}$$

Since $X'_T = X_T$, it is clear that $\{\overline{X}_n, n \geq 0\}$ is a Markov chain with transition probabilities P_{ij} whose initial state is chosen according to a set of stationary probabilities. Now

$$\begin{aligned} P\{\overline{X}_n = j\} &= P\{\overline{X}_n = j | T \leq n\}P\{T \leq n\} + P\{\overline{X}_n = j | T > n\}P\{T > n\} \\ &= P\{X_n = j | T \leq n\}P\{T \leq n\} + P\{\overline{X}_n = j, T > n\}. \end{aligned}$$

Similarly,

$$P_{ij}^n = P\{X_n = j\} = P\{X_n = j | T \leq n\}P\{T \leq n\} + P\{X_n = j, T > n\}.$$

Hence

$$P_{ij}^n - P\{\overline{X}_n = j\} = P\{X_n = j, T > n\} - P\{\overline{X}_n = j, T > n\}$$

implying that

$$\begin{aligned}|P_{ij}^n - P\{\overline{X}_n = j\}| &\leq P\{T > n\} \\ &\leq (1 - \alpha)^{n/N - 1} \qquad \text{(by (9.2.9))}.\end{aligned}$$

But it is easy to verify (say by induction on n) that

$$P\{\overline{X}_n = j\} = \pi_j,$$

and thus we see that

$$|P_{ij}^n - \pi_j| \leq \frac{\beta^n}{1 - \alpha}, \qquad \text{where } \beta = (1 - \alpha)^{1/N}.$$

Hence P_{ij}^n indeed converges exponentially fast to a limit not depending on i. (In addition the above also shows that there cannot be more than one set of stationary probabilities.)

Remark We will use an argument similar to the one given in the above theorem to prove, in Section 9.3, Blackwell's theorem for a renewal process whose interarrival distribution is continuous.

9.3 Hazard Rate Ordering and Applications to Counting Processes

The random variable X has a larger hazard (or failure) rate function than does Y if

$$\lambda_X(t) \geq \lambda_Y(t) \qquad \text{for all } t \geq 0, \tag{9.3.1}$$

where $\lambda_X(t)$ and $\lambda_Y(t)$ are the hazard rate functions of X and Y. Equation (9.3.1) states that, at the same age, the unit whose life is X is more likely to instantaneously perish than the one whose life is Y. In fact, since

$$P\{X > t + s | X > t\} = \exp\left\{-\int_t^{t+s} \lambda(y)\, dy\right\},$$

it follows that (9.3.1) is equivalent to

$$P\{X > t + s | X > t\} \le P\{Y > t + s | Y > t\},$$

or, equivalently,

$$X_t \underset{\text{st}}{\le} Y_t \qquad \text{for all } t \ge 0,$$

where X_t and Y_t are, respectively, the remaining lives of a t-unit-old item having the same distributions as X and Y.

Hazard rate ordering can be quite useful in comparing counting processes. To illustrate this let us begin by considering a delayed renewal process whose first renewal has distribution G and whose other interarrivals have distribution F, where both F and G are continuous and have failure rate functions $\lambda_F(t)$ and $\lambda_G(t)$. Let $\mu(t)$ be such that

$$\max\left(\max_{0\le s\le t} \lambda_F(s),\ \max_{0\le s\le t} \lambda_G(s)\right) \le \mu(t).$$

We first show how the delayed renewal process can be generated by a random sampling from a nonhomogeneous Poisson process having intensity function $\mu(t)$.

Let $S_1, S_2, \ldots$ denote the times at which events occur in the nonhomogeneous Poisson process $\{N(t), t \ge 0\}$ with intensity function $\mu(t)$. We will now define a counting process—which, we will then argue, is a delayed renewal process with initial renewal distribution G and interarrival distribution F—such that events can only occur at times $S_1, S_2, \ldots$. Let

$$I_i = \begin{cases} 1 & \text{if an event of the counting process occurs at time } S_i \\ 0 & \text{otherwise.} \end{cases}$$

Hence, to define the counting process we need specify the joint distribution of the I_i, $i \ge 1$. This is done as follows:

Given $S_1, S_2, \ldots$, take

$$P\{I_1 = 1\} = \lambda_G(S_1)/\mu(S_1) \tag{9.3.2}$$

and, for $i > 1$,

$$P\{I_i = 1 | I_1, \ldots, I_{i-1}\} = \begin{cases} \dfrac{\lambda_G(S_i)}{\mu(S_i)} & \text{if } I_1 = \cdots = I_{i-1} = 0 \\[2ex] \dfrac{\lambda_F(S_i - S_j)}{\mu(S_i)} & \text{if } j = \max\{k\colon\ k < i, I_k = 1\}. \end{cases} \tag{9.3.3}$$

To obtain a feel for the above, let $A(t)$ denote the age of the counting process at time t; that is, it is the time at t since the last event of the counting process that occurred before t. Then $A(S_1) = S_1$, and the other values of $A(S_i)$ are recursively obtained from $I_1, \ldots, I_{i-1}$. For instance, if $I_1 = 0$, then $A(S_2) = S_2$; whereas if $I_1 = 1$, then $A(S_2) = S_2 - S_1$. Then (9.3.2) and (9.3.3) are equivalent to

$$P\{I_i = 1 | I_1, \ldots, I_{i-1}\} = \begin{cases} \dfrac{\lambda_G(S_i)}{\mu(S_i)} & \text{if } A(S_i) = S_i \\ \dfrac{\lambda_F(A(S_i))}{\mu(S_i)} & \text{if } A(S_i) < S_i. \end{cases}$$

We claim that the counting process defined by the I_i, $i \geq 1$, constitutes the desired delayed renewal process. To see this note that for the counting process the probability intensity of an event at any time t, given the past, is given by

$$\begin{aligned} &P\{\text{event in } (t, t+h) | \text{history up to } t\} \\ &\quad = P\{\text{an event of the nonhomogeneous Poisson process occurs in } (t, t+h), \text{ and it is counted} | \text{history up to } t\} \\ &\quad = (\mu(t)h + o(h))P\{\text{it is counted} | \text{history up to } t\} \\ &\quad = \begin{cases} [\mu(t)h + o(h)]\dfrac{\lambda_G(t)}{\mu(t)} = \lambda_G(t)h + o(h) & \text{if } A(t) = t \\ [\mu(t)h + o(h)]\dfrac{\lambda_F(A(t))}{\mu(t)} = \lambda_F(A(t))h + o(h) & \text{if } A(t) < t. \end{cases} \end{aligned}$$

Hence, the probability (intensity) of an event at any time t depends only on the age at that time and is equal to $\lambda_G(t)$ if the age is t and to $\lambda_F(A(t))$ otherwise. But such a counting process is clearly a delayed renewal process with interarrival distribution F and with initial distribution G.

We now use this representation of a delayed renewal process as a random sampling of a nonhomogeneous Poisson process to give a simple probabilistic proof of Blackwell's theorem when the interarrival distribution is continuous.

THEOREM (Blackwell's Theorem)

Let $\{N^*(t), t \geq 0\}$ denote a renewal process with a continuous interarrival distribution F. Then

$$m(t+a) - m(t) \to \frac{a}{\mu} \qquad \text{as } t \to \infty,$$

where $m(t) = E[N^*(t)]$ and μ, assumed finite, is the mean interarrival time.

Proof We will prove Blackwell's theorem under the added simplifying assumption that $\lambda_F(t)$, the failure rate function of F, is bounded away from 0 and ∞. That is, we will assume there is $0 < \lambda_1 < \lambda_2 < \infty$ such that

$$\lambda_1 < \lambda(t) < \lambda_2 \qquad \text{for all } t. \tag{9.3.4}$$

Also let G be a distribution whose failure rate function also lies between λ_1 and λ_2.

Consider a Poisson process with rate λ_2. Let its event times be $S_1, S_2, \ldots$. Now let $I_1^*, I_2^*, \ldots$ be generated by (9.3.3) with $\mu(t) \equiv \lambda_2$, and let $I_1, I_2, \ldots$ be conditionally independent (given $S_1, S_2, \ldots$) of the sequence $I_1^*, I_2^*, \ldots$ and generated by (9.3.3) with $G = F$ and $\mu(t) \equiv \lambda_2$. Thus, the counting process in which events occur at those times S_i for which $I_i^* = 1$—call this process $\{N_0(t), t \geq 0\}$—is a delayed renewal process with interarrival distribution F, and the counting process in which events occur at those times S_i for which $I_i = 1$—call it $\{N(t), t \geq 0\}$—is a renewal process with interarrival distribution F. Let

$$N = \min\{i: \; I_i = I_i^* = 1\}.$$

That is, N is the first event of the Poisson process that is counted by both generated processes. Since independently of anything else each Poisson event will be counted by a given generated process with probability at least λ_1/λ_2, it follows that

$$P\{I_i = I_i^* = 1 | I_1, \ldots, I_{i-1}, I_1^*, \ldots, I_{i-1}^*\} \geq \left(\frac{\lambda_1}{\lambda_2}\right)^2,$$

and hence

$$P\{N < \infty\} = 1.$$

Now define a third sequence $\bar{I}_i$, $i \geq 1$, by

$$\bar{I}_i = \begin{cases} I_i^* & \text{for } i \leq N \\ I_i & \text{for } i \geq N. \end{cases}$$

Thus the counting process $\{\bar{N}(t), t \geq 0\}$ in which events occur at those values of S_i for which $\bar{I}_i = 1$ is a delayed renewal process with initial distribution G and interarrival distribution F whose event times starting at time S_N are identical to those of $\{N(t), t \geq 0\}$.

Letting $N(t, t + a) = N(t + a) - N(t)$ and similarly for $\bar{N}$, we have

$$\begin{aligned} E[\bar{N}(t, t + a)] &= E[\bar{N}(t, t + a) | S_N \leq t] P\{S_N \leq t\} + E[\bar{N}(t, t + a) | S_N > t] P\{S_N > t\} \\ &= E[N(t, t + a) | S_N \leq t] P\{S_N \leq t\} + E[\bar{N}(t, t + a) | S_N > t] P\{S_N > t\} \\ &= E[N(t, t + a)] + (E[\bar{N}(t, t + a) | S_N > t] \\ &\quad - E[N(t, t + a) | S_N > t]) P\{S_N > t\}. \end{aligned}$$

Now it easily follows that $E[N(t, t + a)|S_N > t] \leq \lambda_2 a$ and similarly for the term with $\overline{N}$ replacing N. Hence, since $N < \infty$ implies that $P\{S_N > t\} \to 0$ as $t \to \infty$, we see from the preceding that

$$E[\overline{N}(t, t + a)] - E[N(t, t + a)] \to 0 \qquad \text{as } t \to \infty. \tag{9.3.5}$$

But if we now take $G = F_e$, where $F_e(t) = \int_0^t \overline{F}(y)\, dy/\mu$, it is easy to establish (see the proof of part (i) of Theorem 3.5.2 of Chapter 3) that when $G = F_e$, $E[\overline{N}(t)] = t/\mu$, and so, from (9.3.5),

$$E[N(t, t + a)] \to \frac{a}{\mu} \qquad \text{as } t \to \infty,$$

which completes the proof.

We will also find the approach of generating a renewal process by a random sampling of a Poisson process useful in obtaining certain monotonicity results about renewal processes whose interarrival distributions have decreasing failure rates. As a prelude we define some additional notation.

Definition

For a counting process $\{N(t), t \geq 0\}$ define, for any set of time points T, $N(T)$ to be the number of events occurring in T.

We start with a lemma.

Lemma 9.3.1

Let $N = \{N(t), t \geq 0\}$ denote a renewal process whose interarrival distribution F is decreasing failure rate. Also, let $N_y = \{N_y(t), t \geq 0\}$ denote a delayed renewal process whose first interarrival time has the distribution H_y, where H_y is the distribution of the excess at time y of the renewal process N, and the others have interarrival distribution F. (That is, N_y can be thought of as a continuation, starting at y, of a renewal process having the same interarrival distribution as N.) Then, for any sets of time points $T_1, \ldots, T_n$,

$$(N(T_1), \ldots, N(T_n)) \underset{\text{st}}{\geq} (N_y(T_1), \ldots, N_y(T_n)).$$

Proof Let $N^* = \{N^*(t), t \le y\}$ denote the first y time units of a renewal process, independent of N, but also having interarrival distribution F. We will interpret N_y as the continuation of N^* from time y onward. Let $A^*(y)$ be the age at time y of the renewal process N^*.

Consider a Poisson process with rate $\mu = \lambda(0)$ and let $S_1, S_2, \ldots$ denote the times at which events occur. Use the Poisson process to generate a counting process—call it N—in which events can only occur at times S_i, $i \ge 1$. If I_i equals 1 when an event occurs at S_i and 0 otherwise, then we let

$$P\{I_i = 1 | I_1, \ldots, I_{i-1}\} = \lambda(A(S_i))/\mu,$$

where $A(S_1) = S_1$ and, for $i > 1$, where $A(S_i)$ is the time at S_i since the last counted event prior to S_i, where the Poisson event at S_j is said to be counted if $I_j = 1$. Then, as before, this generated counting process will be a renewal process with interarrival distribution F.

We now define another counting process that also can have events only at times S_i, $i \ge 1$, and we let $\bar{I}_i$ indicate whether or not there is an event at S_i. Let $\bar{A}(t)$ denote the time at t since the last event of this process, or, if there have been no events by t, define it to be $t + A^*(y)$. Let the $\bar{I}_i$ be such that

$$\text{if} \quad I_i = 0, \quad \text{then } \bar{I}_i = 0,$$

$$\text{if} \quad I_i = 1, \quad \text{then } \bar{I}_i = \begin{cases} 1 & \text{with probability } \lambda(\bar{A}(S_i))/\lambda(A(S_i)) \\ 0 & \text{otherwise.} \end{cases}$$

The above is well defined since, as $\bar{I}_i$ can equal 1 only when $I_i = 1$, we will always have that $\bar{A}(t) \ge A(t)$, and since λ is nonincreasing $\lambda(\bar{A}(S_i)) \le \lambda(A(S_i))$. Hence we see

$$\begin{aligned} P\{\bar{I}_i = 1 | \bar{I}_1, \ldots, \bar{I}_{i-1}, I_1, \ldots, I_{i-1}\} &= P\{I_i = 1 | I_1, \ldots, I_{i-1}\} P\{\bar{I}_i = 1 | I_i = 1\} \\ &= \lambda(\bar{A}(S_i))/\mu, \end{aligned}$$

and so the counting process generated by the $\bar{I}_i$, $i \ge 1$—call it N_y—is a delayed (since $\bar{A}(0) = A^*(y)$) renewal process whose initial distribution is H_y. Since events of N_y can only occur at time points where events of N occur ($\bar{I}_i \le I_i$ for all i), it follows that

$$N(T) \ge N_y(T) \qquad \text{for all sets } T,$$

and the lemma follows. ∎

PROPOSITION 9.3.2 Monotonicity Properties of DFR Renewal Processes

Let $A(t)$ and $Y(t)$ denote the age and excess at time t of a renewal process $N = \{N(t), t \ge 0\}$ whose interarrival distribution is DFR. Then both $A(t)$ and $Y(t)$ increase stochastically in t. That is, $P\{A(t) > a\}$ and $P\{Y(t) > a\}$ are both increasing in t for all a.

Proof Suppose we want to show

$$P\{A(t + y) > a\} \geq P\{A(t) > a\}.$$

To do so we interpret $A(t)$ as the age at time t of the renewal process N and $A(t + y)$ as the age at time t of the renewal process N_y, of Lemma 9.3.1. Then letting $T = [t - a, t]$ we have from Lemma 9.3.1

$$P\{N(T) \geq 1\} \geq P\{N_y(T) \geq 1\},$$

or, equivalently,

$$P\{A(t) \leq a\} \geq P\{A(t + y) \leq a\}.$$

The proof for the excess is similar except we now let $T = [t, t + a]$.

Proposition 9.3.2 can be used to obtain some nice bounds on the renewal function of a DFR renewal process and on the distribution of a DFR random variable.

Corollary 9.3.3

Let F denote a DFR distribution whose first two moments are

$$\mu_1 = \int x\, dF(x), \qquad \mu_2 = \int x^2\, dF(x).$$

(i) If $m(t)$ is the renewal function of a renewal process having interarrival distribution F, then

$$\frac{t}{\mu_1} \leq m(t) \leq \frac{t}{\mu_1} + \frac{\mu_2}{2\mu_1^2} - 1,$$

(ii)

$$\overline{F}(t) \geq \exp\left\{-\frac{t}{\mu_1} - \frac{\mu_2}{2\mu_1^2} + 1\right\}.$$

Proof (i) Let $X_1, X_2, \ldots$ denote the interarrival times of a renewal process $\{N(t), t \geq 0\}$ having interarrival distribution F. Now

$$\sum_{i=1}^{N(t)+1} X_i = t + Y(t),$$

where $Y(t)$ is the excess at t. By taking expectations, and using Wald's equation we obtain

$$\mu_1(m(t) + 1) = t + E[Y(t)].$$

But by Proposition 9.3.2 $E[Y(t)]$ is increasing in t, and since $E[Y(0)] = \mu_1$ and (see Proposition 3.4.6 of Chapter 3)

$$\lim_{t\to\infty} E[Y(t)] = \frac{\mu_2}{2\mu_1},$$

we see

$$t + \mu_1 \le \mu_1(m(t) + 1) \le t + \frac{\mu_2}{2\mu_1},$$

or

$$\frac{t}{\mu_1} \le m(t) \le \frac{t}{\mu_1} + \frac{\mu_2}{2\mu_1^2} - 1.$$

(ii) Differentiating the identity $m(t) = \sum_{n=1}^{\infty} F_n(t)$ yields

$$\begin{aligned} m'(t)\,dt &= \sum_{n=1}^{\infty} F_n'(t)\,dt \\ &= \sum_{n=1}^{\infty} P\{n\text{th renewal occurs in } (t, t + dt)\} + o[(dt)] \\ &= P\{\text{a renewal occurs in } (t, t + dt)\} + o[d(t)], \end{aligned}$$

and thus $m'(t)$ is equal to the probability (intensity) of a renewal at time t. But since $\lambda(A(t))$ is the probability (intensity) of a renewal at t given the history up to that time, we thus have

$$\begin{aligned} m'(t) &= E[\lambda(A(t))] \\ &\ge \lambda(t), \end{aligned}$$

where the above inequality follows since λ is decreasing and $A(t) \le t$. Integrating the above inequality yields

$$m(t) \ge \int_0^t \lambda(s)\,ds.$$

Since

$$\overline{F}(t) = \exp\left(-\int_0^t \lambda(s)\,ds\right),$$

we see

$$\overline{F}(t) \ge e^{-m(t)},$$

and the result follows from part (i). ∎

9.4 Likelihood Ratio Ordering

Let X and Y denote continuous nonnegative random variables having respective densities f and g. We say that X is larger than Y in the sense of likelihood ratio, and write

$$X \underset{\text{LR}}{\geq} Y$$

if

$$\frac{f(x)}{g(x)} \leq \frac{f(y)}{g(y)} \qquad \text{for all } x \leq y.$$

Hence $X \geq_{\text{LR}} Y$ if the ratio of their respective densities, $f(x)/g(x)$, is nondecreasing in x. We start by noting that this is a stronger ordering than failure rate ordering (which is itself stronger than stochastic ordering).

PROPOSITION 9.4.1

Let X and Y be nonnegative random variables having densities f and g and hazard rate functions λ_X and λ_Y. If

$$X \underset{\text{LR}}{\geq} Y,$$

then

$$\lambda_X(t) \leq \lambda_Y(t) \qquad \text{for all } t \geq 0.$$

Proof Since $X \geq_{\text{LR}} Y$, it follows that, for $x \geq t$, $f(x) \geq g(x)f(t)/g(t)$. Hence,

$$\begin{aligned}
\lambda_X(t) &= \frac{f(t)}{\int_t^\infty f(x)\,dx} \\
&\leq \frac{f(t)}{\int_t^\infty g(x)f(t)/g(t)\,dx} \\
&= \frac{g(t)}{\int_t^\infty g(x)\,dx} \\
&= \lambda_Y(t).
\end{aligned}$$

Example 9.4(a) If X is exponential with rate λ and Y exponential with rate μ, then

$$\frac{f(x)}{g(x)} = \frac{\lambda}{\mu} e^{(\mu-\lambda)x},$$

and so $X \geq_{\text{LR}} Y$ when $\lambda \leq \mu$.

Example 9.4(b) ***A Statistical Inference Problem.*** A central problem in statistics is that of making inferences about the unknown distribution of a given random variable. In the simplest case, we suppose that X is a continuous random variable having a density function known to be either f or g. Based on the observed value of X, we must decide on either f or g.

A decision rule for the above problem is a function $\phi(x)$, which takes on either value 0 or value 1 with the interpretation that if X is observed to equal x, then we decide on f if $\phi(x) = 0$ and on g if $\phi(x) = 1$. To help us decide upon a good decision rule, let us first note that

$$\int_{x:\phi(x)=1} f(x)\,dx = \int f(x)\phi(x)\,dx$$

represents the probability of rejecting f when it is in fact the true density. The classical approach to obtaining a decision rule is to fix a constant α, $0 \leq \alpha \leq 1$, and then restrict attention to decision rules ϕ such that

$$\int f(x)\phi(x)\,dx \leq \alpha. \tag{9.4.1}$$

Among such rules it then attempts to choose the one that maximizes the probability of rejecting f when it is false. That is, it maximizes

$$\int_{x:\phi(x)=1} g(x)\,dx = \int g(x)\phi(x)\,dx.$$

The optimal decision rule according to this criterion is given by the following proposition, known as the Neyman–Pearson lemma.

Neyman–Pearson Lemma

Among all decision rules ϕ satisfying (9.4.1), the one that maximizes $\int g(x)\phi(x)\,dx$ is ϕ^* given by

$$\phi^*(x) = \begin{cases} 0 & \text{if } f(x)/g(x) \geq c \\ 1 & \text{if } f(x)/g(x) < c, \end{cases}$$

where c is chosen so that

$$\int f(x)\phi^*(x)\,dx = \alpha.$$

Proof Let ϕ satisfy (9.4.1). For any x

$$(\phi^*(x) - \phi(x))(cg(x) - f(x)) \geq 0.$$

The above inequality follows since if $\phi^*(x) = 1$, then both terms in the product are nonnegative, and if $\phi^*(x) = 0$, then both are nonpositive. Hence,

$$\int (\phi^*(x) - \phi(x))(cg(x) - f(x))\,dx \geq 0$$

and so

$$c\left[\int \phi^*(x)g(x)\,dx - \int \phi(x)g(x)\,dx\right] \geq \int \phi^*(x)f(x)\,dx - \int \phi(x)f(x)\,dx$$
$$\geq 0,$$

which proves the result.

If we suppose now that f and g have a monotone likelihood ratio order—that is, $f(x)/g(x)$ is nondecreasing in x—then the optimal decision rule can be written as

$$\phi^*(x) = \begin{cases} 0 & \text{if } x \geq k \\ 1 & \text{if } x < k, \end{cases}$$

where k is such that

$$\int_{-\infty}^{k} f(x)\,dx = \alpha.$$

That is, the optimal decision rule is to decide on f when the observed value is greater than some critical number and to decide on g otherwise.

Likelihood ratio orderings have important applications in optimization theory. The following proposition is quite useful.

PROPOSITION 9.4.2

Suppose that X and Y are independent, with densities f and g, and suppose that

$$X \underset{\text{LR}}{\geq} Y.$$

If $h(x, y)$ is a real-valued function satisfying

$$h(x, y) \geq h(y, x) \qquad \text{whenever } x \geq y,$$

then

$$h(X, Y) \underset{\text{st}}{\geq} h(Y, X).$$

Proof Let $U = \max(X, Y)$, $V = \min(X, Y)$. Then conditional on $U = u$, $V = v$, $u \geq v$, the conditional distribution of $h(X, Y)$ is concentrated on the two points $h(u, v)$ and $h(v, u)$, assigning probability

$$\lambda_1 \equiv P\{X = \max(X, Y), Y = \min(X, Y) | U = u, V = v\}$$
$$= \frac{f(u)g(v)}{f(u)g(v) + f(v)g(u)}$$

to the larger value $h(u, v)$. Similarly, conditional on $U = u$ and $V = v$, $h(Y, X)$ is also concentrated on the two points $h(u, v)$ and $h(v, u)$, assigning probability

$$\lambda_2 \equiv P\{Y = \max(X, Y), X = \min(X, Y) | U = u, V = v\}$$
$$= \frac{g(u)f(v)}{g(u)f(v) + f(u)g(v)}.$$

Since $u \geq v$,

$$f(u)g(v) \geq g(u)f(v),$$

and so, conditional on $U = u$ and $V = v$, $h(X, Y)$ is stochastically larger than $h(Y, X)$. That is,

$$P\{h(X, Y) \geq a | U, V\} \geq P\{h(Y, X) \geq a | U, V\},$$

and the result now follows by taking expectations of both sides of the above.

Remark It is perhaps somewhat surprising that the above does not necessarily hold when we only assume that $X \geq_{\text{st}} Y$. For a counterexample, note that $2x + y \geq x + 2y$ whenever $x \geq y$. However, if X and Y are independent with

$$X = \begin{cases} 3 & \text{with probability } .2 \\ 9 & \text{with probability } .8, \end{cases}$$

$$Y = \begin{cases} 1 & \text{with probability } .2 \\ 4 & \text{with probability } .8, \end{cases}$$

then $X \geq_{st} Y$, but $P\{2X + Y \geq 11\} = .8$ and $P\{2Y + X \geq 11\} = .8 + (.2)(.8) = .96$. Thus $2X + Y$ is not stochastically larger than $2Y + X$.

Proposition 9.4.2 has important applications in optimal scheduling problems. For instance, suppose that n items, each having some measurable characteristic, are to be scheduled in some order. For instance, the measurable characteristic of an item might be the time it takes to complete work on that item. Suppose that if x_i is the measurable characteristic of item i, and if the order chosen is $i_1, \ldots, i_n$, a permutation of $1, 2, \ldots, n$, then the return is given by $h(x_{i_1}, x_{i_2}, \ldots, x_{i_n})$. Let us now suppose that the measurable characteristic of item i is a random variable, say X_i, $i = 1, \ldots, n$. If

$$X_1 \underset{\text{LR}}{\geq} X_2 \underset{\text{LR}}{\geq} \cdots \underset{\text{LR}}{\geq} X_n,$$

and if h satisfies

$$h(y_1, \ldots, y_{i-1}, y_i, y_{i+1}, \ldots, y_n) \geq h(y_1, \ldots, y_i, y_{i-1}, y_{i+1}, \ldots, y_n)$$

whenever $y_i > y_{i-1}$, then it follows from Proposition 9.4.2 that the ordering $1, 2, \ldots, n$ $(n, n-1, \ldots, 1)$ stochastically maximizes (minimizes) the return. To see this consider any ordering that does not start with item 1—say $(i_1, i_2, 1, i_3, \ldots, i_{n-1})$. By conditioning on the values $X_{i_1}, X_{i_3}, \ldots, X_{i_{n-1}}$, we can use Proposition 9.4.2 to show that the ordering $(i_1, 1, i_2, i_3, \ldots, i_{n-1})$ leads to a stochastically larger return. Continuing with such interchanges leads to the conclusion that $1, 2, \ldots, n$ stochastically maximizes the return. (A similar argument shows that $n, n-1, \ldots, 1$ stochastically minimizes return.)

The continuous random variable X having density f is said to have *increasing likelihood ratio* if $\log(f(x))$ is concave, and is said to have *decreasing likelihood ratio* if $\log f(x)$ is convex. To motivate this terminology, note that the random variable $c + X$ has density $f(x - c)$, and so

$$\begin{aligned}
c_1 + X \underset{\text{LR}}{\geq} c_2 + X \qquad & \text{for all } c_1 \geq c_2 \\
\Leftrightarrow \frac{f(x - c_1)}{f(x - c_2)} \uparrow x \qquad & \text{for all } c_1 \geq c_2 \\
\Leftrightarrow \log f(x - c_1) - \log f(x - c_2) \uparrow x \qquad & \text{for all } c_1 \geq c_2 \\
\Leftrightarrow \log f(x) \quad & \text{is concave.}
\end{aligned}$$

Hence, X has increasing likelihood ratio if $c + X$ increases in likelihood ratio as c increases.

For a second interpretation, recall the notation X_t as the remaining life from t onward of a unit, having lifetime X, which has reached the age of t. Now,

$$\begin{aligned}
\overline{F}_t(a) &\equiv P\{X_t > a\} \\
&= \overline{F}(t + a)/\overline{F}(t),
\end{aligned}$$

and so the density of X_t is given by

$$f_t(a) = f(t+a)/\bar{F}(t).$$

Hence,

$$X_s \underset{\text{LR}}{\geq} X_t \quad \text{for all } s \leq t \Leftrightarrow \frac{f(s+a)}{f(t+a)} \uparrow a \quad \text{for all } s \leq t$$

$$\Leftrightarrow \log f(x) \quad \text{is concave.}$$

Therefore, X has increasing likelihood ratio if X_s decreases in likelihood ratio as s increases. Similarly, it has decreasing likelihood ratio if X_s increases in likelihood ratio as s increases.

PROPOSITION 9.4.3

If X has increasing likelihood ratio, then X is IFR. Similarly, if X has decreasing likelihood ratio, then X is DFR.

Proof

$$X_s \underset{\text{LR}}{\geq} X_t \Rightarrow \lambda_{X_s} \leq \lambda_{X_t} \qquad \text{(from Proposition 9.4.1)}$$

$$\Rightarrow X_s \underset{\text{st}}{\geq} X_t.$$

Remarks

(1) A density function f such that $\log f(x)$ is concave is called a Polya frequency of order 2.

(2) The likelihood ratio ordering can also be defined for discrete random variables that are defined over the same set of values. We say that $X \geq_{\text{LR}} Y$ if $P\{X = x\}/P\{Y = x\}$ increases in x.

9.5 Stochastically More Variable

Recall that a function h is said to be convex if for all $0 < \lambda < 1$, x_1, x_2,

$$h(\lambda x_1 + (1 - \lambda)x_2) \leq \lambda h(x_1) + (1 - \lambda)h(x_2).$$

We say that X is more variable than Y, and write $X \geq_v Y$, if

$$E[h(X)] \geq E[h(Y)] \qquad \text{for all increasing, convex } h. \tag{9.5.1}$$

If X and Y have respective distributions F and G, then we also say that $F \geq_v G$ when (9.5.1) holds. We will defer an explanation as to why we call X more variable than Y when (9.5.1) holds until we prove the following results.

PROPOSITION 9.5.1

If X and Y are nonnegative random variables with distributions F and G respectively, then $X \geq_v Y$ if, and only if,

$$\int_a^\infty \overline{F}(x)\,dx \geq \int_a^\infty \overline{G}(x)\,dx \qquad \text{for all } a \geq 0. \tag{9.5.2}$$

Proof Suppose first that $X \geq_v Y$. Let h_a be defined by

$$h_a(x) = (x-a)^+ = \begin{cases} 0 & x \leq a \\ x - a & x > a. \end{cases}$$

Since h_a is a convex increasing function, we have

$$E[h_a(X)] \geq E[h_a(Y)].$$

But

$$\begin{aligned} E[h_a(X)] &= \int_0^\infty P\{(X-a)^+ > x\}\,dx \\ &= \int_0^\infty P\{X > a + x\}\,dx \\ &= \int_a^\infty \overline{F}(y)\,dy. \end{aligned}$$

And, similarly,

$$E[h_a(Y)] = \int_a^\infty \overline{G}(y)\,dy,$$

thus establishing (9.5.2). To go the other way, suppose (9.5.2) is valid for all $a \geq 0$ and let h denote a convex increasing function that we shall suppose is twice differentiable. Since h convex is equivalent to $h'' \geq 0$, we have from (9.5.2)

$$\int_0^\infty h''(a) \int_a^\infty \overline{F}(x)\,dx\,da \geq \int_0^\infty h''(a) \int_a^\infty \overline{G}(x)\,dx\,da. \tag{9.5.3}$$

Working with the left-hand side of the above:

$$\begin{aligned}\int_0^\infty h''(a)\int_a^\infty \bar{F}(x)\,dx\,da &= \int_0^\infty \int_0^x h''(a)\,da\,\bar{F}(x)\,dx\\ &= \int_0^\infty h'(x)\bar{F}(x)\,dx - h'(0)E[X]\\ &= \int_0^\infty h'(x)\int_x^\infty dF(y)\,dx - h'(0)E[X]\\ &= \int_0^\infty \int_0^y h'(x)\,dx\,dF(y) - h'(0)E[X]\\ &= \int_0^\infty h(y)\,dF(y) - h(0) - h'(0)E[X]\\ &= E[h(X)] - h(0) - h'(0)E[X].\end{aligned}$$

Since a similar identity is valid when $\bar{F}$ is replaced by $\bar{G}$, we see from (9.5.3)

(9.5.4) $$E[h(X)] - E[h(Y)] \geq h'(0)(E[X] - E[Y]).$$

The right-hand side of the above inequality is nonnegative since $h'(0) \geq 0$ (h is increasing) and since $E[X] \geq E[Y]$, which follows from (9.5.2) by setting $a = 0$.

Corollary 9.5.2

If X and Y are nonnegative random variables such that $E[X] = E[Y]$, then $X \geq_v Y$ if, and only if,

$$E[h(X)] \geq E[h(Y)] \qquad \text{for all convex } h.$$

Proof Let h be convex and suppose that $X \geq_v Y$. Then as $E[X] = E[Y]$, the inequality (9.5.4), which was obtained under the assumption that h is convex, reduces to

$$E[h(X)] \geq E[h(Y)],$$

and the result is proven.

Thus for two nonnegative random variables having the same mean we have that $X \geq_v Y$ if $E[h(X)] \geq E[h(Y)]$ for all convex functions h. It is for this reason we say that $X \geq_v Y$ means that X has more variability than Y. That is, intuitively X will be more variable than Y if it gives more weight to the extreme values, and one way of guaranteeing this is to require that $E[h(X)] \geq E[h(Y)]$ whenever h is convex. (For instance, since $E[X] = E[Y]$ and since $h(x) = x^2$ is convex, we would have that $\text{Var}(X) \geq \text{Var}(Y)$.)

Corollary 9.5.3

If X and Y are nonnegative with $E[X] = E[Y]$, then $X \geq_v Y$ implies that $-X \geq_v -Y$.

Proof Let h denote an increasing convex function. We must show that

$$E[h(-X)] \geq E[h(-Y)].$$

This, however, follows from Corollary 9.5.2 since the function $f(x) = h(-x)$ is convex.

Our next result deals with the preservation of the variability ordering.

PROPOSITION 9.5.4

If $X_1, \ldots, X_n$ are independent and $Y_1, \ldots, Y_n$ are independent, and $X_i \geq_v Y_i$, $i = 1, \ldots, n$, then

$$g(X_1, \ldots, X_n) \underset{v}{\geq} g(Y_1, \ldots, Y_n)$$

for all increasing convex functions g that are convex in each argument.

Proof Start by assuming that the set of $2n$ random variables is independent. The proof is by induction on n. When $n = 1$ we must show that

$$E[h(g(X_1))] \geq E[h(g(Y_1))]$$

when g and h are increasing and convex and $X_1 \geq_v Y_1$. This follows from the definition of $X_1 \geq_v Y_1$ as the function $h(g(x))$ is increasing and convex since

$$\frac{d}{dx} h(g(x)) = h'(g(x))g'(x) \geq 0,$$

$$\frac{d^2}{dx^2} h(g(x)) = h''(g(x))(g'(x))^2 + h'(g(x))g''(x) \geq 0.$$

Assume the result for vectors of size $n - 1$. Again let g and h be increasing and convex. Now

$$\begin{aligned}
&E[h(g(X_1, X_2, \ldots, X_n))|X_1 = x] \\
&\quad = E[h(g(x, X_2, \ldots, X_n))|X_1 = x] \\
&\quad = E[h(g(x, X_2, \ldots, X_n))] \qquad \text{(by independence of } X_1, \ldots, X_n\text{)} \\
&\quad \geq E[h(g(x, Y_2, \ldots, Y_n))] \qquad \text{(by the induction hypothesis)} \\
&\quad = E[h(g(X_1, Y_2, \ldots, Y_n))|X_1 = x] \qquad \text{(by the independence of } X_1, Y_2, \ldots, Y_n\text{).}
\end{aligned}$$

Taking expectations gives that

$$E[(h(g(X_1, X_2, \ldots, X_n))] \geq E[h(g(X_1, Y_2, \ldots, Y_n))].$$

But, by conditioning on $Y_2, \ldots, Y_n$ and using the result for $n = 1$, we can show that

$$E[h(g(X_1, Y_2, \ldots, Y_n))] \geq E[h(g(Y_1, Y_2, \ldots, Y_n))]$$

which proves the result. Since assuming that the set of $2n$ random variables is independent does not affect the distributions of $g(X_1, \ldots, X_n)$ and $g(Y_1, \ldots, Y_n)$ the result remains true under the weaker hypothesis that the two sets of n random variables are independent.

9.6 Applications of Variability Orderings

Before presenting some applications of variability orderings we will determine a class of random variables that are more (less) variable than an exponential.

Definition

The nonnegative random variable X is said to be *new better than used in expectation* (NBUE) if

$$E[X - a|X > a] \leq E[X] \qquad \text{for all } a \geq 0.$$

It is said to be *new worse than used in expectation* (NWUE) if

$$E[X - a|X > a] \geq E[X] \qquad \text{for all } a \geq 0.$$

If we think of X as being the lifetime of some unit, then X being NBUE (NWUE) means that the expected additional life of any used item is less (greater) than or equal to the expected life of a new item. If X is NBUE and F is the distribution of X, then we say that F is an NBUE distribution, and similarly for NWUE.

PROPOSITION 9.6.1

If F is an NBUE distribution having mean μ, then

$$F \underset{v}{\leq} \exp(\mu),$$

where $\exp(\mu)$ is the exponential distribution with mean μ. The inequality is reversed if F is NWUE.

Proof Suppose F is NBUE with mean μ. By Proposition 9.5.1 we must show that

$$\int_c^\infty \bar{F}(x)\,dx \le \int_c^\infty e^{-x/\mu}\,dx \qquad \text{for all } c \ge 0. \tag{9.6.1}$$

Now if X has distribution F, then

$$\begin{aligned} E[X-a \mid X>a] &= \int_0^\infty P\{X-a>x \mid X>a\}\,dx \\ &= \int_0^\infty \frac{\bar{F}(a+x)}{\bar{F}(a)}\,dx \\ &= \int_a^\infty \frac{\bar{F}(y)}{\bar{F}(a)}\,dy. \end{aligned}$$

Hence, for F NBUE with mean μ, we have

$$\int_a^\infty \frac{\bar{F}(y)}{\bar{F}(a)}\,dy \le \mu$$

or

$$\frac{\bar{F}(a)}{\int_a^\infty \bar{F}(y)\,dy} \ge \frac{1}{\mu},$$

which implies

$$\int_0^c \left(\frac{\bar{F}(a)}{\int_a^\infty \bar{F}(y)\,dy} \right) da \ge \frac{c}{\mu}.$$

We can evaluate the left-hand side by making the change of variables $x = \int_a^\infty \bar{F}(y)\,dy$, $dx = -\bar{F}(a)\,da$ to obtain

$$-\int_\mu^{x(c)} \frac{dx}{x} \ge \frac{c}{\mu}$$

where $x(c) = \int_c^\infty \bar{F}(y)\,dy$. Integrating yields

$$-\log\left(\int_c^\infty \frac{\bar{F}(y)\,dy}{\mu}\right) \ge \frac{c}{\mu}$$

or

$$\int_c^\infty \bar{F}(y)\,dy \le \mu e^{-c/\mu},$$

which proves (9.6.1). When F is NWUE the proof is similar. ∎

9.6.1 Comparison of G/G/1 Queues

The $G/G/1$ system supposes that the interarrival times between customers, X_n, $n \geq 1$, are independent and identically distributed as are the successive service times S_n, $n \geq 1$. There is a single server and the service discipline is "first come first served."

If we let D_n denote the delay in queue of the nth customer, then it is easy to verify (see Section 7.1 of Chapter 7 if you cannot verify it) the following recursion formula:

$$D_1 = 0,$$

$$D_{n+1} = \max\{0, D_n + S_n - X_{n+1}\}. \tag{9.6.2}$$

THEOREM 9.6.2

Consider two G/G/1 systems. The ith, $i = 1, 2$, having interarrival times $X_n^{(i)}$ and service times $S_n^{(i)}$, $n \geq 1$. Let $D_n^{(i)}$ denote the delay in queue of the nth customer in system i, $i = 1, 2$. If

(i) $$E[X_n^{(1)}] = E[X_n^{(2)}]$$

and

(ii) $$X_n^{(1)} \underset{v}{\geq} X_n^{(2)}, \qquad S_n^{(1)} \underset{v}{\geq} S_n^{(2)},$$

then

$$D_n^{(1)} \underset{v}{\geq} D_n^{(2)} \qquad \textit{for all } n.$$

Proof The proof is by induction. Since it is obvious for $n = 1$, assume it for n. Now

$$D_n^{(1)} \underset{v}{\geq} D_n^{(2)} \qquad \text{(by the induction hypothesis)},$$

$$S_n^{(1)} \underset{v}{\geq} S_n^{(2)} \qquad \text{(by assumption)},$$

$$-X_n^1 \underset{v}{\geq} -X_n^2 \qquad \text{(by Corollary 9.5.3)}.$$

Thus, by Proposition 9.5.4,

$$D_n^{(1)} + S_n^{(1)} - X_{n+1}^{(1)} \underset{v}{\geq} D_n^{(2)} + S_n^{(2)} - X_{n+1}^{(2)}.$$

Since $h(x) = \max(0, x)$ is an increasing convex function, it follows from the recursion (9.6.2) and Proposition 9.5.4 that

$$D_{n+1}^{(1)} \underset{v}{\geq} D_{n+1}^{(2)}$$

thus completing the proof.

Let $W_Q = \lim E[D_n]$ denote the average time a customer waits in queue.

Corollary 9.6.3

For a $G/G/1$ queue with $E[S] < E[X]$.

(i) If the interarrival distribution is NBUE with mean $1/\lambda$, then

$$W_Q \leq \frac{\lambda E[S^2]}{2(1 - \lambda E[S])}.$$

The inequality is reversed if the distribution is NWUE.

(ii) If the service distribution is NBUE with mean $1/\mu$, then

$$W_Q \leq \mu\beta(1 - \beta),$$

where β is the solution of

$$\beta = \int_0^\infty e^{-\mu t(1-\beta)}\, dG(t),$$

and G is the interarrival distribution. The inequality is reversed if G is NWUE.

Proof From Proposition 9.6.1 we have that an NBUE distribution is less variable than an exponential distribution with the same mean. Hence in (i) we can compare the system with an $M/G/1$ and in (ii) with a $G/M/1$. The result follows from Theorem 9.6.2 since the right-hand side of the inequalities in (i) and (ii) are respectively the average customer delay in queue in the systems $M/G/1$ and $G/M/1$ (see Examples 4.3(A) and 4.3(B) of Chapter 4).

9.6.2 A Renewal Process Application

Let $\{N_F(t), t \geq 0\}$ denote a renewal process having interarrival distribution F. Our objective in this section is to prove the following theorem.

THEOREM 9.6.4

If F is NBUE with mean μ, then $N_F(t) \leq_v N(t)$, where $\{N(t)\}$ is a Poisson process with rate $1/\mu$, the inequality being reversed when F is NWUE.

Interestingly enough it turns out to be easier to prove a more general result. (We'll have more to say about that later.)

Lemma 9.6.5

Let F_i, $i \geq 1$, be NBUE distributions, each having mean μ, and let G denote the exponential distribution with mean μ. Then for each $k \geq 1$,

$$\sum_{i=k}^{\infty} (F_1 * F_2 * \cdots * F_i)(t) \leq \sum_{i=k}^{\infty} G_{(i)}(t), \tag{9.6.3}$$

where $*$ represents convolution, and $G_{(i)}$ is the i-fold convolution of G with itself.

Proof The proof is by induction on k. To prove it when $k = 1$ let $X_1, X_2, \ldots$ be independent with X_i having distribution F_i and let

$$N^*(t) = \max\left\{n: \sum_1^n X_i \leq t\right\}.$$

Now

$$E\left[\sum_{i=1}^{N^*(t)+1} X_i\right] = E[X]E[N^*(t) + 1] \tag{9.6.4}$$

by Wald's equation, which can be shown to hold when the X_i are independent, nonnegative, and all have the same mean (even though they are not identically distributed). However, we also have that

$$\sum_{i=1}^{N^*(t)+1} X_i = t + \text{time from } t \text{ until } N^* \text{ increases.}$$

But E[time from t until N^* increases] is equal to the expected excess life of one of the X_i and is thus by NBUE, less than or equal to μ. That is,

$$E\left[\sum_1^{N^*(t)+1} X_i\right] \leq t + \mu$$

and, from (9.6.4),

$$E[N^*(t)] \leq t/\mu. \tag{9.6.5}$$

However,

$$\begin{aligned} E[N^*(t)] &= \sum_{i=1}^{\infty} P\{N^*(t) \geq i\} \\ &= \sum_{i=1}^{\infty} P\{X_1 + \cdots + X_i \leq t\} \\ &= \sum_{i=1}^{\infty} (F_1 * \cdots * F_i)(t). \end{aligned}$$

Since the right-hand side of (9.6.3) is, similarly, when $k = 1$, the mean of the exponential renewal process at t (or t/μ), we see from (9.6.5) that (9.6.3) is established when $k = 1$.

Now assume (9.6.3) for k. We have

$$\begin{aligned}
\sum_{i=k+1}^{\infty} (F_1 * \cdots * F_i)(t) &= \sum_{i=k+1}^{\infty} \int_0^t (F_1 * \cdots * F_{i-1})(t - x)\, dF_i(x) \\
&= \int_0^t \sum_{j=k}^{\infty} (F_1 * \cdots * F_j)(t - x)\, dF_{j+1}(x) \\
&\le \int_0^t \sum_{j=k}^{\infty} G_{(j)}(t - x)\, dF_{j+1}(x) \qquad \text{(by the induction hypothesis)} \\
&= \sum_{j=k}^{\infty} (G_{(j)} * F_{j+1})(t) \\
&= \left(\left(\sum_{j=k}^{\infty} G_{(j-1)} * F_{j+1} \right) * G \right)(t) \\
&\le \left(\left(\sum_{j=k}^{\infty} G_{(j)} \right) * G \right)(t) \qquad \text{(by the induction hypothesis)} \\
&= \sum_{j=k}^{\infty} G_{(j+1)}(t) \\
&= \sum_{i=k+1}^{\infty} G_{(i)}(t),
\end{aligned}$$

which completes the proof. ∎

We are now ready to prove Theorem 9.6.4.

Proof of Theorem 9.6.4. By Proposition 9.5.1 we must show that for all $k \ge 1$

$$\sum_{i=k}^{\infty} P\{N_F(t) \ge i\} \le \sum_{i=k}^{\infty} P\{N(t) \ge i\}.$$

But the left-hand side is equal to the left-hand side of (9.6.3) with each F_i equaling F and the right-hand side is just the right-hand side of (9.6.3).

Remark Suppose we would have tried to prove directly that

$$\sum_{i=k}^{\infty} F_{(i)}(t) \le \sum_{i=k}^{\infty} G_{(i)}(t)$$

whenever F is NBUE and G exponential with the same mean. Then we could have tried to use induction on k. The proof when $k = 1$ would have been identical with the one we used in Lemma 9.6.5. However, when we then tried to go from assuming the result for k to proving it for $k + 1$, we would have reached a point in the proof where we had shown that

$$\sum_{i=k+1}^{\infty} F_{(i)} \leq \left(\sum_{j=k}^{\infty} G_{(j-1)} * F \right) * F.$$

However, at this point the induction hypothesis would not have been strong enough to let us conclude that the right-hand side was less than or equal to $(\sum_{j=k}^{\infty} G_{(j)}) * F$. The moral is that sometimes when using induction it is easier to prove a stronger result since the induction hypothesis gives one more to work with.

9.6.3 A Branching Process Application*

Consider two branching processes and let F_1 and F_2 denote respectively the distributions of the number of offspring per individual in the two processes. Suppose that

$$F_1 \underset{\text{v}}{\geq} F_2.$$

That is, we suppose that the number of offspring per individual is more variable in the first process. Let $Z_n^{(i)}$, $i = 1, 2$, denote the size of the nth generation of the ith process.

THEOREM 9.6.6

If $Z_0^{(i)} = 1$, $i = 1, 2$, then $Z_n^{(1)} \geq_{\text{v}} Z_n^{(2)}$ for all n.

Proof The proof is by induction on n. Since it is true for $n = 0$, assume it for n. Now

$$Z_{n+1}^{(1)} = \sum_{j=1}^{Z_n^{(1)}} X_j$$

and

$$Z_{n+1}^{(2)} = \sum_{j=1}^{Z_n^{(2)}} Y_j,$$

where the X_j are independent and have distribution F_1 (X_j representing the number of offspring of the jth person of the nth generation of process 1) and the Y_j are

* To review this model, the reader should refer to Section 4.5 of Chapter 4.

independent and have distribution F_2. Since

$$X_j \underset{v}{\geq} Y_j \qquad \text{(by the hypothesis)}$$

and

$$Z_n^{(1)} \underset{v}{\geq} Z_n^{(2)} \qquad \text{(by the induction hypothesis)},$$

the result follows from the subsequent lemma.

Lemma 9.6.7

Let $X_1, X_2, \ldots$ be a sequence of nonnegative independent and identically distributed random variables, and similarly $Y_1, Y_2, \ldots$. Let N and M be integer-valued nonnegative random variables that are independent of the X_i and Y_i sequences. Then

$$X_i \underset{v}{\geq} Y_i, \qquad N \underset{v}{\geq} M \Rightarrow \sum_{i=1}^{N} X_i \underset{v}{\geq} \sum_{i=1}^{M} Y_i.$$

Proof We will first show that

$$\sum_{i=1}^{N} X_i \underset{v}{\geq} \sum_{i=1}^{M} X_i. \tag{9.6.6}$$

Let h denote an increasing convex function. To prove (9.6.6) we must show that

$$E\left[h\left(\sum_1^N X_i\right)\right] \geq E\left[h\left(\sum_1^M X_i\right)\right]. \tag{9.6.7}$$

Since $N \geq_v M$, and they are independent of the X_i, the above will follow if we can show that the function $g(n)$, defined by

$$g(n) = E[h(X_1 + \cdots + X_n)],$$

is an increasing convex function of n. Since it is clearly increasing because h is and each X_i is nonnegative, it remains to show that g is convex, or, equivalently, that

$$g(n + 1) - g(n) \qquad \text{is increasing in } n. \tag{9.6.8}$$

To prove this let $S_n = \sum_1^n X_i$, and note that

$$g(n + 1) - g(n) = E[h(S_n + X_{n+1}) - h(S_n)].$$

Now,

$$E[h(S_n + X_{n+1}) - h(S_n)|S_n = t] = E[h(t + X_{n+1}) - h(t)]$$
$$= f(t) \text{ (say)}.$$

Since h is convex, it follows that $f(t)$ is increasing in t. Also, since S_n increases in n, we see that $E[f(S_n)]$ increases in n. But

$$E[f(S_n)] = g(n+1) - g(n),$$

and thus (9.6.8) and (9.6.7) are satisfied.

We have thus proven that

$$\sum_1^N X_i \underset{v}{\geq} \sum_1^M X_i$$

and the proof will be completed by showing that

$$\sum_1^M X_i \underset{v}{\geq} \sum_1^M Y_i,$$

or, equivalently, that for increasing, convex h

$$E\left[h\left(\sum_1^M X_i\right)\right] \geq E\left[h\left(\sum_1^M Y_i\right)\right].$$

But

$$E\left[h\left(\sum_1^M X_i\right)\middle| M = m\right] = E\left[h\left(\sum_1^m X_i\right)\right] \qquad \text{(by independence)}$$

$$\geq E\left[h\left(\sum_1^m Y_i\right)\right] \qquad \text{since } \sum_1^m X_i \underset{v}{\geq} \sum_1^m Y_i$$

$$= E\left[h\left(\sum_1^M Y_i\right)\middle| M = m\right].$$

The result follows by taking expectations of both sides of the above. ∎

Thus we have proven Theorem 9.6.6, which states that the population size of the nth generation is more variable in the first process than it is in the second process. We will end this section by showing that if the second (less variable) process has the same mean number of offspring per individual as does the first, then it is less likely at each generation to become extinct.

Corollary 9.6.8

Let μ_1 and μ_2 denote respectively the mean of F_1 and F_2, the offspring distributions. If $Z_0^{(i)} = 1$, $i = 1, 2$, $\mu_1 = \mu_2 = \mu$, and $F_1 \geq_v F_2$, then

$$P\{Z_n^{(1)} = 0\} \geq P\{Z_n^{(2)} = 0\} \qquad \text{for all } n.$$

Proof From Theorem 9.6.6 we have that $Z_n^{(1)} \geq_v Z_n^{(2)}$, and thus from Proposition 9.5.1

$$\sum_{i=2}^{\infty} P\{Z_n^{(1)} \geq i\} \geq \sum_{i=2}^{\infty} P\{Z_n^{(2)} \geq i\},$$

or, equivalently, since

$$E[Z_n] = \sum_{i=1}^{\infty} P\{Z_n \geq i\} = \mu^n,$$

we have that

$$\mu^n - P\{Z_n^{(1)} \geq 1\} \geq \mu^n - P\{Z_n^{(2)} \geq 1\},$$

or

$$P\{Z_n^{(2)} \geq 1\} \geq P\{Z_n^{(1)} \geq 1\},$$

which proves the result.

9.7 Associated Random Variables

The set of random variables $X_1, X_2, \ldots, X_n$ is said to be *associated* if for all increasing functions f and g

$$E[f(\boldsymbol{X})g(\boldsymbol{X})] \geq E[f(\boldsymbol{X})]E[g(\boldsymbol{X})]$$

where $\boldsymbol{X} = (X_1, \ldots, X_n)$, and where we say that h is an increasing function if $h(x_1, \ldots, x_n) \leq h(y_1, \ldots, y_n)$ whenever $x_i \leq y_i$ for $i = 1, \ldots, n$.

PROPOSITION 9.7.1 Independent Random Variables Are Associated

Proof Suppose that $X_1, \ldots, X_n$ are independent. The proof that they are associated is by induction. As the result has already been proved when $n = 1$ (see Proposition 7.2.1), assume that any set of $n - 1$ independent random variables are associated and let f and g be increasing functions. Let $\boldsymbol{X} = (X_1, \ldots, X_n)$. Now,

$$\begin{aligned}E[f(\boldsymbol{X})g(\boldsymbol{X})|X_n = x] &= E[f(X_1,\ldots,X_{n-1},x)g(X_1,\ldots,X_{n-1},x)|X_n = x]\\ &= E[f(X_1,\ldots,X_{n-1},x)g(X_1,\ldots,X_{n-1},x)]\\ &\geq E[f(X_1,\ldots,X_{n-1},x)]E[g(X_1,\ldots,X_{n-1},x)]\\ &= E[f(\boldsymbol{X})|X_n = x]E[g(\boldsymbol{X})|X_n = x]\end{aligned}$$

where the last two equalities follow from the independence of the X_i, and the inequality from the induction hypothesis. Hence,

$$E[f(\boldsymbol{X})g(\boldsymbol{X})|X_n] \geq E[f(\boldsymbol{X})|X_n]E[g(\boldsymbol{X})|X_n],$$

and so,

$$E[f(\boldsymbol{X})g(\boldsymbol{X})] \geq E[E[f(\boldsymbol{X})|X_n]E[g(\boldsymbol{X})|X_n]].$$

However, $E[f(\boldsymbol{X})|X_n]$ and $E[g(\boldsymbol{X})|X_n]$ are both increasing functions of X_n and so Proposition 7.2.1 yields that

$$E[f(\boldsymbol{X})g(\boldsymbol{X})] \geq E[E[f(\boldsymbol{X})|X_n]] \cdot E[E[g(\boldsymbol{X})|X_n]] = E[f(\boldsymbol{X})]E[g(\boldsymbol{X})]$$

which proves the result.

It follows from the definition of association that increasing functions of associated random variables are also associated. Hence, from Proposition 9.7.1 we see that increasing functions of independent random variables are associated.

EXAMPLE 9.7(A) Consider a system composed of n components, each of which is either working or failed. Let X_i equal 1 if component i is working and 0 if it is failed, and suppose that the X_i are independent and that $P\{X_i = 1\} = p_i$, $i = 1, \ldots, n$. In addition suppose that there is a family of component subsets $C_1, \ldots, C_r$ such that the system works if, and only if, at least one of the components in each of these subsets is working. (These subsets are called cut sets and when they are chosen so that none is a proper subset of another they are called minimal cut sets.) Hence, if we let S equal 1 if the system works and 0 otherwise, then

$$S = \prod_{i=1}^{r} Y_i,$$

where

$$Y_i = \max_{j \in C_i} X_j.$$

As $Y_1, \ldots, Y_r$ are all increasing functions of the independent random variables $X_1, \ldots, X_n$, they are associated. Hence,

$$P\{S = 1\} = E[S] = E\left[\prod_{i=1}^{r} Y_i\right] \geq E[Y_1]E\left[\prod_{i=2}^{r} Y_i\right] \geq \cdots \geq \prod_{i=1}^{r} E[Y_i]$$

where the inequalities follow from the association property. Since Y_i is equal to 1 if at least one of the components in C_i works, we obtain that

$$P\{\text{system works}\} \geq \prod_{i=1}^{r} \left\{1 - \prod_{j \in C_i} (1 - p_j)\right\}.$$

Often the easiest way of showing that random variables are associated is by representing each of them as an increasing function of a specified set of independent random variables.

Definition

The stochastic process $\{X(t), t \geq 0\}$ is said to be associated if for all n and $t_1, \ldots, t_n$ the random variables $X(t_1), \ldots, X(t_n)$ are associated.

Example 9.7(b) Any stochastic process $\{X(t), t \geq 0\}$ having independent increments and $X(0) = 0$, such as a Poisson process or a Brownian motion process, is associated. To verify this assertion, let $t_1 < t_2 < \cdots < t_n$. Then as $X(t_1), \ldots, X(t_n)$ are all increasing functions of the independent random variables $X(t_i) - X(t_{i-1})$, $i = 1, \ldots, n$ (where $t_0 = 0$) it follows that they are associated.

The Markov process $\{X_n, n \geq 0\}$ is said to be *stochastically monotone* if $P\{X_n \leq a | X_{n-1} = x\}$ is a decreasing function of x for all n and a. That is, it is stochastically monotone if the state at time n is stochastically increasing in the state at time $n - 1$.

PROPOSITION 9.7.2

A stochastically monotone Markov process is associated.

Proof We will show that $X_0, \ldots, X_k$ are associated by representing them as increasing functions of a set of independent random variables. Let $F_{n,x}$ denote the conditional distribution function of X_n given that $X_{n-1} = x$. Let $U_1, \ldots, U_n$ be independent uniform

(0, 1) random variables that are independent of $\{X_n, n \geq 0\}$. Now, let X_0 have the distribution specified by the process, and for $i = 1, \ldots, n$, define successively

$$X_i = F^{-1}_{i,X_{i-1}}(U_i) \equiv \inf\{x: U_i \leq F_{i,X_{i-1}}(x)\}.$$

It is easy to check that, given X_{i-1}, X_i has the appropriate conditional distribution.

Since $F_{i,X_{i-1}}(x)$ is decreasing in X_{i-1} (by the stochastic monotonicity assumption) we see that X_i is increasing in X_{i-1}. Also, since $F_{i,X_{i-1}}(x)$ is increasing in x, it follows that X_i is increasing in U_i. Hence, $X_1 = F^{-1}_{1,X_0}(U_1)$ is an increasing function of X_0 and U_1; $X_2 = F^{-1}_{2,X_1}(U_2)$ is an increasing function of X_1 and U_2 and thus is an increasing function of X_0, U_1, and U_2; and so on. As each of the X_i, $i = 1, \ldots, n$, is an increasing function of the independent random variables $X_0, U_1, \ldots, U_n$ it follows that $X_1, \ldots, X_n$ are associated.

Our next example is of interest in the Bayesian theory of statistics. It states that if a random sample is stochastically increasing in the value of a parameter having an a priori distribution then this random sample is, unconditionally, associated.

EXAMPLE 9.7(c) Let Λ be a random variable and suppose that, conditional on $\Lambda = \lambda$, the random variables $X_1, \ldots, X_n$ are independent with a common distribution F_λ. If $F_\lambda(x)$ is decreasing in λ for all x then $X_1, \ldots, X_n$ are associated. This result can be verified by an argument similar to the one used in Proposition 9.7.2. Namely, let $U_1, \ldots, U_n$ be independent uniform (0, 1) random variables that are independent of Λ. Then define $X_1, \ldots, X_n$ recursively by $X_i = F^{-1}_\Lambda(U_i)$, $i = 1, \ldots, n$. As the X_i are increasing functions of the independent random variables $\Lambda, U_1, \ldots, U_n$ they are associated.

PROBLEMS

9.1. If $X \geq_{st} Y$, prove that $X^+ \geq_{st} Y^+$ and $Y^- \geq_{st} X^-$.

9.2. Suppose $X_i \geq_{st} Y_i$, $i = 1, 2$. Show by counterexample that it is not necessarily true that $X_1 + X_2 \geq_{st} Y_1 + Y_2$.

9.3. **(a)** If $X \geq_{st} Y$, show that $P\{X \geq Y\} \geq \frac{1}{2}$. Assume independence.

(b) If $P\{X \geq Y\} \geq \frac{1}{2}$, $P\{Y \geq Z\} \geq \frac{1}{2}$, and X, Y, Z are independent, does this imply that $P\{X \geq Z\} \geq \frac{1}{2}$? Prove or give a counterexample.

9.4. One of n elements will be requested—it will be i with probability P_i, $i = 1, \ldots, n$. If the elements are to be arranged in an ordered list, find the arrangement that stochastically minimizes the position of the element requested.

9.5. A random variable is said to have a gamma distribution if its density is given, for some $\lambda > 0$, $\alpha > 0$, by

$$f(t) = \frac{\lambda e^{-\lambda t}(\lambda t)^{\alpha-1}}{\Gamma(\alpha)}, \qquad t \geq 0,$$

where

$$\Gamma(\alpha) = \int_0^\infty e^{-x}x^{\alpha-1}\,dx.$$

Show that this is IFR when $\alpha \geq 1$ and DFR when $\alpha \leq 1$.

9.6. If X_i, $i = 1, \ldots, n$, are independent IFR random variables, show that $\min(X_1, \ldots, X_n)$ is also IFR. Give a counterexample to show that $\max(X_1, \ldots, X_n)$ need not be.

9.7. The following theorem concerning IFR distributions can be proven. *Theorem:* If F and G are IFR, then so if $F * G$, the convolution of F and G. Show that the above theorem is not true when DFR replaces IFR.

9.8. A random variable taking on nonnegative integer values is said to be discrete IFR if

$$P\{X = k | X \geq k\} \text{ is nondecreasing in } k,\ k = 0, 1, \ldots.$$

Show that (a) binomial random variables, (b) Poisson random variables, and (c) negative binomial random variables are all discrete IFR random variables. (*Hint:* The proof is made easier by using the preservation of IFR distribution both under convolution (the theorem stated in Problem 9.7), which remains true for discrete IFR distributions, and under limits.)

9.9. Show that a binomial n, p distribution $B_{n,p}$ increases stochastically both as n increases and as p increases. That is, $\overline{B}_{n,p}$ increases in n and in p.

9.10. Consider a Markov chain with states $0, 1, \ldots, n$ and with transition probabilities

$$P_{ij} = \begin{cases} c_i & \text{if } j = i - 1 \\ 1 - c_i & \text{if } j = i + k(i), \end{cases} \qquad (i = 1, \ldots, n-1; \quad P_{0,0} = P_{n,n} = 1),$$

where $k(i) \geq 0$. Let f_i denote the probability that this Markov chain ever enters state 0 given that it starts in state i. Show that f_i is an increasing function of $(c_1, \ldots, c_{n-1})$. (*Hint:* Consider two such chains, one having $(c_1, \ldots, c_{n-1})$ and the other $(\bar{c}_1, \ldots, \bar{c}_{n-1})$, where $\bar{c}_j \geq c_j$. Suppose both start in state i. Couple the processes so that the next state of the first is no less than that of the second. Then let the first chain run (keeping the second one fixed) until it is either in the same state as the second one or in state n. If it is in the same state, start the procedure over.)

9.11. Prove that a normal distribution with mean μ and variance σ^2 increases stochastically as μ increases. What about as σ^2 increases?

9.12. Consider a Markov chain with transition probability matrix P_{ij} and suppose that $\sum_{j=k}^{\infty} P_{ij}$ increases in i for all k.

(a) Show that, for all increasing functions f, $\sum_j P_{ij} f(j)$ increases in i.

(b) Show that $\sum_{j=k}^{\infty} P_{ij}^n$ increases in i for all k, where P_{ij}^n are the n-step transition probabilities, $n \geq 2$.

9.13. Let $\{N_i(t), t \geq 0\}$, $i = 1, 2$, denote two renewal processes with respective interarrival distributions F_1 and F_2. Let $\lambda_i(t)$ denote the hazard rate function of F_i, $i = 1, 2$. If

$$\lambda_1(t) \geq \lambda_2(s) \qquad \text{for all } s, t,$$

show that for any sets $T_1, \ldots, T_n$,

$$(N_1(T_1), \ldots, N_1(T_n)) \underset{\text{st}}{\geq} (N_2(T_1), \ldots, N_2(T_n)).$$

9.14. Consider a conditional Poisson process (see Section 2.6 of Chapter 2). That is, let Λ be a nonnegative random variable having distribution G and let $\{N(t), t \geq 0\}$ be a counting process that, given that $\Lambda = \lambda$, is a Poisson process with rate λ. Let $G_{t,n}$ denote the conditional distribution of Λ given that $N(t) = n$.

(a) Derive an expression for $G_{t,n}$.

(b) Does $G_{t,n}$ increase stochastically in n and decrease stochastically in t?

(c) Let $Y_{t,n}$ denote the time from t until the next event, given that $N(t) = n$. Does $Y_{t,n}$ increase stochastically in t and decrease stochastically in n?

9.15. For random variables X and Y and any set of values A, prove the coupling bound

$$|P\{X \in A\} - P\{Y \in A\}| \leq P\{X \neq Y\}.$$

9.16. Verify the inequality (9.2.4) in Example 9.2 (E).

9.17. Let $R_1, \ldots, R_m$ be a random permutation of the numbers $1, \ldots, m$ in the sense that for any permutation $i_1, \ldots, i_m$, $P\{R_j = i_j, \text{ for } j = 1, \ldots, m\} = 1/m!$.

(a) In the coupon collectors problem described in Example 9.2(E), argue that $E[N] = \sum_{j=1}^{m} j^{-1} E[1/P_{R_j} | R_j \text{ is the last of types } R_1, \ldots, R_j \text{ to be collected}]$, where N is the number of coupons needed to have at least one of each type.

(b) Argue that the conditional distribution of P_{R_j} given that R_j is the last of types $R_1, \ldots, R_j$ to be obtained, is stochastically smaller than the unconditional distribution of P_{R_j}.

(c) Prove that

$$E[N] \geq \frac{1}{m} \sum_{i=1}^{m} 1/P_i \sum_{j=1}^{m} 1/j.$$

9.18. Let X_1 and X_2 have respective hazard rate functions $\lambda_1(t)$ and $\lambda_2(t)$. Show that $\lambda_1(t) \geq \lambda_2(t)$ for all t if, and only if,

$$\frac{P\{X_1 > t\}}{P\{X_1 > s\}} \leq \frac{P\{X_2 > t\}}{P\{X_2 > s\}}$$

for all $s < t$.

9.19. Let X and Y have respective hazard rate functions $\lambda_X(t) \leq \lambda_Y(t)$ for all t. Define a random variable $\overline{X}$ as follows: If $Y = t$, then

$$\overline{X} = \begin{cases} t & \text{with probability } \lambda_X(t)/\lambda_Y(t) \\ t + X_t & \text{with probability } 1 - \lambda_X(t)/\lambda_Y(t), \end{cases}$$

where X_t is a random variable, independent of all else, with distribution

$$P\{X_t > s\} = \frac{P\{X > t + s\}}{P\{X > t\}}.$$

Show that $\overline{X}$ has the same distribution as X.

9.20. Let F and G have hazard rate functions λ_F and λ_G. Show that $\lambda_F(t) \geq \lambda_G(t)$ for all t if, and only if, there exist independent continuous random variables Y and Z such that Y has distribution G and $\min(Y, Z)$ has distribution F.

9.21. A family of random variables $\{X_\theta, \theta \in [a, b]\}$ is said to be a monotone likelihood ratio family if $X_{\theta_1} \leq_{\text{LR}} X_{\theta_2}$ when $\theta_1 \leq \theta_2$. Show that the following families have monotone likelihood ratio:

(a) X_θ is binomial with parameters n, θ, n fixed.

(b) X_θ is Poisson with mean θ.

(c) X_θ is uniform $(0, \theta)$.

(d) X_θ is gamma with parameters $(n, 1/\theta)$, n fixed.

(e) X_θ is gamma with parameters (θ, λ), λ fixed.

9.22. Consider the statistical inference problem where a random variable X is known to have density either f or g. The Bayesian approach is to postulate a prior probability p that f is the true density. The hypothesis that f were the true density would then be accepted if the posterior probability given the value of X is greater than some critical number. Show that if $f(x)/g(x)$ is nondecreasing in x, this is equivalent to accepting f whenever the observed value of X is greater than some critical value.

9.23. We have n jobs with the ith requiring a random time X_i to process. The jobs must be processed sequentially and the objective is to stochastically maximize the number of jobs that are processed by a (fixed) time t.

(a) Determine the optimal strategy if

$$X_i \underset{\text{LR}}{\leq} X_{i+1}, \qquad i = 1, \dots, n-1.$$

(b) What if we only assumed that

$$X_i \underset{\text{st}}{\leq} X_{i+1}, \qquad i = 1, \dots, n-1?$$

9.24. A stockpile consists of n items. Associated with the ith item is a random variable X_i, $i = 1, \dots, n$. If the ith item is put into the field at time t, its field life is $X_i e^{-\alpha t}$. If $X_i \geq_{\text{LR}} X_{i+1}$, $i = 1, \dots, n-1$, what ordering stochastically maximizes the total field life of all items? Note that if $n = 2$ and the ordering 1, 2 is used, then the total field life is $X_1 + X_2 e^{-\alpha X_1}$.

9.25. Show that the random variables, having the following densities, have increasing likelihood ratio; that is, the log of the densities are concave.

(a) Gamma: $f(x) = \lambda e^{-\lambda x}(\lambda x)^{\alpha-1}/\Gamma(\alpha)$, $\alpha \geq 1$.

(b) Weibull: $f(x) = \alpha\lambda(\lambda x)^{\alpha-1}e^{-(\lambda x)^\alpha}$, $\alpha \geq 1$

(c) Normal truncated to be positive,

$$f(x) = \frac{1}{a\sqrt{2\pi}\sigma} e^{-(x-\mu)^2/2\sigma^2}, \qquad x > 0.$$

9.26. Suppose $X_1, \ldots, X_n$ are independent and $Y_1, \ldots, Y_n$ are independent. If $X_i \geq_v Y_i$, $i = 1, \ldots, n$, prove that

$$E[\max(X_1, \ldots, X_n)] \geq E[\max(Y_1, \ldots, Y_n)].$$

Give a counterexample when independence is not assumed.

9.27. Show that F is the distribution of an NBUE random variable if, and only if,

$$\bar{F}_e(a) \leq \bar{F}(a) \qquad \text{for all } a,$$

where F_e, the equilibrium distribution of F, is defined by

$$F_e(a) = \int_0^a \frac{\bar{F}(x)\,dx}{\mu}$$

with $\mu = \int_0^\infty \bar{F}(x)\,dx$.

9.28. Prove or give a counterexample to the following: If $F \geq_v G$, then $N_F(t) \geq_v N_G(t)$, where $\{N_H(t)\, t \geq 0\}$ is the renewal process with interarrival distribution H, $H = F, G$.

9.29. Let $X_1, \ldots, X_n$ be independent with $P\{X_i = 1\} = P_i = 1 - P\{X_i = 0\}$, $i = 1, \ldots, n$. Show that

$$\sum_{i=1}^n X_i \underset{v}{\leq} \text{Bin}(n, \bar{p}),$$

where $\text{Bin}(n, \bar{p})$ is a binomial random variable with parameters n and $\bar{p} = \sum_{i=1}^n P_i/n$. (*Hint:* Prove it first for $n = 2$.) Use the above to show that $\sum_{i=1}^n X_i \leq_v Y$, where Y is Poisson with mean $n\bar{p}$. Hence, for instance, a Poisson random variable is more variable than a binomial having the same mean.

9.30. Suppose $P\{X = 0\} = 1 - M/2$, $P\{X = 2\} = M/2$ where $0 < M < 2$. If Y is a nonnegative integer-valued random variable such that $P\{Y = 1\} = 0$ and $E[Y] = M$, then show that $X \leq_v Y$.

9.31. Jensen's inequality states that for a convex function f,

$$E[f(X)] \geq f(E[X]).$$

(a) Prove Jensen's inequality.

(b) If X has mean $E[X]$, show that

$$X \underset{v}{\geq} E[X],$$

where $E[X]$ is the constant random variable.

(c) Suppose that there exists a random variable Z such that $E[Z|Y] \geq 0$ and such that X has the same distribution as $Y + Z$. Show that $X \geq_v Y$. In fact, it can be shown (though the other direction is difficult to prove) that this is a necessary and sufficient condition for $X \geq_v Y$.

9.32. If $E[X] = 0$, show that $cX \geq_v X$ when $c \geq 1$.

9.33. Suppose that $P\{0 \leq X \leq 1\} = 1$, and let $\theta = E[X]$. Show that $X \leq_v Y$ where Y is a Bernoulli random variable with parameter θ. (That is, $P\{Y = 1\} = \theta = 1 - P\{Y = 0\}$.)

9.34. Show that $E[Y|X] \leq_v Y$. Use this to show that:

(a) if X and Y are independent then $XE[Y] \leq_v XY$.

(b) if X and Y are independent and $E[Y] = 0$ then $X \leq_v X + Y$.

(c) if X_i, $i \geq 1$, are independent and identically distributed then

$$\frac{1}{n}\sum_{i=1}^{n} X_i \underset{v}{\geq} \frac{1}{n+1}\sum_{i=1}^{n+1} X_i.$$

(d) $X + E[X] \leq_v 2X$.

9.35. Show that if $X_1, \ldots, X_n$ are all decreasing functions of a specified set of associated random variables then they are associated.

9.36. In the model of Example 9.7(A), one can always find a family of component subsets $M_1, \ldots, M_m$ such that the system will work if, and only if, all of the components of at least one of these subsets work. (If none of the M_i is a proper subset of another then they are called minimal path sets.) Show that

$$P\{S = 1\} \leq 1 - \prod_i \left(1 - \prod_{j \in M_i} p_j\right).$$

9.37. Show that an infinite sequence of exchangeable Bernoulli random variables is associated.

REFERENCES

Reference 8 is an important reference for all the results of this chapter. In addition, on the topic of stochastically larger, the reader should consult Reference 4. Reference 1 should be seen for additional material on IFR and DFR random variables. Coupling is a valuable and useful technique, which goes back to Reference 5; indeed, the approach used in Section 9.2.2 is from this reference. The approach leading up to and establishing Proposition 9.3.2 and Corollary 9.3.3 is taken from References 2 and 3. Reference 4 presents some useful applications of variability orderings.

1. R. E. Barlow, and F. Proschan, *Statistical Theory of Reliability and Life Testing*, Holt, New York, 1975.
2. M. Brown, "Bounds, Inequalities and Monotonicity Properties for Some Specialized Renewal Processes," *Annals of Probability*, 8 (1980), pp. 227–240.
3. M. Brown, "Further Monotonicity Properties for Specialized Renewal Processes," *Annals of Probability* (1981).
4. S. L. Brumelle, and R. G. Vickson, "A Unified Approach to Stochastic Dominance," in *Stochastic Optimization Models in Finance*, edited by W. Ziemba and R. Vickson, Academic Press, New York, 1975.
5. W. Doeblin, "Sur Deux Problèmes de M. Kolmogoroff Concernant Les Chaines Dénombrables," *Bulletin Societe Mathematique de France*, 66 (1938), pp. 210–220.
6. T. Lindvall, Lectures on the Coupling Method, Wiley, New York, 1992.
7. A. Marshall and I. Olkin, *Inequalities: Theory of Majorization and Its Applications*, Academic Press, Orlando, FL, 1979.
8. J. G. Shanthikumar and M. Shaked, *Stochastic Orders and Their Applications*, Academic Press, San Diego, CA, 1994.

CHAPTER 10

Poisson Approximations

Introduction

Let $X_1, \ldots, X_n$ be Bernoulli random variables with

$$P\{X_i = 1\} = \lambda_i = 1 - P\{X_i = 0\}, \qquad i = 1, \ldots, n$$

and let $W = \sum_{i=1}^{n} X_i$. The *"Poisson paradigm"* states that if the λ_i are all small and if the X_i are either independent or at most "weakly" dependent, then W will have a distribution that is approximately Poisson with mean $\lambda \equiv \sum_{i=1}^{n} \lambda_i$, and so

$$P\{W = k\} \approx e^{-\lambda}\lambda^k/k!.$$

In Section 10.1 we present an approach, based on a result known as Brun's sieve, for establishing the validity of the preceding approximation. In Section 10.2 we give the Stein-Chen method for bounding the error of the Poisson approximation. In Section 10.3, we consider another approximation for $P\{W = k\}$, that is often an improvement on the one specified above.

10.1 Brun's Sieve

The validity of the Poisson approximation can often be established upon applying the following result, known as *Brun's sieve.*

PROPOSITION 10.1.1

Let W be a bounded nonnegative integer-valued random variable. If, for all $i \geq 0$,

$$E\left[\binom{W}{i}\right] \approx \lambda^i/i!$$

then

$$P\{W = j\} \approx e^{-\lambda}\lambda^j/j!, \qquad j \geq 0.$$

Proof Let I_j, $j \geq 0$, be defined by

$$I_j = \begin{cases} 1 & \text{if } W = j \\ 0 & \text{otherwise.} \end{cases}$$

Then, with the understanding that $\binom{r}{k}$ is equal to 0 for $r < 0$ or $k > r$, we have that

$$\begin{aligned} I_j &= \binom{W}{j}(1-1)^{W-j} \\ &= \binom{W}{j}\sum_{k=0}^{W-j}\binom{W-j}{k}(-1)^k \\ &= \sum_{k=0}^{\infty}\binom{W}{j}\binom{W-j}{k}(-1)^k \\ &= \sum_{k=0}^{\infty}\binom{W}{j+k}\binom{j+k}{k}(-1)^k. \end{aligned}$$

Taking expectations gives,

$$\begin{aligned} P\{W = j\} &= \sum_{k=0}^{\infty} E\left[\binom{W}{j+k}\right]\binom{j+k}{k}(-1)^k \\ &\approx \sum_{k=0}^{\infty}\frac{\lambda^{j+k}}{(j+k)!}\binom{j+k}{k}(-1)^k \\ &= \frac{\lambda^j}{j!}\sum_{k=0}^{\infty}(-\lambda)^k/k! \\ &= e^{-\lambda}\lambda^j/j!. \end{aligned}$$

Example 10.1(a) Suppose that W is a binomial random variable with parameters n and p, where n is large and p is small. Let $\lambda = np$ and interpret W as the number of successes in n independent trials where each is a success with probability p; and so $\binom{W}{i}$

represents the number of sets of i successful trials. Now, for each of the $\binom{n}{i}$ sets of i trials, define an indicator variable X_j equal to 1 if all these trials result in successes and equal to 0 otherwise. Then,

$$\binom{W}{i} = \sum_j X_j$$

and thus

$$E\left[\binom{W}{i}\right] = \binom{n}{i} p^i$$

$$= \frac{n(n-1)\cdots(n-i+1)}{i!} p^i.$$

Hence, if i is small in relation to n, then

$$E\left[\binom{W}{i}\right] \approx \lambda^i/i!.$$

If i is not small in relation to n, then $E\left[\binom{W}{i}\right] \approx 0 \approx \lambda^i/i!$, and so we obtain the classical Poisson approximation to the binomial.

Example 10.1(b) Let us reconsider the matching problem, which was considered in Example 1.3(A). In this problem, n individuals mix their hats and then each randomly chooses one. Let W denote the number of individuals that choose his or her own hat and note that $\binom{W}{i}$ represents the number of sets of i individuals that all select their own hats. For each of the $\binom{n}{i}$ sets of i individuals define an indicator variable that is equal to 1 if all the members of this set select their own hats and equal to 0 otherwise. Then if $X_j, j = 1, \ldots, \binom{n}{i}$ is the indicator for the jth set of i individuals we have

$$\binom{W}{i} = \sum_j X_j$$

and so, for $i \le n$,

$$E\left[\binom{W}{i}\right] = \binom{n}{i} \frac{1}{n(n-1)\cdots(n-i+1)} = 1/i!$$

Thus, from Brun's sieve, the number of matches approximately has a Poisson distribution with mean 1.

Example 10.1(c) Consider independent trials in which each trial is equally likely to result in any of r possible outcomes, where r is large. Let X denote the number of trials needed until at least one of the outcomes has occurred k times. We will approximate the probability that X is larger than n by making use of the Poisson approximation.

Letting W denote the number of outcomes that have occurred at least k times in the first n trials, then

$$P\{X > n\} = P\{W = 0\}.$$

Assuming that it is unlikely that any specified outcome would have occurred at least k times in the first n trials, we will now argue that the distribution of W is approximately Poisson.

To begin, note that $\binom{W}{i}$ is equal to the number of sets of i outcomes that have all occurred k or more times in the first n trials. Upon defining an indicator variable for each of the $\binom{r}{i}$ sets of i outcomes we see that

$$E\left[\binom{W}{i}\right] = \binom{r}{i} p$$

where p is the probability that, in the first n trials, each of i specified outcomes has occurred k or more times. Now, each trial will result in any one of i specified outcomes with probability i/r, and so if i/r is small, it follows from the Poisson approximation to the binomial that Y, the number of the first n trials that result in any of i specified outcomes, is approximately Poisson distributed with mean ni/r. Consider those Y trials that result in one of the i specified outcomes, and categorize each by specifying which of the i outcomes resulted. If Y_j is the number that resulted in the jth of the i specified outcomes, then by Example 1.5(I) it follows that the Y_j, $j = 1, \ldots, i$, are approximately independent Poisson random variables, each with mean $\frac{1}{i}(ni/r) = n/r$. Therefore, from the independence of the Y_j, we see that $p \approx (\alpha_n)^i$, where

$$\alpha_n = 1 - \sum_{j=0}^{k-1} e^{-n/r}(n/r)^j/j!$$

Thus, when i/r is small

$$E\left[\binom{W}{i}\right] \approx \binom{r}{i} \alpha_n^i$$
$$\approx (r\alpha_n)^i/i!$$

In addition, since we have supposed that the probability that any specified outcome occurs k or more times in the n trials is small it follows that α_n (the Poisson approximation to this binomial probability) is also small and so when i/r is not small,

$$E\left[\binom{W}{i}\right] \approx 0 \approx (r\alpha_n)^i/i!$$

Hence, in all cases we see that

$$E\left[\binom{W}{i}\right] \approx (r\alpha_n)^i/i!$$

Therefore, W is approximately Poisson with mean $r\alpha_n$, and so

$$P\{X > n\} = P\{W = 0\} \approx \exp\{-r\alpha_n\}.$$

As an application of the preceding consider the birthday problem in which one is interested in determining the number of individuals needed until at least three have the same birthday. This is in fact the preceding with $r = 365$, $k = 3$. Thus,

$$P\{X > n\} \approx \exp\{-365\alpha_n\}$$

where α_n is the probability that a Poisson random variable with mean $n/365$ is at least 3. For instance, with $n = 88$, $\alpha_{88} = .001952$ and so

$$P\{X > 88\} \approx \exp\{-365\alpha_{88}\} \approx .490.$$

That is, with 88 people in a room there is approximately a 51 percent chance that at least 3 have the same birthday.

10.2 The Stein-Chen Method for Bounding the Error of the Poisson Approximation

As in the previous section, let $W = \sum_{i=1}^{n} X_i$, where the X_i are Bernoulli random variables with respective means λ_i, $i = 1, \ldots, n$. Set $\lambda = \sum_{i=1}^{n} \lambda_i$ and let A denote a set of nonnegative integers. In this section we present an approach, known as the Stein-Chen method, for bounding the error when approximating $P\{W \in A\}$ by $\sum_{i \in A} e^{-\lambda}\lambda^i/i!$.

To begin, for fixed λ and A, we define a function g for which

$$E[\lambda g(W+1) - Wg(W)] = P\{W \in A\} - \sum_{i \in A} e^{-\lambda}\lambda^i/i!$$

This is done by recursively defining g as follows:

$$g(0) = 0$$

and for $j \geq 0$,

$$g(j+1) = \frac{1}{\lambda}\left[I\{j \in A\} - \sum_{i \in A} e^{-\lambda}\lambda^i/i! + jg(j)\right]$$

where $I\{j \in A\}$ is 1 if $j \in A$ and 0 otherwise. Hence, for $j \geq 0$,

$$\lambda g(j+1) - jg(j) = I\{j \in A\} - \sum_{i \in A} e^{-\lambda}\lambda^i/i!$$

and so,

$$\lambda g(W+1) - Wg(W) = I\{W \in A\} - \sum_{i \in A} e^{-\lambda}\lambda^i/i!$$

Taking expectations gives that

$$E[\lambda g(W+1) - Wg(W)] = P\{W \in A\} - \sum_{i \in A} e^{-\lambda}\lambda^i/i! \tag{10.2.1}$$

The following property of g will be needed. We omit its proof.

Lemma 10.2.1

For any λ and A

$$|g(j) - g(j-1)| \leq \frac{1 - e^{-\lambda}}{\lambda} \leq \min(1, 1/\lambda).$$

Since, for $i < j$

$$g(j) - g(i) = g(j) - g(j-1) + g(j-1) - g(j-2) + \cdots + g(i+1) - g(i),$$

we obtain from Lemma 10.2.1 and the triangle inequality that,

$$|g(j) - g(i)| \le |j - i| \min(1, 1/\lambda). \tag{10.2.2}$$

■

To continue our analysis we will need the following lemma.

Lemma 10.2.2

For any random variable R

$$E[WR] = \sum_{i=1}^{n} \lambda_i E[R \mid X_i = 1].$$

Proof

$$\begin{aligned} E[WR] &= E\left[\sum_{i=1}^{n} RX_i\right] \\ &= \sum_{i=1}^{n} E[RX_i] \\ &= \sum_{i=1}^{n} E[RX_i \mid X_i = 1]\lambda_i \\ &= \sum_{i=1}^{n} \lambda_i E[R \mid X_i = 1]. \end{aligned}$$

If, in Lemma 10.2.2, we let $R = g(W)$ then we obtain that

$$\begin{aligned} E[Wg(W)] &= \sum_{i=1}^{n} \lambda_i E[g(W) \mid X_i = 1] \\ &= \sum_{i=1}^{n} \lambda_i E[g(V_i + 1)], \end{aligned} \tag{10.2.3}$$

where V_i is any random variable whose distribution is the same as the conditional distribution of $\sum_{j \ne i} X_j$ given that $X_i = 1$. That is, V_i is any random variable such that

$$P\{V_i = k\} = P\left\{\sum_{j \ne i} X_j = k \mid X_i = 1\right\}.$$

Since $E[\lambda g(W+1)] = \sum_{i=1}^{n} \lambda_i E[g(W+1)]$ we obtain from Equations (10.2.1) and (10.2.3) that,

$$\text{(10.2.4)}\quad \left|P\{W \in A\} - \sum_{i \in A} e^{-\lambda}\lambda^i/i!\right| = \left|\sum_{i=1}^{n} \lambda_i(E[g(W+1)] - E[g(V_i+1)])\right|$$
$$= \left|\sum_{i=1}^{n} \lambda_i E[g(W+1) - g(V_i+1)]\right|$$
$$\leq \sum_{i=1}^{n} \lambda_i E[|g(W+1) - g(V_i+1)|]$$
$$\leq \sum_{i=1}^{n} \lambda_i \min(1, 1/\lambda) E[|W - V_i|]$$

where the final inequality used Equation (10.2.2). Therefore, we have proven the following.

THEOREM 10.2.3

Let V_i be any random variable that is distributed as the conditional distribution of $\sum_{j \neq i} X_j$ given that $X_i = 1$, $i = 1, \ldots, n$. That is, for each i, V_i is such that

$$P\{V_i = k\} = P\left\{\sum_{j \neq i} X_j = k \,|\, X_i = 1\right\}, \qquad \text{for all } k.$$

Then for any set A of nonnegative integers

$$\left|P\{W \in A\} - \sum_{i \in A} e^{-\lambda}\lambda^i/i!\right| \leq \min(1, 1/\lambda) \sum_{i=1}^{n} \lambda_i E[|W - V_i|].$$

Remark The proof of Theorem 10.2.3 strengthens our intuition as to the validity of the Poisson paradigm. From Equation (10.2.4) we see that the Poisson approximation will be precise if W and V_i have approximately the same distribution for all i. That is, if the conditional distribution of $\sum_{j \neq i} X_j$ given that $X_i = 1$ is approximately the distribution of W, for all i. But this will be the case if

$$\sum_{j \neq i} X_j | X_i = 1 \overset{d}{\approx} \sum_{j \neq i} X_j \overset{d}{\approx} \sum_{j} X_j,$$

where $\overset{d}{\approx}$ means "approximately has the same distribution." That is, the dependence between the X_j must be weak enough so that the information that a specified one of them is equal to 1 does not much affect the distribution of the sum of the others, and also $\lambda_i = P\{X_i = 1\}$ must be small for all i so that $\sum_{j \neq i} X_j$ and $\sum_j X_j$ approximately have the same distribution. Hence, the Poisson approximation will be precise provided the λ_i are small and the X_i are only weakly dependent.

EXAMPLE 10.2(A) If the X_i are independent Bernoulli random variables then we can let $V_i = \sum_{j \neq i} X_j$. Therefore,

$$E[|W - V_i|] = E[X_i] = \lambda_i$$

and so,

$$\left| P\{W \in A\} - \sum_{i \in A} e^{-\lambda}\lambda^i/i! \right| \leq \min(1, 1/\lambda) \sum_{i=1}^{n} \lambda_i^2 .$$

Since the bound on the Poisson approximation given in Theorem 10.2.3 is in terms of any random variables V_i, $i = 1, \ldots, n$, having the distributions specified, we can attempt to couple V_i with W so as to simplify the computation of $E[|W - V_i|]$. For instance, in many cases where the X_j are negatively correlated, and so the information that $X_i = 1$ makes it less likely for other X_j to equal 1, there is a coupling that results in V_i being less than or equal to W. If W and V_i are so coupled in this manner, it follows that

$$E[|W - V_i|] = E[W - V_i] = E[W] - E[V_i].$$

EXAMPLE 10.2(B) Suppose that m balls are placed among n urns, with each ball independently going into urn i with probability p_i, $i = 1, \ldots, n$. Let X_i be the indicator for the event that urn i is empty and let $W = \sum_{i=1}^{n} X_i$ denote the number of empty urns. If m is large enough so that each of the

$$\lambda_i = P\{X_i = 1\} = (1 - p_i)^m$$

is small, then we might expect that W should approximately have a Poisson distribution with mean $\lambda = \sum_{i=1}^{n} \lambda_i$.

As the X_i have a negative dependence, since urn i being empty makes it less likely that other urns will also be empty, we might suppose that we can construct a coupling that results in $W \geq V_i$. To show that this is indeed possible, imagine that the m balls are distributed according to the specified probabilities and let W denote the number of empty urns. Now take each of the balls that are in urn i and redistribute them, independently, among the other urns according to the probabilities $p_j/(1 - p_i)$, $j \neq i$. If we now let V_i denote the number of urns j, $j \neq i$, that are empty, then it is easy to see that V_i has the appropriate distribution specified in Theorem 10.2.3. Since $W \geq V_i$ it follows that

$$\begin{aligned} E[|W - V_i|] &= E[W - V_i] \\ &= E[W] - E[V_i] \\ &= \lambda - \sum_{j \neq i} \left(1 - \frac{p_j}{1 - p_i}\right)^m. \end{aligned}$$

Thus, for any set A,

$$\begin{aligned} &\left|P\{W \in A\} - \sum_{i \in A} e^{-\lambda}\lambda^i/i!\right| \\ &\leq \min(1, 1/\lambda) \sum_{i=1}^{n} \lambda_i \left[\lambda - \sum_{j \neq i} \left(1 - \frac{p_j}{1 - p_i}\right)^m\right]. \end{aligned}$$

Added insight about when the error of the Poisson approximation is small is obtained when there is a way of coupling V_i and W so that $W \geq V_i$. For in this case we have that

$$\begin{aligned} \sum_i \lambda_i E[|W - V_i|] &= \sum_i \lambda_i E[W] - \sum_i \lambda_i E[V_i] \\ &= \lambda^2 - \sum_i \lambda_i (E[1 + V_i] - 1) \\ &= \lambda^2 + \lambda - \sum_i \lambda_i E\left[1 + \sum_{j \neq i} X_j \middle| X_i = 1\right] \\ &= \lambda^2 + \lambda - \sum_i \lambda_i E[W | X_i = 1] \\ &= \lambda^2 + \lambda - E[W^2] \\ &= \lambda - \text{Var}(W), \end{aligned}$$

where the next to last equality follows by setting $R = W$ in Lemma 10.2.2.

Thus, we have shown the following.

PROPOSITION 10.2.4

If, for each i, $i = 1, \ldots, n$, there is a way of coupling V_i and W so that $W \geq V_i$, then for any set A

$$\left|P\{W \in A\} - \sum_{i \in A} e^{-\lambda}\lambda^i/i!\right| \leq \min(1, 1/\lambda)[\lambda - \text{Var}(W)].$$

Thus, loosely speaking, in cases where the occurrence of one of the X_i makes it less likely for the other X_j to occur, the Poisson approximation will be quite precise provided $\text{Var}(W) \approx E[W]$.

10.3 Improving the Poisson Approximation

Again, let $X_1, \ldots, X_n$ be Bernoulli random variables with $\lambda_i = P\{X_i = 1\}$. In this section we present another approach to approximating the probability mass function of $W = \sum_{i=1}^{n} X_i$. It is based on the following proposition.

PROPOSITION 10.3.1

With V_i distributed as the conditional distribution of $\sum_{j \neq i} X_j$ given that $X_i = 1$,

(a) $P\{W > 0\} = \sum_{i=1}^{n} \lambda_i E[1/(1 + V_i)]$.

(b) $P\{W = k\} = \dfrac{1}{k}\sum_{i=1}^{n} \lambda_i P\{V_i = k - 1\}, \qquad k \geq 1.$

Proof Both parts follow from Lemma 10.2.2. To prove (a), let

$$R = \begin{cases} 1/W & \text{if } W > 0 \\ 0 & \text{if } W = 0. \end{cases}$$

Then from Lemma 10.2.2, we obtain that

$$P\{W > 0\} = \sum_{i=1}^{n} \lambda_i E[1/W \mid X_i = 1] = \sum_{i=1}^{n} \lambda_i E[1/(1 + V_i)]$$

which proves (a). To prove (b), let

$$R = \begin{cases} 1/k & \text{if } W = k \\ 0 & \text{otherwise,} \end{cases}$$

and apply Lemma 10.2.2 to obtain

$$P\{W = k\} = \sum_{i=1}^{n} \lambda_i \frac{1}{k} P\{W = k \mid X_i = 1\} = \frac{1}{k} \sum_{i=1}^{n} \lambda_i P\{V_i = k - 1\}.$$

Let

$$a_i = E[V_i] = \sum_{j \neq i} E[X_j \mid X_i = 1].$$

Now, if the a_i are all small and if, conditional on $X_i = 1$, the remaining X_j, $j \neq i$, are only "weakly" dependent then the conditional distribution of V_i will approximately be Poisson with mean a_i. Assuming this is so, then

$$P\{V_i = k - 1\} \approx \exp\{-a_i\} a_i^{k-1}/(k-1)!, \qquad k \geq 1$$

and

$$\begin{aligned} E[1/(1 + V_i)] &\approx \sum_j (1 + j)^{-1} \exp\{-a_i\} a_i^j / j! \\ &= \frac{1}{a_i} \exp\{-a_i\} \sum_j a_i^{j+1}/(j+1)! \\ &= \exp\{-a_i\}(\exp\{a_i\} - 1)/a_i \\ &= (1 - \exp\{-a_i\})/a_i. \end{aligned}$$

Using the preceding in conjunction with Proposition 10.3.1 leads to the following approximations:

$$\begin{aligned} &P\{W > 0\} \approx \sum_{i=1}^{n} \lambda_i (1 - \exp\{-a_i\})/a_i, \\ &P\{W = k\} \approx \frac{1}{k} \sum_{i=1}^{n} \lambda_i \exp\{-a_i\} a_i^{k-1}/(k-1)!, \qquad k \geq 1. \end{aligned} \tag{10.3.1}$$

If the X_i are "negatively" dependent then given that $X_i = 1$ it becomes less likely for the other X_j to equal 1. Hence, V_i is the sum of smaller mean

Bernoullis than is W and so we might expect that its distribution is even closer to Poisson than is that of W. In fact, it has been shown in Reference [4] that the preceding approximation tends to be more precise than the ordinary Poisson approximation when the X_i are negatively dependent, provided that $\lambda < 1$. The following examples provide some numerical evidence for this statement.

EXAMPLE 10.3(A) ***(Bernoulli Convolutions).*** Suppose that X_i, $i = 1, \ldots, 10$, are independent Bernoulli random variables with $E[X_i] = i/1000$. The following table compares the new approximation given in this section with the usual Poisson approximation.

k	$P\{W = k\}$	New Approx.	Usual Approx.
0	.946302	.946299	.946485
1	.052413	.052422	.052056
2	.001266	.001257	.001431
3	.000018	.000020	.000026

EXAMPLE 10.3(B) Example 10.2(B) was concerned with the number of multinomial outcomes that never occur when each of m independent trials results in any of n possible outcomes with respective probabilities $p_1, \ldots, p_n$. Let X_i equal 1 if none of the trials result in outcome i, $i = 1, \ldots, n$, and let it be 0 otherwise. Since the nonoccurrence of outcome i will make it even less likely for other outcomes not to occur there is a negative dependence between the X_i. Thus we might expect that the approximations given by Equation (10.3.1), where

$$a_i = \sum_{j \neq i} E[X_j | X_i = 1] = \sum_{j \neq i} \left(1 - \frac{p_j}{1 - p_i}\right)^m$$

are more precise than the straight Poisson approximations when $\lambda < 1$.

The following table compares the two approximations when $n = 4$, $m = 20$, and $p_i = i/10$. W is the number of outcomes that never occur.

k	$P\{W = k\}$	New Approx.	Usual Approx.
0	.86690	.86689	.87464
1	.13227	.13230	.11715
2	.00084	.00080	.00785

EXAMPLE 10.3(C) Again consider m independent trials, each of which results in any of n possible outcomes with respective probabilities $p_1, \ldots, p_n$, but now let W denote the number of outcomes

that occur at least r times. In cases where it is unlikely that any specified outcome occurs at least r times, W should be roughly Poisson. In addition, as the indicators of W are negatively dependent (since outcome i occurring at least r times makes it less likely that other outcomes have also occurred that often) the new approximation should be quite precise when $\lambda < 1$.

The following table compares the two approximations when $n = 6$, $m = 10$, and $p_i = 1/6$. W is the number of outcomes that occur at least five times.

k	$P\{W = k\}$	New Approx.	Usual Approx.
0	.9072910	.9072911	.91140
1	.0926468	.0926469	.08455
2	.0000625	.0000624	.00392

Proposition 10.3.1(a) can also be used to obtain a lower bound for $P\{W > 0\}$.

Corollary 10.3.2

$$P\{W > 0\} \geq \sum_{i=1}^{n} \lambda_i/(1 + a_i).$$

Proof Since $f(x) = 1/x$ is a convex function for $x \geq 0$ the result follows from Proposition 10.3.1(a) upon applying Jensen's inequality.

The bound given in Corollary 10.3.2 can be quite precise. For instance, in Example 10.3(A) it provides the upper bound $P\{W = 0\} \leq .947751$ when the exact probability is .946302. In Examples 10.3(B) and 10.3(C) it provides upper bounds, $P\{W = 0\} \leq .86767$ and $P\{W = 0\} \leq .90735$, that are less than the straight Poisson approximations.

Problems

10.1. A set of n components, each of which independently fails with probability p, are in a linear arrangement. The system is said to fail if k consecutive components fail.

(a) For $i \leq n + 1 - k$, let Y_i be the event that components $i, i + 1, \ldots, i + k - 1$ all fail. Would you expect the distribution of $\sum_i Y_i$ to be approximately Poisson? Explain.

(b) Define indicators X_i, $i = 1, \ldots, n + 1 - k$, such that for all $i \neq j$ the events $\{X_i = 1\}$ and $\{X_j = 1\}$ are either mutually exclusive or independent, and for which

$$P(\text{system failure}) = P\left\{X + \sum_i X_i > 0\right\},$$

where X is the indicator for the event that all components are failed.

(c) Approximate P(system failure).

10.2. Suppose that m balls are randomly chosen from a set of N, of which n are red. Let W denote the number of red balls selected. If n and m are both small in comparison to N, use Brun's sieve to argue that the distribution of W is approximately Poisson.

10.3. Suppose that the components in Problem 10.1 are arranged in a circle. Approximate P(system failure) and use Brun's sieve to justify this approximation.

10.4. If X is a Poisson random variable with mean λ, show that

$$E[Xh(X)] = \lambda E[h(X + 1)]$$

provided these expectations exist.

10.5. A group of $2n$ individuals, consisting of n couples, are randomly arranged at a round table. Let W denote the number of couples seated next to each other.

(a) Approximate $P\{W = k\}$.

(b) Bound this approximation.

10.6. Suppose that N people toss their hats in a circle and then randomly make selections. Let W denote the number of the first n people that select their own hat.

(a) Approximate $P\{W = 0\}$ by the usual Poisson approximation.

(b) Approximate $P\{W = 0\}$ by using the approximation of Section 10.3.

(c) Determine the bound on $P\{W = 0\}$ that is provided by Corollary 10.3.2.

(d) For $N = 20$, $n = 3$, compute the exact probability and compare with the preceding bound and approximations.

10.7. Given a set of n vertices suppose that between each of the $\binom{n}{2}$ pairs of vertices there is (independently) an edge with probability p. Call the sets of vertices and edges a graph. Say that any set of three vertices constitutes a triangle if the graph contains all of the $\binom{3}{2}$ edges connecting these vertices. Let W denote the number of triangles in the graph. Assuming that p is small, approximate $P\{W = k\}$ by

(a) using the Poisson approximation.

(b) using the approximation in Section 10.3.

10.8. Use the approximation of Section 10.3 to approximate P(system failure) in Problem 10.3. Also determine the bound provided by Corollary 10.3.2.

10.9. Suppose that W is the sum of independent Bernoulli random variables. Bound the difference between $P\{W = k\}$ and its approximation as given in Equation (10.3.1). Compare this bound with the one obtained in Example 10.2(A) for the usual Poisson approximation.

References

For additional results on the Poisson paradigm and Brun's sieve, including a proof of the limit theorem version of Proposition 10.1.1, the reader should see Reference 2. For learning more about the Stein-Chen method, we recommend Reference 3. The approximation method of Section 10.3 is taken from Reference 4. Many nice applications regarding Poisson approximations can be found in Reference 1.

1. D. Aldous, *Probability Approximations via the Poisson Clumping Heuristic*, Springer-Verlag, New York, 1989.
2. N. Alon, J. Spencer, and P. Erdos, *The Probabilistic Method*, Wiley, New York, 1992.
3. A. D. Barbour, L. Holst, and S. Janson, *Poisson Approximations*, Clarendon Press, Oxford, England, 1992.
4. E. Peköz and S. M. Ross, "Improving Poisson Approximations," *Probability in the Engineering and Informational Sciences*, Vol. 8, No. 4, December, 1994, pp. 449–462.

Answers and Solutions to Selected Problems

Chapter 1

1.5. **(a)** $$P_{N_1,\ldots,N_{r-1}}(n_1,\ldots,n_{r-1}) = \frac{n!}{\prod_{i=1}^{r} n_i!}\prod_{i=1}^{r} P_i^{n_i},$$

where $n_i = 0, \ldots, n$ and $\sum_{i=1}^{r} n_i = n$.

(b) $E[N_i] = nP_i, \qquad E[N_i^2] = nP_i - nP_i^2 + n^2P_i^2,$

$E[N_j] = nP_j, \qquad E[N_j^2] = nP_j - nP_j^2 + n^2P_j^2,$

$E[N_iN_j] = E[E[N_iN_j|N_j]],$

$$E[N_iN_j|N_j = m] = mE[N_i|N_j = m] = m(n - m)\frac{P_i}{1 - P_j}$$

$$= \frac{nmP_i - m^2P_i}{1 - P_j},$$

$$E[N_iN_j] = \frac{nE[N_j]P_i - E[N_j^2]P_i}{1 - P_j} = \frac{n^2P_jP_i - nP_iP_j + nP_j^2P_i - n^2P_j^2P_i}{1 - P_j}$$

$$= \frac{n^2P_jP_i(1 - P_j) - nP_iP_j(1 - P_j)}{1 - P_j} = n^2P_iP_j - nP_iP_j,$$

$\mathrm{Cov}(N_i, N_j) = E[N_iN_j] - E[N_i]E[N_j] = -nP_iP_j, \quad i \neq j.$

(c) Let $I_j = \begin{cases} 1 & \text{if outcome } j \text{ never occurs} \\ 0 & \text{otherwise.} \end{cases}$

$E[I_j] = (1 - P_j)^n, \qquad \mathrm{Var}[I_j] = (1 - P_j)^n(1 - (1 - P_j)^n),$

$E[I_iI_j] = (1 - P_i - P_j)^n, \qquad i \neq j,$

Number of outcomes that do not occur $= \sum_{j=1}^{r} I_j$,

$$E\left[\sum_{j=1}^{r} I_j\right] = \sum_{j=1}^{r} (1 - P_j)^n,$$

$$\text{Var}\left[\sum_{i=1}^{r} I_j\right] = \sum_{1}^{r} \text{Var}[I_j] + \sum\sum_{i \neq j} \text{Cov}[I_i, I_j],$$

$$\begin{aligned}\text{Cov}[I_i I_j] &= E[I_i I_j] - E[I_i]E[I_j] \\ &= (1 - P_i - P_j)^n - (1 - P_i)^n(1 - P_j)^n,\end{aligned}$$

$$\begin{aligned}\text{Var}\left[\sum_{j=1}^{r} I_j\right] &= \sum_{j=1}^{r} (1 - P_j)^n(1 - (1 - P_j)^n) \\ &\quad + \sum\sum_{i \neq j} [(1 - P_i - P_j)^n - (1 - P_i)^n(1 - P_j)^n].\end{aligned}$$

1.6. **(a)** Let $I_j = \begin{cases} 1 & \text{if there is a record at time } j \\ 0 & \text{otherwise.} \end{cases}$

$$N_n = \sum_{j=1}^{n} I_j,$$

$$E[N_n] = \sum_{j=1}^{n} E[I_j] = \sum_{j=1}^{n} \frac{1}{j},$$

$$\text{Var}[N_n] = \sum_{j=1}^{n} \text{Var}[I_j] = \sum_{j=1}^{n} \frac{1}{j}\left(1 - \frac{1}{j}\right), \text{ since the } I_j \text{ are independent.}$$

(b) Let $T = \min\{n: n > 1 \text{ and a record occurs at } n\}$.

$$T > n \Leftrightarrow X_1 = \text{largest of } X_1, X_2, \ldots, X_n,$$

$$E[T] = \sum_{n=1}^{\infty} P\{T > n\} = \sum_{n=1}^{\infty} \frac{1}{n} = \infty,$$

$$P\{T = \infty\} = \lim_{n \to \infty} P\{T > n\} = 0.$$

(c) Let T_y denote the time of the first record value greater than y. Let XT_y be the record value at time T_y.

$$\begin{aligned}P\{XT_y > x \mid T_y = n\} &= P\{X_n > x \mid X_1 < y, X_2 < y, \ldots, X_{n-1} < y, X_n > y\} \\ &= P\{X_n > x \mid X_n > y\} \\ &= \begin{cases} 1 & x < y \\ \overline{F}(x)/\overline{F}(y) & x > y. \end{cases}\end{aligned}$$

Since $P\{XT_y > x \mid T_y = n\}$ does not depend on n, we conclude T_y is independent of XT_y.

1.10. Suppose that the outcome of each contest is independent, with either contestant having probability 1/2 of winning. For any set S of k contestants let $A(S)$

be the event that no member of S^c beats every member of S. Then

$$P\left\{\bigcup_S A(S)\right\} \leq \sum_S P\{A(S)\} = \binom{n}{k}[1 - (1/2)^k]^{n-k},$$

where the equality follows since there are $\binom{n}{k}$ sets of size k and $P\{A(S)\}$ is the probability that each of the $n - k$ contestants not in S lost at least one of his or her k matches with players in S. Hence, if $\binom{n}{k}[1 - (1/2)^k]^{n-k} < 1$ then there is a positive probability that none of the events $A(S)$ occurs.

1.14. **(a)** Let Y_j denote the number of 1's that occur between the $(j - 1)$st and jth even number, $j = 1, \ldots, 10$. By conditioning on whether the first appearance of either a 1 or an even number is 1, we obtain

$$\begin{aligned} E[Y_j] &= E[Y_j|\text{even before } 1]3/4 + E[Y_j|1 \text{ before even}]1/4 \\ &= (1 + E[Y_j])1/4, \end{aligned}$$

implying that

$$E[Y_j] = 1/3.$$

As,

$$X_1 = \sum_{j=1}^{10} Y_j$$

it follows that

$$E[X_1] = 10/3.$$

(b) Let W_j denote the number of 2's that occur between the $(j - 1)$st and jth even number, $j = 1, \ldots, 10$. Then,

$$\begin{aligned} E[W_j] &= E[W_j|\text{even number is } 2]1/3 + E[W_j|\text{even number is not } 2]2/3 \\ &= 1/3 \end{aligned}$$

and so,

$$E[X_2] = E\left[\sum_{j=1}^{10} W_j\right] = 10/3.$$

(c) Since each outcome that is either 1 or even is even with probability 3/4, it follows that $1 + Y_j$ is a geometric random variable with mean 4/3.

Hence, $\sum_{j=1}^{10} (1 + Y_j) = 10 + X_1$ is a negative binomial random variable with parameters $n = 10$ and $p = 3/4$. Therefore,

$$P\{X_1 = i\} = P\{\text{Neg Bin}(10, 3/4) = 10 + i\} = \binom{9+i}{9} (3/4)^{10}(1/4)^i.$$

(d) Since there are 10 outcomes that are even, and each is independently 2 with probability 1/3, it follows that X_2 is a binomial random variable with parameters $n = 10$, $p = 1/3$. That is,

$$P\{X = i\} = \binom{10}{i} (1/3)^i(2/3)^{10-i}.$$

1.17. **(a)**
$$\begin{aligned} F_{i,n}(x) &= P\{i\text{th smallest} \le x | X_n \le x\}F(x) \\ &\quad + P\{i\text{th smallest} \le x | X_n > x\}\bar{F}(x) \\ &= P\{X_{i-1,n-1} \le x\}F(x) + P\{X_{i,n-1} \le x\}\bar{F}(x). \end{aligned}$$

(b)
$$\begin{aligned} F_{i,n-1}(x) &= P\{X_{i,n-1} \le x | X_n \text{ is among } i \text{ smallest}\}\, i/n \\ &\quad + P\{X_{i,n-1} \le x | X_n \text{ is not among } i \text{ smallest}\}(1 - i/n) \\ &= P\{X_{i+1,n} \le x | X_n \text{ is among } i \text{ smallest}\} i/n \\ &\quad + P\{X_{i,n} \le x | X_n \text{ is not among } i \text{ smallest}\}(1 - i/n) \\ &= P\{X_{i+1,n} \le x\} i/n + P\{X_{i,n} \le x\}(1 - i/n) \end{aligned}$$

where the final equality follows since whether or not X_n is among the i smallest of $X_1, \ldots, X_n$ does not change the joint distribution of $X_{i,n}$, $i = 1, \ldots, n$.

1.21. $P\{N = 0\} = P\{u_1 < e^{-\lambda}\} = e^{-\lambda}$. Suppose that

$$\begin{aligned} P\{N = n\} &= P\{u_1 \ge e^{-\lambda}, u_1u_2 \ge e^{-\lambda}, \ldots, u_1 \cdots u_n \ge e^{-\lambda}, u_1 \cdots u_{n+1} < e^{-\lambda}\} \\ &= e^{-\lambda}\lambda^n/n! \end{aligned}$$

Hence,

$$\begin{aligned} P\{N = n + 1\} &= \int_0^1 P\{u_1 \ge e^{-\lambda}, \ldots, u_1 \cdots u_{n+1} \ge e^{-\lambda}, u_1 \cdots u_{n+2} < e^{-\lambda} | u_1 = x\}\, dx \\ &= \int_{e^{-\lambda}}^1 P\left\{u_2 \ge \frac{e^{-\lambda}}{x}, \ldots, u_2 \cdots u_{n+1} \ge \frac{e^{-\lambda}}{x}, u_2 \cdots u_{n+2} < \frac{e^{-\lambda}}{x}\right\} dx \\ &= \int_{e^{-\lambda}}^1 e^{-(\lambda + \log x)} \frac{(\lambda + \log x)^n}{n!}\, dx, \end{aligned}$$

where the last equality follows from the induction hypothesis since $e^{-\lambda}/x = e^{-(\lambda + \log x)}$. From the above we thus have

$$P\{N = n + 1\} = \frac{1}{n!}\int_{e^{-\lambda}}^{1} \frac{e^{-\lambda}(\lambda + \log x)^n}{x}\,dx$$

$$= \frac{e^{-\lambda}}{n!}\int_{0}^{\lambda} y^n\,dy \qquad (\text{by } y = \lambda + \log x)$$

$$= \frac{e^{-\lambda}\lambda^{n+1}}{(n+1)!},$$

which completes the induction.

1.22. $\mathrm{Var}(X|Y) = E[(X - E(X|Y))^2|Y]$

$$= E[X^2 - 2XE(X|Y) + (E(X|Y))^2|Y]$$

$$= E[X^2|Y] - 2[E(X|Y)]^2 + [E(X|Y)]^2$$

$$= E[X^2|Y] - (E[X|Y])^2,$$

$$\mathrm{Var}(X) = E[X^2] - (E[X])^2$$

$$= E(E[X^2|Y]) - (E[E[X|Y]])^2$$

$$= E[\mathrm{Var}(X|Y) + (E[X|Y])^2] - (E[E[X|Y]])^2$$

$$= E[\mathrm{Var}(X|Y)] + E[(E[X|Y])^2] - (E[E[X|Y]])^2$$

$$= E[\mathrm{Var}(X|Y)] + \mathrm{Var}(E[X|Y]).$$

1.34. $P\{X_1 < X_2|\min(X_1, X_2) = t\} = \dfrac{P\{X_1 < X_2, \min(X_1, X_2) = t\}}{P\{\min(X_1, X_2) = t\}}$

$$= \frac{P\{X_1 = t, X_2 > t\}}{P\{X_1 = t, X_2 > t\} + P\{X_2 = t, X_1 > t\}}$$

$$= \frac{P\{X_1 = t\}P\{X_2 > t\}}{P\{X_1 = t\}P\{X_2 > t\} + P\{X_2 = t\}P\{X_1 > t\}},$$

$$P\{X_2 = t\} = \lambda_2(t)P\{X_2 > t\},$$

$$P\{X_1 > t\} = \frac{P\{X_1 = t\}}{\lambda_1(t)},$$

$$P\{X_2 = t\}P\{X_1 > t\} = \frac{\lambda_2(t)}{\lambda_1(t)}P\{X_1 = t\}P\{X_2 > t\}.$$

Hence

$$P\{X_1 < X_2|\min(X_1, X_2) = t\} = \frac{1}{1 + \dfrac{\lambda_2(t)}{\lambda_1(t)}}$$

$$= \frac{\lambda_1(t)}{\lambda_1(t) + \lambda_2(t)},$$

1.37. Let I_n equal 1 if a peak occurs at time n, and let it be 0 otherwise; and note, since each of X_{n-1}, X_n, or X_{n+1} is equally likely to be the largest of the three, that $E[I_n] = 1/3$. As $\{I_2, I_5, I_8, \ldots\}$, $\{I_3, I_6, I_9, \ldots\}$, $\{I_4, I_7, I_{10}, \ldots\}$ are all independent and identically distributed sequences, it follows from the strong law of large numbers that the average of the first n terms in each sequence converges, with probability 1, to 1/3. But this implies that, with probability 1, $\lim_{n\to\infty} \sum_{i=1}^{n} I_{i+1}/n = 1/3$.

1.39. $E[T_1] = 1$.

For $i > 1$,

$$E[T_i] = 1 + 1/2(E[\text{time to go from } i-2 \text{ to } i] = 1 + 1/2(E[T_{i-1}] + E[T_i])$$

and so,

$$E[T_i] = 2 + E[T_{i-1}], \qquad i > 1.$$

Hence,

$$E[T_2] = 3$$
$$E[T_3] = 5$$
$$E[T_i] = 2i - 1, \qquad i = 1, \ldots, n.$$

If $T_{0,n}$ is the expected number of steps to go from 0 to n, then

$$E[T_{0,n}] = E\left[\sum_{i=1}^{n} T_i\right] = 2n(n+1)/2 - n = n^2.$$

Chapter 2

2.6. Letting N denote the number of components that fail, the desired answer is $E[N]/(\mu_1 + \mu_2)$, where

$$E[N] = \sum_{k=\min(n,m)}^{n+m-1} k\left[\binom{k-1}{n-1}\left(\frac{\mu_1}{\mu_1+\mu_2}\right)^n\left(\frac{\mu_2}{\mu_1+\mu_2}\right)^{k-n} + \binom{k-1}{m-1}\left(\frac{\mu_1}{\mu_1+\mu_2}\right)^{k-m}\left(\frac{\mu_2}{\mu_1+\mu_2}\right)^{m}\right],$$

and where $\binom{k-1}{i} = 0$ when $i > k - 1$.

2.7. Since $(S_1, S_2, S_3) = (s_1, s_2, s_3)$ is equivalent to $(X_1, X_2, X_3) = (s_1, s_2 - s_1, s_3 - s_2)$, it follows that the joint density of S_1, S_2, S_3 is given by

$$f(s_1, s_2, s_3) = \lambda e^{-\lambda s_1}\lambda e^{-\lambda(s_2-s_1)}\lambda e^{-\lambda(s_3-s_2)}$$
$$= \lambda^3 e^{-\lambda s_3}, \qquad 0 < s_1 < s_2 < s_3.$$

2.8. (a) $$P\left\{\frac{-\log U_i}{\lambda} \le x\right\} = P\left\{\log\left(\frac{1}{U_i}\right) \le \lambda x\right\}$$
$$= P\{1/U_i \le e^{\lambda x}\}$$
$$= P\{U_i \ge e^{-\lambda x}\}$$
$$= 1 - e^{-\lambda x}.$$

(b) Letting X_i denote the interarrival times of a Poisson process, then $N(1)$—the number of events by time 1—will equal that value of n such that

$$\sum_1^n X_i < 1 < \sum_1^{n+1} X_i,$$

or, equivalently, that value of n such that

$$-\sum_1^n \log U_i < \lambda < -\sum_1^{n+1} \log U_i,$$

or, equivalently,

$$\sum_1^n \log U_i > -\lambda > \sum_1^{n+1} \log U_i$$

or,

$$\prod_1^n U_i > e^{-\lambda} > \prod_1^{n+1} U_i.$$

The result now follows since $N(1)$ is a Poisson random variable with mean 1.

2.9. (a) $P_s\{\text{winning}\} = P\{1 \text{ event in } [s, T]\} = \lambda(T - s)e^{-\lambda(T-s)}$.

(b) $\frac{d}{ds} P_s\{\text{winning}\} = \lambda e^{-\lambda T}[(T - s)\lambda e^{\lambda s} - e^{\lambda s}]$. Setting this equal to 0 gives that the maximal occurs at $s = T - 1/\lambda$.

(c) Substituting $s = T - 1/\lambda$ into part (a) gives the maximal probability e^{-1}.

2.14. Let $N_{i,j}$ denote the number of people that get on at floor i and off at floor j. Then N_{ij} is Poisson with mean $\lambda_i P_{ij}$ and all N_{ij}, $i \ge 0, j \ge i$, are independent. Hence:

(a) $E[0_j] = E\left[\sum_i N_{ij}\right] = \sum_i \lambda_i P_{ij}$;

(b) $0_j = \sum_i N_{ij}$ is Poisson with mean $\sum_i \lambda_i P_{ij}$;

(c) 0_j and 0_k are independent.

2.15. (a) N_i is negative binomial. That is,

$$P\{N_i = k\} = \binom{k-1}{n_i - 1} P_i^{n_i}(1 - P_i)^{k-n_i}, \qquad k \ge n_i.$$

(b) No.

(c) T_i is gamma with parameters n_i and P_i.

(d) Yes.

(e)
$$
\begin{aligned}
E[T] &= \int_0^\infty P\{T > t\}\, dt \\
&= \int_0^\infty P\{T_i > t, i = 1, \ldots, r\}\, dt \\
&= \int_0^\infty \left(\prod_{i=1}^r P\{T_i > t\} \right) dt \qquad \text{(by independence)} \\
&= \int_0^\infty \left(\prod_{i=1}^r \int_t^\infty \frac{P_i e^{-P_i x} (P_i x)^{n_i - 1}}{(n_i - 1)!}\, dx \right) dt.
\end{aligned}
$$

(f) $T = \sum_{i=1}^N X_i$, where X_i is the time between the $(i - 1)$st and ith flip. Since N is independent of the sequence of X_i, we obtain

$$
\begin{aligned}
E[T] &= E[E[T|N]] \\
&= E[NE[X]] \\
&= E[N] \quad \text{since } E[X] = 1.
\end{aligned}
$$

2.22. Say that an entering car is type 1 if it will be between a and b at time t. Hence a car entering at times s, $s < t$, will be type 1 if its velocity V is such that $a < (t - s)V < b$, and so it will be type 1 with probability

$$
F\left(\frac{b}{t-s}\right) - F\left(\frac{a}{t-s}\right).
$$

Thus the number of type-1 cars is Poisson with mean

$$
\lambda \int_0^t \left(F\left(\frac{b}{t-s}\right) - F\left(\frac{a}{t-s}\right) \right) ds.
$$

2.24. Let v be the speed of the car entering at time t, and let $t_v = L/v$ denote the resulting travel time. If we let G denote the distribution of travel time, then as $T \equiv L/X$ is the travel time when X is the speed, it follows that $G(x) = \bar{F}(L/x)$. Let an event correspond to a car entering the road and say that the event is counted if that car encounters the one entering at time t. Now, independent of other cars, an event occurring at time s, $s < t$, will be counted with probability $P\{s + T > t + t_v\}$, whereas one occurring at time s, $s > t$, will be counted with probability $P\{s + T < t + t_v\}$. That is, an event occurring at time s is, independently of other events, counted with probability $p(s)$, where

$$
p(s) = \begin{cases} \bar{G}(t + t_v - s) & \text{if } s < t; \\ G(t + t_v - s) & \text{if } t < s < t + t_v; \\ 0 & \text{otherwise.} \end{cases}
$$

Hence, the number of encounters is Poisson with mean

$$\lambda \int_0^\infty p(s)\,ds = \lambda \int_0^t \overline{G}(t + t_v - s)\,ds + \lambda \int_t^{t+t_v} G(t + t_v - s)\,ds$$
$$= \lambda \int_{t_v}^{t+t_v} \overline{G}(y)\,dy + \lambda \int_0^{t_v} G(y)\,dy.$$

To choose the value of t_v that minimizes the preceding, differentiate to obtain:

$$\frac{d}{dt_v}\left\{\lambda \int_0^\infty p(s)\,ds\right\} = \lambda[\overline{G}(t + t_v) - \overline{G}(t_v) + G(t_v)].$$

Setting this equal to 0 gives, since $\overline{G}(t + t_v) \approx 0$ when t is large, that

$$\overline{G}(t_v) = G(t_v)$$

or,

$$G(t_v) = 1/2.$$

Thus, the optimal travel time t_v^* is the median of the distribution, which implies that the optimal speed $v^* = L/t_v^*$ is such that $G(L/v^*) = 1/2$. As $G(L/v^*) = \overline{F}(v^*)$ this gives the result.

2.26. It can be shown in the same manner as Theorem 2.3.1 that, given $S_n = t$, $S_1, \ldots, S_{n-1}$ are distributed as the order statistics from a set of $n - 1$ uniform $(0, t)$ random variables. A second, somewhat heuristic, argument is:

$$S_1, \ldots, S_{n-1} | S_n = t$$
$$= S_1, \ldots, S_{n-1} | N(t^-) = n - 1, N(t) = n$$
$$= S_1, \ldots, S_{n-1} | N(t^-) = n - 1 \qquad \text{(by independent increments)}$$
$$= S_1, \ldots, S_{n-1} | N(t) = n - 1.$$

2.41. **(a)** It cannot have independent increments since knowledge of the number of events in any interval changes the distribution of Λ.

(b) Knowing $\{N(s), 0 \leq s \leq t\}$ is equivalent to knowing $N(t)$ and the arrival times $S_1, \ldots, S_{N(t)}$. Now (arguing heuristically) for $0 < s_1 < \cdots < s_n < t$,

$$P\{\Lambda = \lambda, N(t) = n, S_1 = s_1, \ldots, S_n = s_n\}$$
$$= P\{\Lambda = \lambda\}P\{N(t) = n | \Lambda = \lambda\}P\{S_1 = s_1, \ldots, S_n = s_n | \Lambda, N(t) = n\}$$
$$= dG(\lambda)e^{-\lambda t}\frac{(\lambda t)^n}{n!}\frac{n!}{t^n}. \qquad \text{(by Theorem 2.3.1)}$$

Hence,

$$P\{\Lambda \in (\lambda, \lambda + d\lambda) | N(t) = n, S_1 = s_1, \ldots, S_n = s_n\}$$
$$= \frac{e^{-\lambda t}(\lambda t)^n\,dG(\lambda)}{\int_0^\infty e^{-\lambda t}(\lambda t)^n\,dG(\lambda)}.$$

Hence the conditional distribution of Λ depends only on $N(t)$. This is so since, given the value of $N(t)$, $S_1, \ldots, S_{N(t)}$ will be distributed the same (as the order statistics from a uniform $(0, t)$ population) no matter what the value of Λ.

(c) $P\{\text{time of first event after } t \text{ is greater than } t + s | N(t) = n\}$

$$= \frac{\int_0^\infty e^{-\lambda s} e^{-\lambda t} (\lambda t)^n \, dG(\lambda)}{\int_0^\infty e^{-\lambda t} (\lambda t)^n \, dG(\lambda)}.$$

(d) $$\lim_{h \to 0} \int_0^\infty \frac{1 - e^{-\lambda h}}{h} \, dG(\lambda) = \int_0^\infty \lim_{h \to 0} \left(\frac{1 - e^{-\lambda h}}{h} \right) dG(\lambda)$$

$$= \int_0^\infty \lambda \, dG(\lambda).$$

(e) Identically distributed but not independent.

2.42. **(a)** $$P\{N(t) = n\} = \int_0^\infty e^{-\lambda t} \frac{(\lambda t)^n}{n!} \alpha e^{-\alpha\lambda} \frac{(\alpha\lambda)^{m-1}}{(m-1)!} \, d\lambda$$

$$= \frac{\alpha^m t^n (m + n - 1)!}{n!(m-1)!(\alpha + t)^{m+n}} \int_0^\infty (\alpha + t) e^{-(\alpha+t)\lambda} \frac{((\alpha + t)\lambda)^{m+n-1}}{(m + n - 1)!} \, d\lambda$$

$$= \binom{m + n - 1}{n} \left(\frac{\alpha}{\alpha + t} \right)^m \left(\frac{t}{\alpha + t} \right)^n.$$

(b) $$P\{\Lambda = \lambda | N(t) = n\} = \frac{P\{N(t) = n | \Lambda = \lambda\} P\{\Lambda = \lambda\}}{P\{N(t) = n\}}$$

$$= \frac{e^{-\lambda t} \frac{(\lambda t)^n}{n!} \alpha e^{-\alpha\lambda} \frac{(\alpha\lambda)^{m-1}}{(m-1)!}}{\binom{m + n - 1}{n} \left(\frac{\alpha}{\alpha + t} \right)^m \left(\frac{t}{\alpha + t} \right)^n}$$

$$= (\alpha + t) e^{-(\alpha+t)\lambda} \frac{((\alpha + t)\lambda)^{m+n-1}}{(m + n - 1)!}.$$

(c) By part (d) of Problem 2.41, the answer is the mean of the conditional distribution which by (b) is $(m + n)/(\alpha + t)$.

CHAPTER 3

3.7. The renewal equation is, for $t \leq 1$,

$$m(t) = t + \int_0^t m(t - s) \, ds = t + \int_0^t m(y) \, dy.$$

Differentiating yields that

$$m'(t) = 1 + m(t).$$

Letting $h(t) = 1 + m(t)$ gives

$$h'(t) = h(t) \qquad \text{or} \qquad h(t) = ce^t.$$

Upon evaluating at $t = 0$, we see that $c = 1$ and so,

$$m(t) = e^t - 1.$$

As $N(1) + 1$, the time of the first renewal after 1, is the number of interarrival times that we need add until their sum exceeds 1, it follows that the expected number of uniform (0, 1) random variables that need be summed to exceed 1 is equal to e.

3.17. $g = h + g * F$

$$= h + (h + g * F) * F = h + h * F + g * F_2$$

$$= h + h * F + (h + g * F) * F_2 = h + h * F + h * F_2 + g * F_3$$

$$\vdots$$

$$= h + h * F + h * F_2 + \cdots + h * F_n + g * F_{n+1}.$$

Letting $n \to \infty$ and using that $F_n \to 0$ yields

$$g = h + h * \sum_{n=1}^{\infty} F_n = h + h * m.$$

(a) $P(t) = \int_0^\infty P\{\text{on at } t | Z_1 + Y_1 = s\}\, dF(s)$

$$= \int_0^t P(t - s)\, dF(s) + \int_t^\infty P\{Z_1 > t | Z_1 + Y_1 = s\}\, dF(s)$$

$$= \int_0^t P(t - s)\, dF(s) + P\{Z_1 > t\}.$$

(b) $g(t) = \int_0^\infty E[A(t) | X_1 = s]\, dF(s)$

$$= \int_0^t g(t - s)\, dF(s) + \int_t^\infty t\, dF(s)$$

$$= \int_0^t g(t - s)\, dF(s) + t\bar{F}(t).$$

$$P(t) \to \frac{\int_0^\infty P\{Z_1 > t\}\, dt}{\mu_F} = \frac{E[Z]}{E[Z] + E[Y]},$$

$$g(t) \to \frac{\int_0^\infty t\bar{F}(t)\, dt}{\mu} = \frac{\int_0^\infty t \int_t^\infty dF(s)\, dt}{\mu} = \frac{\int_0^\infty \int_0^s t\, dt\, dF(s)}{\mu}$$

$$= \frac{\int_0^\infty s^2\, dF(s)}{2\mu} = \frac{E[X^2]}{2E[X]}.$$

3.24. Let T denote the number of draws needed until four successive cards of the same suit appear. Assume that you continue drawing cards indefinitely and any time that the last four cards have all been of the same suit say that a renewal has occurred. The expected time between renewals is then

$$E[\text{time between renewals}] = 1 + \frac{3}{4}(E[T] - 1).$$

The preceding equation being true since if the next card is of the same suit then the time between renewals is 1, and if it is of a different suit then it is like the first card of the initial cycle. By Blackwell's theorem,

$$E[\text{time between renewals}] = (\lim P\{\text{renewal at } n\})^{-1} = (1/4)^{-3} = 64.$$

Hence,

$$E[T] = 85.$$

3.27. $E[R_{N(t)+1}] = \int_0^t E[R_{N(t)+1} | S_{N(t)} = s]\bar{F}(t-s)\, dm(s) + E[R_{N(t)+1} | S_{N(t)} = 0]\bar{F}(t)$

$$= \int_0^t E[R_1 | X_1 > t - s]\bar{F}(t-s)\, dm(s) + E[R_1 | X_1 > t]\bar{F}(t)$$

$$\to \int_0^\infty E[R_1 | X_1 > t]\bar{F}(t)\, dt/\mu$$

$$= \int_0^\infty \int_t^\infty E[R_1 | X_1 = s]\, dF(s)\, dt/\mu$$

$$= \int_0^\infty \int_0^s dt\, E[R_1 | X_1 = s]\, dF(s)/\mu$$

$$= \int_0^\infty sE[R_1 | X_1 = s]\, dF(s)/\mu$$

$$= E[R_1 X_1]/\mu,$$

where $\mu = E[X_1]$. We have assumed that $E[R_1 X_1] < \infty$, which implies that

$$E[R_1 | X_1 > t]\bar{F}(t) \to 0 \text{ as } t \to \infty.$$

$E[X^2] > E^2[X]$ since $\text{Var}(X) > 0$ except when X is constant with probability 1.

3.33. **(a)** Both processes are regenerative.

(b) If T is the time of a cycle and N the number of customers served, then

$$V = E\left[\int_0^T V(s)\, dx\right] \Big/ E[T], \qquad W_Q = \frac{E[D_1 + \cdots + D_N]}{E[N]}.$$

Imagine that each pays, at any time, at a rate equal to the customer's remaining service time. Then

$$\text{reward in cycle} = \int_0^T V(s)\, ds.$$

Also, letting Y_i denote the service time of customer i,

$$\text{reward in cycle} = \sum_{i=1}^N \left[D_i Y_i + \int_0^{Y_i} (Y_i - t)\, dt\right]$$

$$= \sum_{i=1}^N D_i Y_i + \sum_{i=1}^N \frac{Y_i^2}{2}.$$

Hence,

$$E[\text{reward in cycle}]$$
$$= E\left[\sum_{i=1}^{N} D_i Y_i\right] + E\left[\sum_{i=1}^{N} \frac{Y_i^2}{2}\right]$$
$$= E\left[\sum_{i=1}^{N} D_i Y_i\right] + \frac{E[N]E[Y^2]}{2} \qquad \text{(by Wald's equation).}$$

Now

$$\frac{E\left[\sum_{i=1}^{N} D_i Y_i\right]}{E[N]}$$
$$= \lim_{n\to\infty} \frac{E[D_1Y_1 + \cdots + D_nY_n]}{n}$$
$$= E[Y] \lim_{n\to\infty} \frac{E[D_1 + \cdots + D_n]}{n} \quad \text{since } D_i \text{ and } Y_i \text{ are independent}$$
$$= E[Y]W_Q.$$

Therefore from the above

$$E[\text{reward in cycle}] = E[N]\left(E[Y]W_Q + \frac{E[Y^2]}{2}\right).$$

Equating this to $E[\int_0^T V(s)\,ds] = VE[T]$ and using the relationship $E[T] = E[N]/\lambda$ establishes the identity.

3.34. Suppose $P\{X < Y\} = 1$. For instance, $P\{X = 1\} = 1$, and Y is uniform $(2, 3)$ and $k = 3$.

3.35. **(a)** Regenerative.

(b) $E[\text{time during cycle } i \text{ packages are waiting}]/\mu_F$, where

$$E[\text{time } i \text{ packages are waiting}]$$
$$= \int_0^\infty E[\text{time} \mid \text{cycle of length } x]\,dF(x)$$
$$= \int_0^\infty \sum_{j=i}^{\infty} E[\text{time} \mid \text{length is } x, N(x) = j]e^{-\lambda x}\frac{(\lambda x)^j}{j!}\,dF(x)$$
$$= \int_0^\infty \sum_{j=i}^{\infty} \frac{x}{j+1} e^{-\lambda x}\frac{(\lambda x)^j}{j!}\,dF(x).$$

The last equality follows from Problem 2.17 of Chapter 2.

3.36. $\lim \int_0^t r(X(s))\, ds/t = E\left[\int_0^T r(X(s))\, ds\right] \Big/ E[T]$

$$= \frac{E\left[\sum_j r(j) \cdot (\text{amount of time in } j \text{ during } T)\right]}{E[T]}$$

$$= \sum_j r(j) P_j.$$

Chapter 4

4.10. **(a)** $\alpha_i \equiv 1 - (1 - p)^i$.

(b) No.

(c) No.

(d) Yes, the transition probabilities are given by $P\{X_{n+1} = k, Y_{n+1} = j - k | X_n = i, Y_n = j\} = \binom{j}{k} \alpha_i^k (1 - \alpha_i)^{j-k}, \quad 0 \le k \le j.$

4.13. Suppose $i \leftrightarrow j$ and i is positive recurrent. Let m be such that $P_{ij}^m > 0$. Let N_k denote the kth time the chain is in state i and let

$$I_k = \begin{cases} 1 & \text{if } X_{N_k+m} = j \\ 0 & \text{otherwise.} \end{cases}$$

By the strong law of large numbers

$$\sum_{k=1}^n \frac{I_k}{n} \to P_{ij}^m > 0.$$

Hence

$$\lim_{n\to\infty} \frac{\text{number of visits to } j \text{ by time } N_n + m}{n} \frac{n}{N_n + m} \ge P_{ij}^m \frac{1}{E[T_{ii}]} > 0,$$

where T_{ii} is the time between visits to i. Hence j is also positive recurrent. If i is null recurrent, then as $i \leftrightarrow j$, and recurrence is a class property, j is recurrent. If j were positive recurrent, then by the above so would be i. Hence j is null recurrent.

4.14. Suppose i is null recurrent and let C denote the class of states communicating with i. Then all states in C are null recurrent implying that

$$\lim_{n\to\infty} P_{ij}^n = 0 \qquad \text{for all } j \in C.$$

But this is a contradiction since $\sum_{j\in C} P_{ij}^n = 1$ and C is a finite set. For the same reason not all states in a finite-state chain can be transient.

4.16. **(a)** Let the state be the number of umbrellas she has at her present location. The transition probabilities are

$$P_{0,r} = 1, \quad P_{i,r-i} = 1 - p, \quad P_{i,r-i+1} = p, \qquad i = 1, \dots, r.$$

(b) The equations for the limiting probabilities are

$$\begin{aligned}
\pi_r &= \pi_0 + \pi_1 p, \\
\pi_j &= \pi_{r-j}(1-p) + \pi_{r-j+1} p, \qquad j = 1, \dots, r-1, \\
\pi_0 &= \pi_r(1-p).
\end{aligned}$$

It is easily verified that they are satisfied by

$$\pi_i = \begin{cases} \dfrac{q}{r+q} & \text{if } i = 0 \\[2ex] \dfrac{1}{r+q} & \text{if } i = 1, \dots, r, \end{cases}$$

where $q = 1 - p$.

(c) $p\pi_0 = \dfrac{pq}{r+q}$.

4.17. Let $I_n(j) = \begin{cases} 1 & \text{if } X_n = j \\ 0 & \text{otherwise.} \end{cases}$ Then

$$\begin{aligned}
\pi_j &= \lim_n \frac{E\left[\sum_{k=1}^{n} I_k(j)\right]}{n} \\
&= \lim_n E\left[\sum_{k=1}^{n} \sum_i I_{k-1}(i) I_k(j)\right] \Big/ n \\
&= \lim_n \sum_{k=1}^{n} \sum_i E[I_{k-1}(i) I_k(j)]/n \\
&= \lim_n \sum_{k=1}^{n} \sum_i E[I_{k-1}(i)] P_{ij}/n \\
&= \lim_n \sum_i P_{ij} \sum_{k=1}^{n} E[I_{k-1}(i)]/n \\
&= \sum_i P_{ij} \lim_n E\left[\sum_{k=1}^{n} I_{k-1}(i)\right] \Big/ n \\
&= \sum_i \pi_i P_{ij}.
\end{aligned}$$

4.20. Let $X_{ij}(n)$ equal 1 if the nth transition out of state i is into state j, and let it be 0 otherwise. Also, let N_i denote the number of time periods the chain is in state i before returning to 0. Then, for $j > 0$

$$m_j = E\left[\sum_i \sum_{n=1}^{N_i} X_{ij}(n)\right] = \sum_i E\left[\sum_{n=1}^{N_i} X_{ij}(n)\right].$$

But, by Wald's equation

$$E\left[\sum_{n=1}^{N_i} X_{ij}(n)\right] = E[N_i]P_{ij} = m_i P_{ij}.$$

For a second proof, by regarding visits to state 0 as cycles, it follows that π_j, the long-run proportion of time in state j, satisfies

$$\pi_j = m_j/\mu_{00}.$$

Hence, from the stationarity equations $\pi_j = \sum_i \pi_i P_{ij}$ we obtain for $j > 0$ that

$$m_j = \sum_i m_i P_{ij}.$$

4.21. This Markov chain is positive recurrent if, and only if, the system of equations

$$y_0 = y_1 q_1,$$
$$y_j = y_{j+1} q_{j+1} + y_{j-1} p_{j-1}, \qquad j \geq 1,$$

possesses a solution such that $y_j \geq 0$, $\sum_j y_j = 1$. We may now rewrite these equations to obtain

$$y_0 = y_1 q_1,$$
$$y_{j+1} q_{j+1} - y_j p_j = y_j q_j - y_{j-1} p_{j-1}, \qquad j \geq 1.$$

From the above it follows that

$$y_{j+1} q_{j+1} = y_j p_j, \qquad j \geq 0.$$

Hence,

$$y_{j+1} = y_0 \frac{p_0 \cdots p_j}{q_1 \cdots q_{j+1}}, \qquad j \geq 0.$$

Therefore, a necessary and sufficient condition for the random walk to be a positive recurrent is for

$$\sum_{j=0}^{\infty} \frac{p_0 \cdots p_j}{q_1 \cdots q_{j+1}} < \infty.$$

4.30. Since $P\{X_i - Y_i = 1\} = P_1(1 - P_2)$ and $P\{X_i - Y_i = -1\} = (1 - P_1)P_2$ if we only look when a pair X_i, Y_i occurs for which $X_i - Y_i \neq 0$, then what is observed is a simple random walk with

$$P = \frac{P_1(1 - P_2)}{P_1(1 - P_2) + P_2(1 - P_1)}.$$

Hence from gambler's ruin results

$$\begin{aligned} P\{\text{error}\} &= P\{\text{down } M \text{ before up } M\} \\ &= 1 - \frac{1 - (1 - (q/p)^M)}{1 - (q/p)^{2M}} = \frac{(q/p)^M(1 - (q/p)^M)}{1 - (q/p)^{2M}} = \frac{(q/p)^M}{1 + (q/p)^M} \\ &= \frac{1}{(p/q)^M + 1} = \frac{1}{1 + \lambda^M}. \end{aligned}$$

By Wald's equation

$$E\left[\sum_{i=1}^{N} (X_i - Y_i)\right] = E[N](P_1 - P_2),$$

or,

$$E[N](P_1 - P_2) = M\frac{\lambda^M}{1 + \lambda^M} - \frac{M}{1 + \lambda^M} = \frac{M(\lambda^M - 1)}{1 + \lambda^M}.$$

4.40. Consider the chain in steady state, and note that the reverse process is a Markov chain with the transition probabilities $P^*_{ij} = \pi_j P_{ji}/\pi_i$. Now, given that the chain has just entered state i, the sequence of (reverse) states until it again enters state i (after T transitions) has exactly the same distribution as Y_j, $j = 0, \ldots, T$. For a more formal argument, let $i_0 = i_n = 0$. Then,

$$\begin{aligned} P\{\mathbf{Y} = (i_0, \ldots, i_n)\} &= P\{\mathbf{X} = (i_n, \ldots, i_0)\} = \prod_{k=1}^{n} P_{i_k, i_{k-1}} \\ &= \prod_{k=1}^{n} P^*_{i_{k-1}, i_k} \pi_{i_{k-1}}/\pi_{i_k} = \prod_{k=1}^{n} P^*_{i_{k-1}, i_k}. \end{aligned}$$

4.43. For any permutation $i_1, i_2, \ldots, i_n$ of $1, 2, \ldots, n$, let $\pi(i_1, i_2, \ldots, i_n)$ denote the limiting probability under the one-closer rule. By time reversibility, we have

$$(*) \quad P_{i_{j+1}}\pi(i_1, \ldots, i_j, i_{j+1}, \ldots, i_n) = P_{i_j}\pi(i_1, \ldots, i_{j+1}, i_j, \ldots, i_n)$$

for all permutations. Now the average position of the element requested can be expressed as

$$\text{average position} = \sum_i P_i E[\text{position of element } i]$$

$$= \sum_i P_i \left[1 + \sum_{j \neq i} P\{\text{element } j \text{ precedes element } i\}\right]$$

$$= 1 + \sum_i \sum_{j \neq i} P_i P\{e_j \text{ precedes } e_i\}$$

$$= 1 + \sum_{i<j} [P_i P\{e_j \text{ precedes } e_i\} + P_j P\{e_i \text{ precedes } e_j\}]$$

$$= 1 + \sum_{i<j} [P_i P\{e_j \text{ precedes } e_i\} + P_j(1 - P\{e_j \text{ precedes } e_i\})]$$

$$= 1 + \sum_{i<j}\sum (P_i - P_j) P\{e_j \text{ precedes } e_i\} + \sum_{i<j}\sum P_j.$$

Hence, to minimize the average position of the element requested, we would want to make $P\{e_j \text{ precedes } e_i\}$ as large as possible when $P_j > P_i$ and as small as possible when $P_i > P_j$. Now under the front-of-the-line rule

$$P\{e_j \text{ precedes } e_i\} = \frac{P_j}{P_j + P_i}$$

since under the front-of-the-line rule element j will precede element i if, and only if, the last request for either i or j was for j. Therefore, to show that the one-closer rule is better than the front-of-the-line rule, it suffices to show that under the one-closer rule

$$P\{e_j \text{ precedes } e_i\} > \frac{P_j}{P_j + P_i} \qquad \text{when } P_j > P_i.$$

Now consider any state where element i precedes element j, say $(\ldots, i, i_1, \ldots, i_k, j, \ldots)$. By successive transpositions using (*), we have

$$\pi(\ldots, i, i_1, \ldots, i_k, j, \ldots) = \left(\frac{P_i}{P_j}\right)^{k+1} \pi(\ldots, j, i_1, \ldots, i_k, i, \ldots).$$

Now when $P_j > P_i$, the above implies that

$$\pi(\ldots, i, i_1, \ldots, i_k, j, \ldots) < \frac{P_i}{P_j} \pi(\ldots, j, i_1, \ldots, i_k, i, \ldots).$$

Letting $\alpha(i, j) = P\{e_i \text{ precedes } e_j\}$, we see by summing over all states for which i precedes j and by using the above that

$$\alpha(i, j) < \frac{P_i}{P_j} \alpha(j, i),$$

which, since $\alpha(i, j) = 1 - \alpha(j, i)$, yields

$$\alpha(j,i) > \frac{P_j}{P_j + P_i}.$$

4.46. **(a)** Yes.

(b) Proportion of time *in* j $= \pi_j / \sum_{i=0}^{N} \pi_i$, $0 \le j \le N$.

(c) Note that by thinking of visits to i as renewals

$$\pi_i(N) = (E[\text{number of } Y\text{-transitions between } Y\text{-visits to } i])^{-1}$$

$$\pi_j(N) = \frac{E[\text{number of } Y\text{-visits to } j \text{ between } Y\text{-visits to } i]}{E[\text{number of } Y\text{-transitions between } Y\text{-visits to } i]}$$

$$= \frac{E[\text{number of } X\text{-visits to } j \text{ between } X\text{-visits to } i]}{1/\pi_i(N)}.$$

(d) For the symmetric random walk, the Y-transition probabilities are

$$P_{i,i+1} = \tfrac{1}{2} = P_{i,i-1}, \qquad i = 1, \ldots, N-1,$$
$$P_{00} = \tfrac{1}{2} = P_{01}, \qquad P_{NN} = \tfrac{1}{2} = P_{N,N-1}.$$

This transition probability matrix is doubly stochastic and thus

$$\pi_i(N) = \frac{1}{N+1}, \qquad i = 0, 1, \ldots, N.$$

Hence, from (b),

$$E[\text{time the } X \text{ process spends in } j \text{ between visits to } i] = 1.$$

(e) Use Theorem 4.7.1.

CHAPTER 5

5.3. **(a)** Let $N(t)$ denote the number of transitions by t. It is easy to show in this case that

$$P\{N(t) \ge n\} \le \sum_{j=n}^{\infty} e^{-Mt} \frac{(Mt)^j}{j!}$$

and thus $P\{N(t) < \infty\} = 1$.

5.5. Rather than imagining a single Yule process with $X(0) = i$, imagine i independent Yule process each with $X(0) = 1$. By conditioning on the sizes of each of these i populations at t, we see that the conditional distribution of the k births is that of k independent random variables each having distribution given by (5.3.2).

5.8. We start by showing that the probability of 2 or more transitions in a time t is $o(t)$. Conditioning on the next state visited gives

$$P\{\geq 2 \text{ transitions by } t | X_0 = i\} = \sum_j P_{ij} P\{T_i + T_j \leq t\},$$

where T_i and T_j are independent exponentials with respective rates v_i and v_j; they represent the times to leave i and j, respectively. (Note that T_i is independent of the information that j is the state visited from it.) Hence,

$$\begin{aligned}
&\lim_{t \to 0} P\{\geq 2 \text{ transitions by } t | X_0 = i\}/t \\
&\quad = \lim_{t \to 0} \sum_j P_{ij} P\{T_i + T_j \leq t\}/t \\
&\quad \leq \lim_{t \to 0} \left[\sum_{j \leq M} P_{ij} P\{T_i + T_j \leq t\}/t + \sum_{j > M} P_{ij} P\{T_i \leq t\}/t \right] \\
&\quad = \lim_{t \to 0} \left[\sum_{j \leq M} P_{ij} P\{T_i + T_j \leq t\}/t + \frac{1 - e^{-v_i t}}{t} \left(1 - \sum_{j \leq M} P_{ij}\right) \right].
\end{aligned}$$

Now

$$\begin{aligned}
P\{T_i + T_j \leq t\} &\leq P\{T + T' \leq t\}, \qquad T, T' = \text{independent exponentials with rates } v = \max(v_i, v_j) \\
&= P\{N(t) \geq 2\}, \qquad N(t) = \text{Poisson process with rate } v \\
&= o(t)
\end{aligned}$$

and

$$\frac{1 - e^{-v_i t}}{t} \to v_i \qquad \text{as } t \to 0.$$

Hence, from the above

$$\lim_{t \to 0} P\{\geq 2 \text{ transitions by } t | X_0 = i\}/t \leq v_i \left(1 - \sum_{j \leq M} P_{ij}\right) \qquad \text{for all } M.$$

Letting $M \to \infty$ now gives the desired result.

Now,

$$\begin{aligned}
P_{ii}(t) &= P\{X(t) = i | X(0) = i\} \\
&= P\{X(t) = i, \text{ no transitions by } t | X(0) = i\} \\
&\quad + P\{X(t) = i, \text{ at least 2 transitions by } t | X(0) = i\} \\
&= e^{-v_i t} + o(t).
\end{aligned}$$

Similarly for $i \neq j$,

$$\begin{aligned}
P_{ij}(t) = P\{&\text{first state visited is } j, \text{ transition time } \leq t | X(0) = i\} \\
&+ P\{X(t) = j, \text{ first state visited } \neq j | X(0) = i\}.
\end{aligned}$$

Hence,

$$P_{ij}(t) - P_{ij}(1 - e^{-v_i t}) \leq P\{\geq 2 \text{ transitions by } t\} = o(t)$$

or

$$P_{ij}(t) = v_i P_{ij} t + o(t).$$

5.14. With the state equal to the number of infected members of the population, this is a pure birth process with birth rates $\lambda_k = k(n-k)\lambda$. The expected time to go from state 1 to state n is

$$\sum_{k=1}^{n-1} 1/\lambda_k = \sum_{k=1}^{n-1} 1/\{k(n-k)\lambda\}.$$

5.23. Let q^x and P^x denote the transition rate and stationary probability functions for the $X(t)$ process and similarly q^y, P^y for the $Y(t)$ process. For the chain $\{(X(t), Y(t)), t \geq 0\}$, we have

$$q_{(i,j),(i',j)} = q^x_{i,i'},$$

$$q_{(i,j),(i,j')} = q^y_{j,j'}.$$

We claim that the limiting probabilities are

$$P_{i,j} = P^x_i P^y_j.$$

To verify this claim and at the same time prove time reversibility, all we need do is check the reversibility equations. Now

$$\begin{aligned} P_{i,j} q_{(i,j),(i',j)} &= P^x_i P^y_j q^x_{ii'} \\ &= P^x_{i'} P^y_j q^x_{i',i} \qquad \text{(by time reversibility of } X(t)\text{)} \\ &= P_{i',j} q_{(i',j),(i,j)}. \end{aligned}$$

Since the verification for transitions of the type (i, j) to (i, j') is similar, the result follows.

5.34. **(a)** $P_i \Big/ \sum_{j \in B} P_j$.

(b) $$P\{X(t) = i \mid X(t) \in B, X(t^-) \in G\} = \frac{P\{X(t) = i, X(t^-) \in G\}}{P\{X(t) \in B, X(t^-) \in G\}}$$

$$= \frac{\sum_{j \in G} P\{X(t^-) = j\} P\{X(t) = i \mid X(t^-) = j\}}{\sum_{j \in G} P\{X(t^-) = j\} P\{X(t) \in B \mid X(t^-) = j\}}$$

$$= \frac{\sum_{j \in G} P_j q_{ji}}{\sum_{j \in G} \sum_{k \in B} P_j q_{jk}}.$$

(c) Let T denote the time to leave state i and T' the additional time after leaving i until G is entered. Using the independence of T and T', we obtain upon conditioning on the state visited after i,

$$\begin{aligned}\tilde{F}_i(s) &= E[e^{-s(T+T')}] \\ &= E[e^{-sT}]E[e^{-sT'}] \\ &= \frac{v_i}{v_i + s}\sum_j E[e^{-sT'}|\text{next is } j]P_{ij}.\end{aligned}$$

Now

$$E[e^{-sT'}|\text{next is } j] = \begin{cases} 1 & \text{if } j \in G \\ \tilde{F}_j(s) & \text{if } j \in B, \end{cases}$$

which proves (c).

(d) In any time t, the number of transitions from G to B must equal to within 1 the number from B to G. Hence the long-run rate at which transitions from G to B occur (the left-hand side of (d)) must equal the long-run rate of ones from B to G (the right-hand side of (d)).

(e) It follows from (c) that

$$(s + v_i)\tilde{F}_i(s) = \sum_{j\in B} \tilde{F}_j(s)q_{ij} + \sum_{j\in G} q_{ij}.$$

Multiplying by P_i and summing over all $i \in B$ yields

$$\begin{aligned}\sum_{i\in B} P_i(s + v_i)\tilde{F}_i(s) &= \sum_{i\in B}\sum_{j\in B} P_i\tilde{F}_j(s)q_{ij} + \sum_{i\in B}\sum_{j\in G} P_iq_{ij} \\ &= \sum_{j\in B}\tilde{F}_j(s)\sum_{i\in B} P_iq_{ij} + \sum_{i\in G}\sum_{j\in B} P_iq_{ij} \\ &= \sum_{j\in B}\tilde{F}_j(s)\left[v_jP_j - \sum_{i\in G} P_iq_{ij}\right] + \sum_{i\in G}\sum_{j\in B} P_iq_{ij},\end{aligned}$$

where the last equality follows from

$$v_jP_j = \sum_i P_iq_{ij}$$

and the next to last from (d). From this we see that

$$s\sum_{i\in B} P_i\tilde{F}_i(s) = \sum_{i\in G}\sum_{j\in B} P_iq_{ij}(1 - \tilde{F}_j(s)).$$

(f)

$$E[e^{-sT_v}] = \frac{\sum_{i\in B}\tilde{F}_i(s)\sum_{j\in G} P_jq_{ji}}{\sum_{j\in G}\sum_{k\in B} P_jq_{jk}} \text{ (from (b)).}$$

(g) From (f), we obtain upon using (e)

$$s\sum_{i\in B} P_i\tilde{F}_i(s) = \left(\sum_{i\in G}\sum_{j\in B} P_i q_{ij}\right)(1 - E[e^{-sT_v}]).$$

Dividing by s and then letting $s \to 0$ yields

$$\sum_{i\in B} P_i = \sum_{i\in G}\sum_{j\in B} P_i q_{ij} E[T_v].$$

(h)

$$E[e^{-sT_x}] = \frac{\sum_{i\in B} P_i\tilde{F}_i(s)}{\sum_{j\in B} P_j} \qquad \text{(from (a))}$$

$$= \frac{\sum_{i\in G}\sum_{j\in B} P_i q_{ij}(1 - \tilde{F}_j(s))}{s\sum_{j\in B} P_j} \qquad \text{(from (e))}$$

$$= \frac{\sum_{i\in G}\sum_{j\in B} P_i q_{ij}(1 - E[e^{-sT_v}])}{s\sum_{j\in B} P_j} \qquad \text{(from (f))}$$

$$= \frac{1 - E[e^{-sT_v}]}{sE[T_v]} \qquad \text{(by (g)).}$$

(i) If $P\{T_x \le t\}$ is as given, then

$$E[e^{-sT_x}] = \int_0^\infty e^{-st} P\{T_v > t\}\, dt / E[T_v]$$

$$= \int_0^\infty e^{-st} \int_t^\infty dF_{T_v}(y)\, dt / E[T_v]$$

$$= \int_0^\infty \int_0^y e^{-st}\, dt\, dF_{T_v}(y) / E[T_v]$$

$$= \frac{1 - E[e^{-sT_v}]}{sE[T_v]}.$$

Hence, from (h) the hypothesized distribution yields the correct Laplace transform. The result then follows from the one-to-one correspondence between transforms and distributions.

(j)

$$E[T_x] = \int_0^\infty t\, dF_{T_x}(t)$$

$$= \int_0^\infty t \int_t^\infty dF_{T_v}(y)\, dt / E[T_v]$$

$$= \int_0^\infty \int_0^y t\, dt\, dF_{T_v}(y) / E[T_v]$$

$$= \frac{E[T_v^2]}{2E[T_v]}.$$

$E[T_v^2] \ge (E[T_v])^2$ since $\mathrm{Var}(T_v) \ge 0$.

5.35. $\{R(t), t \geq 0\}$ is a two-state Markov chain that leaves state 1(2) to enter state 2(1) at any exponential rate $\lambda_1 q(\lambda_2 p)$. Since $P\{R(0) = 1\} = p$, it follows from Example 5.8(A) (or 5.4(A)) that

$$P\{R(t) = 1\} = pe^{-\bar{\lambda}t} + (1 - e^{-\bar{\lambda}t})\lambda_2 p/\bar{\lambda},$$

where $\bar{\lambda} = \lambda_1 q + \lambda_2 p$. Hence

$$(*) \qquad \int_0^t P\{R(s) = 1\}\, ds = \frac{pq(\lambda_1 - \lambda_2)}{\bar{\lambda}^2}(1 - e^{-\bar{\lambda}t}) + \frac{\lambda_2 pt}{\bar{\lambda}}.$$

To prove (b), let $\Lambda(t) = \lambda_{R(t)}$ and write for $\varepsilon > 0$

$$N(t) = \sum_{n=1}^{t/\varepsilon} [N(n\varepsilon) - N((n-1)\varepsilon)] + o(\varepsilon).$$

Now,

$$E[N(n\varepsilon) - N((n-1)\varepsilon) | \Lambda((n-1)\varepsilon)] = \Lambda((n-1)\varepsilon)\varepsilon + o(\varepsilon).$$

Thus,

$$E[N(n\varepsilon) - N((n-1)\varepsilon)] = \varepsilon E[\Lambda((n-1)\varepsilon)] + o(\varepsilon).$$

Hence, from the above,

$$E[N(t)] = E\left[\sum_{n=1}^{t/\varepsilon} \varepsilon\Lambda((n-1)\varepsilon) + \frac{to(\varepsilon)}{\varepsilon}\right].$$

Now let $\varepsilon \to 0$ to obtain

$$\begin{aligned} E[N(t)] &= E\left[\int_0^t \Lambda(s)\, ds\right] \\ &= \sum_{i=1}^{2} \lambda_i \int_0^t P\{R(s) = i\}\, ds \\ &= \frac{pq(\lambda_1 - \lambda_2)^2}{\bar{\lambda}^2}(1 - e^{-\bar{\lambda}t}) + \frac{\lambda_1\lambda_2 t}{\bar{\lambda}}, \end{aligned}$$

where the last equality follows from (*). Another (somewhat more heuristic) argument for (b) is as follows:

$$P\{\text{renewal in } (s, s+h)\} = hE[\Lambda(s)] + o(h),$$

and so,

$$E[\Lambda(s)] = P\{\text{renewal in } (s, s+h)\}/h + o(h)/h.$$

Letting $h \to 0$ gives

$$E[\Lambda(s)] = m'(s),$$

and so,

$$m(t) = \int_0^t E[\Lambda(s)]\, ds.$$

CHAPTER 6

6.2. $\text{Var } Z_n = \text{Var}\left(\sum_1^n X_i\right)$

$$= \sum_1^n \text{Var}(X_i) + \sum\sum_{i \neq j} \text{Cov}(X_i, X_j).$$

But, for $i < j$,

$$\begin{aligned}
\text{Cov}(X_i, X_j) &= E[X_i X_j] \\
&= E[(Z_i - Z_{i-1})(Z_j - Z_{j-1})] \\
&= E[E[(Z_i - Z_{i-1})(Z_j - Z_{j-1})|Z_1, \ldots, Z_i]] \\
&= E[(Z_i - Z_{i-1})(E(Z_j|Z_1, \ldots, Z_i) - E[Z_{j-1}|Z_1, \ldots, Z_i])] \\
&= E[(Z_i - Z_{i-1})(Z_i - Z_i)] \\
&= 0.
\end{aligned}$$

6.7.
$$\begin{aligned}
E[S_n^2|S_1^2, \ldots, S_{n-1}^2] &= E[(S_{n-1} + X_n)^2|S_1^2, \ldots, S_{n-1}^2] \\
&= E[S_{n-1}^2|S_1^2, \ldots, S_{n-1}^2] + 2E[X_n S_{n-1}|S_1^2, \ldots, S_{n-1}^2] \\
&\quad + E[X_n^2|S_1^2, \ldots, S_{n-1}^2] \\
&= S_{n-1}^2 + 2E(X_n)E[S_{n-1}|S_1^2, \ldots, S_{n-1}^2] + E[X_n^2] \\
&= S_{n-1}^2 + \sigma^2.
\end{aligned}$$

Hence,

$$E[S_n^2 - n\sigma^2|S_1^2, \ldots, S_{n-1}^2] = S_{n-1}^2 - (n-1)\sigma^2.$$

6.8. (a)
$$\begin{aligned}
&E[X_n + Y_n|X_i + Y_i, i = 1, \ldots, n-1] \\
&\quad = E[X_n|X_i + Y_i, i = 1, \ldots, n-1] \\
&\quad\quad + E[Y_n|X_i + Y_i, i = 1, \ldots, n-1]. \qquad (*)
\end{aligned}$$

Now,

$$\begin{aligned}
&E[X_n|X_i + Y_i, i = 1, \ldots, n-1] \\
&\quad = E[E[X_n|X_i + Y_i, X_i, i = 1, \ldots, n-1]|X_i + Y_i, i = 1, \ldots, n-1] \\
&\quad = E[E[X_n|X_i, i = 1, \ldots, n-1]|X_i + Y_i, i = 1, \ldots, n-1] \\
&\quad\quad \text{(by independence)} \\
&\quad = E[X_{n-1}|X_i + Y_i, i = 1, \ldots, n-1].
\end{aligned}$$

Similarly,

$$E[Y_n|X_i + Y_i, i = 1, \ldots, n-1] = E[Y_{n-1}|X_i + Y_i, i = 1, \ldots, n-1].$$

Hence, from (*)

$$\begin{aligned} &E[X_n + Y_n|X_i + Y_i, i = 1, \ldots, n-1] \\ &= E[X_{n-1} + Y_{n-1}|X_i + Y_i, i = 1, \ldots, n-1] \\ &= X_{n-1} + Y_{n-1}. \end{aligned}$$

The proof of (b) is similar.

Without independence, both are false. Let U_i, $i \geq 1$, be independent and equally likely to be either 1 or -1. Set $X_n = \sum_{i=1}^{n} U_i$, and set $Y_1 = 0$, and for $n > 1$, $Y_n = \sum_{i=1}^{n-1} U_i$. It is easy to see that neither $\{X_n Y_n, n \geq 0\}$ nor $\{X_n + Y_n, n \geq 0\}$ is a martingale.

6.20. Let $x_{i_1}, x_{i_2}, \ldots, x_i, \ldots, x_{i_n}$ denote the shortest path and let $h(\boldsymbol{x})$ denote its length. Now if $\boldsymbol{y}$ differs from $\boldsymbol{x}$ by only a single component, say the ith one, then the shortest path connecting all the $\boldsymbol{y}$ points is less than or equal to the length of the path $x_{i_1}, x_{i_2}, \ldots, y_i, \ldots, x_{i_n}$. But the length of this path differs from $h(\boldsymbol{x})$ by at most 4. Hence, $h/4$ satisfies the conditions of Corollary 6.3.4 and so, with $T = h(\boldsymbol{X})$

$$P\{|T/4 - E[T]/4| \geq b\} \leq 2 \exp\{-b^2/2n\}.$$

Letting $a = 4b$ gives the result.

CHAPTER 7

7.5. $0 \leq \operatorname{Var}\left(\sum_1^n X_i\right) = n \operatorname{Var}(X_1) + n(n-1) \operatorname{Cov}(X_1, X_2)$

$\therefore \quad \operatorname{Var}(X_1) + (n-1)\operatorname{Cov}(X_1, X_2) \geq 0 \qquad$ for all n

$\Rightarrow \operatorname{Cov}(X_1, X_2) \geq 0.$

For a counterexample in the finite case, let X_1 be normal with mean 0 and let $X_2 = -X_1$.

7.9. Use the fact that $f(x) = e^{\theta x}$ is convex. Then, by Jensen's inequality,

$$1 = E[e^{\theta X}] \geq e^{\theta E[X]}.$$

Since $E[X] < 0$, the above implies that $\theta > 0$.

CHAPTER 8

8.3. It follows from (8.1.4) that, starting at some given time, the conditional distribution of the change in the value of Brownian motion in a time $s - t_1$ given that the process changes by $B - A$ in a time $t_2 - t_1$ is normal with mean $(B - A)(s - t_1)/(t_2 - t_1)$ and variance $(s - t_1) \times (t_2 - s)/(t_2 - t_1)$, $t_1 < s < t_2$. Hence, given $X(t_1) = A$, $X(t_2) = B$, $X(s)$ is normal with mean and variance

$$E[X(s)|X(t_1) = A, X(t_2) = B] = A + (B - A)\frac{s - t_1}{t_2 - t_1},$$

$$\text{Var}(X(s)|X(t_1) = A, X(t_2) = B) = \frac{(s - t_1)(t_2 - s)}{t_2 - t_1}.$$

8.4. For $s \leq t$,

$$\text{Cov}(X(s), X(t)) = (s + 1)(t + 1)\,\text{Cov}(Z(t/(t + 1)), Z(s/(s + 1))).$$

Since $\{Z(t)\}$ has the same probability law as $\{W(t) - tW(1)\}$, where $\{W(t)\}$ is Brownian motion, we see that

$$\begin{aligned}
\text{Cov}(X(s), X(t)) &= (s + 1)(t + 1)\left[\text{Cov}\left(W\left(\frac{t}{t + 1}\right), W\left(\frac{s}{s + 1}\right)\right)\right. \\
&\quad - \frac{t}{t + 1}\text{Cov}\left(W(1), W\left(\frac{s}{s + 1}\right)\right) \\
&\quad - \frac{s}{s + 1}\text{Cov}\left(W\left(\frac{t}{t + 1}\right), W(1)\right) \\
&\quad \left. + \frac{st}{(s + 1)(t + 1)}\text{Cov}(W(1), W(1))\right] \\
&= s(t + 1) - st - st + st \\
&= s, \qquad s \leq t.
\end{aligned}$$

Since it is easy to see that $\{X(t)\}$ is Gaussian with $E[X(t)] = 0$, it follows that it is Brownian motion.

8.6. $\mu = \lambda$, $c = \sqrt{1/2\lambda}$.

8.7. All three have the same density.

8.14. $\frac{1}{6}$.

8.15. $f(x) = \dfrac{A - x}{\mu} + (B + A)\dfrac{e^{-2\mu A} - e^{-2\mu x}}{\mu(e^{2\mu B} - e^{-2\mu A})}$.

8.16. **(b)**

$$\begin{aligned}
E[T_x|X(h)] &= E[h + T_{x-X(h)}] + o(h) \\
&= h + \frac{x - X(h)}{\mu} + o(h) \qquad \text{since } E[T_x] = \frac{x}{\mu}.
\end{aligned}$$

$$\begin{aligned}
\text{Var}(T_x|X(h)) &= \text{Var}(h + T_{x-X(h)}) + o(h) \\
&= g(x - X(h)) + o(h).
\end{aligned}$$

Hence, from the conditional variance formula,

$$\begin{aligned} g(x) &= \mathrm{Var}(E[T_x|X(h)]) + E[\mathrm{Var}(T_x|X(h))] \\ &= \frac{h}{\mu^2} + E[g(x - X(h))] + o(h) \\ &= \frac{h}{\mu^2} + E\left[g(x) - X(h)g'(x) + \frac{X^2(h)}{2}g''(x) + \cdots\right] + o(h) \\ &= \frac{h}{\mu^2} + g(x) - \mu h g'(x) + \frac{h}{2}g''(x) + o(h). \end{aligned}$$

Upon dividing by h and letting $h \to 0$,

$$0 = -\mu g'(x) + g''(x)/2 + 1/\mu^2.$$

(c) $$\begin{aligned} \mathrm{Var}(T_{x+y}) &= \mathrm{Var}(T_x + T_{x+y} - T_x) \\ &= \mathrm{Var}(T_x) + \mathrm{Var}(T_{x+y} - T_x) \qquad \text{(by independence)} \\ &= \mathrm{Var}(T_x) + \mathrm{Var}(T_y). \end{aligned}$$

So,

$$g(x + y) = g(x) + g(y)$$

implying that,

$$g(x) = cx.$$

(d) By (c), we see that

$$g(x) = cx$$

and by (b) this implies

$$g(x) = x/\mu^3.$$

8.17. $\min(1, 5x/4)$.

Chapter 9

9.8. (a) Use the representation

$$X = \sum_{i=1}^{n} X_i,$$

where $X_1, \ldots, X_n$ are independent with

$$P\{X_i = 1\} = 1 - P\{X_i = 0\} = p, \qquad i = 1, \ldots, n.$$

Since the X_i are discrete IFR, so is X from the convolution result.

(b) Let X_n be the binomial with parameters n, p_n, where $np_n = \lambda$. By preservation of the IFR property under limits, the result follows since X_n converges to a Poisson random variable with mean λ.

(c) Let X_i, $i \geq 1$, be independent geometric random variables—that is,

$$P\{X_i = k\} = p(1 - p)^{k-1}, \qquad k \geq 1.$$

Since

$$P\{X_i = k | X_i \geq k\} = p,$$

it follows that X_i is discrete IFR, $i \geq 1$. Hence $\sum_1^r X_i$ is IFR, which gives the result.

9.12. Let T_i denote the next state from i in the Markov chain. The hypothesis yields that $T_i \leq_{st} T_{i+1}$, $i \geq 1$. Hence (a) follows since $\sum_j P_{ij} f(s) = E[f(T_i)]$. Prove (b) by induction on n as follows: writing P_i and E_i to mean probability and expectation conditional on $X_0 = i$, we have

$$\begin{aligned} P_i\{X_n \geq k\} &= E_i[P_i\{X_n \geq k | X_1\}] \\ &= E_i[P_{X_1}\{X_{n-1} \geq k\}]. \end{aligned}$$

Now the induction hypothesis states that $P_i\{X_{n-1} \geq k\}$ is increasing in i and thus $P_{X_1}\{X_{n-1} \geq k\}$ is an increasing function of X_1. From (a), X_1 is stochastically increasing in the initial state i, and so for any increasing function $g(X_1)$—in particular for $g(X_1) = P_{X_1}\{X_{n-1} \geq k\}$—we have

$$E_i[g(X_1)] \uparrow i.$$

9.15. Start with the inequality

$$I\{X \in A\} - I\{Y \in A\} \leq I\{X \neq Y\},$$

which follows because if the left-hand side is equal to 1 then X is in A but Y is not, and so the right-hand side is also 1. Take expectations to obtain

$$P\{X \in A\} - P\{Y \in A\} \leq P\{X \neq Y\}.$$

But reversing the roles of X and Y establishes that

$$P\{Y \in A\} - P\{X \in A\} \leq P\{X \neq Y\}$$

which proves the result.

9.20. Suppose Y, Z are independent and $Y \sim G$, $\min(Y, Z) \sim F$. Using the fact that the hazard rate function of the minimum of two independent random variables is equal to the sum of their hazard rate functions gives

$$\lambda_G(t) = \lambda_Y(t) \leq \lambda_Y(t) + \lambda_Z(t) = \lambda_F(t).$$

To go the other way suppose $\lambda_F(t) \geq \lambda_G(t)$. Let Y have distribution F and define Z to be independent of Y and have hazard rate function

$$\lambda_Z(t) = \lambda_F(t) - \lambda_G(t).$$

The random variables Y, Z satisfy the required conditions.

9.29. Let $n = 2$. Then

$$P\{X_1 + X_2 \geq 2\} = P_1 P_2 \leq \left(\frac{P_1 + P_2}{2}\right)^2 = P\{\text{Bin}(2, \bar{p}) \geq 2\}$$

$$\sum_{i=1}^{2} P\{X_1 + X_2 \geq i\} = P_1 P_2 + P_1 + P_2 - P_1 P_2 = \sum_{i=1}^{2} P\{\text{Bin}(2, \bar{p}) \geq i\}$$

showing that

$$X_1 + X_2 \underset{\text{v}}{\leq} \text{Bin}(2, \bar{p}).$$

Now consider the n case and suppose $P_1 \leq P_2 \leq \cdots \leq P_n$. Letting f be increasing and convex, we see, using the result for $n = 2$, that

$$E\left[f\left(\sum_{i=1}^{n} X_i\right) \middle| X_2, \ldots, X_{n-1}\right] \leq E\left[f\left(\bar{X}_1 + \bar{X}_n + \sum_{2}^{n-1} X_i\right) \middle| X_2 \cdots X_{n-1}\right],$$

where $\bar{X}_1, \bar{X}_n$ are independent of all else and are Bernoulli random variables with

$$P\{\bar{X}_1 = 1\} = P\{\bar{X}_n = 1\} = \frac{P_1 + P_n}{2}.$$

Taking expectations of the above shows

$$\sum_{i=1}^{n} X_i \underset{\text{v}}{\leq} \bar{X}_1 + \bar{X}_n + \sum_{i=2}^{n-1} X_i.$$

Repeating this argument (and thus continually showing that the sum of the present set of X's is less variable than it would be if the X having the largest P and the X having the smallest P were replaced with two X's having the average of their P's) and then going to the limit yields

$$\sum_{1}^{n} X_i \underset{\text{v}}{\leq} \text{Bin}(n, \bar{p}).$$

Now write

$$\sum_{1}^{n} X_i = \sum_{1}^{n} X_i + \sum_{n+1}^{n+m} X_i,$$

where $X_i \equiv 0$, $n + 1 \le i \le n + m$. Then from the preceding

$$\sum_1^n X_i \le \text{Bin}\left(n + m, \sum_1^n \frac{P_i}{n + m}\right).$$

Letting $m \to \infty$ yields

$$\sum_1^n X_i \underset{v}{\le} \text{Poisson}\left(\sum_1^n P_i\right).$$

Note: We are assuming here that the variability holds in the limit.

9.32. Let f be a convex function. A Taylor series expansion of $f(Xc)$ about X gives

$$f(cX) = f(X) + f'(X)(c - 1)X + f''(Z)(cX - Z)^2/2,$$

where $X < Z < cX$. Taking expectations and using the convexity of f gives

$$E[f(cX)] \ge E[f(X)] + (c - 1)E[Xf'(X)].$$

Now the functions $g(X) = X$ and $g(X) = f'(X)$ are both increasing in X for $X \ge 0$ (the latter by the convexity of f) and so by Proposition 7.2.1.

$$E[Xf'(X)] \ge E[X]E[f'(X)] = 0,$$

which shows that $E[f(cX)] \ge E[f(X)]$.

9.34. Let f be convex. Then

$$E[f(Y)] = E[E[f(Y)|X]] \ge E[f(E[Y|X])]$$

where the inequality follows from Jensen's inequality applied to a random variable whose distribution is the conditional distribution of Y given X.

(a) follows from the preceding since,

$$XY \underset{v}{\ge} E[XY|X] = XE[Y|X].$$

(b) $X + Y \underset{v}{\ge} E[X + Y|X] = X + E[Y|X] = X + E[Y] = X.$

(c) $\displaystyle\sum_{i=1}^n X_i \underset{v}{\ge} E\left[\sum_{i=1}^n X_i \,\middle|\, \sum_{i=1}^{n+1} X_i\right] = \frac{n}{n+1}\sum_{i=1}^{n+1} X_i.$

(d) Let X and Y be independent and identically distributed. Then,

$$2X \underset{v}{\ge} X + Y \underset{v}{\ge} E[X + Y|X] = X + E[Y|X] = X + E[X]$$

where the first inequality follows from (c).

Index

Applied Probability and Statistics (Continued)

ELANDT-JOHNSON and JOHNSON · Survival Models and Data Analysis
EVANS, PEACOCK, and HASTINGS · Statistical Distributions, *Second Edition*
FISHER and VAN BELLE · Biostatistics: A Methodology for the Health Sciences
FLEISS · The Design and Analysis of Clinical Experiments
FLEISS · Statistical Methods for Rates and Proportions, *Second Edition*
FLEMING and HARRINGTON · Counting Processes and Survival Analysis
FLURY · Common Principal Components and Related Multivariate Models
GALLANT · Nonlinear Statistical Models
GLASSERMAN and YAO · Monotone Structure in Discrete-Event Systems
GROSS and HARRIS · Fundamentals of Queueing Theory, *Second Edition*
GROVES · Survey Errors and Survey Costs
GROVES, BIEMER, LYBERG, MASSEY, NICHOLLS, and WAKSBERG · Telephone Survey Methodology
HAHN and MEEKER · Statistical Intervals: A Guide for Practitioners
HAND · Discrimination and Classification
*HANSEN, HURWITZ, and MADOW · Sample Survey Methods and Theory, Volume I: Methods and Applications
*HANSEN, HURWITZ, and MADOW · Sample Survey Methods and Theory, Volume II: Theory
HEIBERGER · Computation for the Analysis of Designed Experiments
HELLER · MACSYMA for Statisticians
HINKELMAN and KEMPTHORNE: · Design and Analysis of Experiments, Volume 1: Introduction to Experimental Design
HOAGLIN, MOSTELLER, and TUKEY · Exploratory Approach to Analysis of Variance
HOAGLIN, MOSTELLER, and TUKEY · Exploring Data Tables, Trends and Shapes
HOAGLIN, MOSTELLER, and TUKEY · Understanding Robust and Exploratory Data Analysis
HOCHBERG and TAMHANE · Multiple Comparison Procedures
HOCKING · Methods and Applications of Linear Models: Regression and the Analysis of Variance
HOEL · Elementary Statistics, *Fifth Edition*
HOGG and KLUGMAN · Loss Distributions
HOLLANDER and WOLFE · Nonparametric Statistical Methods
HOSMER and LEMESHOW · Applied Logistic Regression
HØYLAND and RAUSAND · System Reliability Theory: Models and Statistical Methods
HUBERTY · Applied Discriminant Analysis
IMAN and CONOVER · Modern Business Statistics
JACKSON · A User's Guide to Principle Components
JOHN · Statistical Methods in Engineering and Quality Assurance
JOHNSON · Multivariate Statistical Simulation
JOHNSON and KOTZ · Distributions in Statistics
Continuous Univariate Distributions—2
Continuous Multivariate Distributions
JOHNSON, KOTZ, and BALAKRISHNAN · Continuous Univariate Distributions, Volume 1, *Second Edition;* Volume 2, *Second Edition*
JOHNSON, KOTZ, and KEMP · Univariate Discrete Distributions, *Second Edition*
JUDGE, GRIFFITHS, HILL, LÜTKEPOHL, and LEE · The Theory and Practice of Econometrics, *Second Edition*
JUDGE, HILL, GRIFFITHS, LÜTKEPOHL, and LEE · Introduction to the Theory and Practice of Econometrics, *Second Edition*
JUREČKOVÁ and SEN · Robust Statistical Procedures: Aymptotics and Interrelations
KADANE · Bayesian Methods and Ethics in a Clinical Trial Design
KADANE and SCHUM · A Probabilistic Analysis of the Sacco and Vanzetti Evidence
KALBFLEISCH and PRENTICE · The Statistical Analysis of Failure Time Data
KASPRZYK, DUNCAN, KALTON, and SINGH · Panel Surveys

*Now available in a lower priced paperback edition in the Wiley Classics Library.

Applied Probability and Statistics (Continued)

KISH · Statistical Design for Research

*KISH · Survey Sampling

LANGE, RYAN, BILLARD, BRILLINGER, CONQUEST, and GREENHOUSE · Case Studies in Biometry

LAWLESS · Statistical Models and Methods for Lifetime Data

LEBART, MORINEAU., and WARWICK · Multivariate Descriptive Statistical Analysis: Correspondence Analysis and Related Techniques for Large Matrices

LEE · Statistical Methods for Survival Data Analysis, *Second Edition*

LePAGE and BILLARD · Exploring the Limits of Bootstrap

LEVY and LEMESHOW · Sampling of Populations: Methods and Applications

LINHART and ZUCCHINI · Model Selection

LITTLE and RUBIN · Statistical Analysis with Missing Data

MAGNUS and NEUDECKER · Matrix Differential Calculus with Applications in Statistics and Econometrics

MAINDONALD · Statistical Computation

MALLOWS · Design, Data, and Analysis by Some Friends of Cuthbert Daniel

MANN, SCHAFER, and SINGPURWALLA · Methods for Statistical Analysis of Reliability and Life Data

MASON, GUNST, and HESS · Statistical Design and Analysis of Experiments with Applications to Engineering and Science

McLACHLAN · Discriminant Analysis and Statistical Pattern Recognition

MILLER · Survival Analysis

MONTGOMERY and MYERS · Response Surface Methodology: Process and Product in Optimization Using Designed Experiments

MONTGOMERY and PECK · Introduction to Linear Regression Analysis, *Second Edition*

NELSON · Accelerated Testing, Statistical Models, Test Plans, and Data Analyses

NELSON · Applied Life Data Analysis

OCHI · Applied Probability and Stochastic Processes in Engineering and Physical Sciences

OKABE, BOOTS, and SUGIHARA · Spatial Tesselations: Concepts and Applications of Voronoi Diagrams

OSBORNE · Finite Algorithms in Optimization and Data Analysis

PANKRATZ · Forecasting with Dynamic Regression Models

PANKRATZ · Forecasting with Univariate Box-Jenkins Models: Concepts and Cases

PORT · Theoretical Probability for Applications

PUTERMAN · Markov Decision Processes: Discrete Stochastic Dynamic Programming

RACHEV · Probability Metrics and the Stability of Stochastic Models

RÉNYI · A Diary on Information Theory

RIPLEY · Spatial Statistics

RIPLEY · Stochastic Simulation

ROSS · Introduction to Probability and Statistics for Engineers and Scientists

ROUSSEEUW and LEROY · Robust Regression and Outlier Detection

RUBIN · Multiple Imputation for Nonresponse in Surveys

RYAN · Statistical Methods for Quality Improvement

SCHUSS - Theory and Applications of Stochastic Differential Equations

SCOTT · Multivariate Density Estimation: Theory, Practice, and Visualization

SEARLE · Linear Models

SEARLE · Linear Models for Unbalanced Data

SEARLE · Matrix Algebra Useful for Statistics

SEARLE, CASELLA, and McCULLOCH · Variance Components

SKINNER, HOLT, and SMITH · Analysis of Complex Surveys

STOYAN, KENDALL, and MECKE · Stochastic Geometry and Its Applications, *Second Edition*

STOYAN and STOYAN · Fractals, Random Shapes and Point Fields: Methods of Geometrical Statistics